Edited by
Georg J. Schmitz and Ulrich Prahl

Handbook of Software Solutions for ICME

Edited by
Georg J. Schmitz and Ulrich Prahl

Handbook of Software Solutions for ICME

WILEY-VCH
Verlag GmbH & Co. KGaA

Editors

Dr. Georg J. Schmitz
ACCESS e.V.
Intzestr. 5
52072 Aachen
Germany

Dr. Ulrich Prahl
RWTH Aachen University
Steel Institute (IEHK)
Intzestr. 1
52072 Aachen
Germany

Library of Congress Card No.: applied for

British Library Cataloguing-in-Publication Data
A catalogue record for this book is available from the British Library.

Bibliographic information published by the Deutsche Nationalbibliothek
The Deutsche Nationalbibliothek lists this publication in the Deutsche Nationalbibliografie; detailed bibliographic data are available on the Internet at <http://dnb.d-nb.de>.

Print ISBN: 978-3-527-33902-0
ePDF ISBN: 978-3-527-69359-7
ePub ISBN: 978-3-527-69358-0
Mobi ISBN: 978-3-527-69357-3
oBook ISBN: 978-3-527-69356-6

Cover Design Formgeber, Mannheim, Germany
Typesetting SPi Global, Chennai, India
Printing and Binding Markono Print Media Pte Ltd., Singapore

Printed on acid-free paper

Contents

List of Contributors

Laurent Adam
MSC Software Belgium
S.A. – e-Xstream engineering
Axis Parc. Rue Emile Francqui 9
1435 Mont-Saint-Guibert
Belgium

Carlos Agelet de Saracibar
International Center for
Numerical Methods in
Engineering, CIMNE
Edificio C1, Campus Norte UPC
c/ Gran Capitán, s/n
8034 Barcelona
Spain

Ilkin Alkhasli
RWTH Aachen University
Surface Engineering Institute
Kackertstr. 15
52072 Aachen
Germany

Markus Apel
ACCESS e.V.
Intzestr. 5
52072 Aachen
Germany

Sander Arnout
InsPyro
Kapeldreef 60
3001 Leuven
Belgium

Önder Babur
Eindhoven University of
Technology
De Zaale, 5600 MB Eindhoven
The Netherlands

Markus Bambach
Chair for Mechanical Design and
Manufacturing
Brandenburg University of
Technology Cottbus-Senftenberg
Konrad-Wachsmann-Allee 17
03046 Cottbus
Germany

Ralph Bernhardt
Simufact Engineering GmbH
Tempowerkring 19
21079 Hamburg
Germany

Ulrike Beyer
Fraunhofer-Institut für
Werkzeugmaschinen und
Umformtechnik
Reichenhainer Straße 88
9126 Chemnitz
Germany

Olle Blomberg
University of Copenhagen
Faculty of Humanities
Department of Media
Cognition and Communication
Karen Blixens Vej 4
2300 Copenhagen S
Denmark

Kirsten Bobzin
RWTH Aachen University
Head of Institute
Surface Engineering Institute
Kackertstr. 15
52072 Aachen
Germany

Thomas Böhlke
Karlsruhe Institute of
Technology (KIT)
Institute of Engineering
Mechanics
Kaiserstr. 10
76131 Karlsruhe
Germany

Romain Boman
University of Liège
LTAS-MN2L, Campus
Universitaire du Sart-Tilman
B52/3, Chemin des chevreuils 1
4000 Liège
Belgium

Günter Bräuer
Fraunhofer Institute for Surface
Engineering and Thin Films IST
Bienroder Weg 54e
38108 Braunschweig
Germany

Christoph Broeckmann
RWTH Aachen University
Institute for Materials
Application in Mechanical
Engineering (IWM)
Augustinerbach 4
52062 Aachen
Germany

Philippe Bussetta
São Paulo State University
Department of Mechanical
Engineering
Campus de Bauru
São Paulo SP 01049-010
Brazil

Juan Carlos Cajas
Barcelona Supercomputing
Center – Centro Nacional de
Supercomputación
Nexus II Building, c/ Jordi
Girona, 29
8034 Barcelona
Spain

Miguel Cervera
International Center for
Numerical Methods in
Engineering, CIMNE
Edificio C1, Campus Norte UPC
c/ Gran Capitán, s/n
8034 Barcelona
Spain

Geng Chen
RWTH Aachen University
Institute for Materials
Application in Mechanical
Engineering (IWM)
Augustinerbach 4
52062 Aachen
Germany

Long-Qing Chen
The Pennsylvania State University
Department of Materials Science and Engineering
Materials Research Institute
N229 Millennium Science Complex
University Park, PA 16802
USA

Qing Chen
Thermo-Calc Software AB
Norra Stationsgatan 93
11364 Stockholm
Sweden

Michele Chiumenti
International Center for Numerical Methods in Engineering, CIMNE
Edificio C1, Campus Norte UPC
c/ Gran Capitán, s/n
8034 Barcelona
Spain

Abel Coll
International Center for Numerical Methods in Engineering, CIMNE
Edificio C1, Campus Norte UPC
c/ Gran Capitán, s/n
8034 Barcelona
Spain

André Costa e Silva
Universidade Federal Fluminense
Escola de Engenharia Industrial Metalúrgica
Av dos Trabalhadores 420
Volta Redonda 27255-125
Brazil

Jean-Claude Crivello
East Paris Institute of Chemistry and Materials Science
2-8 Rue Henri Dunant
94320 Thiais
France

Pooyan Dadvand
International Center for Numerical Methods in Engineering, CIMNE
Edificio C1, Campus Norte UPC
c/ Gran Capitán, s/n
8034 Barcelona
Spain

Johan Deconinck
Vrije Universiteit Brussel
Faculty of Engineering
Department of Electrotechnical Engineering
Pleinlaan 2
1050 Brussels
Belgium

Charles-André Gandin
MINES ParisTech
CEMEF – CNRS
1, rue Claude Daunesse
CS 10207
06904 Sophia Antipolis
France

László Gránásy
Wigner Research Centre for Physics
Konkoly-Thege 29-33
1121 Budapest
Hungary

Gildas Guillemot
MINES ParisTech, CEMEF
1, rue Claude Daunesse
CS 10207
06904 Sophia Antipolis
France

Theo Hack
Airbus Group Innovations
TX2 – Metallic Technologies
Surface Engineering
81663 München
Germany

Klaus Hack
GTT Technologies Gesellschaft
für Technische Thermochemie
und -physik mbH
Kaiserstrasse 103
52134 Herzogenrath
Germany

Seyed Masood Hafez Haghighat
Department of Microstructure
Physics and Alloy Design
Max-Planck Institute for Iron
Research
Max-Planck-Str. 1
40237 Düsseldorf
Germany

Bengt Hallstedt
RWTH Aachen University
Institute for Materials
Application in Mechanical
Engineering (IWM)
Augustinerbach 4
52062 Aachen
Germany

Mauricio Hanzich
HPC Software Engineering
Group Leader
Barcelona Supercomputing
Center
CASE Department
Nexus II Building, c/ Jordi
Girona, 29, Office 311
8034 Barcelona
Spain

Adham Hashibon
Fraunhofer Institute for
Mechanics of Materials (IWM)
Wöhlerstraße 11
79108 Freiburg
Germany

Tore Haug-Warberg
Norwegian University of Science
and Technology
Department of Chemical
Engineering
Høgskoleringen 1
7491 Trondheim
Norway

Joaquin A. Hernández Ortega
International Center for
Numerical Methods in
Engineering, CIMNE
Edificio C1, Campus Norte UPC
c/ Gran Capitán, s/n
8034 Barcelona
Spain

Daniel Höche
Department of Corrosion and
Surface Technology
Institute of Materials Research
Helmholtz-Zentrum Geesthacht
Zentrum für Material- und
Küstenforschung GmbH
Max-Planck-Strasse 1
21502 Geesthacht
Germany

Stephan Hojda
RWTH Aachen University
Institute of Metal Forming (IBF)
Intzestraße 10
52056 Aachen
Germany

Christian Hopmann
RWTH Aachen University
Institute of Plastics Processing
(IKV) in Industry and the Skilled
Crafts
Pontstr. 49
52062 Aachen
Germany

Stephan Horstmann
Universität Paderborn
Fakultät für Maschinenbau
Warburger Straße 100
33098 Paderborn
Germany

Guillaume Houzeaux
Physical and Numerical
Modelling Group Leader
Barcelona Supercomputing
Center
CASE Department
Nexus II Building
c/ Jordi Girona
29, Office 3A
8034 Barcelona
Spain

Martin Hunkel
Stiftung Institut für
Werkstofftechnik
Badgasteiner Str. 3
28359 Bremen
Germany

Bethany Huseby
Thermo-Calc Software AB
Norra Stationsgatan 93
11364 Stockholm
Sweden

Jean-Marc Joubert
East Paris Institute of Chemistry
and Materials Science
2-8 Rue Henri Dunant
94320 Thiais
France

Ursula R. Kattner
National Institute of Standards
and Technology
Materials Science and
Engineering Division
100 Bureau Dr
Gaithersburg, MD 20899
USA

Martin Keunecke
Center for Tribological Coatings
Fraunhofer Institute for Surface
Engineering and Thin Films IST
Bienroder Weg 54e
38108 Braunschweig
Germany

Pavel A. Korzhavyi
Department of Materials Science
and Engineering
KTH Royal Institute of
Technology
Brinellvägen 23
10044 Stockholm
Sweden

Markus Krömer
RWTH Aachen University
Gear Department
Laboratory for Machine Tools
and Production Engineering
(WZL)
Steinbachstr. 19
52074 Aachen
Germany

Henrik Larsson
Thermo-Calc Software AB
Norra Stationsgatan 93
11364 Stockholm
Sweden

Gottfried Laschet
ACCESS e.V.
Intzestr. 5
52072 Aachen
Germany

Joonho Lee
Korea University
Department of Materials Science
and Engineering
145 Anam-ro
Seongbuk-gu, Seoul 02841
Republic of Korea

Maxime Lesueur
Purdue University
School of Engineering Education
701 W Stadium Ave.
West Lafayette, IN 47907
USA

Thomas Frederik Linke
RWTH Aachen University
Surface Engineering Institute
Kackertstr. 15
52072 Aachen
Germany

Mauricio Lobos
Karlsruhe Institute of
Technology (KIT)
Institute of Engineering
Mechanics
Kaiserstr. 10
76131 Karlsruhe
Germany

Cláudio M. Lousada
Department of Materials Science
and Engineering
KTH Royal Institute of
Technology
Brinellvägen 23
10044 Stockholm
Sweden

Ignacio Martin-Bragado
Formerly, IMDEA Materials
Institute
Eric Kandel 2
28906 Madrid
Spain

and

Present address,
Synopys Inc.
690 E. Middlefield Road
Mountain View, CA 94043
USA

Thomas Melzig
Low Pressure Plasma Processes
Fraunhofer Institute for Surface
Engineering and Thin Films IST
Bienroder Weg 54e
38108 Braunschweig
Germany

Gerson Meschut
Universität Paderborn
Fakultät für Maschinenbau
Warburger Straße 100
33098 Paderborn
Germany

Roland Müller
RWTH Aachen University
High Performance Cutting
Department
Fraunhofer Institute for
Production Technology IPT
Steinbachstr. 17
52074 Aachen
Germany

Sebastian Münstermann
RWTH Aachen University
Steel Institute (IEHK)
Group for Integrity of Materials
and Structures
Intzestr. 1
52072 Aachen
Germany

Ludovic Noels
University of Liege
Aerospace and Mechanical
Engineering Department
Computational & Multiscale
Mechanics of Materials (CM3)
Allée de la découverte 9
4000 Liège
Belgium

Stefan Oberpeilsteiner
University of Applied Sciences
Upper Austria
Stelzhamerstr. 23
4600 Wels
Austria

Mehmet Öte
RWTH Aachen University
Surface Engineering Institute
Kackertstr. 15
52072 Aachen
Germany

Miguel Ángel Pasenau de Riera
International Center for
Numerical Methods in
Engineering, CIMNE
Edificio C1, Campus Norte UPC
c/ Gran Capitán, s/n
8034 Barcelona
Spain

Bořek Patzák
Czech Technical University
Faculty of Civil Engineering
Department of Mechanics
Thakurova 7
16629 Prague
Czech Republic

Stephan Petersen
GTT Technologies Gesellschaft
für Technische Thermochemie
und -physik mbH
Kaiserstrasse 103
52134 Herzogenrath
Germany

Andreas Pflug
Low Pressure Plasma Processes
Fraunhofer Institute for Surface
Engineering and Thin Films IST
Bienroder Weg 54e
38108 Braunschweig
Germany

Frigyes Podmaniczky
Wigner Research Centre for
Physics
Konkoly-Thege 29-33
1121 Budapest
Hungary

Jean-Philippe Ponthot
University of Liège
LTAS-MN2L, Campus
Universitaire du Sart-Tilman
B52/3, Chemin des chevreuils 1
4000 Liège
Belgium

Ulrich Prahl
RWTH Aachen University
Steel Institute (IEHK)
Intzestr. 1
52072 Aachen
Germany

Thomas Reiter
University of Applied Sciences
Upper Austria
Stelzhamerstr. 23
4600 Wels
Austria

Markus A. Reuter
Helmholtz-Institut Freiberg für
Ressourcentechnologie
Chemnitzer Str. 40
09599 Freiberg
Germany

Franz Roters
Max-Planck-Institut für
Eisenforschung GmbH
Max-Planck-Str. 1
40237 Düsseldorf
Germany

Dietmar Salaberger
University of Applied Sciences
Upper Austria
Stelzhamerstr. 23
4600 Wels
Austria

Lothar Schäfer
Chemical Vapour Deposition
Fraunhofer Institute for Surface
Engineering and Thin Films IST
Bienroder Weg 54e
38108 Braunschweig
Germany

Georg J. Schmitz
ACCESS e.V.
Intzestr. 5
52072 Aachen
Germany

Javier Segurado
Polytechnic University of Madrid
and IMDEA Materials Institute
Eric Kandel 2
28906 Madrid
Spain

Jan Seyfarth
MSC Software GmbH
Am Moosfeld 13
81829 Mûnchen
Germany

Mohamed Sharaf
Formerly, RWTH Aachen
University
IEHK Steel Institute
Intzestr. 1
52072 Aachen
Germany

Michael Siemers
Low Pressure Plasma Processes
Fraunhofer Institute for Surface
Engineering and Thin Films IST
Bienroder Weg 54e
38108 Braunschweig
Germany

Sree Koundinya Sistla
RWTH Aachen University
Institute for Materials
Application in Mechanical
Engineering (IWM)
Augustinerbach 4
52062 Aachen
Germany

Ganesh Soni
MSC Software Corporation India
Pvt Ltd
6th Floor MrApex, Baner Road
Pune 411045
India

Robert Spatschek
Research Center Jülich
Institute for Energy and Climate
Research IEK-2
Wilhem-Johnen-Str.
52428 Jülich
Germany

Marcel Spekowius
RWTH Aachen University
Institute of Plastics Processing
(IKV) in Industry and the Skilled
Crafts
Seffenter Weg 201
52074 Aachen
Germany

Roberto Spina
Politecnico di Bari
Department of Mechanics
Mathematics and Management
Viale Japigia 182
70126 Bari
Italy

André Teixeira
RWTH Aachen University
High Performance Cutting
Department
Fraunhofer Institute for
Production Technology IPT
Steinbachstr. 17
52074 Aachen
Germany

Chung Van Nguyen
RWTH Aachen University
Institute for Materials
Application in Mechanical
Engineering (IWM)
Augustinerbach 4
52062 Aachen
Germany

Dieter Weichert
RWTH Aachen University
Institute of General Mechanics
(IAM)
Templergraben 64
52062 Aachen
Germany

Ling Wu
University of Liege
Aerospace and Mechanical
Engineering Department
Computational & Multiscale
Mechanics of Materials (CM3)
Allée de la découverte 9
4000 Liège
Belgium

Daniela Zander
Chair of Corrosion and Corrosion Protection
RWTH Aachen University
Faculty of Georesources and Materials Engineering
Division of Materials Science and Engineering
Intzestrasse 5
52072 Aachen
Germany

Fan Zhang
CompuTherm
437 S Yellowstone Dr
Madison, WI 53719
USA

Preface

Integrated Computational Materials Engineering: Past, Present, and Future

The present *Handbook on Software Solutions for Integrated Computational Materials Engineering* is probably best introduced by giving a look back onto the roots and by providing an outlook into the future. Based on the editor's own professional experience, we have a quick look at the situations ±30 years from now.

1986: The Historical Ground

Materials were often characterized by optical microscopy and respective microstructures were recorded on black and white glossy prints. These were collected in microstructure catalogs. First personal computers with 80 286 processors and a Windows 3.1 operating system entered into the research practice of noncomputer experts. Floppy disks had a storage capability of a few hundred kilobytes, and a 10 MB hard disk was already considered as advanced equipment. The Internet was in the early nascent state. Digitizing the glossy prints at that time was the first step to automatic image processing and to subsequent statistical evaluation of microstructures. Finite element method (FEM) modeling of entire components on large computers filling a whole room entered into applications. The foundations for computational thermodynamics were already laid. Materials processing, however, could still be considered rather as a skill or as an art than as a science at that time.

2016: The Present Status

The evolution of computational capabilities during the last decades has triggered a tremendous progress. The development of simulation models proceeded on all time and length scales, and a huge variety of simulation tools being nowadays available has been compiled in this book. Even complex simulations may sometimes be developed and be run on a standard multicore laptop computer. Data storage in the terabyte region is usual even in private use. Microstructure features increasingly are digitally recorded in 3D and sometimes even in 4D. FEM modeling has replaced experimental efforts to a large extent and often only final validation proceeds "physically," for example, in crash tests. Computational thermodynamics have further matured into a spatially resolved description of phase transformations based on the phase-field concept, which nowadays allows the simulation of microstructure evolution even in complex technical alloy systems. Materials engineering thus has transformed from being skill-based toward being a science. The complex interplay of atomistic processes, thermodynamics, processing conditions, microstructure evolution, materials and component properties, component functionality, and component performance has been identified to be only accessible via a combination of different simulation tools in an ICME-type, holistic approach. The current major challenge seems to facilitate communication between the different model worlds and communities.

2046: The Future Vision

All software tools and experimental devices in the area of ICME have a common communication standard similar to jpeg formats for pictures in 2016. 3D and 4D simulation data with highest spatial resolution can easily be exchanged. Metadata will collect all information about origin, precision, validation, and many other aspects of the data. Data will be stored in the cloud or on powerful exabyte local devices. Simulations running for weeks in 2016 will run in hours. Well-calibrated surrogate models with a lower precision will run within seconds and provide assistance in business decisions. Models will be available to simultaneously describe all known phenomena affecting the properties of any material. Individual results can easily be integrated into suitable common data structures and can easily be retrieved. A new community of holistically educated "ICME engineers" has entered their professional life and takes responsibility in leading positions.

The design of new materials and components will essentially be based on simulations. It will be optimized with respect to a desired functionality and performance obeying constraints given, for example, by their manufacturing processes, ecological footprint, and economic impact. The prediction of materials and component properties will be possible along their entire production and service life cycle. Simulations will bridge interfaces between inorganic, organic, and biological materials and even encompass human tissue. The morphogenesis in complex biological systems can be tackled by simulations.

The editors are eager to contribute to further shaping the necessary developments, and they are also curious to see whether their vision might really come true by 2046.

Editing this book has been performed within the ICMEg project and has received funding from the European Union Seventh Framework Programme

(FP7/2007-2011) under grant agreement n° 6067114 (ICMEg) and from the Cluster of Excellence "Integrative Production Technologies for High-Wage Countries" being funded by the Deutsche Forschungsgemeinschaft (DFG). Chapter 8 was compiled with help of the European Multiscale Modelling Cluster, which received funding from the European Union Seventh Framework Programme (FP7/2007-2011) under grant agreements n° 604005 (SimPhoNy) and n° 604279 (MMP). Thanks are due to the Commission's project officers, to the coordinators and coworkers of the cluster projects, and to the members of the European Materials Modelling Council for numerous stimulating discussions.

Our thanks also go to 93 authors from 15 countries, who volunteered and contributed their expertise and took quite an effort to make this book real. We do hope that it will become a valuable documentation for anybody – whether "process engineer" or "simulation guy" – seeking a holistic view on things.

Aachen,
January 2016

Georg J. Schmitz and Ulrich Prahl

1
Introduction

Georg J. Schmitz and Ulrich Prahl

1.1
Motivation

Integrated Computational Materials Engineering (ICME) – by its name and its nature – draws on the combination and the simultaneous or consecutive use of a variety of software and modeling tools. This simple phrase immediately raises a number of further questions:

How to combine tools?
How to select suitable tools?
How to decide on a specific tool?
And many more ....

Eventually, before answering these questions, some even more direct issues arise:

Which tools are available at all?
How to become aware about suitable tools?

Thus, there is obviously a need for something like the "yellow pages of software solutions for ICME" or a similar "one-stop shop" like institution. Such kind of yellow pages listing – even if being very comprehensive – would be quite boring to read (and also to write ...) and probably would even be outdated after a short time in view of new codes emerging and old codes being discontinued.

The ICMEg project partners [1] therefore decided to extend the scope of this book beyond the "yellow pages" and to include also a general introduction and overviews to the different fields, models, and software tools. The book content thus eventually evolved into an "overview" of overviews. All contributions are as generic as possible and references are mainly limited to "further reading" and refer to textbooks and tutorials for the different fields and review articles. The book also provides an "overview" of tutorials, reviews, and textbooks for the different fields.

Major scope, however, is to "name" phenomena, models, descriptors, and other terms and to arrange them in an overall context structure. Looking up the details

Handbook of Software Solutions for ICME, First Edition. Edited by Georg J. Schmitz and Ulrich Prahl.

Figure 1.1 A large variety of simulation tools is available around the globe. Not all codes in general have their own websites and logos. There is an even bigger number of particularly academic tools that are hidden and waiting to be exploited.

behind the different "names" is left to the reader and nowadays can often best be achieved by drawing on the Internet.

In summary, the ultimate motivation for the authors to write this book is that it could become a standard tutorial for future ICME engineers, which by nature of ICME need to have a holistic education, a general background, and a "bird's eye view" on things.

The motivation for this book is making the first steps toward providing a thematically structured directory of the huge and heterogeneous variety of state-of-the-art models (Figure 1.1). It is thus also particularly suited for young scientists and engineers seeking an overview of modern simulation tools in the area of Computational Materials Science and ICME.

1.2 What is ICME?

"Integrated computational materials engineering (ICME) as an emerging discipline aiming to integrate computational materials science tools into a holistic system will accelerate materials development, transform the engineering design optimization process, and unify design and manufacturing" [2].

Looking at the names, a definition of ICME has been attempted in a previous book [3] based on the analysis of the ingredients I, C, M, and E. An ambiguity has been identified with respect to the term "E" – engineering – which is applied to a product/component in Computational Engineering/Integrated Computational Engineering (CE/ICE) and to a specific material in Computational Materials/Computational Materials Engineering (CM/CME). This ambiguity can be resolved by putting the focus of ICME on:

Engineering the properties of a component as a function of
the local properties of the material inside the component
and along its entire production and service life cycle

"I" in this context especially means integrating along the process chain (time, history), integrating across the scales (space, structures), integrating several models/tools, and integrating real and virtual worlds. "E" refers to engineering of technical alloy systems, engineering under industrial boundary conditions, and engineering of materials in components during manufacture and under operational load.

Technically, ICME is an approach for solving advanced engineering problems related to the design of new materials, processes, and products by combining individual materials and process models. Is ICME just a synonym for the coupling/linking of simulation tools by data exchange? From a systems point of view, the coupling of individual models and/or software codes across length scales and along material processing chains leads to highly complex metamodels.

ICME thus is more than just linking/coupling tools. The global optimum of a process chain might – and actually will – differ from a chain of individually optimized process steps (Figure 1.2).

ICME is also not only about exchange of some data between different simulation tools but further requires information contingency in view of subsequent processes downstream the value chain (Figure 1.3).

ICME currently is already known to combine, to address, and to exploit "processing–microstructure" relationships and "microstructure–property" relationships. An emerging area is the "microstructure–processing" relationship, which investigates how processing is affected by the initial microstructure or how the microstructure affects the robustness of a process – in other words, how the history of a component affects its response to processing and how it defines its properties for operation.

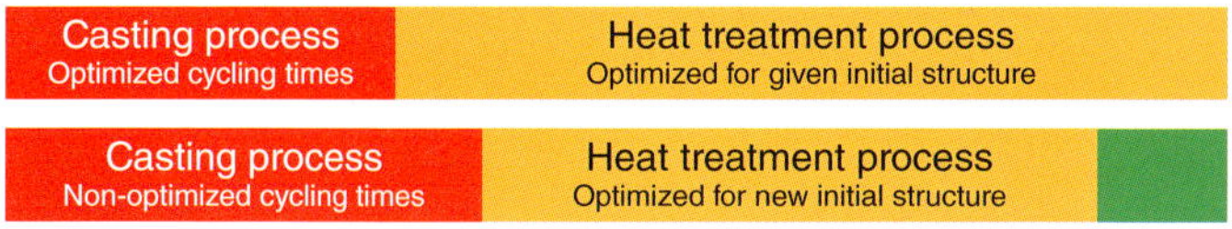

Figure 1.2 Example for a global optimum. Slightly suboptimal casting cycles with samples remaining at higher temperature for some more time may allow for shorter heat treatment times and thus result in an overall shorter production cycle.

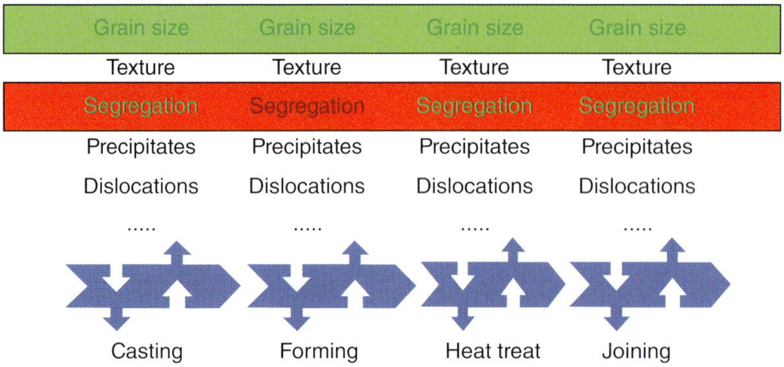

Figure 1.3 The importance of information contingency for the example of a process chain. While, for example, the grain size is simulated in all tools, the segregation of alloy elements is not modeled in the "forming" step. Segregation however becomes important again in subsequent processes.

In summary, ICME is an emerging discipline spanning various disciplines from materials technology, mechanical engineering, chemical and physical science, information technology, and numerical and mathematical science, for which a generic structural framework has to be elaborated, established, and maintained.

1.3 Industrial Needs for ICME

When discussing the industrial needs for ICME, two perspectives have to be differentiated. On the one hand, there are the commercial software providers trying to provide their software solutions to as many customers as possible. On the other hand, any industrial user of software tools is interested in exploiting software solutions to design new materials and production processes for components with tailored performance. The interests and needs of these two communities have to be discussed independently.

The interests of commercial software providers in ICME essentially relate to providing software solutions, to continuously developing new functionalities, to providing data along with their models, to making models faster and/or more robust, to making reliable/predictive models, and eventually to earning money by selling their software solutions and/or their simulation-based consulting competence.

Commercial software industry needs to identify, meet, and anticipate the needs of industrial users of their models and codes. This especially includes anticipating the potential needs not even yet being identified by these users themselves. The situation somehow corresponds to a fire brigade, which in general is not needed but anybody is happy to have it in case of a fire. The better the software provider meets the industrial user's needs for specific application tasks, the higher the user's motivation will be to further use (and thus to pay for ...) the provider's software solution.

Even before any type of simulation was available, people were already able to construct airplanes. Nowadays – using simulations – they can do it faster, cheaper, better, and also with less ecologic impact. Currently, however, materials data entering respective finite element method (FEM) simulations are still often estimates based on similar materials, being isotropic and often revealing no temperature dependency. Large safety margins thus have still to be considered making the airplane heavier than needed. Better understanding and knowledge of materials and their processing will open pathways to new designs and even lighter airplanes.

From the application point of view, the industrial user of ICME is interested in solutions to his actual, real problems in ongoing production processes. He aims at optimizing current production sequences and the value of his products or at obtaining an improved understanding and control of materials and processes along the production chain. He aims at improving his products and processes in terms of cost and time and to increase the planning quality toward predictability of process chains to decrease waste and recycling material. Additionally he will develop new materials, new processes, and new products and will exploit emerging new options and applications. Eventually, simulation shall support a faster time to market of new material and process solutions by minimizing risks. This especially holds for "first-time-right" products where a classical trial-and-error approach bears unacceptable financial risks. Additional benefit of ICME can be generated for industrial users by using simulation results originating from their suppliers to improve their own processes or by providing a simulation history as an added value of products to their customers.

Industrial users of ICME are interested in designing their product as efficient as possible. ICME will be applied to reduce the design effort for new products/processes/materials in terms of costs and time. A specific requirement during modern simulation tasks within industrial design process concerns the configurable combination of different tools from different providers instead of monolithic solutions. A "plug-and-play"-type combination of tools in workflows and open simulation platforms is the key vision of future ICME. A future standardized data exchange will drastically decrease the efforts of software providers with respect to providing and maintaining a large number of import and export functionalities for their tools.

Though the benefits of using advanced material simulations are widely accepted, the application of the ICME approach within industrial settings still is a challenge in terms of complexity, capacity, and specific knowledge needed. Analyzing the needs of industry for ICME the following general conclusions can be drawn:

Conclusion 1

Cheap, fast, readily available, and reliable solutions are needed to tackle complex topics of technical interest.

Future markets will not relate to *products* and their *properties* but rather to *functionalities* and *performance*. An airplane manufacturer – or the airline as its final customer – is not really interested in the turbine but in procuring and having "a propulsion *functionality*" performing best for an estimated operational scenario. Best *performance* could, for example, be a long operational period with a minimum of interrupts in operation or a low price with a minimum of operational costs (fuel), meeting environmental constraints and laws and many other conditions.

Topics and materials of interest are exemplarily depicted for processing and properties of bulk materials/components essentially being made from metallic alloys. Typical customers for metallic alloys are the aerospace industry (superalloys, light materials), the steel industry (steel), power utilities and respective industry (superalloys, steels, etc.), automotive industry (steels, Al alloys, and Mg alloys), electronics industry (solders, semiconductors, solar silicon), biomedical devices (Ti alloys), and other industrial branches. The alloys of interest, in general, comprise a large number of alloy elements each of them purposely added in a well-defined amount to fulfill a specific functionality (Figure 1.4). The specification of the exact amounts of alloy elements by now is a long-lasting and expensive – in terms of both time and money – task. The development and qualification of new materials and their processing in the past took years to decades. These development cycles will be drastically shortened by future ICME-type approaches.

In other areas, the relevance of treating numerous chemical elements results from the small size of components being at the scale of the diffusion length such as in electronic solder configuration, where the composition of the solder joint is determined by both the solder composition comprising, for example, some melting point depressants, and the coatings on the components to be joined (Figure 1.5):

Conclusion 2

Models and tools are needed for complex multicomponent and multiphase materials and their processing.

Fe	C	Mn	Si	P	Cr	Ni
Bulk material, defines temperature stability range	Workability, hardness, wear	Binds sulphur, hardness, cold workability, machinability	Electromagnetic properties, surface properties, hardness, and so on	Hardness, hot workability	Corrosion resistance, wear/heat resistance	Toughness, formability, thermal expansion matching
Bulk	0.003–2.1%	0.02–27%	0.01–6%	0.01–0.6%	0.01–13%	0.01–12%

Figure 1.4 The technical metallic alloy steel consists of numerous alloying elements with very different amounts and specific contribution to the overall performance and processing of the material.

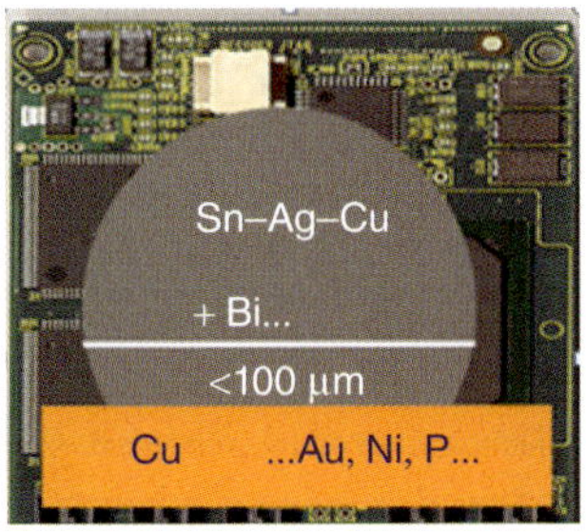

Figure 1.5 Steadily increasing miniaturization requires consideration of the effects of boundaries dissolving their chemical elements, for example, into a solder ball. A ternary Sn–Ag–Cu solder ball thus rapidly turns into a complex multicomponent, multiphase alloy system.

Advanced technical materials are not only a composition of various ingredients (alloy composition) but also consist of various phases of different crystallographic nature and/or local chemical composition. In steels and also in aluminum alloys, hard and brittle precipitates and inclusions are embedded in a more ductile matrix controlling the dislocation mobility. For example, steel may consist of two or more crystalline phases (e.g., ferrite, martensite, and austenite in parallel) being present in specific size, shape, and orientation. Recent developments aim at decreasing the structural length of single components (nanosized microstructures) and at exploiting metastable phases revealing complex deformation mechanisms during loading (e.g., transformation-induced phase transformation or inhomogeneous dislocation gliding) (Figure 1.6).

The costs of components are determined by the raw material, the production costs, and the costs arising during service life including repair and recycling. It is not only about the alloy composition. The *processing history* plays a decisive role in achieving the desired properties. During production, the material undergoes various process steps like casting and solidification, hot or cold forming, annealing, machining, joining, and coating. Each process during the entire production

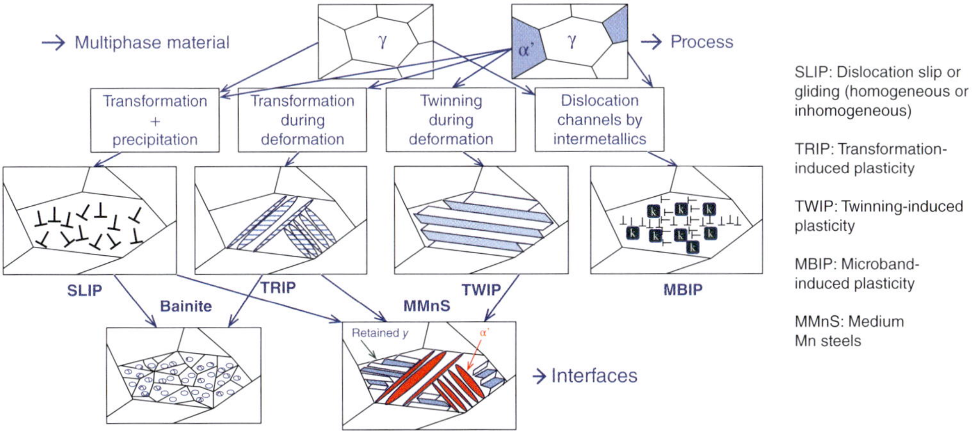

Figure 1.6 Modern steel development aims at tailored properties by adjusting a multiphase microstructure by applying deformation mechanisms based on metastable phases and by decreasing the structural length scale of microstructural components.

chain affects the microstructure and thus the properties of the material. Also, the process itself is affected by the properties being controlled by the microstructure. Thus, there is a strong dependence between processing and properties during the production and also during the final operation of the component. It should be mentioned that the macroscopic properties of the component in general are determined by mechanisms occurring at the length scale of nanometers or even below.

Damascus steel can be considered as a thousand-year-old nanomaterial although nobody really knew about "nano" at that time. Knowledge about the process parameters for the repeated and well-defined application of numerous individual forming and heat-treatment steps eventually leading to the superior properties of Damascus steel took generations to evolve on the basis of trial-and-error approaches.

Conclusion 3

Models are needed to track the evolution of materials and components along their entire production and service life cycle.

Modern simulation techniques will drastically shorten the development times for new materials and their processing. Respective developments will become much more targeted. However, the vision of "materials design out of the computer" by now is still only in reach for few examples. Nevertheless, it is widely accepted that simulating processes and materials evolution during processes significantly supports the design process. Some modern materials and components can only be produced using quantitative simulations of their microstructure, their processing, and their performance.

Eventually, simulation results have to be validated against measurements, existing data, and real – sometimes even financial – arguments increasing the confidence into the model, in its predictive capabilities and in its commercial value. Such validation can be based on data provided and shared by companies and academia alike. This validation includes information about the quality of calculated results including the uncertainty to be expected due to the use of nonexact input parameters and also due to the limits of the models. In particular, any user has to be aware of such uncertainties and limitations if he is applying models for situations or questions that are not within the – hopefully well-defined – input parameter space.

Conclusion 4

Validation is needed requiring improved interaction between virtual and real worlds.

Validation does not only include proving that models are able to predict materials properties correctly in a quantitative manner. The modeling-based design approach itself has to be validated respectively has to be evaluated against a traditional design approach based on trial and error combined with tailored experiments. Thus, it is mandatory to know the quantitative predictability of models and to balance the benefit against the simulation costs. Only if modeling costs and efforts can be turned into an added value, for example, by improved production results, industry will continue to invest into simulations. To decrease simulation costs, it will be necessary to integrate all modeling components into a single work flow in a way that all tools are independent, but mutually connected.

This financial validation is a major industrial requirement. All industrial applications of ICME will focus on the delivery of the outcome. The following four key questions have to be discussed: (i) What is the outcome and the contribution of simulated results to the design or optimization process? (ii) How much does a specific tool and its application cost? (iii) How much does a specific calculation cost? (iv) What is the risk associated with feeding the virtual simulation results into real production?

Quantifying the "risk" becomes a priority, especially for SME use. A desirable output would be a predictive tool running hypothetical scenarios, which include both the technical and the market inputs into a holistic decision-making process by optimizing component performance and cost efficiency in parallel.

Conclusion 5

Multicriteria optimization of the workflow is needed to balance all business.

Improving the knowledge about processes and materials is not the final aim of an industrial ICME user. Eventually, simulations shall support the industrial user in making his business decisions and in optimizing his products. Simulations are an investment for a company in terms of software tools, IT infrastructure, and a highly skilled work force.

In terms of business decisions, the optimal solution from a technical point of view may not be the most effective solution from an economic point of view. The same holds for simulations during the design process. The perfect, quantitative, and predictive simulation approach is not always needed. Often simpler approaches offering a sufficient depth of understanding in a shorter response time are more appropriate to meet a specific design objective. This implies the need for nonmonolithic software solutions and for an effective adaptation of various software tools into a single simulation workflow. Different types of software solutions have to be available to tackle the same topic with different degrees of accuracy. These may reach from fast solutions providing quick estimates for decision making up to highly precise approaches requiring substantial simulation times for complex trouble shooting.

1.4 Present ICME

A major target of ICME is a holistic simulation approach comprising entire process chains for engineering components. ICME addresses not only the macroscopic component and the process scale, but also the local material properties of individual components on the microscopic scale. ICME not only encompasses

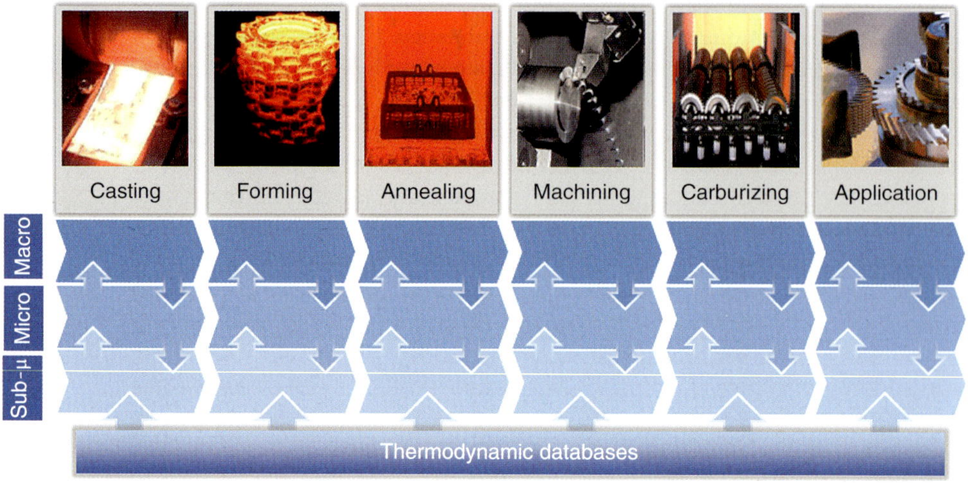

Figure 1.7 ICME aims to combine multiscale and through-process simulation toward a thorough description of all mechanisms being relevant for the production and performance of technical materials.

the material state for a given time at various length scales, but it also integrates the evolution of local material properties along the entire manufacturing process and eventually during service life (Figure 1.7). This concept promises better predictions and control of performance and service time of individual components or component assemblies. In this context, material microstructures and their evolution represent central issues in ICME as the microstructure is the carrier of the material properties.

Currently, numerous software solutions have reached a level allowing for valuable contributions to modern engineering tasks within knowledge-driven production models.

Simulations on the macroscopic scale of the manufacturing processes already are widely accepted and successfully used in industry on a Technology Readiness Level (TRL, according to the European Commission definition) of 8–9. Microstructure evolution codes and even more detailed electronic, atomistic, and mesoscopic models by now have reached a TRL of approximately 4–5 or even less. Respective approaches thus essentially are used in very specialized laboratories. However, the benefits of respective simulation tools and especially the needs for their mutual coupling and linking are widely accepted. The need for describing and simulating materials in more detail and at higher resolution is clearly formulated from an industrial perspective.

A major gap has especially been identified between continuum mesoscale models and world of discrete models (electronic/atomistic/mesoscopic). The usability, compatibility, and interoperability between these two model worlds have been formulated to be one of the major challenges for future ICME.

In spite of ICME currently emerging as a new and powerful discipline, coupling of different software tools is, however, still in its infancy and represents an issue consuming significant effort in terms of time and workforce if a coupling is realized at all. The combination of models by now is mainly achieved by manual transformation of the output of a simulation to form the input to a subsequent one. This subsequent simulation is either performed at a different length scale or constitutes a subsequent step along the process chain. Concerning the return of investment, the industry still complains on the significant amount of engineering efforts spent in data conversion and result management at the interfaces between different heterogeneous software tools. And indeed, there are several serious hindering factors: the missing of general standards for data format and platforms (in particular, the user friendliness of lower scale models is limited in most cases), the limited simulation knowledge at user side, expensive IT capacities, and no license schemes for short term but challenging calculation projects have to be mentioned.

1.5 Scope of this Book

The book covers essentially process chain for metals and alloys and to some extent also plastics, composites, and bulk components for structural applications (including coatings). These specific process chains could be composed based on the experience of the contributing authors. The developed concepts and descriptions, however, are all meant to be generic and may in future be extended to other fields actually not yet being covered, for example, pharmaceuticals, smart or functional materials, electronic or optical materials, textiles, and human tissue.

The book does not address (at least not explicitly):

- Hardware requirements
- Processing of nanoparticles, powder, molecules, and so on
- Subsystems of ICME like ICE not comprising materials as a central ingredient
- ICME politics

Scopes of this book are to:

1) *Serve as a tutorial for ICME engineers and scientists*
   Currently, there are only few educational schemes of "ICME engineers and scientists". ICME by now, in general, is performed by "self-learning" by people who recognized the importance of combining models and tools to achieve better performance of their materials, processes/products, and software tools. There is a need to at least basically acquaint engineers and scientists in a specific field with the basic concepts, problems, and solutions of all other field downstream or upstream the production chain. This handbook shall allow skilled engineers and scientists to get such an overview of adjacent fields. It shall create awareness about the possibilities and the needs/requirements of the other fields and possible limitations.

This overview shall further encompass the phenomena, models, and tools across all length scales relevant for the production process and for the properties of the materials and eventually the final product.

2) *Provide an introduction to different processes along the production chain including the lifetime of components*
For each of the process steps along the value chain like casting, forming, heat treatment, coating, joining, machining, corrosion, and recycling, an introduction is given even substructuring the particular field (e.g., casting: continuous casting, die casting, investment casting, etc.) and highlighting the relevant phenomena for each of these process steps. For each of these fields, current trends and actual developments are also shortly highlighted.
3) *Provide a general overview about concepts, methods, models, and tools across all scales*
The book addresses different concepts and methodologies like numerical methods, pre- and postprocessing of simulation results, thermodynamics, electronics, atomistics, mesoscopic, and discrete models as well as continuum models for both microstructure and processing scales.
4) *Provide references for further reading*
For each field, a number of references like review articles are provided for further reading, which themselves may be considered as "standard handbooks" or tutorials, reviews, or key publications for the field.
5) *Provide a structured directory of some prominent software tools in the different fields*
This book provides a directory of globally available software tools in the area of ICME. However, it does not simply draw on an alphabetic list-type directory. In contrast, it provides a structured compilation of these tools. The structure in this context is provided by the processes along the value chain during the production of a component.
6) *Discuss the origin of different data, initial conditions, and boundary conditions necessary to run the tools*
This discussion allows the identification of possible interactions between different models, for example, by providing data by one model for another model at the same or at a different scale. A similar strategy is applied for data on initial and boundary conditions. Further needs for new models to create required data may emerge from this discussion. Another topic under this discussion is the harmonization of information exchange between real and virtual worlds.
7) *Discuss current concepts to combine simulation tools in platform-type approaches*
The identified simulation tools are further discussed with respect to their integration into simulation platforms and ICME types of operation. This discussion especially focuses on interfaces and data formats being provided for import/export of information and also covers aspects of workflow management.

8) *Identify missing tools and functionalities and formulate suggestion for further development*
   In view of covering an entire life cycle of a component by simulations, also "missing tools" and "missing functionalities" are identified and suggestions for their development are outlined.

1.6 Structure of the Book

The structure of this book is "top-down" instead of "bottom-up" and thus follows the philosophy of "starting from the outcome." This outcome may be a certain consumer-desirable property of a specific product or of a specific component. Based on the respective requirements the physical, processing, and additional requirements allowing meeting the objective and tools for their simulation can be identified. Based on the identified requirements at each length scale, models at smaller scales eventually are either needed or allow for refining "down to the atoms" wherever this seems necessary or beneficial. This strategy is based on the conclusion that processing and properties cannot be treated independently but are strongly correlated. Materials cannot be tailored without considering their processing, and processes can only be optimized if materials behavior during processing is well understood in a quantitative manner.

A future strategy in ICME may be to investigate and consider only those scales contributing relevant information for understanding the given problem – a strategy of scale hopping instead of multiscale modeling. Even at a specific length scale, a fast and robust model providing estimates might be given preference over a sophisticated and very detailed model. This choice eventually will depend on the risks associated with the given task and the time and capacity being available to tackle it.

This chapter provides an introduction to the field and the motivation for future readers. Following an overview on "Integrated Computational Materials Engineering (ICME)," the potential benefits to industrial users are shortly highlighted. A section on industrial needs for ICME is followed by a short review of the current status on ICME. The introduction concludes with an outline of the book.

Chapter 2 details the variety of processes, simulations, models, tools, and phenomena occurring at the scale of component and being relevant to its production and its use. Following a short overview over a typical process chain, this chapter starts from primary shaping processes, for example, by solidification of a melt and then follows the entire manufacturing cycle, comprising the fundamental processes of forming, heat treatment, joining, coating, and machining. Eventually, this chapter also addresses the service life cycle in terms of corrosion, fatigue, and recycling.

Understanding of component behavior during processing or operation often requires understanding the microstructures of the materials building up the component. The evolution of microstructures is not only affected by the process conditions but by phenomena at all length scales.

Chapter 3 summarizes different models and tools to describe the evolution of microstructures. Following an overview and a specification of the term microstructure, a categorization is provided to describe microstructures at different levels of detail. Subsequent to the identification of the major phenomena affecting microstructure evolution, an overview of the different modeling approaches is provided. A section on nucleation modeling is followed by sections on models/methods describing diffusion/reaction processes and precipitation phenomena. Models and methods for the description of the spatiotemporal evolution of microstructures based on cell, lattice, or field representations are outlined. These include cellular automata and Potts Hamiltonian lattice models solved by Monte Carlo methods, as well as phase-field and multiphase-field models, phase-field crystal models, and crystal plasticity models. The chapter concludes with a structured list of software tools, specifically comprising tools for the analysis of 3D digital experimental microstructures.

Major requirements for any type of microstructure simulation are thermodynamic data and kinetic data as well as other parameters/properties of the phases present in the microstructure. Also the properties of the interfaces play a significant role. Respective data can be obtained from thermodynamic/kinetic databases and calculations as depicted in Chapter 4 or from electronic/atomistic/mesoscopic models as described in Chapter 5.

Chapter 4 provides an overview and an introduction to thermodynamics and thermodynamic modeling. Following a discussion on the role of thermodynamic modeling in ICME, the CALPHAD approach is explained. Based on its history and the minimization of the Gibbs energy as theoretical basis, the crystallography and models of phases, the development of CALPHAD databases, their formats, and future extensions are highlighted. The chapter concludes with sections on "use of thermodynamics at larger scales" and on "deriving thermodynamics/materials properties from small scale models." This section addresses the future of determination of effective properties in ICME where first principles' DFT methods are used to predict, for example, thermoelastic properties of materials.

Chapter 5 focuses on mesoscopic, atomistic, and electronic models following the overall "top-down" structure of the book. Following an overview on discrete and semidiscrete mesoscopic models in materials science, the specification of some definitions and an introduction to atomistic simulations like kinetic Monte Carlo and molecular dynamics is provided. A section on electronic structure methods addresses Hartree–Fock theory and post-Hartree–Fock methods and density functional theory (DFT) in its different approximations (local density approximation, "LDA," generalized gradient approach, GGA). Further sections address actual developments like meta-GGA methods, hybrid DFT–Hartree–Fock approaches, and van der Waals-corrected DFT. In addition to providing a structured overview of approximately 80 different simulation tools in the area of electronic, atomistic, and mesoscopic simulations, the chapter further provides a number of links to online sources for potentials, force fields, and effective cluster interactions.

Chapter 6 goes back to larger scales and is devoted to the determination of effective properties and provides an overview of different approaches like finite-element-based homogenization methods, mean-field homogenization methods, and virtual testing approaches. Finite-element-based homogenization methods apply the method of homogenization on the continuum description of the microstructure of the material – resulting, for example, from phase-field models – being discretized into a finite element description. Details of this approach such as sensitivity of the obtained properties to the underlying microstructure and to numerical influences, for example, size of representative volume elements (RVEs), mesh density, and element type are highlighted and discussed on the basis of examples. Mean-field homogenization methods are explained based on example(s) for fiber-reinforced materials and address especially nonlinear and evolving material properties. A section on screening and virtual testing of material properties highlights the potential of usage of computationally derived effective properties to obtain tailor-made materials from the virtual stage and is presented for an application to a polycrystalline material and cover the subsequent usage for different microstructures.

Chapter 7 deals with numerical methods for the computational simulation of a wide range of multiphysics and multiscale problems within an ICME framework. Following an overview on preprocess and space discretization methods, a large section provides an overview of different kinematic descriptions, solution methods for coupled problems, and numerical methods for the solution of a wide variety of engineering problems such as the finite difference method (FDM), FEM, boundary element method (BEM), discrete element method (DEM), mesh-free methods (MFMs), particle finite element method (PFEM), extended finite element method (XFEM), isogeometric analysis (IGA) method, and model order reduction (MOR) methods. This overview is followed by a description of numerical methods for contact problems, such as Lagrange multipliers, penalty and augmented Lagrangian, direct elimination, mortar, and IGA methods. Further sections address postprocessing and visualization methods, and numerical methods for mapping and data transfer in the FEM simulation of multiphysics/multiscale problems using either Lagrangian or arbitrary Lagrangian–Eulerian (ALE) formalisms. Data transfer algorithms are discussed for the numerical simulation of a single process using Lagrangian or ALE formalisms and requiring remeshing operations. Sections on reduced-order models and on different issues related to high-performance computing (HPC) and parallelization precede a list of software tools for pre- and postprocessing.

Chapter 8 addresses "software platforms" and "software integration," which in industry currently essentially correspond to integrating continuum models. Respective developments are dominated by structural mechanics or computational fluid dynamics tools based on finite elements or finite volume numerical schemes as part of product life management or computer-aided engineering (PLM/CAE) environments. However, advances in nanotechnology require descriptions of processes occurring on the smaller scales, which can be provided by electronic, atomistic, and mesoscopic models and continuum models applied to the mesoscale. This demand requires extending the tools and approaches of ICME to include tools on the smaller scales as well and to make this variety of tools "interoperable." Following an overview of different integration approaches (object-oriented, component-based, service-oriented, data-centric, model-based, and ontology-based approaches), existing standards for integration and different available approaches to coupling and linking are briefly highlighted.

Chapter 9 discusses the needs for future developments required toward the objective of covering an entire life cycle of a component by simulations. In view of covering an entire life cycle of a component by simulations, also "missing tools" and "missing functionalities" and needs for interoperability are identified and some suggestions for future developments into these directions are outlined.

This book eventually provides a *directory of globally available software tools* in the area of ICME. However, it does not simply draw on an alphabetic list-type directory. In contrast, it provides a structured compilation of relevant models and tools. The structure in this context is provided by the processes along the value chain during the production of a component from a given material as depicted in the individual chapters. Almost all chapters identify the most relevant simulation tools and discuss them with respect to their underlying models, the origin of the required data, the necessary initial and boundary conditions, and eventually with respect to their integration into simulation platforms and ICME types of interoperation. A structured table of software solutions for each field marks the end of each chapter.

These tables are based on and compiled from a survey of modeling tools in a number of resources:

- Precompiled lists of software codes, for example, the TMS [4], or a recent Japanese activity [5]
- A list of software tools being used in research project funded by the European Commission with a focus on electronic/atomistic/mesoscopic models [6]
- Software repositories such as GitHub [7], nanoHub [8], or SourceForge [9]
- Software identified during the 1st International Workshop on Software Solutions for ICME
- Software companies in different fields being compiled on the basis of personal acquaintance and by Internet search
- A private homepage [10] providing a survey of public domain and commercial mesh generators
- The NIST Materials Data Digital Library [11]
- Two TMS studies on ICME [12, 13].

The resulting list – eventually comprising approximately 350 different software tools – was then further complemented by the knowledge of the authors of the individual chapters. Each of the almost 70 authors from at least 12 countries in 3 continents is a renowned scientist or engineer in his/her field making their comments on the different tools a highly valuable piece of information and advice.

The focus of the book definitely is on structural mechanics applications. In other areas, for example, electronic design automation, the Technology CAD (Technology computer-aided design or TCAD) is an ICME-type approach of modeling semiconductor fabrication and semiconductor device operation. The modeling of the fabrication is termed Process TCAD, while the modeling of the device operation is termed Device TCAD. The software tools available in this field are widely documented in [14]. Some thematic overlap between structural and electronic materials processing already occurs in the domains of crystal growth and thin-film deposition.

In the end, the book has actually become what it was originally intended to be: "the yellow pages for software solutions for ICME." But the book has evolved even further into something the editors and authors would like it to become: "a tutorial for young students in the emerging field of ICME" and a handbook for scientists and engineer wanting to get a quick but sound impression on adjacent fields of expertise.

References

1. ICMEg (2013) Integrated Computational Materials Engineering expert group – a Coordination and Support action of the European Commission, www.icmeg.eu (accessed 25 April 2016).
2. National Research Council (2008) *Integrated Computational Materials Engineering: A Transformational Discipline for Improved Competitiveness and National Security*, National Academic Press, Washington, DC. ISBN: 0-309-12000-4
3. Schmitz, G.J. and Prahl, U. (eds) (2012) *Integrative Computational Materials Engineering – Concepts and Applications of a Modular Simulation Platform*, Wiley-VCH Verlag GmbH, Weinheim. ISBN: 978-3-527-33081-2
4. TMS http://www.tms.org/cyberportal/showData.aspx?tooltype=properties#grid (accessed 25 April 2016).
5. MateriApps http://ma.cms-initiative.jp/en (accessed 25 April 2016)
6. European Commission Review of Materials Modelling being Performed in European Projects, http://ec.europa.eu/research/industrial_technologies/e-library.cfm (accessed 25 April 2016).
7. GitHub, Inc. https://github.com/ (accessed 25 April 2016).
8. NanoHUB https://nanohub.org/resources/tools (accessed 25 April 2016).
9. Sourceforge http://sourceforge.net/ (accessed 25 April 2016).
10. www.robertschneiders.de/meshgeneration (accessed 25 April 2016).
11. NIST MATDL www.MatDl.org (accessed 25 April 2016).
12. TMS (2013) Implementing ICME in the Aerospace, Automotive, and Maritime Industries, http://www.tms.org/ICMEstudy/ (accessed 25 April 2016).
13. TMS (2015) Modeling Across Scales: A Roadmapping Study for Connecting Materials Models and Simulations Across Length and Time Scales, http://www.tms.org/multiscalestudy/ (accessed 25 April 2016).
14. TCAD Central http://www.tcadcentral.com/index.php/CommercialTCAD (accessed 25 April 2016).

2
Modeling at the Process and Component Scales

Handbook of Software Solutions for ICME, First Edition. Edited by Georg J. Schmitz and Ulrich Prahl.

2.1
Overview of Processing Methods and Process Chains

Ralph Bernhardt and Georg J. Schmitz

The processing of materials and their targeted shape and property modification have a long-standing tradition in human history. Entire historical eras are named after a material, whose production and processing dominated a time period and determined technological advancement – examples are the “Stone Age” or the “Iron Age.” In general, many of the ensuing principles and approaches discussed in this book can be applied to technologically relevant engineering materials for solids. Whether for glass, plastic, concrete, wood, or metal, each material class has a specialized subdivision in materials sciences, materials engineering, and manufacturing technology. What they all have in common is mankind’s ongoing search for the most suitable method and approach to efficiently manufacture products made from the different materials, tailored to the individual requirements.

Today, the field of materials sciences and materials engineering already stands for new innovations in socially relevant, cutting-edge fields such as energy, climate and environmental protection, resource optimization, and mobility as well as the health sector, security, and communication. Current studies underline the exceptional direct or indirect dependency of all technological innovations on materials. Materials science and materials engineering belong – next to information technology – to the key technologies of the twenty-first century [1].

The insights gained in materials science enable the manufacture of high-tech materials with new or improved properties and also the improvement of existing and the development of new processes. The material selection, the structural design, the manufacturing process, and the workload and operational conditions influence a component’s properties. This encompasses the entire life cycle of a component up to recycling or material recovery, including the development of entirely new manufacturing processes.

Covering all possible material classes in this chapter would hardly be possible and also not serious. This particular chapter thus focuses on essential fundamental principles and approaches to the production, processing, and treatment of metallic materials as well as their modeling at the process scale, representative for all other relevant types of engineering materials.

Handbook of Software Solutions for ICME, First Edition. Edited by Georg J. Schmitz and Ulrich Prahl.

This introductory chapter briefly depicts the history of primary shaping and forming technologies of different metals. A second historical excursion relates to the development of modeling of manufacturing processes.

To widen the readers' view and to demonstrate the generic nature of the concepts being discussed in this book, the appendix of this introductory chapter discusses a very special process chain for a very special nonmetallic product: asparagus.

2.1.1
History of Metalworking

The beginnings of metalworking can be traced all the way back to the fifth century B.C. Depending on the historical source, either Persia (now Iraq) or the Mediterranean region is identified as its origin. It is proven that during the "Copper Age," gold, silver, and copper were already melted, cast, and cold-hammered. Craftsmen of that time already discovered that alloying copper with, for example, arsenic increases the material hardness, making it more suitable for the manufacture of weapons. About 3000 B.C., the first alloy system was established with the discovery of bronze. In the wake of this discovery, sheet pressing tools were introduced, the marking stamp was invented, and needles or rather profiled wires were manufactured. The spread of forming technology over the course of the Iron Age and throughout the Roman Empire – until about the thirteenth century – led to the first heyday of metalworking between the fourteenth and eighteenth centuries. Hydropower was increasingly used as a power unit driving forging hammers; the first presses and rolling mills found their way into manufacturing.

The great significance of metal tools and commodities is reflected in the contents and definition of the "Artes Mechanicae" of the seven mechanical arts of the Middle Ages [2, 3].

Further milestones on the way to mechanized and automated mass production of metal products were the invention of the steam engine by James Watt and the improvement of steel quality through new processing methods in the middle of the nineteenth century. Only these new advancements in steel production could supply the durable tools required for metalworking. With them, the technologies still relevant today such as die forging or rolling were able to gain their industrial and economic importance. After the 1960s, the industrial era was succeeded by the information age of today. Also known as the "silicon era," modern important advancements have occurred in the field of computer-aided modeling of technological processes. Next to metallic materials (ferrous and nonferrous metals), silicon has gained a particular importance as a metalloid. Metallic materials and iron-based materials such as steel continue to be of immense value as engineering materials. Around 1.6 billion tons of raw steel were produced in the year 2014. For 2015 alone, 1.5 billion tons of finished steel products (excluding nonferrous metals) were projected (see [2–6]).

2.1.2 History of Modeling of Manufacturing Processes

For centuries, the production and processing of materials was marked by empirical and heuristic approaches. Exploiting observed or accidentally discovered phenomena and effects was predominant. For example, the hardening of steel, a 3000-year-old technology, was "long enveloped by superstition and a mystery even to experienced masters" [7].

First systematic approaches to explain and to model materials already existed in the form of the particle model by Demokrit (around 400 B.C.) in old Greece. Similar to the works of Leonardo da Vinci (1452–1519) and Galileo Galilei (1564–1642) about determining and analyzing hardness on metal wire and structures, these discoveries were not systematically used to proactively develop new technologies. The book of metallurgy *De re metallica libri XII*, published in 1556 by Georgius Agricola, provided a substantial compilation of "recipes," yet this approach to the contemporary technologies was of a more descriptive and heuristic character (Figure 2.1.1).

During the seventeenth and eighteenth centuries, the philosophers of nature began to address the subject. They catalyzed the breakthrough of the old Greek concept of atomic structure of matter and discovered "crystallinity." With the transition from alchemy to chemistry and from metaphysics to physics at the end of the eighteenth century, the foundations for the systematic, mathematical–physical description of material properties and material processing were laid. The beginnings, for example, of the plasticity theory were born. Names such as Coulomb, Tresca, Levy, Huber, von Mises, Hencky, Poisson, Cauchy, Lamé, Green, Young,

Figure 2.1.1 Steelmaking and blacksmithing in the Middle Ages – woodcuts from the book *De re metallica libri XII* (Georgius Agricola) – showing the interior view of workshops with men at various stages of the metallurgical process. (Wikimedia-commons.)

and many others stand for some of the groundbreaking models and theories still supported today. A good overview of the historic evolution of the theory of plasticity can be found in [8].

The development of mathematical methods and algorithms for solving complex equations and systems of equations in particular for partial differential equations enabled an increasing depth and importance of modeling results. Numerical methods became an indispensable tool for scientists as a supportive branch of mathematics. The importance of numerics was raised to new heights along with the invention of the computer in the 1930s. Computers and numerics provided the two basic ingredients for the introduction of the finite element method (FEM) in the 1940s. Today, FEM is one of the most internationally widespread software methodologies for numerically solving partial differential equations in almost all engineering fields, in both research and industrial applications.

Solving problems with the help of the FEM can boast a 70-year tradition today. Limited to linear elastic problems in the beginning, the still valid concept of a stiffness matrix and element assembly to also solve nonlinear problems has established itself since the 1960s – more precisely with the introduction of NASTRAN in 1965 by John Davidson. While the development of elaborate and complex models was reserved for an elite circle of highly specialized computer experts in the early days, the method now has been established as a commercial tool in the manufacturing industry since the mid-1980s. While the first applications were limited to 2D problems and rather simple geometries, nowadays even complex 3D applications with a high level of detail can be calculated on conventional desktop computers.

The digital revolution and along with it the development of measuring and characterization technologies down to the individual atoms have made tremendous progress during the last decades. They are nowadays both prerequisites and driving forces for increasingly more detailed modeling of industrial manufacturing processes. Respective simulations have made the much quoted "lightweight design" possible in the first place. Integrated Computational Materials Engineering (ICME) is emerging as a main pillar of this overall concept as this integrated model approach not only integrates different length and timescales but also simulation and experiment and fundamental research with applications.

Crucial for the significant improvement and innovation in functionality, reliability, life span, and business efficiency of products is the integral approach, meaning the interlinking of information from different length scales and tracking of evolution of material properties over several process steps. A number of mechanisms influence and therefore determine material responses to workloads. ICME is the method of choice to quantitatively describe the relationships between process steps, material microstructure, material properties, and component behavior.

Future ICME will achieve the continuous cross-linking of production and application-related mechanisms, which the materials undergo on different length and timescales. Thus, changes in the material properties can be tracked and numerically described throughout the entire component manufacturing process. Based on these observations, weak spots in the process chain as well as during the lifetime can be determined and eliminated.

2.1.3
Overview of Processing Methods

A first rough classification of manufacturing processes following the German Standard DIN 8580 [9] is depicted in Figure 2.1.2. Similar classifications can also be found in [10]. Manufacturing processes are defined as processes in which products (outputs) are generated from other goods (inputs). Manufacturing technology defines manufacturing processes as processes for the production of geometrically defined solid bodies. This includes semifinished products or components of technical structures.

In general, several manufacturing processes have to be combined to proceed from raw parts to semifinished products and eventually to finished products such as machines, devices, tools, vehicles, and other single- or multipart objects/components and systems. The aforementioned DIN standard comprises – besides the geometric form – also the targeted modification of the properties of the manufactured product. The manufacturing sequence of a product is illustrated for the example of a vise in Figure 2.1.3.

Modern sustainable economies have to take into account not only the manufacture of a product but also its efficient and economic use during lifetime operation and eventually its recycling. Recycling is the process of reusing wastes/waste materials to generate secondary raw materials. Recycling is defined as "every recovery process through which waste products are processed to products, materials or substances either for their original purpose or for other uses. It includes the processing of organic materials, but not energetic recovery or the processing of materials purposed as fuel or back filling" [4].

This chapter covers all processes along the entire life cycle of a typical product in individual subchapters starting from primary shaping (2.2), via forming (2.3),

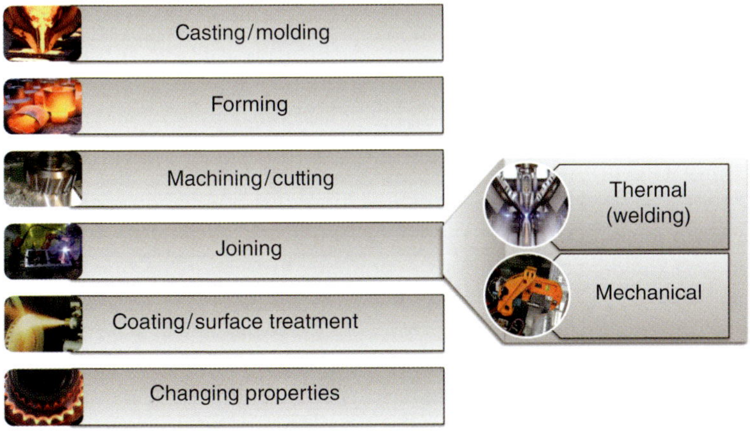

Figure 2.1.2 Main methods of manufacturing processes according to German Norm DIN standard 8580.

1. Base body: primary shaping: melting (formless substance) → casting (primary shaping) → work piece (vise body) and semi-finished parts for other components
2. Crank handle: Forming: raw part (billets) → rolling on a roll stand (forming) → component (Round material for the lever)
3. Vise spindle: Cutting (by machining): semi-finished product (round material) → turning, threading → component (Spindle)
4. Hardened vise jaws: Changing material properties: raw material (steel) → Hardening of the jaws (changing material properties) → component material (hardened steel)
5. Painting: coating: raw component → varnish application (coating) → painted component
6. Final assembly: Joining: individual parts → assembly (joining) → finished part (vise)

Figure 2.1.3 Manufacturing steps for components of a product for the example of a vise after [9]. (Benutzer https://commons.wikimedia.org/wiki/File:Schraubstock-800.jpg#. Used under CC BY 3.0 license https://creativecommons.org/licenses/by-sa/3.0/deed.en.)

heat treatment (2.4), joining (2.5), coating with thick (2.6) and thin (2.7) films, machining (2.8), fatigue (2.9), corrosion (2.10), and eventually recycling (2.11).

2.1.4
Processes and Process Chains

A technical *process* in the sense of manufacturing a product after DIN 66201 can be summarized as the entirety of processes in a system in which the state of the product is specifically altered. This change in state can entail an alteration of the geometric form and/or of the properties of the product. DIN 66201 also states: "A technical process is a process, the physical quantities of which can be determined and influenced with technical means."

A correlation between the process, the microstructure of the finished product, and accordingly its properties exists in all manufacturing processes:

> Process influences microstructure and microstructure determines properties.

The interpretation of this statement means that any processing of a material by any of the manufacturing processes automatically results in a change of its microstructure and therefore in the (local) properties. These changes may be small and even almost nondetectable or may be dramatic in some cases, for example, phase transitions in the materials are evoked by the process.

One of the oldest and globally well-known *process chains* in the field of metal forming was already applied to the manufacturing of Damascus steel in order to achieve a targeted geometric form and a specific combination of properties [11]. The origins of Damascus steel date back to the year 600 B.C. and took place in modern-day India and Sri Lanka. Essentially, after an initial primary shaping process (melting and casting), a multistage hot forming process (forging) is

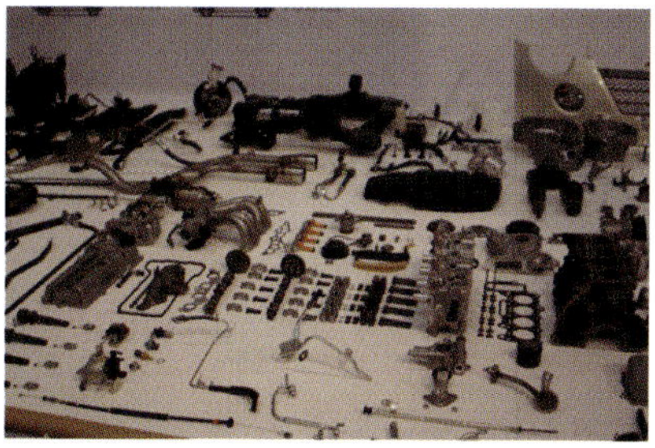

Figure 2.1.4 "Disassembly of a car." (Paul Hudson https://commons.wikimedia.org/wiki/File:Disassembly_of_a_car_(3).jpg. Used under CC BY 2.0 license https://creativecommons.org/licenses/by/2.0/deed.en.)

linked with a joining process (forge welding) with intermediary heat treatments (quenching and tempering) and cutting operations. The basic idea is to join two single bars of materials of varying hardness, heat them, and create a form by forging. The early craftsmen already understood that the joining of two different materials results in a product that is not only very hard and durable (hard steel) but also possesses sufficient toughness (soft steel). Furthermore it was perceived that if the process is repeated very often, the continuous combination of forming, folding, and heating results in a very fine microstructure in terms of the number of "folded layers."

Recently it was discovered that "the forge masters of the time used trial and error of different methods to create nanostructures, which constituted the key to the particular hardness. Periodic heating and forming of the steel fostered the creation of carbon nanotubes, which in turn contribute to the formation of microscopically thin fibers of iron carbide" [12]

The necessity of linking different individual processes and repeatedly applying them in a process chain therefore are an evident prerequisite to reach the superior properties of the Damascus steel.

It is obvious that the combination of processes results in an arbitrary number of different process chains with an arbitrary number of complex combinations. Even each of the individual processes in such a chain reveals an almost infinite option to adjust its parameter set such as temperature–time profiles. Different process chains can further be combined to form a final sequence, which typically happens when individual components are assembled into a system. These components are joined and assembled in a specific assembly sequence to form a system – each component representing the process chain in itself, which has been run to manufacture it (Figure 2.1.4)

2.1.5 Benefits of Modeling Process Chains

"Customers neither buy products nor features, but benefits" – this statement can be ubiquitously found in standard literature about marketing and communications. "But benefits are determined through properties" – only this clarification explains the necessity for the current shift in the manufacturing industry from product-oriented to property-oriented development and manufacturing.

As already discussed in Chapter 1, "properties" in this context are now increasingly replaced by the demand for "functionality" (e.g., determined by a number of properties) and even further by the demand for "performance," meaning the provision of a functionality for a given life cycle.

Manufacturers are increasingly facing the challenge of reliably manufacturing new products with high-dimensional accuracy and only revealing small tolerances for the local properties. They further need to operate their processes and process chains not only economically but also ecologically in terms of energy and resource efficiency [13]. Possible solutions are customized and tailored materials, specified for the component requirements, and adjusted processes for their manufacturing and processing. The focus is on the *entire* process chain, which is generally heterogeneous in terms of the process owners in the supply chain and also in terms of the variety of process types.

To meet the demands for individually adjustable product properties at the end of the process chain, the complex relationship between material design (e.g., the alloy composition for metallic materials) and the desired local creation of different microstructures and phase components (e.g., fiber form and orientation in fiber composites) and the specific process parameters have to be analyzed along the entire process chain and across the different scales of relevance. In the end, locally occurring effects in microstructure of different length scales determine the distribution of properties at the component level. Properties in this context are not limited to strength requirements. The terms "life cycle" and "performance characteristics" also include, fatigue, abrasion, and corrosion – here the focus is on the machining and heat treatment processes as well as the methods of surface treatment and joining.

All the aforementioned statements have a common and central denominator: the importance of the materials used and their (local) properties in compliance with their operational performance. For a targeted design and robust adjustment of the required component properties, the early knowledge and design of the following parameters are indispensable:

- Type, proportion, distribution (segregation), and mode of action respectively purpose of alloy elements
- Phase fractions and their local distribution and properties of the phases
- Grain size and their (statistical) distribution and derivable properties
- Type, proportion, size, shape, and distribution of precipitates
- Existence of preferred property orientations (anisotropy effects)
- Consideration of micro-/macrodefects/damage (porosity, inclusions, blowholes)
- Undesired inclusions (e.g., oxides in casting processes)

In order to control these parameters, appropriate process designs have to be developed in advance, and robust process situations must be found for every single process step. Taking into account the history of the previous manufacturing steps is the only way to reach this objective. The biggest challenge is that the aforementioned factors are often controlled by effects in the micro- and nanoscale and are in most cases not homogeneously but stochastically distributed over the component.

Using the example of a safety-critical automotive component, the outstanding importance of a virtual design approach can be clearly demonstrated: Redesign of such a component first requires selection of a suitable alloy, which has to be ordered from suppliers of semifinished products in an amount of typically 5 tons in business practice. Samples being produced by the desired production process are then subjected to durability tests and – according to the client's specification – to static and/or cyclic torsion or bending tests. Structural analyses, the creation of entire tool sets, and a prototype production under near-series conditions drive the costs for one loop to the six-digit euro range. A similar reasoning also holds for the approval of new aerospace components [14].

2.1.6 Available Modeling Tools at Component Scale

There is a large variety of available software codes and modeling tools at component scale. Subchapters 2.2–2.11 provide a list with respective tools in the different fields. Independent of the dedicated application field and methods being used, three types of software can be distinguished:

- General-purpose (GP) codes
- Commercial codes
- Proprietary codes

GP codes – as the name implies – are not limited (or only to a certain extent) in terms of applicable problems. They enable users to define a wide range of problems with respect to manufacturing processes at the component scale. Their biggest advantage is the unbeatable flexibility. On the other hand, they demand for very experienced users. Handling of simulations may become very complex and time consuming. In some cases such codes would be able to model manufacturing processes, but modeling and computational efforts to spend make it unattractive to use the software.

Commercial codes mostly make use of the same or similar solver solutions like GP codes, but they are specialized for a single main application – in the manufacturing areas this can be commercial codes for primary shaping or plastic forming or welding or heat treatment or surface treatment or machining applications. Specialization in this context means the best fit to the specific areas of application.

Proprietary codes are most of the time highly specialized solutions to cover one given process type or process chain of a company. They have nearly no flexibility to solve other problems but may provide excellent output for the very specific application. Such codes are not available on the market.

All types of software packages shall predict – for the selected manufacturing processes – integral quantities like forces, moments, work, and energy.

Furthermore they shall predict the viability of the process design approach, the existence of potential failure sources, and the evolution and distribution of typical scalar field variables like temperatures, strains, strain rates, and others.

The microstructure of a material has a huge impact on the properties of the component after each process stage and is itself influenced by the process. Modeling tools being designed to compute manufacturing processes at the component level mostly provide averaged output for microstructure features like phase transformation rates and phase fractions, grain growth, and grain size along with some information on recrystallized fraction and in few cases information about diffusion-driven processes. Methods and models to describe the microstructures and their evolution in more detail are discussed in Chapter 3.

When combining different codes, one has to differentiate carefully between the so-called incremental (fully coupled) and offline calculations (decoupled, postvariables). Fully coupled approaches considering the dynamic change of the material conditions (e.g., workability) are expensive with respect to computational efforts. Aspects of numerically coupling different modeling tools are discussed in Chapters 7 and 8.

Within the existing variety of software codes, no single software vendor can provide all the necessary answers. In order to virtually describe the entire manufacturing process of a component, thus several solutions are needed to cover the different process types by suitable modeling approaches. In the existing heterogeneous world of software solutions, it is, however, still hardly possible to consider the manufacturing history of the previous process steps and hence their impact on initial and boundary conditions with respect to the material state and the component properties. This will be one of the major challenges of introducing ICME approaches to a broader industrial community [15].

The following subchapters provide an introduction to the different individual processes in the process chain according to the categorization depicted in Figure 2.1.2 and provide lists of software tools currently available in the individual fields: casting – forming – joining – heat treatment – machining – coating – operations – corrosion – recycling.

References

1. *acatech* bezieht Position – Nr. 3: Materialwissenschaft und Werkstofftechnik in Deutschland. Empfehlungen zu Profilierung, Lehre und Forschung, Hrsg.: Deutsche Akademie der Technikwissenschafte ISSN: 1863–1738, ISBN: 978-3-8167-7913-1, www.acatech.de (accessed 27 April 2016).
2. Prof. Dr.-Ing. Uwe Glatzel Vorlesungsskripte: Aufbau und Eigenschaften Metalle, http://www.metalle.uni-bayreuth.de/de/download/teaching_downloads/Vorl_Einf_Mawi_Metalle/index.html (accessed 27 April 2016).
3. Wikipedia, The Free Encyclopedia Artes Mechanicae, http://en.wikipedia.org/wiki/Artes_Mechanicae (accessed 27 April 2016).
4. von Doege, E. and Behrens, B.-A. (2010) *Handbuch Umformtechnik: Grundlagen, Technologien, Maschinen*, Springer-Verlag. ISBN: 978-3-642-04249-2
5. Stahl http://www.stahl-online.de/index.php/statistiken/ (accessed 27 April 2016).

6. Wikipedia, The Free Encyclopedia History of Metalworking, http://en.wikipedia.org/wiki/Metalworking (accessed 27 April 2016).
7. Berns, H. (2002) Die Geschichte des Härtens, Härterei Gerster, Egerkingen.
8. Osakada, K. (2008) History of plasticity and metal forming analysis, Osaka University, Japan. ICTP 2008, The 9th International Conference on Technology of Plasticity.
9. Wikipedia, Die freie Enzyklopädie Fertigungsverfahren, Bearbeitungsstand: 12 January 2015, 12:32 UTC, http://de.wikipedia.org/w/index.php?title=Fertigungsverfahren&oldid=137704782 (Abgerufen: 23 January 2015, 10:41 UTC).
10. Wikipedia, The Free Encyclopedia (5 January 2015) List of Manufacturing Processes http://en.wikipedia.org/w/index.php?title=List_of_manufacturing_processes&oldid=641057116 (accessed 23 January 2015).
11. Wikipedia http://en.wikipedia.org/w/index.php?title=Damascus_steel&oldid=643676977 (accessed 27 April 2016).
12. Reibold, M. *et al.* (0000) *Nature*, Bd. **444**, S. 286.
13. Ehrlenspiel, K., Kiewert, A., and Lindemann, U. (2007) *Kostengünstig Entwickeln und Konstruieren*. 6, überarb. u. korr. Aufl., Springer, Berlin.
14. Göhler, T., Maiwald-Immer, T., Fischersworring-Bunk, A., and Oberwinkler, B. (2014) Communication standards between ICME tools – a crucial success factor for industrial application. 1st International Workshop on Software Solutions for ICME, Aachen/Rolduc, June 24–27, 2014.
15. ICMEg (2014) 1st International Workshop on Software Solutions for ICME, Aachen/Rolduc, June 24–27, 2014, www.icmeg.info (accessed 27 April 2016).

Appendix

The "Asparagus Process Chain"

This appendix may widen the view of the reader and remind that material modeling and processing of materials is not only limited to metals and alloys but also comprises plastics, ceramics, biomaterials, drugs, pharmaceuticals, biological tissue, and many others. Thus the manufacturing chain for an organic product – asparagus – is briefly depicted in the following, highlighting the influence of modifications of the process chain. The process chain for the production of an asparagus meal – without any claim for completeness – reads as follows:

> Seeding – growing – harvesting – buying/selling – peeling – cooking – coating – assembly – cutting – eating/quality assessment – recycling

Seeding and growing asparagus need the right soil, humidity, a suitable temperature–time profile, no light, and other boundary conditions not being further detailed here. Some farmers take advantage of using waste heat, for example, from power utilities, allowing them to grow their *product faster* and to *harvest it earlier*. They thus gain a competitive advantage by modifying the process conditions and shortening the production cycle as they can sell the product earlier than their competitors.

When selling their product to the end user, often a *very productive peeling service* is offered, where the peeled skin in general remains at the marketplace and is deposited as waste. In the best case it is recycled to form new soil. There may be reasons for the end user to ask for the skin parts as further detailed below. At home the customer starts to further process the asparagus toward the final product, the meal. In case he/she did not get the asparagus peeled at the marketplace, he has to perform this action now.

Cooking is performed in different ways by different people. Traditionally the peeled asparagus rods are either (i) placed into boiling water for a defined period of time resulting in the cooked asparagus meal or (ii) wrapped into an aluminum foil and placed into an oven to be cooked in their own vapor pressure. Method (ii) results in a better – more intense – taste as the taste-making molecules are not diluted in the boiling water. Alternative (iii) is to boil the peeled skin before boiling the asparagus itself. This preprocess saturates the boiling water and the taste of the asparagus remains intense instead of being diluted by the boiling water.

The actual choice of the cooking process eventually depends on the availability of the required infrastructure (oven and aluminum foil vs. cooking pot and water) and on the preferences of the final customer – maybe he would like some butter *coating* on top or the asparagus *assembled* with a slice of ham?

When eating the asparagus, the anisotropic nature of its fibrous microstructure has a strong impact on the cutting process required to make it fit into the mouth. This fibrous nature is very pronounced in the skin regions of the asparagus, and the preceding peeling process thus has a direct impact on the cutting process.

The taste itself – besides their concentration – depends on the molecular, atomic, and electronic structures of the "taste-providing molecules" and their interaction with the sensors in the customer's mouth representing his QA system.

Some lessons to be learned from this – admittedly exotic – example are that temperature–time profiles (e.g., during growth and also during cooking), the ambient conditions and their evolution (both during growth and during cooking/baking), the composition of the product (here the concentration of taste-providing molecules), and the sequence of its processing all significantly determine the final taste: the functionality of the product. In summary, the properties of a product are determined by the history of its entire production chain. The recycling of the skins to form new soil implies ecological sustainability of the overall life cycle of the asparagus.

2.2
Primary Shaping Processes

Christoph Broeckmann, Christian Hopmann, Georg J. Schmitz, Sree Koundinya Sistla, Marcel Spekowius, Roberto Spina, and Chung Van Nguyen

2.2.1
Overview

Most materials for engineering applications are used in their solid state, and in general a specific and well-defined shape is required to make them a functional component.

The basic characteristic of primary shaping processes is to provide a shape to a component for the first time from a fluid (gases, melts, and solutions). Also powders are considered as a fluid in this context. The primary shape does not necessarily correspond to the final shape, and often subsequent forming and machining processes are required to further process the semifinished parts like slabs, rods, and blankets. The importance of primary shaping processes results also from their capability of generating complex shaped parts requiring only minor subsequent processing.

Casting is one of the primary shaping processes which further include sintering, curing of polymers, or also the most recent methods of additive manufacturing (AM). Furthermore, all methods of single crystal growth such as that from a vapor phase of a solution belong to this category. According to DIN 8580, primary shaping is the first major group of the processing technologies, and the following subcategories exist:

Primary shaping

- From the liquid phase (e.g., metallic melt)
- From the viscous state (e.g., polymer melts)
- From pastes (e.g., ceramic slurries)
- From the powder state (e.g., pressing of ceramics)
- From chips
- From fibrous state (e.g., textiles)

Handbook of Software Solutions for ICME, First Edition. Edited by Georg J. Schmitz and Ulrich Prahl.

Condensation from gas or vapor phase
Electrolytic deposition from the ionized state

This chapter gives a brief introduction to crystal growth processes and exemplarily addresses the primary shaping of metallic melts, polymer melts, and ceramics.

2.2.1.1
Solidification and Crystal Growth

The formation of crystals is a well-known phenomenon occurring in nature. In general, respective natural crystals grow by evaporation of solutes. The macroscopic shape of such crystals is not determined by any mold but by the structure and the symmetry of the bonds between the individual atoms forming the crystal.

This also holds for some technical processes aiming at the creation of single crystals, for example, the Czochralski process. Here a seed is withdrawn from a melt at a very slow rate, allowing the melt to crystallize at its interface. The resulting crystals reveal the macroscopic symmetry of the underlying atomic bonds. Typical applications of the Czochralski process are found in the semiconductor, optics, electronics, and photovoltaics industries.

Oxide crystals are often grown in a Verneuil-type process, where molten droplets of the material are deposited onto an existing crystal and solidified, thus adapting the crystal structure of the base material.

Several types of crystal growth processes lead not only to single crystal structures but also to single crystals revealing a predefined macroscopic shape. Examples are the Bridgman process and the vertical gradient freeze (VGF) method. In both methods a molten material is placed in a mold providing a cavity of the desired shape. This mold is then transferred through a temperature gradient providing a directional heat flux. This directional heat flux leads to a directional solidification of the material. Grains not being oriented favorably with respect to the imposed temperature gradient vanish, and eventually only one grain prevails. To shorten this grain selection process or to generate a desired crystal orientation, often helix-type grain selectors or a seeding crystal is used. The Bridgman process is frequently used in the aerospace and power industries for the manufacture of single crystalline turbine blades.

Also the metallurgy of solidifying electronic solders and purification/homogenization methodologies such as zone melting, electroslag remelting (ESR), and vacuum arc remelting (VAR) belong to the area of solidification in a wider sense.

2.2.2
Casting

Casting is one of the oldest processes for the production of metallic goods. Almost any metallic product in everyday life at the beginning of its life cycle is created by casting technologies. The alloys being cast range from steels, cast iron, and light metal alloys (Al based, Mg based) to superalloys, titanium alloys, and many others. The basic method of casting seems easy at a first glance – just pouring a liquid melt into a mold. However, the details of modern casting technologies for high-tech products are manifold and sophisticated.

Different types of casting technologies may be identified. On the one hand, there are moldless casting technologies like continuous casting for the production of semifinished parts such as slabs, rods, or profiles. Such parts, in general, are subject to subsequent processing steps. On the other hand, there are casting technologies directly providing a complex shape to the products. These may be distinguished into permanent mold technologies such as die casting or pressure die casting and lost mold processes such as sand casting or investment casting. All these methods reveal qualitative differences, and all find their typical applications for which they are suited best in terms of fulfilling requirements and meeting cost targets.

Going back to the basic technology of just getting a melt into a mold, solidifying it in the mold, and eventually removing the mold, a number of phenomena have to be considered. When filling the mold with the melt, an escape path for the air/atmosphere within the mold has to be provided. This can either be achieved by vacuum casting methods or by providing a porous mold like in sand casting or investment casting processes. For die casting processes, special ventlines have to be foreseen. Another major issue is to assure that solidified parts within the casting do not block further melt flow that is meant to fill unfilled regions. Significant density differences between the liquid melt and the solid – which, in general, reveals a higher density – lead to the requirement of a so-called feeding system, providing a compensation of the volume changes during solidification.

A typical sequence for the generation of shaped cast parts reads as follows: preparation of models, design of the mold including feeding and gating systems, preparation of the mold, generation of the melt, casting, solidification, heat treatment, removal of the mold, removal of the feeding gating/system, and surface treatment. In the case of hollow cast parts, additional process steps arise for the design of a core, for the preparation of the core, for the placement alignment of the core and its alignment with the mold, and eventually for the removal of the core from the cast part.

The target of a casting process at first is a well-filled cast part not revealing any porosity or surface defects and matching the desired shape tolerances.

Another important target of casting processes is the control of the microstructure in the cast part and eventually of the properties of the cast part. In some cases the evolving microstructure also directly affects the stability of the process such as hot tearing susceptibility or tendency of freckle formation.

Means to influence the microstructure via the process conditions are the cooling rates (where higher cooling rates in general lead to smaller average grain sizes revealing better mechanical properties and to smaller secondary dendrite arm spacings), temperature gradients leading to directional/textured or even single crystalline structures, and seeding technologies either aiming at grain refinement when using multiple seeding particles (so-called inoculants) or aiming at single crystalline structures when using a single well-defined seed or a grain selection mechanism by geometrical constraints in the mold. The microstructure can further be influenced by temperature–time profiles before and after removal of the cast part from the mold and/or by the application of external fields such as ultrasound or electromagnetic fields.

An important aspect to control the microstructure in the cast part is the metallurgy of the melt. Even minor amounts of dopants within the alloy may considerably influence the formation sequence and topology of precipitates and/or secondary phases. Such phases and their topology are either detrimental or beneficial with respect to the envisaged application. Examples are eutectic silicon refinement in Al alloys; control of shape and size of graphite in cast iron; amount and distribution of carbides, nitrides, and other precipitates in steels and superalloys; and many other still not exploited phenomena.

The local formation of such phases strongly depends on the presence and the local concentration of the different alloy elements at different places within the cast part. The distribution of the alloy elements in the cast part is called macrosegregation and plays an important role in the local properties of the materials and the overall properties of the component as well as in further processing in ICME-type process chains.

To obtain more homogeneous properties of the cast part, some applications require additional processes aiming at the reduction/minimization of the macrosegregation. Examples are the remelting processes depicted in the following.

VAR and ESR processes are used for the production of high-quality ingots of superalloys, high-performance steels, and titanium alloys. Major scopes of these processes are the reduction of macrosegregation, the removal of gas phases, and the generation of textured structures. Due to the complexity of the processes and high capital investments and operating costs, a trial-and-error-based approach for process improvement is very expensive, making them an interesting topic for ICME-type optimization approaches. In VAR and ESR, typically cast rods of the materials serve as electrodes being molten and redeposited either by an arc in vacuum (VAR) or by melting the electrode by an electric current passing the molten droplets through a slag and depositing them on a newly formed solid in the case of ESR.

2.2.3 Plastics Processing

Plastics is a general term for macromolecular materials that are created synthetically or based on natural products. They contain polymers and additional materials that are added to influence the material properties, for example, to allow an easier processing. The polymers are made of repetitive units called monomers that are connected via chemical bonds. In general, there are two different types of chemical reactions to connect the monomers to each other and form the polymer chains:

1) Chain-growth polymerization: The chain-growth polymerization denotes a chemical reaction in which unsaturated monomers add onto the polymer chain one by one. Popular materials that are created with this type of process are polyethylene (PE), polypropylene (PP), and polystyrene (PS).
2) Step-growth polymerization: The step-growth polymerization is a chemical reaction in which bi- or multifunctional monomers react to form polymer chains. The step-growth polymerization is usually further subdivided into polyaddition and polycondensation based on the products that are formed during the reaction. The polyaddition denotes processes in which only polymers are formed. If other side products are formed (e.g., water, ammoniac, alcohols, etc.), the process is called polycondensation. Typical products are polyurethanes (PURs) for polyaddition and polyamides (PAs) for polycondensation.

If the polymer is not only made of a single type of monomer, it is called a copolymer.

Pure polymers are barely used in plastics processing because they are often difficult to process and do not match the material requirements. Thus different substances are added to the polymers to form a plastic material. Usually these additional substances are distinguished into reinforcements, fillers, and additives. Reinforcements are materials that are added to influence the final mechanical properties. The most important ones are fibers usually made of glass, carbon, or aramid, which significantly improve the mechanical properties of the material. These fiber-reinforced plastics are widespread in the car and aviation industry. Fillers are materials that are added to influence other properties other than the mechanical ones (such as tribological or electrical properties) or to lower the material costs; for example, calcium carbonate ($CaCO_3$), which can be found as a major component of chalk, is often added in film production to decrease the price of PE films. Because of the higher Young's modulus of the calcium carbonate, it is acting as a slight reinforcement. In contrast to filler and reinforcements, additives are added in a much lower concentration (1 – 5%). They are often used as stabilizer, as antiblocking or bonding agent, or as aid for processing.

Plastics are separated in three different groups called thermoplastics, elastomers, and thermosets. The distinguishing factor between these groups is their temperature-dependent melting/solidification and cross-linking behavior. In general, thermoplastics are not cross-linked and can be melted and solidified multiple times in a reversible process, if the temperature does not exceed a limit where decomposition takes place. Thermoplastics can be further distinguished into amorphous and semicrystalline materials. In contrast to the amorphous materials, which solidify in an isotropic state (assuming there is no external deformation of the melt), the molecular chains of semicrystalline materials can solidify in periodic patterns and form crystalline superstructures. Typical semicrystalline materials are PP, PE, and polyoxymethylene (POM), and typical amorphous materials are polycarbonate (PC), PS, and polymethyl methacrylate (PMMA). In contrast to thermoplastics, the polymer chains of elastomers are weakly cross-linked. Elastomers are widely known as rubber materials. After the vulcanization process, they are not meltable and decompose or combust at high temperatures. Elastomers are usually used if an elastic load transmission, damping of vibrations, or elastic sealing is required. Typical elastomers are natural rubber (NR) and polybutadiene (BR = butadiene rubber). Materials with a densely cross-linked structure are called thermosets. The material properties of these materials are almost constant in a wide temperature range up to the decomposition temperature. They have a good environmental resistance and are often used in fiber-reinforced plastics. A lot of matrix materials that are used in lightweight design plastic components are thermosets such as epoxy resins (EP).

The two most important manufacturing processes for plastic materials are extrusion and injection molding (IM). The extrusion process is used to manufacture continuous semifinished products that can be further processed, such as plates, films, cables, tubes, or complex profiles. Based on the final product, there are many different extrusion processes available, such as film casting and blow film extrusion for the manufacturing of films, profile extrusion for tubes and more complex profiles, or fiber spinning for the manufacturing of textiles and fibers. All these processes have in common that the plastic material is put in an extruder machine where it is continuously melted and homogenized. The extruder is connected to a mold which is responsible for a homogeneous melt delivery to further process units, such as a calibrating unit or a quality assurance unit. Most of today's machines are additionally equipped with a melt filter to clean the melt from pollutions. The exact design of the mold and the following process units are defined by the final product and require an application-dependent construction. Because of the continuous manufacturing process, online systems for process control and quality assurance are important for the extrusion processes. Thus most extrusion facilities use online measurements of quality characteristics to dynamically adjust the process parameters.

In contrast to the continuous extrusion processes, IM represents a discontinuous manufacturing process. A screw-shaped plunger is used to inject molten plastics in a mold cavity. The filling of the cavity is usually followed by a holding phase in which further material is pressed into the cavity to compensate for the material shrinkage during solidification. After solidification the mold is opened and the final part is ejected. Because of the high viscosity from plastic melts, it is necessary to inject them with high pressures of up to 2000 bar. This in combination with the mostly thin-walled components (usual wall thicknesses of injection-molded parts are from 1 to 4 mm) leads to complicated deformation histories of the melt. Additionally, the temperature gradients usually vary in the melt because of the geometrical cavity design or the cooling channel layout. These issues make the prediction of the final properties of the injection-molded parts barely possible in early states of the mold construction. Thus it is quite common that the molds undergo multiple cost-intense adjustments until they are finally used for series production. Besides extrusion processing and IM, there are several other manufacturing processes available on the market. The most important of these are pressing, blow molding, and thermoforming. Compared to IM, pressing is much simpler and cheaper and is usually used for the production of thermoset or elastomer parts in small numbers. It is further used if the material cannot be processed with another manufacturing process, for example, because its viscosity is too high. The blow molding is a method to produce hollow parts at low prices. There are three different types of blow molding methods that are called extrusion blow molding, injection blow molding, and injection stretch blow molding. What they all have in common is that a preform is inserted in a mold where it is inflated and pressed against the mold walls. After the part is cooled down, it is ejected from the mold. The most widely known application for the blow molding process is the manufacturing of drinking bottles often made of polyethylene terephthalate (PET). Thermoforming is used to manufacture three-dimensional (3D) parts from films or plates that are usually extruded before. The semifinished product is heated up and stretched over a mold that has the shape of the final product. It is usually used for the manufacturing of packaging articles.

2.2.4 Sintering

Besides metals and plastics mentioned earlier, ceramic materials play an important role in many applications. They are often used, for example, for high-temperature insulation or as electrical isolators, as functional ceramics, and many more. The high melting point of ceramics, however, limits or even fully prevents their processing by melting and casting processes as detailed earlier. Ceramics and other high-melting-point materials such as refractory metals thus are often shaped by sintering processes.

Sintering has been a well-established component fabrication technique since time immemorial. Some of the early products also include earthen ware, bricks, jewelry, and other artifacts. The applications and the materials used in sintering have evolved over the years.

Sintering is a heat treatment applied to a powder compact at temperatures lower than the melting point of the constituent material in order to impart strength and integrity. The process of sintering generally leads to denser products and also results in elevated mechanical and electrical properties. At the microstructural level, sinter bonding becomes evident as the necks at the particle interactions grow in size. Atomic-level interactions are responsible for the sintering of particles which try to eliminate the surface energies of the interacting particles (Figure 2.2.1).

The entire sintering process can be divided into various stages based on the geometrical transformations the grains undergo during the densification process. When the powders are mixed and poured into the die before green compaction, the particles have a point contact with a random distribution between each other. The initial stage of sintering is characterized with the growth of the sinter bond

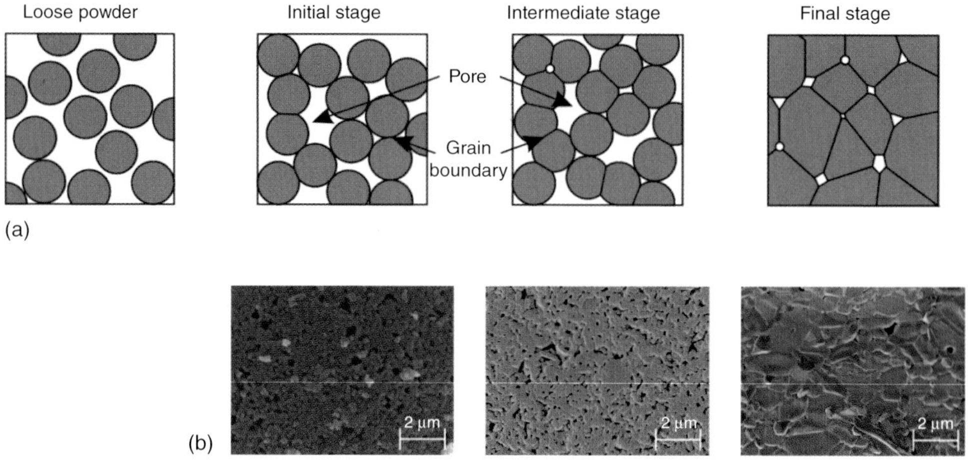

Figure 2.2.1 Schematic representation of a sintering process (a) and sintering of aluminum oxide powder (b).

from the initial single point contact. In the intermediate stage, the pore volume starts to reduce, and the major densification of the compact takes place. This stage is also characterized by pore rounding and increase in grain size. The final stage of sintering is characterized by closure of pores, pore isolation, and migration.

The sintering processes can be broadly classified into pressureless and pressure-assisted sintering [1]. Solid-state sintering and liquid-phase sintering are the most commonly practiced techniques in pressureless sintering. The two processes are distinguished based on the bonding of particles occurring below or above the melting point of the constituent material. The more modern and novel processes include pressure-assisted sintering which aid in the densification and coarsening of the sintering compact. The pressure-assisted sintering techniques can be further classified into low-stress and high-stress processes based on the magnitude of the pressure applied during the process. Creep and viscous flow constitute the major mechanisms in low-stress pressure-assisted processes, and plastic flow (PF) is majorly responsible for sintering with higher stresses.

Hot isostatic pressing (HIP) is a near-net-shape process as it involves heating and pressurizing a powder compact, with a hot gas, sealed in a flexible container. As HIP includes compaction at elevated temperatures, a pressurized inert gas medium is used to pressurize the powder capsule or container, and the plastic mediums used in CIP are replaced with materials that can withstand the elevated temperatures. Full density complex shaped parts or components can be achieved by HIP.

Reactive sintering is a sintering process that forms a sinter compound due to a reaction caused due to heating of a powder compact. These are self-propagating high-temperature reactions, which are sometimes difficult to control, leading to the production of a porous body. Certain intermetallic and ceramic alloys and composites such as borides, carbides, nitrides, oxides, aluminides, and silicates are directly formed from elemental powders using reactive sintering. A controlled reactive sintering process is possible by inducing an inert environment and by restricting the initial peak temperature induced to the powder material. Densification and maintenance of a compact shape are enhanced by preheating the compact and by adding inert particles.

The most recent developments have been spark plasma sintering (SPS) and field-assisted sintering techniques (FAST) where heating of the compact by a furnace is replaced by heating through passage of electric current. This process is advantageous over conventional sintering processes as rapid heating rates are achieved with shorter processing time. The introduction of an electric current enhances the sintering kinetics, and therefore high densities can be achieved at lower temperatures or in shorter times as compared to conventional sintering methods. In this way, full or near-full densification may be realized with minimal undesirable microstructural changes. The usage of a controlled atmosphere is eliminated in FAST. Numerous materials including nanostructured materials, composite materials, and ceramics can be sintered using this novel technique.

Typical holding times in sintering are usually between 20 and 60 min under a controlled and protected atmosphere. The main purpose of using a controlled environment is to control the oxidation that might occur during heating in air. Other additional functions of usage of protected atmospheres include reduction, carburization, and removal of lubricants and binding materials. The most commonly used furnace atmospheres include endothermic gas (hydrogen, nitrogen, or carbon monoxide), dissociated ammonia (DA), hydrogen, hydrogen–nitrogen mixtures, and vacuum.

Sintered iron-based materials made from iron powders, alloy steel powders, or mixtures of iron powder with graphite and other metal powders form the largest segment of the components manufactured by powder metallurgy. Copper-based materials such as brass, bronze, and nickel silver are ranked second in terms of commercial applicability of components produced by sintering in the powder metallurgy industry. Components produced with sintering of powdered aluminum and aluminum alloy are also common nowadays and substitute conventional materials with attractive physical and mechanical properties. Commercially available aluminum powder alloy compositions consist of mixtures made with atomized aluminum powders and powders of various alloying elements such as copper, silicon, zinc, and magnesium. Cemented carbides are generally produced by liquid-phase sintering or HIP achieving 100% theoretical densities. Some of the commonly manufactured cemented carbides by sintering include mixtures of carbides and iron-group metals such as tungsten carbide/cobalt, tungsten carbide/titanium carbide/tantalum carbide/niobium carbide/cobalt, or titanium carbide/molybdenum carbide/nickel. Refractory metals are a class of metals that have atomic numbers greater than 30 in the periodic table. This class of metals has higher melting points and use similar powder metallurgy production processes and applications. Some of the metals include tungsten, molybdenum, chromium, tantalum, niobium, and rhenium. These metals show elevated mechanical properties and are highly resistant to heat and wear when compared to conventional metals. Sintered nickel-based materials enjoy wide commercial usage due to their unique properties, such as thermal expansion, corrosion resistance, wear resistance, electrical conductivity, and magnetic permeability and higher mechanical strength at low and elevated temperatures. Most of the ceramic components are manufactured by sintering, and the most commonly used materials are alumina, titania, zirconia, and so on and are generally manufactured by pressure-assisted sintering techniques [2].

2.2.5
Additive Manufacturing

An emerging technology branch for the generation of primary shaped products is the method of additive manufacturing (AM) – also called 3D printing. 3D printing of plastics, polymers, and waxes in the meantime can even be performed at home, and respective printers are readily available. 3D plastic products may enter into the processing of metals, for example, as models for an investment casting process.

AM of metals is still found in the area of R&D with first technological products being successfully demonstrated. One of the key characteristics of AM is that it does not require any mold. The desired geometry is realized by a controlled path of a laser beam subsequently and locally melting powder layers, for example, in case the of selective laser melting (SLM) processes. Respective AM processes combine different processing technologies and respective phenomena, for example, preparation and sintering of powders, melting and solidification, laser welding, deposition welding, and others. This complexity and the interaction and control of process parameters make AM a highly interesting area for ICME approaches.

2.2.6
Typical Applications of Simulations in Primary Shaping Processes

2.2.6.1
Casting

Objectives for modeling of casting processes are the optimization of cast parts and their processing, the optimization of feeding/gating system design, the estimation of local solidification time, the avoidance of porosity formation, the provision of estimates for phase fractions and for local microstructure characteristics based on criteria functions, and many others.

The optimization of the geometry typically targets an optimal position of the feeding/gating system to avoid the formation of pores within the cast part and to minimize waste respectively circulation material or to obtain a cast part revealing a minimum of distortions. Further targets are to simulate both surface and internal stresses and to predict if they could cause cracking, hot tearing, or formation of other defects. In ingot casting simulations indicate where the shrinkage of the cooling metal will produce voids. Simulations are often the only way to know that they might occur. In continuous casting applications, simulations frequently help to assess the risk of breakouts, thus preventing accidents, delays, and equipment damage.

The outputs of typical casting simulations are, for example, an optimized geometry of the casting system which comprises the component plus the feeding and gating systems, optimized filling sequences such as fluid flow over time, temperature, and velocity fields as function of time. Further optimization parameters are the fraction of phases being present at the different locations in the cast part, criteria functions for microstructure features such as secondary dendrite arm spacing "SDAS," the distribution of alloy elements in the cast part, that is, the "macrosegregation," thermomechanical stresses and strains and many other beneficial information.

2.2.6.2 Plastics Processing

In recent years, the plastics processing companies have shown an increasing interest in the use of enhanced simulation tools to speed up and lower the costs of the design and the realization of prototypes and products. One of the main fields of application for modern simulation tools in plastics processing is the IM, especially the mold design. It is important for the mold designer to know about the filling and cooling behavior of the melt in the cavity at an early state of the design process. It allows him/her to identify weak spots in the mold construction such as hotspots or air traps and to adequately adjust the design. The cavity filling can be modeled as a multiphase flow problem in which the melt itself, the air in the cavity, and the solidified parts of the melt represent three different phases. Another important application for the IM simulation is the prediction of the final part geometry after the ejection. During the manufacturing, the melt undergoes a complex deformation and temperature history which results in a distribution of residual stresses in the part. After ejection the part is free of any mold constraints, and the compensation of the residual stresses leads to a distortion of the component. If the distortions are undesired, they usually imply a lowering of the product quality that has to be compensated by an adjustment of the mold. To avoid these cost-intense modifications, the prediction of shrinkage and warpage is one of the major goals of today's software providers and research institutes. However, nowadays, it is only possible to make a qualitative prediction of the final distortion. A quantitative computation for any material/cavity combination is still part of current research.

Another important field of application for plastics processing is the structural analysis. The increasing complexity of the components in combination with the demand for lower costs results in much higher requirements for the predictability of the mechanical behavior. Structural components made of plastic materials are barely accepted by the customers when they cannot be computed. Because of the close relation between the manufacturing process and final part properties, it is usually necessary to take inhomogeneous material properties into account. If the simulation tool of the manufacturing process is not including a sufficient module for the structural analysis, this has to be done in integrative simulation chains. Typical applications for integrative simulations are the application of custom-made material models or the calculation of special part properties (such as the acoustic behavior).

Simulation methods for the extrusion process exist but are not used as widely as in IM. The most common application is the construction of the die, for example, a spiral mandrel die or a flat thin die for the manufacturing of films. The goal of the die design is a homogeneous velocity distribution at the die exit to allow a robust manufacturing process. During the design process, it is necessary to compute and compare a lot of different designs which usually makes time-consuming finite element analysis unsuitable. However, very often, the task for the die is standardized and well defined. Thus specialized simulation tools have been developed in which standardized and well-defined extrusion die concepts are used to find suitable flow channel geometries for a specific material/process combination. These tools are based on the network theory in which the flow channel is approximated by a 2D network of geometrically simple elements that account for the computation of the local flow. An analogy between electrical engineering quantities (such as current, voltage, or resistance) and hydrodynamic quantities (such as volume flow, pressure, or flow resistance) is used to calculate the characteristic of the entire flow channel. These calculations can be done in a few seconds allowing the comparison of hundreds of designs in a short time.

2.2.6.3 Sintering

Development of numerical models and simulations of various stages of sintering has been in progress from the middle of the last century. Based on the different scales of modeling, sintering models can be classified in the following manner (Figures 2.2.2 and 2.2.3).

On the component level, numerical simulations of sintering are performed to study parameters such as densification, sintering stresses, grain growth, and geometrical tolerances on the components produced. Predicting a uniform

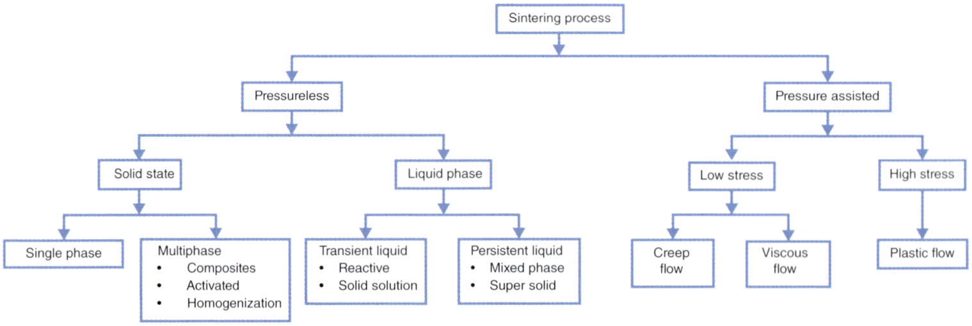

Figure 2.2.2 Classification of sintering processes. (Redrawn from [1].)

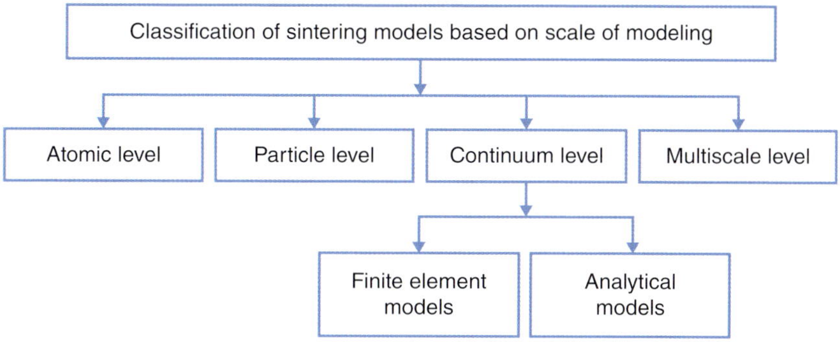

Figure 2.2.3 Classification of sintering models based on scale of modeling. (Redrawn from [3].)

densification and close geometrical tolerances is of utmost importance to the industry, as close prediction of these parameters helps in the reduction of material and man power wastage.

Certain analytical approaches are available for the prediction of densification and grain growth of a powder compact but are restricted on their usage as they fail to predict the geometrical changes a component undergoes during sintering. One of the first analytical models was developed by Ashby [4]. He constructed diagrams which help to predict the dominant transport mechanism and the rate of sintering for a particular temperature, grain, and neck size. The interpretation was carried out for metals and ceramics. A master sintering curve was developed by Johnson and Su [5]. It characterizes the sintering behavior of a powder compact regardless of the heating profiles. This procedure helps to plan sintering strategies to a certain extent with the aforementioned drawback.

The finite element [3] approach is one of the approaches most widely used for component-level modeling and simulation of sintering. In this approach the compact is divided into small finite elements which are connected to each other and transmit the stresses, forces, and temperatures through the continuum, and hence, most of the sintering simulations are coupled thermomechanical simulations. Numerical solvers solve equations via various numerical methods and integration schemes applied to each element. Since the material under consideration is a porous material, constitutive equations are required to implement the sintering behavior. This is carried out by defining state variables that are programmed with the constitutive relations. These relations for sintering are best represented and understood through stress–strain rate relations. In the case of sintering, the powder compact behaves as a continuum body containing porosities, of which the evolution can be described by an internal state variable correlated to the relative density.

The finite element analysis helps us predict the stresses, strains, evolution of grain size, and relative density along with shape and dimensions of the powder component over the entire time period of the sintering process. These entities are calculated as functions of time, temperature, and location and can be used to characterize the microstructure. Finite element models are one of the most commonly used component-scale models for sintering, and these models built for sintering can be broadly classified into phenomenological models and physical models. The phenomenological models for sintering, developed by Olevsky [6], are based on generalized viscous flow of porous bodies, linear viscous incompressible material with voids, incompressible nonlinear viscous material, and nonlinear viscous porous material. Abouaf *et al.* [7] developed a constitutive model for simulating HIP. The phenomenological model for HIP is built on the concept of relating the viscoplastic part of a constitutive equation to the rate of variation of porosity. Due to the application of high pressure in HIP, several mechanisms such as particle rearrangement, plastic deformation of the particle, formation of spherical pores, isolation of pores, and so on occur simultaneously. Due to the limitation on the predictive power of the phenomenological models, physical models were built by considering the effects of various physical diffusion mechanisms (grain boundary (GB) diffusion, surface diffusion (SD), volume diffusion, PF, etc.) on physical sintering parameters such as grain size, neck size, temperature, and sintering stresses at various stages of sintering. A comprehensive micromechanical model for sintering was built by Riedel and coworkers [8, 9, 10]. A relation between the strain rates and the macroscopic stresses was derived by equating the stresses on the GB with the macroscopic stresses.

2.2.7 Phenomena to be Modeled

2.2.7.1 Casting/Crystal Growth

The major phenomena to be modeled in casting processes can be categorized according to the different phases being involved (melt, gas, solid mold, solid core) and eventually by their interaction:

a) Fluid dynamics of the melt and the ambient gas phase
   Fluid flow plays an important role in any casting process. The fluids, in general, are multicomponent liquids, and they move in a multiphase flow simultaneously revealing gas–liquid, liquid–solid, or even liquid–solid–gas phases. The resulting flow patterns can either be forced flow during mold filling, forced convection in special casting processes such as centrifugal casting, natural convection during cooling, and gas flow through porous media in sand casting. The outer part of the mold, in general, is subject to a convective cooling boundary condition.
   The respective flow patterns strongly influence the local temperature distribution and also the local distribution of the alloy elements, that is, the macrosegregation.
b) Thermomechanics of the mold, of the core, and of the solidified cast part

Thermal expansion of all solid parts is important in view of the large temperature intervals being covered by casting processes. Mold and core have to reveal a sufficient strength at high temperatures to withstand the metallostatic pressure and further forces acting on both mold and core. Shrinkage during cooling leads to distortions of the cast part and to the formation of eigenstrains. In particular, the large density difference between the liquid melt and the solidified body leads to volume shrinkage upon solidification, causing a formation of gaps between the mold and the cast part.

c) Cooling
Cooling of the cast part proceeds by radiation, convection, and heat conduction. Radiation dominates the cooling process at temperatures above 600°C. At lower temperatures heat conduction becomes increasingly important. An important issue is the formation of "air gaps" which appear between the metal and the mold as the metal begins to cool. The formation of these gaps is due to density differences between the solid and the melt. They significantly influence the heat transfer coefficient and accordingly the subsequent cooling process.

d) Phase transformations
The major phase transformation in a casting process is solidification. Other phase transitions such as solid-state transformations or precipitation phenomena occurring at lower temperatures do not influence the casting process itself. The formation of tiny precipitates in the melt, however, may significantly influence the nucleation of the solid phase and thus the transformation temperature. Also the hot tearing susceptibility of the cast part may be influenced by such precipitates. The phase transformation is especially associated with the release of latent heat which has to be accounted for in any casting simulations. Other important aspects of any phase transformation are the redistribution of solute between the different phases and the generation of elastic stresses/strains in the cast part.

e) Diffusion phenomena
Diffusion phenomena – especially in the liquid melt in most casting processes – locally compete with or even are dominated by macroscopic flow phenomena. They determine the temperature, the sequence, and the amount of phases being formed. In many cases the Scheil model can be used to obtain estimates for the compositions of the solidifying solids and of the melt as well as for the phase fractions being formed. In contrast to thermodynamics equilibrium calculations (assuming infinite diffusion in both liquid and solid), the Scheil model assumes infinite diffusion in the liquid and no diffusion in the solid. For fast diffusing atoms (e.g. carbon), a modified Scheil model assumes also infinite diffusion in the solid. Increasingly modeling of diffusion requires finite values for diffusion in both the solid and the liquid (see Chapter 3).

f) Wetting phenomena
Wetting phenomena at interfaces being present in the process play a major role. Capillarity effects associated with wetting are especially important for the filling of thin-walled parts.

g) Influences of external fields
   A number of external fields have been shown to significantly influence casting processes. Besides the temperature fields, which can be considered as generic for casting processes, especially ultrasound is often applied to influence the nucleation of the first solid phases. Ultrasound often further finds an application when removing ceramic molds. Electromagnetic fields are frequently used in inductive melting, in the control of fluid flow in the melt, and eventually in the control of the positions of secondary phases within the melt.

2.2.7.2 Plastics Processing

For the simulation of thermoplastic materials, the flow and cooling process the melt undergoes has to be calculated. Thereby, a lot of factors such as the unsteady melt flow with a moving flow front, the contact between the non-Newtonian viscoelastic melt and the air inside the cavity, the phase change during the process, and the heat transfer have to be considered in a model. In general, a manufacturing process for plastics can be classified by verifying if one or more of these basic phenomena exist:

1) Handling of particulate solids or mixtures. The manipulation of the base materials and additives (shapes and states) and the way to preshape them before processing (pellets, film, etc.) are important steps to assure stable and repetitive processing conditions.
2) Heat softening and melting. The phenomenon of reaching a well-defined heat softening or melting temperature strictly depends on the heating rate, the presence of impurities (additives, fillers, etc.), the type of polymer (e.g., amorphous or semicrystalline), and the molecular weight distribution.
3) Melt deformation. Once the polymer is in a softening or melt state, a die or a mold can be used to shape the plastic into the final product. The molecules undergo a pressure, temperature, and velocity profile in which they can be aligned in flow or stretch direction.
4) Pressurization and pumping of the plastic melt. Some processes require that additional materials enter into the mold to compensate shrinkage effects.
5) Solidification or cross-linking. The material can be solidified by cooling below glass transition temperature or melt temperature (thermoplastics) or by a chemical reaction (thermosets).

2.2.7.3 Sintering

The manner of mass flow in the sintering process can be understood by studying the effects of various transport phenomena that occur during the movement of vacancies against the driving force in sintering and the collection of the vacancies that are treated as pores. The creation of pores, pore reduction, and isolation are key mechanisms to the sintering behavior. The transport phenomena have been broadly classified by German [1] into surface transport and bulk transport mechanisms.

In the surface mechanisms the vacancies move along the surface of the grain and are prominent during the initial stages of sintering as the surface areas are large compared to the later stages. The different surface mechanisms are SD and evaporation–condensation (E–C). The surface mechanisms contribute more to necking and less to densification. When the vacancies move along the surface of the grain, it is termed as SD. In E–C diffusion an atom evaporates to condense on another surface, thus causing the atoms to move across pore spaces.

On the other hand, bulk transport mechanisms are mainly responsible for the densification of the powder compact. The different bulk transport mechanisms are GB diffusion, PF and volume diffusion, or bulk diffusion (BD). When the vacancies move along the GB, the mechanism is termed as GB diffusion, and when they move through the lattice interior, the mechanism is termed as volume diffusion or BD. PF is the dislocation movement of mass under applied force that causes the material to flow.

2.2.8 Basic Equations to be Solved

2.2.8.1 Casting/Plastics Processing

The basic equations to be solved can be generalized for plastics and metals if similar processes are investigated such as IM and high-pressure die casting (HPDC). The differences are mainly related to the values of the material data and/or typical phenomena which occur during each process. Using this assumption, three main groups of models/equations can be identified:

1) Flow models
The Navier–Stokes equation is employed to describe pressure-driven flow, coupled to the Fourier heat conduction equation. In this way the flow is correctly computed by considering the heat fluxes between the melt and the mold. Multiphase flows in the laminar regime are commonly used to represent the simultaneous coexistence of air and material into the cavity. The model can be extended to include additional features such as the k/ε turbulence (for metals) or the application of pressure or vacuum.
2) Solidification models
The Fourier heat conduction equation is used for both classes of materials. The models include the solidification laws which specify the phase transformation enthalpies (melt heat, glass transition) and important effects such as the different precipitation of solidification phases (metals) or crystallization (polymers) caused by supercooling. In addition, values calculated during the

flow simulation, such as particle orientation, viscosity changes, cooling rate, and differential temperature gradient, allow the computation of the final microstructures. In addition the mechanical contact between the material and the mold is crucial to be modeled to assure the correct computation of the heat contact resistance.

3) Stress models
The formation of residual stresses and distortions is very complex because it is affected by different process conditions such as melt temperature, mold temperature, pressure or vacuum conditions, and flow time. The correct prediction of temperature and solidification, as well as an accurate material model, is the basis to simulate stresses and distortions. These phenomena are induced by shrinkage constraints in the mold as well as the part geometry including the gating system. Residual stresses are primary (independent of geometry and occurrence during cooling) or secondary (depending on the geometry of the component). In modeling the development of residual stresses, the mechanical model describes the evolution of thermal (and mechanical) strains due to thermal (and mechanical) forces. Stress relaxation effects play an important role, and thermoviscoelastic models with volume relaxation are commonly used. The thermal, mechanical, and material models are not independent but can show strong interaction in terms of thermomechanical equilibria.
4) Particular models
High temperatures in metal casting and the alloy composition being quite complex in general require further models. At temperatures exceeding approximately 600 °C, particularly radiation becomes an important issue, and materials' spectral emissivities and shielding effects have to be considered. The complex alloy chemistry requires composition-dependent models and data on thermodynamic equilibria/phase equilibria and further models describing phase transformations. At the process scale, frequently empirical models or criteria functions are used to describe phase transformations. An example is the Johnson–Mehl–Avrami–Kolmogorov (JMAK) approach describing the transitions between two arbitrary states on the basis of a function, whose parameters in general are fitted respectively calibrated based on experimental findings.

2.2.8.2 Sintering

The resulting work of various models on sintering was the construction of a constitutive strain rate equation in tensor form. The strain rate tensor equations are functions of the deviatoric stress tensor, some material parameters, and some sintering parameters such as sintering temperatures, grain size, hydrostatic stresses, and relative density. The densification equation or the rate of densification used in these models is constructed based on the mass conservation equation and is dependent on various mechanisms such as creep, diffusion, and grain growth. Additional equations that support the strain rate equation for physical models are equations of grain growth built on the Hillert law, the sintering stress equation which is a function of specific surface energy and grain size, and equations for shear and bulk viscosities that take all the mass transport mechanisms into consideration. In finite element simulations these equations are calculated for every integration point and are functions of time and position.

Based on the theories of plastic and nonlinear viscous deformation of porous bodies, Olevsky *et al.* [6, 11, 12] developed a phenomenological equation for sintering where the strain rate of the sintered material is represented as the sum of response items of free sintering and the viscous material:

$$\sigma_{ij} = \frac{\sigma(W)}{W}\left[\varphi\dot{e}_{ij} + \left(\psi - \frac{1}{3}\varphi\right)\dot{e}\delta_{ij}\right] + P_{\mathrm{L}}\delta_{ij}, \tag{2.2.1}$$

where σ_{ij} is the external applied stress, the first term represents the material resistance, and the second term corresponds to the influence of sintering stresses. W is the strain rate, e is shrinkage rate, and φ and ψ are the bulk and shear moduli and are functions of porosity of the material, and P_{L} is the effective sintering stress.

The comprehensive macroscopic strain rate tensor equation was derived and used by Riedel and coworkers for the solid-state sintering [8–10, 13]:

$$\dot{\varepsilon}_{ij} = \frac{\acute{\sigma}_{ij}}{2G} + \delta_{ij}\frac{\sigma_{\mathrm{m}} - \sigma_{\mathrm{s}} + \Delta p}{3K}, \tag{2.2.2}$$

where microscopic strain rate tensor $\dot{\varepsilon}_{ij}$, is treated as a linear function of deviatoric stress tensor $\acute{\sigma}_{ij}$, hydrostatic stress σ_m, sintering stress σ_s, gas overpressure Δp, shear G, and bulk viscosities K defined as follows. The physical sintering model takes into account different diffusive transports of matter as mentioned in the previous section and grain coarsening and also makes a clear distinction between the mechanisms taking place in different stages of sintering:

$$\dot{\varepsilon}_{ij} = \frac{\acute{\sigma}_{ij}}{2G} + \delta_{ij}\frac{\sigma_m - \sigma_s + \Delta p}{3K}. \tag{2.2.3}$$

In the same model the bulk viscosity K and shear viscosity G have been defined in the following manner. The values are directly proportional to sintering temperature T, third power of grain radius R, a phenomenological parameter U for particle rearrangement, normalized viscosities $k_1\, and k_2$, and $g_1\, and g_2$ for bulk and shear viscosities, respectively, at open and closed pore conditions. They are inversely proportional to the atomic volume of the particle Ω and GB thickness times GB diffusion coefficient δD_b, as GB diffusion is treated as a dominant mass transport mechanism in the aforementioned model. A detailed summary of the constitutive equations has been explained in [9, 10]:

$$K = \frac{kTR^3}{\Omega\delta D_b}((1-\theta)k_1 + \theta k_2)U. \tag{2.2.4}$$

$$G = \frac{kTR^3}{\Omega\delta D_b}((1-\theta)g_1 + \theta g_2)U. \tag{2.2.5}$$

A densification model was developed by Abouaf and coworkers [7] for HIP. Similar to the macroscopic model developed by Riedel for sintering, the HIP model also constitutes of a strain rate $\dot{\varepsilon}_{ij}$ equation dependent on relative density D, equivalent stress σ_{eq}, hydrostatic stress σ_{kk}, and rheological parameters $c\,and\,f$, which are known as porosity coefficients and are determined by performing uniaxial compression tests on fully dense and porous materials [14]. Details of the constitutive equation can be found in [7, 14–16]:

$$\dot{\varepsilon}_{ij} = A\sigma_{\rm eq}^{n-1}\left[\frac{3}{2}cS_{ij} + f\sigma_{kk}\delta_{ij}\right], \tag{2.2.6}$$

$$f = 0.16\left(\frac{1-D}{D-D_0}\right)^{1.1}, \tag{2.2.7}$$

$$c = 1 + 5.23\left(\frac{1-D}{D-D_0}\right)^{0.914}, \tag{2.2.8}$$

where S_{ij} is the deviatoric stress tensor, A is the Dorn constant, n is the creep exponent of the fully dense compact under power law creep, and D_0 is the initial relative density of the powder compact.

2.2.9
Initial and Boundary Conditions

2.2.9.1
Casting

Initial and boundary conditions for a casting simulation typically are defined by the process conditions and by the geometry to be simulated. The most relevant process parameters during casting processes are the following: composition of the alloy, superheat of the melt, preheat temperature of the mold, imposed external cooling rates, casting atmosphere, pressure of the melt, flow velocity of the melt stream, orientation of the setup with respect to the gravity vector, pulling speed in the case of continuous casting processes, and many others.

An important role takes the geometry of the cast part (final component including the required feeding/gating systems) and of the mold (control of cooling rates, thermomechanical stresses) and the mutual arrangement of multiple cast carts in view of shading effects.

2.2.9.2
Plastics Processing

Initial and boundary conditions complete the statement of the problem. It is important to note that two fluids (polymer melt and air) usually coexist. For this reason different boundary conditions are set depending on the material. Usually, on the entry section, the melt velocity and temperature are imposed by a flow rate as a function of the parameters of the plasticizing unit (e.g., velocity of the screw). The air does not move at the beginning of the simulation. Because of thickness changes in the flow channel or material freezing during the flow front advancement, the polymer velocity does not remain constant but undergoes several variations. This leads to flow fields in which elongational and shearing deformations of the melt occur. It should be noted that a lot of simulation models assume a homogeneous flow field in which these changes should not be too high. In the case of the filling of a cavity, a free boundary condition is often imposed on the unfilled part which allows air movement, whereas no-slip boundary conditions (zero velocity at wall) are imposed on the filled parts. At the end of the cavity, the polymer velocity is zero, while the air can escape from the mold (venting). Concerning temperature and pressure boundary conditions, the temperature at the beginning is equal to the melt temperature, while the temperature of the air is equal to the environment temperature.

2.2.9.3 Sintering

The initial conditions of powder compact are very important parameters to simulate sintering process. The important initial conditions are initial temperature, grain size, and relative density. The constituent material properties also play an important role. Basic models assume a uniform initial temperature and relative density distribution. But more realistic simulations also consider nonuniform distribution for accurate prediction of densification. Based on the type of modeling, initial conditions also have phenomenological parameters such as particle rearrangement and physical parameters such as activation energies and preexponential factors for GB diffusion, SD, and volume diffusion. These parameters are generally determined by additional experiments. Boundary conditions influence the simulation during the sintering process. Friction, thermal properties such as conduction, specific heat, coefficient of expansion, and radiation and geometrical restrictions in the form of tools, when forming complex geometries, also influence the sintering simulations. Pressure-assisted sintering models are sensitive to the pressure cycles. In sinter forming, pressure is applied after the maximum temperature is achieved and is released during the cooling phase. In HIP, the pressure profile is similar to the temperature profile in terms of heating, cooling, and constant temperature stages. In the case of SPS/FAST, the powder compact is subjected to an electrical field, and in this case the additional boundary conditions would be the voltage applied and the frequency of the pulse that is applied.

2.2.10 Required Data and their Origin

2.2.10.1 Casting

The data entering into casting simulations can be categorized into geometry data, material data, and process data.

Geometry data at a first glance refer to the geometry of the final component at its nominal operating temperature. The design of a mold cavity that eventually allows fulfilling this specification is already a complex engineering problem. However, a cast part requires further features such as a feeding and a gating system that allows filling the cavity without creating any porosity. Specification of position and size of these features is a typical optimization problem in casting process simulations. Very thin-walled regions in the mold cavity may cause problems in mold filling in view of capillary effects.

Further geometry data refer to the external geometry of the mold. Regions revealing a thicker mold wall will lead to lower cooling rates. Artificial placement of materials with a high heat capacity into the mold material in contrast will allow for a local increase of the cooling rate influencing the local microstructure.

As casting in general proceeds at high temperatures, radiation effects are relevant. The external geometry of the mold and its mutual arrangement with a number of other molds or within a furnace thus play an important role in terms of shading and view factors.

Following material parameters and the temperature dependence of mold materials(s), furnace materials and molten/partially solidified alloy play a vital role in the mold filling process:

> Liquidus and solidus temperatures of the alloy, viscosity of the melt, fraction solid as a function of temperature, densities, specific heat capacities, enthalpy of fusion, thermal conductivities, heat transfer coefficients, spectral emissivities, thermal expansion coefficients, diffusion data, and many more.

While a number of these parameters in the meantime can – even for complex alloys – be determined from thermodynamic calculations if suitable databases are available (see Chapter 4), especially kinetic data such as thermal conductivities by now are still often experimentally determined. Models at smaller scales (see Chapter 5) may allow the calculation/prediction of respective values for complex alloys at high temperatures in future.

Process data in general enter into the simulations as initial and boundary conditions and are specified in the respective sections. They are either based on experience or subject of the optimization of the process itself.

2.2.10.2
Sintering

On a component scale, the required data for sintering are material data and model data. Young's modulus, density, specific heat, coefficient of thermal expansion, conductivity, and so on are some of the material data of the constituent material that are required and are determined by some simple laboratory experiments. Powder characteristics such as particle size distribution, particle shape, dihedral angles, and packing densities are also important data as they influence the densification of the powder compact and are generally determined by microstructure image analysis. On the other hand, based on the model being used, there are some model parameters that are needed to simulate the respective process model. Other data that are required are initial grain size which is determined using microstructure image analysis, initial relative density which is determined before the start of an experiment, and temperature, pressure, and strains which are recorded during the simulations. Accurate furnace temperature prediction and the environment used in the sintering experiments are crucial for the prediction of sintering properties in simulations.

Material data	
Young's modulus	Universal testing machine (UTM)
Density	Dimension and weight or X-ray/gamma ray attenuation
Specific heat	Differential scanning calorimeter (DSC)
Coefficient of thermal expansion	Dilatometer
Thermal conductivity	Laser flash analyzer and heat flow meters
Electrical conductivity	Eddy current measurement
Model data	
Activation energies	Chemical analysis
Pre-exponential factors	Chemical analysis
Dihedral angle	Microstructure image analysis
Surface energies	Contact angle measurement, gas adsorption/permeability
Pore size	Quantitative microscopy
Rheological parameters	Respective sintering experiments

2.2.10.3
Plastics Processing

A crucial point for the accuracy of simulation models for plastic materials is the modeling of the material properties and the capability to link them to the process. Usually this is done by formulating the material properties as a function of the process parameters such as temperature, pressure, and shear rate of the melt flow. However, the specific formulation depends on the material system and can be quite complicated. Especially semicrystalline thermoplastics require the consideration of the crystallization process. Some approximations calculate a crystallization degree and include it in the interpolation for the material properties. Unfortunately this neglects microeffects such as the local distribution of crystalline superstructures. Typical material properties that are needed for the simulation of manufacturing processes with thermoplastic materials are density, viscosity, specific heat capacity, temperature conductivity, and bulk modulus.

These are usually obtained by measurements in the laboratory that are used to fit the material models. Because the measurements barely represent the real manufacturing conditions of high pressures, melt deformations, and cooling gradients, the models have to be extrapolated which often implies inaccuracies. The following measurements of the plastic material are usually required:

Material data	Parameter	Instrument
Rheology	η, fitting coefficients of the used model (power law, Carreau, cross-CLW)	Rotational, capillarity, slit die rheometers
Thermal analysis	Thermal conductivity	Laser flash analyzer, comparative cut bar, flux gauge, guarded hot plate, and hot wire
	Specific heat	DSC
	Melt, glass transition, vicat, ejection temperatures	DSC
pvT	Density, fitting coefficients of the used model (Tait, Schmitz, Renner)	pvT apparatus
Mechanic	Young's modulus, Poisson's ratio	Tensile test machine
	Coefficient of thermal expansion	Dilatometer

Besides the material properties, the simulation also requires process data. The required parameters strongly vary between the different manufacturing processes. However, what the processes all have in common is that they require the melt temperature and the cooling/heating conditions of the environment.

2.2.11 Simulation Codes in the Area of Primary Shaping

The following table summarizes a large number of existing simulation codes in the area of primary shaping. The codes for this software classification follow the overall scheme of this book, that is, 2xxx for process scale codes and in more detail 22xx primary shaping codes for this chapter.

The codes for primary shaping 22xx have been further classified into the following subcategories:

2210 Crystal growth/solidification
2220 Casting
2221 Flow solver
2222 Solidification
223× Sintering
224× Additive manufacturing
225× Plastics processing
226× Remelting/refining ESR/VAR

Crystal growth codes

SW cat	Name of code	Website of software	Type of code	Short description/typical applications/ phenomena being treated
2210	CGSim	http://www.str-soft.com/products/CGSim/		The CGSim (Crystal Growth Simulator) code is specialized for the simulation of Czochralski (Cz), liquid-encapsulated Czochralski (LEC), vapor pressure–controlled Czochralski (VCz), and Bridgman growth
2210	FEMAG Soft	https://www.femagsoft.com/		The FEMAG Czochralski (FEMAG/CZ) software is dedicated to the simulation of the Czochralski process (Cz, MCz, VCz, Kyropoulos) The FEMAG Directional Solidification (FEMAG/DS) is dedicated to the simulation of the directional solidification process (DS, HEM, VB, VGF). The FEMAG Floating Zone (FEMAG/FZ) is dedicated to the simulation of the floating zone process (FZ, PFZ)
2210	CrysMAS CrysVUn,	http://download.iisb.fraunhofer.de/downloads/Manual/index.html		CrysMAS is a program for the global numerical simulation of crystal growth processes in complex furnaces with axial or translational symmetry CrysMAS is able to perform stationary and time-dependent simulations for furnaces heated by resistance as well as by inductive heaters. Fluid flow, inductive heating, different kinds of treated electromagnetic fields, and alloy solidification can be considered

Crystal growth codes

SW cat	Name of code	Website of software	Type of code	Short description/typical applications/ phenomena being treated
2210	Krist Mag	http://www.matheon.de/research/show_project.asp?id=37		Research project on crystal growth in magnetic fields
2210	Leo Crystal			LeoCrystal is a software performing numerical Monte Carlo modeling of reaction on the surface of the growing crystal for illustrative educational and research purposes in the area of crystallization phenomena It provides estimates for the influence of effective geometrical sizes of structure elements of the crystal and thermodynamic parameters of the process on the topology and kinetics of the crystallization
2220	Any Casting	www.anycastsoftware.com	Commercial	AnyCasting (casting) is based on the FDM and allows analyzing the filling and solidification pattern during casting process such as die casting, sand casting, investment casting, and others AnyTX (thermal stress) draws on the FEM to predict the displacement and thermal stress distribution in the casting and in the mold during solidification and cooling ConCast (continuous casting) allows for the analysis of the continuous casting process especially for steel and aluminum AnyFoam is a simulation software for analyzing the foaming molding process for polyurethane

Crystal growth codes

SW cat	Name of code	Website of software	Type of code	Short description/typical applications/ phenomena being treated
2220	CAD mold	http://www.simcon-worldwide.com/pages/en/products/cadmould.php?lang=EN	Commercial	Automatic geometry generation and meshing. Very high resolution in areas of high gradients (e.g., temperatures and shear velocities across the wall thickness). Exact simulation of volumetric part areas and areas with discontinuous wall thicknesses
2220	SRIF Cast	http://www.foundryworld.com/uploadfile/20094260585877.pdf	Commercial	
2220	Castech	http://www.castech.fi/	Commercial	CastCAE – casting simulation software form•Z – 3D CAD and rendering engine CastDESIGN – risering design software CastCHECK – component analysis software CastSPEX – quality specification software
2220	EKKcap cast	http://ekkinc.com/index.html	Commercial	EKKcapcast software includes automatic finite element mesh generation and solidification, thermal, fluid flow, stress, and porosity analysis tools
2220	Click2 Cast	Quantech.es	Commercial	Click2Cast has possible applications in a wide variety of industries, such as mold making, automotive, naval, aerospace, train, electrical appliance, furniture and casting industries

Casting codes

SW cat	Name of code	Website of software	Type of code	Short description/typical applications/ phenomena being treated
2220	Finite Solutions	http://finite .solutions/	Commercial	SOLIDCast; FLOWCast, a full-featured computational fluid dynamics (CFD) simulation of flow. Based on solution of the Navier – Stokes equations for fluid flow. With FLOWCast, users can view progressive temperature, fluid velocity, and fluid pressure during the fill from any angle of view; OPTICast builds on the user's initial design, then modifies it, and runs simulations to achieve an optimum result. An easy-to-use design tool, OPTICast maximizes yield and quality while freeing design engineers from the repetitive task of trial-and-error design
2221	FLOW-3D	http://www .flow3d.de/	Commercial	FLOW-3D is a powerful and highly accurate CFD software giving valuable insight into many physical flow processes during casting and water resource engineering. It realistically simulates flows with free surfaces, moving bodies in exact geometries with effective meshing
2221	Fluent	http://www .ansys.com/ Products/ Simulation+ Technology/ Fluid+ Dynamics/ Fluid+ Dynamics+ Products/ ANSYS+ Fluent	Commercial	ANSYS Fluent software contains the broad physical modeling capabilities needed to model flow, turbulence, heat transfer, and reactions for industrial applications. Special models that give the software the ability to model in-cylinder combustion, aeroacoustics, turbomachinery, and multiphase systems have served to broaden its reach

Crystal growth codes

SW cat	Name of code	Website of software	Type of code	Short description/typical applications/ phenomena being treated
2220	JSCast	www.jscast.com.cn/en.as		3D integrated CAE system for casting optimization with a wide variety of specific applications. Misrun and gas porosity prediction (mixed mesh, back pressure); shrinkage prediction in ductile cast iron, casting deformation simulation porosity prediction; melt flow in short sleeve (multistep injection, parallel shot, distance-plunger speed setting); smoother and faster rendering when rotating or moving geometry models with OpenGL; and faster and easier to identify defect locations with CT-Scan animation and arbitrary sectioning
2221	Kratos	www.cimne.com/kratos/	Open source	Kratos is a framework for building multidisciplinary finite element programs. It provides several tools for fast implementation of finite element applications: multiphysics CFD, CSD, thermally coupled problems, and particle problems
2220	MAGMA	www.magmasoft.de	Commercial	MAGMA 5 is designed to predict total casting quality by simulating mold filling, solidification and cooling, and, optionally, stresses/strains as well as microstructure formation and property distributions in casting manufacturing processes With its modular design, MAGMA 5 contributes to cost reduction along the complete process chain of a cast component: from conceptual to final component design, tooling layout, and prototyping all the way through to process optimization, reliable production, and quality management

Casting codes

SW cat	Name of code	Website of software	Type of code	Short description/typical applications/ phenomena being treated
2220	MAVIS-FLOW	www.alpha cast-software.	Commercial	MAVIS-FLOW 7 for Windows combines the leading-edge solidification simulation technology of MAVIS with the benefits of the familiar Windows platforms, enabling incredible ease of use and functionality. MAVIS-FLOW 7 includes a Navier–Stokes mold filling simulator bringing even more power within reach of the foundryman. With improved graphics output, easy navigation menus, and reduced simulation times, the MAVIS-FLOW 7 software represents a significant step forward for casting design on the desktop
2260	MeltFlow	http://inres .com/index .html	Commercial	MeltFlow-VAR* and MeltFlow-ESR* utilize advanced CFD techniques developed specifically for a detailed analysis of the electromagnetic (DC in VAR, AC in ESR), fluid flow, heat transfer, and alloy element redistribution phenomena occurring in the VAR and ESR processes. They enable prediction of pool evolution during the entire process and the thermal history, distributions of alloy element concentrations, dendrite arm spacings, and freckle formation probabilities in the solidified ingot

Casting codes

SW cat	Name of code	Website of software	Type of code	Short description/typical applications/ phenomena being treated
2220	Nova Cast	www.nova castfoundry.se/	Commercial	Foundry Tech™ is a comprehensive toolbox for design of horizontal gating and feeding systems. NovaMethod™ is a complete casting geometry preparation tool with complete CAD functionality, especially designed for foundries. NovaFlow&Solid™ CV is a complete casting process simulation system that uses an advanced and unique meshing technology and control volume, combining reliable results with high calculation speeds. Adding NovaStress™ allows to calculate the stresses. MetalMaster™ helps foundries to calculate and plan for the most economical charge composition. ATASmetstar™, a metallurgical process control system, allows to analyze, optimize, and stabilize the metallurgical foundry process using advanced thermal analysis
2220	Pro CAST	https://www .esi-group .com/software-services/virtual-manufacturing/ casting/procast-quikcast	Commercial	ProCAST is based on finite element technology and offers modules and tools to predict distortions and residual stresses. It can address more specific processes such as semisolid, core blowing, centrifugal, lost foam, and continuous casting SALSA 3D helps calculate size and gating and running systems for high-pressure die casting process. It assists in designing gates, runners, and overflows by controlling the maximum gate velocity and thickness based on the available pressure machine

Casting codes

SW cat	Name of code	Website of software	Type of code	Short description/typical applications/ phenomena being treated
2220	Quick Cast	https://www .esi-group.com/ software-services/ virtual-manufacturing/ casting/ procast-quikcast	Commercial	See also ProCAST QuikCast addresses the basics of any casting process: filling, solidification, and porosity prediction including core blowing and semisolid modeling. It accounts for air back pressure, filters, mold roughness, thermal exchanges, die coatings, and gravity to accurately simulate most casting processes ranging from gravity sand casting to high- and low-pressure die casting
2220	SIMTEC	www. simtec-inc.com	Commercial	SIMTEC provides a full line of finite element analysis (FEA), which includes fluid flow, the cooling down and solidification, residual stress as well as stress coming from an exterior load, and a visualization of the resulting distortion
2220	STAR-Cast	www. starcast.org	Commercial	The prediction of misrun formation is based on STAR-Cast's unique, fully coupled computational continuum mechanics approach, featuring a multiphase mold filling module. Concurrent computation of fluid flow, temperature, and liquid–solid transition. Modeling of dipping-type ceramic shell molds is crucial to accurate simulation of investment casting processes. STAR-Cast provides a high-performance tool for calculating the outer surface of the virtual shell in close correlation to the real shape of the mold. For this reason, a dedicated material database is key to assuring quality of simulation

Casting codes

SW cat	Name of code	Website of software	Type of code	Short description/typical applications/ phenomena being treated
2221	STAR-CD	www,cd-adapco.com	Commercial	STAR-CCM+®, STAR-CD®, Optimate™, SPEED™, STAR-Cast Battery Design Studio®, Battery Simulation Module™, DARS/HEEDS
2220	SUTCAST	www. sutcast.com	Commercial	SUTCAST simulates the molten metal of any casting alloy into sand or permanent molds. Sand casting modules for cast iron, steel casting, nonferrous alloys, and more. Permanent mold modules, low-pressure die casting (LPDC), gravity die casting
2220	THERCAST	http://www.transvalor.com/en/cmspages/thercast.6.html	Commercial	THERCAST® is based on a nonstationary 3D finite element method software system that simulates ingot casting and continuous casting of steel and other metals. Includes algorithms for the liquid state of the metal being cast, modeling both arbitrary Lagrangian–Eulerian (ALE) and Navier–Stokes behavior to account for convection and other liquid metal phenomena. Simulates "mushy zone" thermoviscoplastic Norton–Hoff behavior as the metal begins to solidify and then solid or thermoelastic-plastic behavior when phenomena such as air gaps become more significant. Predicts surface and internal stresses as cause for cracking, hot tearing, or other defects

Casting codes

SW cat	Name of code	Website of software	Type of code	Short description/typical applications/ phenomena being treated
2220	Vulcan	www. quantech.es	Commercial	Vulcan is a simulation software for casting process optimization to obtain error-free first castings, gravity sand and permanent mold casting, low-pressure die casting, high-pressure die casting, filling of the mold, temperature field analysis, flow turbulences, evolution of the solidification, thermomechanical analysis, and the centrifugal casting modules
2220	WinCast	www.rwp-simtec.de	Commercial	WinCast® expert allows the simulation of all typical casting processes: sand casting, pressure die casting, investment casting, permanent mold casting, continuous casting, welding, heat treatment, and others
2221	deal.II	http://www.dealii.org/	Open source	Fluid flow

Plastics processing codes

SW cat	Name of code	Website of software	Type of code (academic/ commercial) installations	Short description/typical applications/phenomena being treated
2250	MOREX	www.aixtrusion.de	Commercial	MOREX is a program for the analysis of compounding processes in corotating closely intermeshing twin-screw extruders Typical results are melt temperature, melt pressure, current viscosity, dissipated driving power, and heating and cooling power exchanged with the barrel
2250	Sheet CoolAix	www.aixtrusion.de	Commercial	Film and sheet extrusion – simulation of cooling processes calculates, for example, temperature distribution across the thickness of film or sheet; temperature distribution along the extrusion direction; heat flow from the plastic material to the cooling rolls; and heat flow to the environment
2250	Erkon	www.aixtrusion.de	Commercial	The program calculates the cooling process of blow-molded parts and works on the basis of an enhanced, convective cooling model. Convective boundary conditions or boundary conditions of the third kind are used as thermal boundary conditions for solving the energy equation and allow a realistic calculation of the cooling at the inner side of the molded part. At the outer side, the mold design (arrangement of the cooling channels, mold material, etc.) and the operating point (throughput and temperature of the cooling water) are taken into account

Plastics processing codes

SW cat	Name of code	Website of software	Type of code (academic/ commercial) installations	Short description/typical applications/phenomena being treated
2250	Moldex	http://www.moldex3d.com/en/	Commercial	Moldex3D is a CAE product for the plastic injection molding industry. Moldex3D can help to simulate the widest application range of injection molding processes to optimize product design and manufacturability, shorten time to market, and maximize product ROI CAD-embedded preprocessing, automatic 3D mesh engine, high-resolution 3D mesh technology, high-performance parallel computing
2250	Moldflow	http://www.moldflow.eu/	Commercial	Autodesk® Moldflow® plastic injection molding simulation software provides tools that help manufacturers validate and optimize the design of plastic parts and injection molds and study the plastic injection molding process
2250	WenVerti	www.aixtrusion.de	Commercial	Design of spiral mandrel dies for the extrusion processing. The software allows the fast calculation and optimization of the die design using the network theory
2250	Vertigo	www.aixtrusion.de	Commercial	Design of flat sheet dies for the extrusion film casting

Plastics processing codes

SW cat	Name of code	Website of software	Type of code (academic/ commercial) installations	Short description/typical applications/phenomena being treated
2250	B-SIM	http://www.t-sim.com/	Commercial	B-SIM is a software package for simulation of blow molding. B-SIM predicts the final wall thickness distribution based on the specified processing parameters (the pressure level, the speed of tools, the parison/preform initial temperature distribution, etc). B-SIM can also predict image distortion and predistortion based on sleeve shrinkage process
2250	T-SIM	http://www.t-sim.com/	Commercial	T-SIM is a software package for simulation of thermoforming. T-SIM predicts the final wall thickness distribution based on the specified processing parameters (the pressure level, the speed of tools, the sheet temperature distribution, etc.). T-SIM can also predict image distortion and predistortion
2250	BlowView	http://www.pacesimulations.com/	Commercial	BlowView simulates the full blow molding process. It allows the calculation of the preform temperature distribution during heating and extrusion, the wall thickness distribution, and the stresses

Plastics processing codes

SW cat	Name of code	Website of software	Type of code (academic/ commercial) installations	Short description/typical applications/phenomena being treated
2250	FormView	http://www.pacesimulations.com/	Commercial	This software technology focuses on thermoforming (sheet reheat, forming, and solidification) and includes the prediction of the thermoforming with multiple plugs and the thermoforming of profile heated preforms or sheets. The core program solves the nonisothermal solid mechanics constitutive equations specifically for the individual phases of the thermoforming process (sheet reheating, sheet formation, part solidification, and warpage)
2250	Sigmasoft	http://www.sigma-soft.de/	Commercial	Sigmasoft is a software tool for the simulation of the injection molding process. It uses real 3D simulation of flow, heat flux, and warpage including all details of the process
2250	SIMCON	http://www.simcon-worldwide.com/	Commercial	Cadmould is developed and sold by SIMCON Kunststofftechnische Software GmbH. It has become a widely recognized and approved software tool in the area of part and mold development for injection molding applications

Sintering codes

SW cat	Name of code	Website of software	Type of code (academic/ commercial) installations	Short description/typical applications/phenomena being treated
2230	FORM-2D	http://www.qform3d.com/?go=main	Commercial	FORM-2D (now QForm) is a fully automated mathematic simulation. The software also calculated tool stresses and deformations efficiently. FORM-2D carried some very essential innovative ideas, which predetermined the development in this field for the years to come The software can be used for the calculation of pore structure parameters and properties of a powder component [17]
2230	PM solver			PMsolver/Compaction-3D is die compaction-oriented CAE/FEM software, which is based on implicit time integration and a rigid viscoplastic algorithm. It was used to design a compaction schedule to obtain a uniform green density distribution, and the effect of loading schedule on distortion during sintering was also compared to experimental results [18]
2230	CASTEM 2000	http://www.cecalc.ula.ve/documentacion/tutoriales/castem/castem3.html	Open source	CASTEM is a multipurpose finite element code developed at the CEA. Domains of applications are structural mechanics, fluids mechanics, and thermal and magnetic analyses. As cited in [19], CASTEM2000 was used in [20], and the models were built with viscoplastic compressible model

Sintering codes

SW cat	Name of code	Website of software	Type of code (academic/ commercial) installations	Short description/typical applications/phenomena being treated
2230	JAS3D		Commercial	JAS3D is a three-dimensional finite element program originally designed to solve Lagrangian quasistatic nonlinear mechanics problems and subsequently extended to include both implicit and explicit dynamics. A temperature-dependent skeleton viscosity function within the Skorohod–Olevsky viscous sintering (SOVS) model is used to predict the sintering behavior [21]
2230	PreCAD		Commercial	PreCAD is an FEM software with design and solver modules developed at CEA/CEREM. It allows the user to design both the part to be produced from powder and all the tooling used in the process and allows to simulate the consolidation of powders by HIP, cold compaction, or sintering using a macromechanical approach. In order to help optimize processes such as die compaction, sintering, and hot isostatic pressing, CEA/CEREM has developed a computer-assisted design and modeling tool named PreCAD. In the case of hot isostatic pressing HIP, this tool is used to optimize the design of the initial mold in order to obtain a very accurate part after compaction [22]

Sintering codes

SW cat	Name of code	Website of software	Type of code (academic/ commercial) installations	Short description/typical applications/phenomena being treated
2230	CIMLib	http://www.cemef.mines-paristech.fr/sections/recherche/equipes-recherche/calcul-intensif	Commercial	The scientific finite element library (CIMLib), developed at CEMEF, is the base for numerical applications. A numerical strategy to simulate the mass transport by surface and lattice diffusion into a granular packing during sintering has been studied as a mechanical problem of coupling the grain elastic behavior to the fluid behavior of the surrounding phase [23]
2230	DEFORM	http://www.deform.com/	Commercial	DEFORM is an engineering software that enables designers to analyze metal forming, heat treatment, machining, and mechanical joining processes on the computer rather than the shop floor using trial and error. Insert geometries have been modeled using 2D axisymmetric modeling to study shrinkage [24]

Sintering codes

SW cat	Name of code	Website of software	Type of code (academic/ commercial) installations	Short description/typical applications/phenomena being treated
2230	ABAQUS	http://www.3ds.com/products-services/simulia/products/abaqus/	Commercial	ABAQUS (with user subroutines such as CREEP, UMAT, and VUMAT) is a software suite developed to address challenging nonlinear problems, large-scale linear dynamics applications, and routine design simulations. A micromechanical model developed by Riedel was modified for sinter forming, and a constitutive model was implemented with VUMAT. The simulations were performed with alumina on a section of a gear wheel [13]. A comprehensive model of solid-state sintering was presented which combines models on various stages of sintering and grain coarsening. The model was implemented with UMAT, and a face seal was simulated [10]
2230	LS-DYNA	http://www.lstc.com/products/ls-dyna	Commercial	LS-DYNA is a general-purpose finite element program capable of simulating complex real-world problems, used by the automobile, aerospace, construction, military, manufacturing, and bioengineering industries. A constitutive model for calcium carbonate applied to high-pressure compaction is presented. The plastic response of powder is nonlinear and described in a rate-independent cap plasticity model. User-defined material routine has been used and implemented in LS-DYNA [25]

Sintering codes

SW cat	Name of code	Website of software	Type of code (academic/ commercial) installations	Short description/typical applications/phenomena being treated
2230	MSC Marc	http://www.mscsoftware.com/product/marc	Commercial	MSC Marc is a powerful, general-purpose nonlinear finite element analysis solution to accurately simulate the response of products under static, dynamic, and multiphysics loading scenarios. A finite element model has been used in MSC Marc to study the sintering process of polytetrafluoroethylene (PTFE) cylinders in order to predict residual thermal stresses; both solid (rods) and hollow (billets) blocks were studied [26]
2230	ANSYS Multi-physics	http://www.ansys.com/Products/Simulation+Technology/Multiphysics	Commercial	ANSYS Multiphysics software offers a comprehensive product solution for both Multiphysics and single-physics analysis. A study considered the hot pressing and the sinter forging of a disk. The model equations were converted from FORTRAN codes to ANSYS codes [27]
2230	COMSOL Multi-physics	http://www.comsol.com/comsol-multiphysics	Commercial	COMSOL Multiphysics is a general-purpose software platform, based on advanced numerical methods, for modeling and simulating physics-based problems. The constitutive equations used here are based on the continuum theory of sintering in the framework of a linear viscous material behavior. The model describes the gravity influence on porosity evolution and shrinkage inhomogeneity. Simulations of densification, shape distortion, and porosity gradients are presented in [28]

References

1. German, R.M. (1996) *Sintering Theory and Practice*, Wiley-VCH Verlag GmbH.
2. ASM International (1998) *ASM Handbook – Powder Metal Technologies and Applications*, vol. **7**, ASM International.
3. Niu, W. and Pan, J. (2010) in: *Sintering of Advanced Materials*, Chapter 4 (ed. Z.Z. Fang), Woodhead Publishing, pp. 86–109.
4. Ashby, M.F. (1974) A first report on sintering diagrams. *Acta Metall.*, **22** (3), 275–289.
5. Su, H. and Johnson, D.L. (1996) Master sintering curve: a practical approach to sintering. *J. Am. Ceram. Soc.*, **79** (12), 3211–3217.
6. Olevsky, E.A. (1998) Theory of sintering: from discrete to continuum. *Mater. Sci. Eng., R*, **23** (2), 41–100.
7. Abouaf, M., Chenot, J., Raisson, G., and Bauduin, P. (1988) Finite element simulation of hot isostatic pressing of metal powders. *Int. J. Numer. Methods Eng.*, **25** (1), 191–212.
8. Riedel, H. (1990) A constitutive model for the finite-element simulation of sintering--distortions and stresses. *Ceram. Powder Sci.*, **III**, 619–630.
9. Riedel, H. and Blug, B. (2002) *Multiscale Deformation and Fracture in Materials and Structures* (eds T.-J. Chuang and J. W. Rudnicki), Springer, pp. 49–70.
10. Kraft, T. and Riedel, H. (2004) Numerical simulation of solid state sintering: model and application. *J. Eur. Ceram. Soc.*, **24** (2), 345–361.
11. Olevsky, E.A., Timmermans, G., Shtern, M.B., and Froyen, L. (1997) The permeable element method for modelling of deformation processes in porous and powder materials: theoretical basis and checking by experiments. *Powder Technol.*, **93** (2), 127–141.
12. Olevsky, E., German, R., and Upadhyaya, A. (2000) Effect of gravity on dimensional change during sintering—II. Shape distortion. *Acta Mater.*, **48** (5), 1167–1180.
13. Reiterer, M., Kraft, T., and Riedel, H. (2004) Manufacturing of a gear wheel made from reaction bonded alumina—numerical simulation of the sinterforming process. *J. Eur. Ceram. Soc.*, **24**, 239–246.
14. Bouaziz, O., Dellis, C., and Stutz, P. (1997) Creation of a material data file for modeling HIPing of an austenitic stainless steel. Proceedings of the International Workshop on Modeling of Metal Powder Forming Process, Grenoble, France.
15. Bose, A. and Eisen, W. (2003) *Hot Consolidation of Powders and Particulates*, MPIF, Princeton, NJ.
16. Jeon, Y. and Kim, K. (1999) Near-net-shape forming of 316L stainless steel powder under hot isostatic pressing. *Int. J. Mech. Sci.*, **41** (7), 815–830.
17. Chernozhukova, I., Leshchinsky, V., and Weinert, H. (2000) Modeling of multi-operational cold forging. Proceedings of the 2000 International Conference on Powder Metallurgy & Particulate Materials Part1, New York.
18. Chung, S.-H., Kwon, Y.-S., Hyun, C.-M., Kim, K., Kim, M.-J., and German, R.M. (2004) Analysis and design of a press-and-sinter process for fabrication of precise tungsten carbide cutting tools. *Adv. Powder. Metall. Part. Mater.*, **8**, 26–39.
19. Sornin, D. and Couvrat, M. (2010) Numerical simulation of hot extrusion consolidation of PM ferritic ODS stainless steels. PM2010 World Congress Proceedings, Vol. 2, Florence.
20. Abouaf, M. and Chenot, J. (1986) Modélisationnumérique de la déformation à chaud de poudresmétalliques. *J. Mécan.Theor. Appl.*, **5** (1), 121–140.
21. Reiterer, M., Ewsuk, K.G., and Argüello, J.G. (2006) An arrhenius-type viscosity function to model sintering using the Skorohod–Olevsky viscous sintering model within a finite-element code. *J. Am. Ceram. Soc.*, **89** (6), 1930–1935.
22. Baccino, R. and Moret, F. (2000) Numerical modeling of powder metallurgy processes. *Mater. Des.*, **21**, 359–364.
23. Bruchon, J., Pino-Munoz, D., Valdivieso, F., and Drapier, S. (2012) Finite element simulation of mass transport during sintering of a granular packing. Part I.

Surface and lattice diffusions. *J. Am. Ceram. Soc.*, **95** (8), 2398–2405.

24. Shamasundar, S., Siddegowda, M., and Chigurupati, P. (2002) Modeling liquid phase sintering of hard metal powder compacts. Advances in Powder Metallurgy & Particulate Materials, Orlando, FL.
25. Pär, J., Häggblad, H.-Å., and Berg, S. (2012) Modelling ultra high pressure compaction of powder. *Tech. Mech.*, **32** (2-5), 287–302.
26. Andena, L., Rink, M., and Polastri, F. (2004) Simulation of PTFE sintering: thermal stresses and deformation behavior. *Polym. Eng. Sci.*, **44** (7), 1368–1378.
27. Camacho-Montes, H., García-Casillas, P.E., Rodriguez-Ramos, R., Fuentes-Montero, M.E., and Fuentes-Cobas, L.E. (2008) Simulation of the stress-assisted densification behavior of a powder compact: effect of constitutive laws. *J. Am. Ceram. Soc.*, **91** (3), 836–845.
28. Alvarado-Contreras, J.A., Olevsky, E.A., and German, R.M. (2013) Modeling of gravity-induced shape distortions during sintering of cylindrical specimens. *Mech. Res. Commun.*, **50**, 8–11.

Further Reading

Crystal Growth/Casting

Bührig-Polaczek, A. and Träger, H. (eds) (2012) Foundry technology, in *Ullmann's Encyclopedia of Industrial Chemistry*, Wiley-VCH Verlag GmbH, Weinheim, doi: 10.1002/14356007.a12_035.pub2.

Campbell, J. (2012) *Complete Casting Handbook -Metal Casting Processes, Techniques and Design*, Elsevier. ISBN: 978-1-85617-809-9

Rudolph, P. (ed.) (2014) *Handbook of Crystal Growth: Bulk Crystal Growth*, 2nd edn, vol. **2**, Elsevier. ISBN: 9780444633033

Plastics

Doi, M. and Edwards, S.F. (1988) *The Theory of Polymer Dynamics*, Oxford University Press, Oxford.

Greener, J. and Wimberger-Friedl, R. (2006) *Process, Materials, and Applications.* Hanser Publications, Cincinnati, OH, ISBN: 9781569904008

Michler, G.H. and Balta-Calleja, F.J. (2012) *Nano- and Micromechanics of Polymers*, Carl Hanser Verlag GmbH & Co. KG, München.

Zhou, H. (ed.) (2013) *Computer Modeling for Injection Molding: Simulation, Optimization, and Control.* John Wiley & Sons, Inc., ISBN: 978-0-470-60299-7

2.3
Forming Processes

Stephan Hojda and Markus Bambach

2.3.1
Overview: Manufacturing Process Forming

Metal forming comprises all manufacturing processes in which the shape, material properties, and/or surface quality of a workpiece is deliberately altered while the volume is preserved (German Institute for Standardization DIN 8582). Metal forming processes can be categorized by:

- The state of stress in the zone of plastic deformation (DIN 8582)
- The temperature during deformation (cold, warm, hot)
- The type of the semifinished product being processed (bulk or sheet)
- The time dependence of the forming operation (transient or steady state)

The aforementioned classes focus on the component scale. It is also possible to categorize forming processes according to the machines, tools, and process kinematics used or their influence on microstructure and properties. Process chains can be separated into the production of a semifinished product and the downstream processing to the final component. Both of these steps can, but not necessarily have to, involve forming operations. An overview of different metal forming operations is given in Figure 2.3.1.

2.3.2
Phenomena Occurring during Forming Processes

The primary purpose of metal forming is to change the shape of a workpiece using plastic deformation processes. As such, metal forming processes belong to the class of field problems with the displacement or velocity field being the primary solution variable. A metal forming problem is defined by the four standard balance equations used in continuum mechanics, that is, the balance of mass, momentum, angular momentum, and energy, together with a suitable constitutive equation or set of constitutive equations as well as boundary (and, if applicable, initial) conditions. The constitutive equation connects configuration and source variables, that is, strains and stresses, for which appropriate measures

Handbook of Software Solutions for ICME, First Edition. Edited by Georg J. Schmitz and Ulrich Prahl.

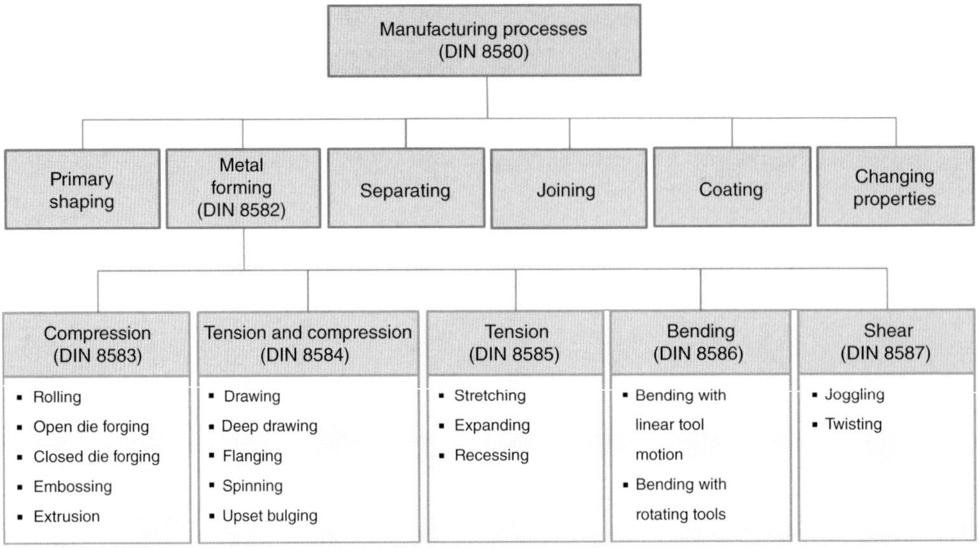

Figure 2.3.1 Metal forming processes according to German Institute for Standardization [1].

need to be defined. In addition, the second law of thermodynamics must not be violated. Hence, the continuum mechanics of solid bodies builds the primary framework for treating metal forming processes (cf., e.g., the classic textbooks on plasticity and metal forming [2–4]).

However, metal forming processes have evolved tremendously in recent years. On the one hand, borders between manufacturing processes have been opened, for example, by combinations of primary shaping and forming processes (semisolid forming [5] or joining by forming processes [6, 7]. Besides that, physical phenomena other than the mechanical action of a solid tool have been used for metal forming, for example, electromagnetic forming [8]; hence, metal forming processes tend to be multiphysic problems in many cases nowadays.

Some typical and mostly nonlinear phenomena that occur depending on the type of metal forming problem at hand are summarized in the following table.

Inside the workpiece	**At the workpiece surface or workpiece–tool interface**
Finite strain (elastic–plastic) deformation Strain hardening (isotropic, kinematic, etc.) Unloading and recovery of elastic deformation Anisotropy and texture evolution	• Generation of new free surface areas due to increase in surface area • Changes in surface roughness
Dissipation due to plastic deformation	Contact and friction
Damage	Wear
Heat generation and conduction	Heat transfer
Recrystallization and grain growth	
Phase transformation, segregation, and precipitation	Chemical reactions (oxidation, bonding, etc.)

These phenomena can be traced back to underlying physical phenomena such as lattice defect formation and motion, cohesion and adhesion in between surfaces, crystal lattice vibration, recrystallization, and recovery [9, 10] or treated from the viewpoint of continuum mechanics. A few details of the multitude of phenomena occurring in metal forming operations are given here, but the interested reader has to be referred to the references given for an in-depth treatment of the phenomena.

2.3.2.1 Finite Strain Deformation

Large deformations or finite strains need a mathematical description that is different from that of small deformation, which covers a range of up to a 10% strain and 10° rotations [11]. Some forming processes such as bending combine small deformations with large rotations [12], while others such as open die forging comprise both large deformations and large rotations. Due to this variety of capabilities, several measures for deformation, strain, and stress have been developed and discussed according to the specific backgrounds in continuum mechanics [11, 12]. However, most of these issues are solved in commercial FE software packages without any user intervention.

2.3.2.2 Strain Hardening

Strain hardening refers to the evolution of flow stress, that is, the stress necessary to evoke plastic deformation, with strain. Plastic flow occurs when the stress state reaches the yield surface, a surface bounding the domain of elastic and plastic deformation in stress space [2]. When the loading is increased after reaching the yield surface, strain hardening occurs. At room temperature, strain hardening is typically subdivided into isotropic and kinematic strain hardening [13]. When the loading direction remains unchanged, isotropic strain hardening is observed, that is, the flow stress increases monotonically with strain, while kinematic strain hardening refers to a change in flow stress that occurs when the material is unloaded after prior plastic deformation (e.g., in tension) and plastically deformed under a load acting in opposite direction (e.g., compression). If the direction of the loading is changed in a general way (i.e., not precisely reversed), cross hardening is observed [14]. Finite deformation also leads to changes in the orientation distribution of grains in the workpiece which affects the shape of the yield surface. Various models for plastic anisotropy have been developed; see, for example, Bunge *et al.* [15] for an overview. At elevated temperature, recovery and recrystallization can occur, which reduce the strain hardening rate considerably. Plastic deformation becomes time dependent at elevated temperatures, and the workpiece is then treated as a viscoplastic material [16].

2.3.2.3
Contact

One of the most challenging issues in forming simulations is the description of contact [12]. This holds especially for highly dynamic processes. Different numerical methods have been developed to describe contact and friction numerically:

- The Lagrange multiplier method
- The penalty method
- The augmented Lagrangian method
- The perturbed Lagrangian method

The most critical difference between the Lagrange multiplier method and the penalty method is that an exact contact is achieved with the Lagrange multiplier method, while a penetration of surfaces that is minimized is the basis for the penalty method.

2.3.2.4
Friction

The standard friction model in the mechanics of solid materials is the Coulomb friction model which states that the frictional shear stress is proportional to the normal contact stress. In metal forming, the shear stress can easily reach the shear flow stress of the softer contact partner (typically the workpiece) such that the shear stress being transmitted no longer increases linearly with contact pressure but remains almost constant (friction factor model) [17].

2.3.2.5
Instability and Damage

Metal forming is usually concerned with ductile material states. Since no produced metal is free from defects such as inclusions of second-phase particles, large plastic deformation leads to the formation, growth, and coalescence of microvoids which may ultimately lead to failure of the workpiece [18]. Alternatively, under tension loading, the workpiece may decrease its cross section at a rate which cannot be compensated by strain hardening, such that necking, a special type of instability, occurs [19].

2.3.2.6
Heat Transfer

The heat transfer has to be considered in all hot forming operations. It is obvious that the stress and strain state during a hot forming process will strongly depend on the temperature evolution. Also in cold forming processes, temperature can rise considerably due to dissipation. A detailed description of thermal boundary conditions is given in the following section.

2.3.3
Modeling and Simulation Methods

Analytical solutions for metal forming problems are only available for a few academic examples. To solve metal forming problems of industrial relevance, the use of numerical solution methods is necessary. Most commonly, the finite element method (FEM) is employed, which was first developed in the 1950s for static analysis and later extended to applications in various other fields, especially to nonlinear phenomena such as metal forming simulations. In the early years, the material behavior was assumed to be rigid plastic; see, for example, [20]. Through research into the numerical solution of plasticity problems, the field of computational plasticity [21–23] developed, and algorithms for the solution of elastic–plastic problems were devised. Various applications of other solution methods such as finite volume and meshless methods can be found in the literature and to some extent also in commercial codes [24, 25]. The main characteristic of all numerical methods is that the original field problem given by the balance equations, the constitutive equation, and initial and boundary conditions is discretized in space and time. This means that the solution is no longer sought at infinitely many points in space and time but at individual points. Also, the original balance equations are no longer fulfilled in the "strong" sense but only approximately. In terms of the FEM, this is typically referred to as the "weak" form of the balance equations. Hence, one always has to keep in mind that a finite element (FE) solution is only equivalent to the solution of the original problem in the weak sense and that the solution is affected by discretization errors.

With the increasing availability of computing capacity, it became possible to set up more sophisticated thermomechanical problems, for example, incorporating complex material models, including models for phase changes and eigenstress during quenching after a forming operation. A coupling of the mechanical and the thermal problem was established to perform such simulations. A review on numerical simulations of sheet metal forming is given by Banabic and Tekkaya [26].

Before the advent of numerical methods, some fundamental solution methods for plasticity problems were developed in the eighteenth and nineteenth centuries. In the beginning of the twentieth century, the models were brought into application and further developed, for example, for forging and rolling process. The development up to the 1980s is summarized by Osakada [27]. Some of the resolution methods, which apply simplifying assumptions to obtain approximate solutions, are known as elementary theory, slip line method, and upper bound method. They are still in use in the industry to enable online process control due to the low computational costs [28].

Most of the material models used in metal forming do not consider the spatial microstructure of the material but use mean-field approaches with empirical or semiempirical material models such as the Johnson–Mehl–Avrami–Kolmogorov (JMAK) equation. In [29], it is shown that grain size predictions can become unstable over a sufficiently large number of process steps when JMAK models are used, which is a severe restriction for Integrated Computational Materials Engineering (ICME) simulations. In the 1990s the material itself became the focus of intensive research activities. During this time, the crystal plasticity finite element method (CPFEM) was developed; see [30] for a detailed review. Representative volume elements or unit cells were increasingly used to simulate the local inter- and intragrain characteristics under mechanical and/or thermal loading. These techniques allowed access to complex processes such as the kinetics of recrystallization after deformation or friction in between the tool and the workpiece for different surface topologies. Current trends show that in addition to the methods described earlier, artificial intelligence is an emerging field. Mainly driven by the availability and standardization of processing data, the target values obtained under certain boundary conditions can be described using neuronal networks or stochastic approaches. A general review of these techniques from 2001 to 2011 is given by Liao *et al.* [31], while Harding *et al.* [32] focus on manufacturing processes in their review.

2.3.4 Typical Applications of Forming Simulations

In the industrial environment, the driving force for innovation are savings in time and cost as well as an improvement of a final or semifinished product's quality. Therefore, during the design of a new product, different geometries and process routes are tested with simulations. Although the benefit of metal forming simulations often cannot be measured in terms of money savings, it is obvious that the development of new tools and processes was mainly driven by trial and error or based on experience in the past, while today simulations are state of the art even in the early stages of component design [33]. These simulations cannot always reduce the expenditure spent for prototypes, but it can at least lead to an educated guess on how to solve a certain problem.

Forming simulations cover a wide range of effects and phenomena and are used for various target values. One focus is the feasibility of a certain process. For example, the simulation can answer the question whether a given forming press has enough force or power to perform a certain forming operation with a given material, tool geometry, and kinematic. Another aspect of interest may be whether fracture or damage occurs in the workpiece when forming limits are reached or in the tools due to wear or fatigue. Yet another goal is the prediction of final product properties due to a specific treatment during the forming operation. These applications are only a selection of challenges that can be solved by simulations, but it is evident that several scales and parts of a process chain are involved and that a perspective isolated to a single processing step is often not sufficient. The challenges for metal forming simulations can be summarized as the calculation of the following:

- **Global process quantities:** The overall response of the system including workpiece, machine, and environment (e.g., temperature, forces)

- **Local process quantities:** The local response of the workpiece in a forming process (e.g., material flow, temperature field, stress distribution)
- **Product properties:** The prediction of final product properties (e.g., geometry, microstructure, hardness, surface properties)
- **Interaction of process, tool, and machine:** The change in process conditions due to the change of the properties of the tools (e.g., wear) or the machine (e.g., clearance, elasticity)
- **Robustness of a manufacturing process:** The reproducibility of a certain process under the perturbation of processing conditions

The aforementioned simulations focus on the understanding of already established processes, and another aspect is the design of new process chains, new tools, and the proof of suitability of new forming machines.

When simulations and methods are developed in the research community, it lasts 10–15 years until they are used in the industry [33]. It is worth mentioning that the early developments of the discipline of ICME are deeply coupled to metal forming problems, with the main focus being on rolling processes. These approaches were initially called through-process modeling (TPM) and integrate simulations along whole manufacturing chains and over several length scales [34–36]. While most of the early scholarly activities under the use of the FEM were done by research institutes, for example, the National Aeronautics and Space Administration (NASA), today several commercial software codes exist that are delivered and adapted to the industrial customers' need [37]. Computer-aided engineering (CAE) for single processes and process chains is continuously being developed to higher performance and take on an ever more holistic view of manufacturing process chains, hence reflecting the idea of ICME to some extent already.

2.3.5 Initial and Boundary Conditions

Most forming processes define an initial boundary value problem (IBVP) which means that for a full definition of the problem the balance equations for mass, linear and angular momentum, and energy, the material behavior and the mechanical and thermal conditions have to be considered (Figure 2.3.2). As mechanical boundary conditions, translations and rotations, forces, and torque can be applied to rigid and deformable bodies or regions of them. Additional mechanical boundary conditions are introduced by contacting surfaces and their tangential and normal behavior. In some software codes boundary conditions are related to physical phenomena like inertia effects and applied automatically. Mechanical initial conditions can be a velocity field, an (admissible) stress field, or an initial amount of strain hardening.

Special attention is required for friction. The character of most forming operations strongly depends on the tribological system which comprises friction, surface structure, and lubrication. These phenomena are represented by a friction law

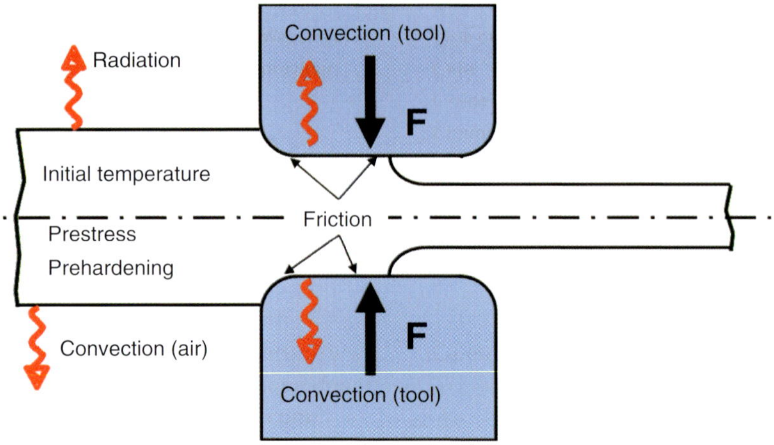

Figure 2.3.2 Typical initial and boundary conditions that have to be defined in a forming simulation as the example of an open die forging process.

in the forming simulation (cf. Section 2.3.2). New developments have been made and are reviewed in [38]. However, the surface interactions of nonideal systems have not been fully understood and are currently still being researched.

While the initial condition for the thermal problem is a temperature distribution in the body, thermal boundary conditions can be subdivided into four types [39]:

1) A temperature that is applied to the boundary (Dirichlet type)
2) A heat flux that is applied to the boundary (Neumann type)
3) A convective boundary condition or mixed boundary condition (Robin type)
4) Equal fluxes at the surface due to conduction in the material and convection/radiation at the surface (Stefan type).

2.3.6
Required Data and their Origin

A crucial aspect for a forming simulation is the input data, which comprise the density, the elastic, and plastic material behavior (i.e., Young's modulus, Poisson ratio, flow stress, strain hardening law, anisotropy, etc.). These properties have to be supplied primarily for the workpiece, but if the tools are also treated as deformable bodies, their material properties have to be defined as well. In contact problems, the friction coefficient or friction factor has to be specified. For thermal problems, also the heat capacity, the thermal conductivity, and the thermal boundary conditions (heat transfer coefficient, emissivity) are needed for workpiece and tool material. It is worth noting that most material data and boundary conditions depend on temperature, that is, they have to be provided for the entire temperature range occurring in the simulation. The friction behavior also depends on the

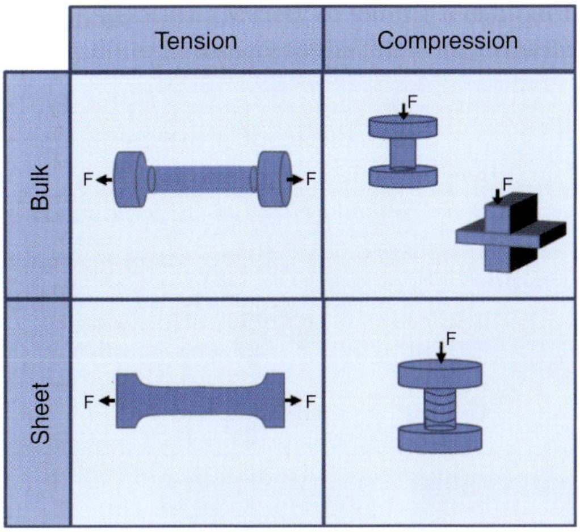

Figure 2.3.3 Testing procedures to evaluate material properties for bulk and sheet metal.

relative sliding velocity and pressure, and the plastic material response depends on the strain rate and temperature.

Aside from the theoretical possibility of *ab initio* calculations [40] which can be used to retrieve some of the input data required, the state of the art is the use of databases for thermomechanical properties or the determination of the properties through lab-scale experiments.

As mentioned earlier, the stress state during the forming operation is very important for the material behavior. Therefore different tests exist to determine the properties of a sample under tension or compression. Not every test is applicable to every sheet and bulk geometry, i.e. tests for sheet and bulk metal forming exist. There are a huge number of testing procedures available. While some are standardized (like the tensile test), there are some which have not yet been subject to standardization. Here, only the frequently used tests are described. For more insight into material testing for forming simulations, refer to [41, 42].

Common tests to access the flow curve of a metal for forming simulations are given in Figure 2.3.3.

It is obvious that some of these tests require the knowledge about the boundary conditions in the experimental procedure. The elastic response of the testing device, the friction behavior for compression testing, the heat transfer coefficient, and the emissivity for experiments at elevated temperatures have to be taken into account. To minimize the uncertainties, multistage approaches for input data determination have been developed [43, 44]. A scheme of such a multistage process is given in Figure 2.3.4. The basic idea is to derive process-specific boundary conditions in a cascade of subsequent (inverse) problems. At first the process-independent properties such as mass density ρ, heat capacity c_p, and the thermal conductivity λ are calculated based on the CALPHAD method, derived from experiments or taken from the literature data. In the next stages

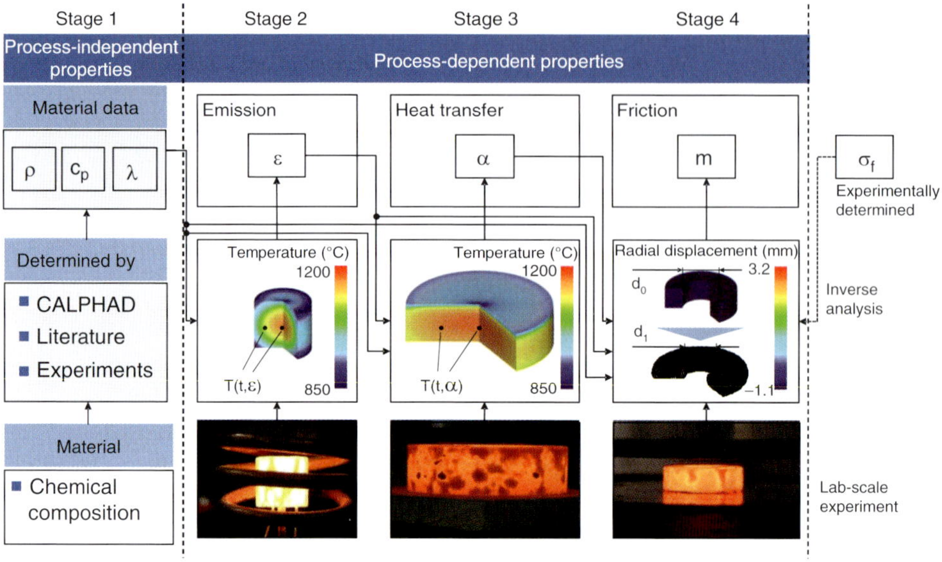

Figure 2.3.4 Multistage approach to determine material properties (flow curve) with respect to the processing conditions (translated from [43]).

the emissivity ε, the heat transfer coefficient α, and the friction factor m are determined by an inverse FE calculation in which only a particular parameter (ε, α, and m) is fitted to the measured value of a corresponding experiment (temperature and radial displacement). The determined boundary conditions from previous stages are used as input for the simulation in the next stage. For the final stage (friction factor m), the flow curve (σ_f) has to be supplied for the process-dependent boundary conditions in the experiment.

2.3.7
Numerical Aspects

The user of a state-of-the-art problem-dependent commercial FE solver normally does not have to change numerical settings in the software. Typically, the software package, containing a preprocessor to set up the model and a postprocessor to view the simulation results, is delivered with templates that enable the user to choose a certain process such as "sheet metal stamping" or "hot forging." By this choice most of the numerical settings are preset to a new default that matches best for the procedure to be simulated. However, special cases or the usage of a general-purpose solvers require manual settings such as choosing appropriate FE, an appropriate mesh size, or even numerical details affecting the convergence behavior. Therefore, it is recommendable for the advanced user to have a sound background on numerical issues. A review on the state of the art in solving methods for nonlinear partial differential equations (PDEs) is given by Tadmor [45].

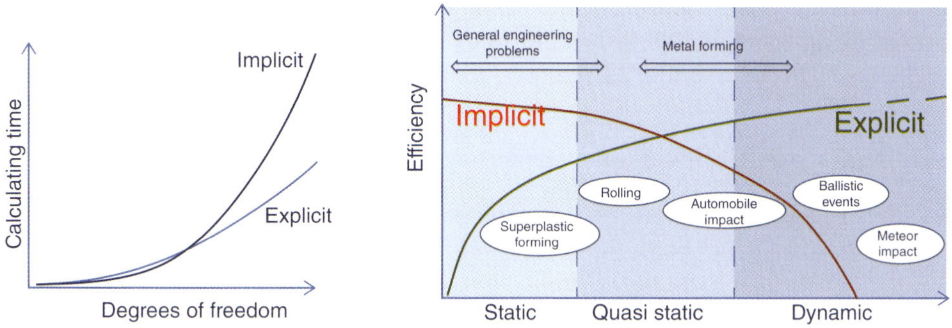

Figure 2.3.5 Implicit and explicit solver performance for varying degrees of freedom and dynamic character of the problem.

Simo and Hughes [13] give a detailed overview on computational inelasticity, and Wriggers [46] gives an overview of nonlinear FE, which is very useful for metal forming problems.

An important distinction can be made between implicit and explicit algorithms to perform the time integration. This choice is sometimes a choice between software codes. The choice whether a problem is solved using explicit or implicit time integration has great impact on the stability and computational costs of a calculation. While the implicit solver searches for an equilibrium solution iteratively, the explicit solver, typically a central difference method, solves a dynamic problem which is rearranged such that the unknowns at time $i + 1$ can be computed directly from known values at time step i and before, whereby no check for equilibrium takes place. Dynamic processes can be handled with a higher efficiency by explicit solvers, while quasistatic problems are often solved implicitly. Computational costs and efficiency of implicit and explicit solutions with respect to degrees of freedom are given in Figure 2.3.5.

2.3.8 Software Codes

As of today, a full list of available software codes would be hard to assemble. Several proprietary and open-source software codes are developed and vanish over time. Here only the most important commercial codes are presented that are commonly used to perform metal forming simulations. While some codes are specific to bulk metal forming or sheet metal forming, others can be applied to a variety of problems and are therefore called multipurpose solvers. In the last years most software providers started to provide full software suites that comprise all preprocessing, postprocessing, and optimization procedures in a single environment. With such software, it is possible to provide the user with an easy-to-use environment with capacities for analyzing entire process chains. Examples for these systems are HyperWorks by Altair or the AutoForm product family in the field of sheet metal

Table 2.3.1 Frequently used commercial FE codes for metal forming simulation.

SW cat	Name of code	Website of software	Type of code	Short description/typical applications/ phenomena being treated
2320	Abaqus	http://www.3ds.com/products-services/simulia/products/abaqus/	Commercial	Multipurpose code with implicit and explicit solvers
2320	ANSYS	http://www.ansys.com/de_de	Commercial	
2330	AutoForm	http://www.autoform.com/de/	Commercial	Easy-to-handle sheet metal simulation for the automotive industry
2320	COMSOL	https://www.comsol.de/	Commercial	Multiphysics software with huge capabilities to extend the software
2340	DEFORM	http://www.deform.com/	Commercial	Bulk metal forming software
2340	FORGE	http://www.transvalor.com/de/	Commercial	Bulk metal forming software with focus on forging
2330	FormingSuite	http://www.forming.com/	Commercial	Sheet metal forming software
2320	HyperWorks	http://www.altairhyperworks.de/	Commercial	Multipurpose suite with interfaces to several forming simulation software tools
2330	LS-DYNA	http://www.dynamore.de/de/produkte/dyna	Commercial	Sheet metal forming software with implicit and explicit solvers
2340	MSC products (Simufact)	http://www.simufact.de/	Commercial	Bulk metal forming software
2330	PAM-STAMP	https://www.esi-group.com/de/software-services/virtual-manufacturing/sheet-metal-forming/stamping-solution-pam-stamp	Commercial	Sheet metal forming software with focus on stamping operations
2340	QForm	http://www.qform3d.com/	Commercial	Bulk metal forming software
2330	Stampack	http://www.maw-cax.de/produkte/stampack/	Commercial	Sheet metal forming software

forming, whereas DEFORM by SFTC, FORGE by Transvalor, MSC Simufact, and QForm cover bulk metal forming processes. Today the statement of Banabic and Tekkaya [26] still holds that sheet metal forming simulations are ahead of 3D bulk metal forming simulations in terms of complexity and capacity although bulk metal forming was the starting point for the FEM in metal forming. Software categories are as follows: 2300 general forming simulation, 2320 multipurpose codes, 2330 sheet metal forming, and 2340 bulk metal forming (Table 2.3.1).

References

1. DIN 8582–1:2003–09. (2003–09) *Manufacturing Processes Forming - Classification; Subdivision, Terms and Definitions, Alphabetical Index.*
2. Hill, R. (1950) *The Mathematical Theory of Plasticity*, Oxford University Press, London.
3. Kobayashi, S., Oh, S.-I., and Altan, T. (1989) *Metal Forming and the Finite-Element Method*, Oxford University Press, Oxford.
4. Lubliner, J. (2005) *Plasticity Theory*, Courier Corporation.
5. Hirt, G. and Kopp, R. (2009) *Thixoforming. Semi-Solid Metal Processing*, Wiley-VCH Verlag GmbH, Weinheim.
6. Groche, P., Wohletz, S., Brenneis, M., Pabst, C., and Resch, F. (2014) Joining by forming—a review on joint mechanisms, applications and future trends. *J. Mater. Process. Technol.*, **214** (10), 1972–1994. doi: 10.1016/j.jmatprotec.2013.12.022
7. Mori, K., Bay, N., Fratini, L., Micari, F., and Tekkaya, A.E. (2013) Joining by plastic deformation. *CIRP Ann. Manuf. Technol.*, **62** (2), 673–694. doi: 10.1016/j.cirp.2013.05.004
8. Psyk, V., Risch, D., Kinsey, B.L., Tekkaya, A.E., and Kleiner, M. (2011) Electromagnetic forming—a review. *J. Mater. Process. Technol.*, **211** (5), 787–829. doi: 10.1016/j.jmatprotec.2010.12.012.
9. Gottstein, G. (2005) *Physical Foundations of Materials Science*, Springer-Verlag.
10. Schey, J.A. (1983) *Tribology in Metalworking: Friction, Lubrication and Wear*, American Society for Metals, Metals Park, OH.
11. Bertram, A. (2011) *Elasticity and Plasticity of Large Deformations – An Introduction*, 3rd edn, Springer-Verlag.
12. Belytschko, T., Liu, W.K., and Moran, B. (2006) *Nonlinear Finite Elements for Continua and Structures*, John Wiley & Sons, Ltd.
13. Simo, J.C. and Hughes, T.J.R. (1998) *Computational Inelasticity*, vol. 7, Springer-Verlag.
14. Wang, J., Levkovitch, V., and Svendsen, B. (2006) Modeling and simulation of directional hardening in metals during non-proportional loading. *J. Mater. Process. Technol.*, **177** (1), 430–432. doi: 10.1016/j.jmatprotec.2006.04.063.
15. Bunge, H.-J., Pöhlandt, K., Tekkaya, A.E., Banabic, D., and Banabic, D. (2000) *Formability of Metallic Materials*, Springer, Berlin, Heidelberg.
16. Perzyna, P. (1966) Fundamental problems in viscoplasticity. *Adv. Appl. Mech.*, **9**, 243–377. doi: 10.1016/S0065-2156(08)70009-7.
17. Bay, N. (1987) Friction stress and normal stress in bulk metal-forming processes. *J. Mech. Work. Technol.*, **14**, 203–223.
18. Lemaitre, J. and Chaboche, J.-L. (Hg.) (1990) *Mechanics of Solid Materials*, Cambridge University Press, Cambridge.
19. Considère, A. (1885) Mémoire sur l'emploi du fer et de l'acier dans les constructions. *Ann. Ponts Chaussées I Sem.*, **9**, 574.
20. Chen, C.C. and Kobayashi, S. (1978) Rigid plastic finite element analysis of ring compression, in *Winter Annual Meeting of the American Society of Mechanical Engineers*, San Francisco, CA, 10–15 December, 1978 (eds H. Armen and R.F. Jones), New York, pp. 163–174, doi: 10.1007/978-1-4613-2880-3_14.

21. Dunne, F. and Petrinic, N. (2005) *Introduction to Computational Plasticity*, Oxford University Press, Oxford.
22. Han, W., Reddy, D., and Tinsley, J. (1995) *Computational Plasticity: The Variational Basis and Numerical Analysis*, North-Holland Publishing Company.
23. Owen, D.R.J. and Hinton, E. (1980) *Finite Elements in Plasticity: Theory and Practice*, Pineridge Press Limited.
24. Chen, J.-S., Roque, C., Pan, C., and Button, S. (1998) Analysis of metal forming process based on meshless method. *J. Mater. Process. Technol.*, **80–81**, 642–646. doi: 10.1016/S0924-0136(98)00171-X.
25. Idelsohn, S., Onate, E., Calvo, N., and Del Pin, F. (2003) The meshless finite element method. *Int. J. Numer. Methods Eng.*, **58** (6), 893–912. doi: 10.1002/nme.798.
26. Banabic, D. and Tekkaya, E. (2006) Numerical simulation and material models of aluminium sheet forming, in *Virtual Fabrication of Aluminium Products. Microstructural Modeling in Industrial Aluminium Fabrication Processes* (Hg. J. Hirsch), Wiley-VCH Verlag GmbH, Weinheim.
27. Osakada, K. (2008) History of plasticity and metal forming analysis. ICTP 2008 (The 9th International Conference of Technology and Plasticity), pp. 22–43.
28. Lenard, J.G., Pietrzyk, M., and Cser, L. (1999) *Mathematical and Physical Simulation of the Properties of Hot Rolled Products*, Elsevier.
29. Bambach, M. and Seuren, S. (2015) On instabilities of force and grain size predictions in the simulation of multipass hot rolling processes. *J. Mater. Process. Technol.*, **216**, 95–113. doi: 10.1016/j.jmatprotec.2014.07.018.
30. Roters, F., Eisenlohr, P., Bieler, T.R., and Raabe, D. (2010) Crystal plasticity finite element methods, in *Materials Science and Engineering*, Wiley-VCH Verlag GmbH, Weinheim.
31. Liao, S.-H., Chu, P.-H., and Hsiao, P.-Y. (2012) Data mining techniques and applications – a decade review from 2000 to 2011. *Expert Syst. Appl.*, **39**, 11303–11311. doi: 10.1016/j.eswa.2012.02.063.
32. Harding, J.A., Shahbaz, M., Srinivas, S., and Kusiak, A. (2006) Data mining in manufacturing: a review. *J. Manuf. Sci. Eng.*, **128**, 969–976. doi: 10.1115/1.2194554.
33. Roll, K. (2008) *Simulation of Sheet Metal Forming – Necessary Developments in the Future*. Daimler AG; Werk Sindelfingen, PWT/VAS, Germany. Keynote Vorträge des LS-DYNA Anwenderforum. LS-DYNA Anwenderforum, DYNAmore GmbH, Bamberg.
34. Gottstein, G. (2007) *Integral Materials Modeling. Towards Physics-Based Through-Process Models*, Wiley-VCH Verlag GmbH, Weinheim.
35. Hirsch, J. and Karhausen, K.F. (2011) History of ICME in the European aluminium industry, in *1st World Congress on Integrated Computational Materials Engineering. ICME. ICME. Seven Springs, PA, July 10–14, 2011. TMS* (Hg. J. Allison, P. Collins, and G. Spanos), John Wiley & Sons, Inc., Hoboken, NJ, pp. 203–210.
36. Raabe, D., Chen, L.-Q., Barlat, F., and Roters, F. (Hg.) (2004) *Continuum Scale Simulation of Engineering Materials, Fundamentals. Microstructures – Process Applications*, Wiley-VCH Verlag GmbH, Berlin.
37. Roll, K. (2007) Advanced simulation techniques – exceeding reality? Materials Science and Technology, Detroit, MI, September 16–20, 2007.
38. Berger, E.J. (2002) Friction modeling for dynamic system simulation. *Appl. Mech. Rev.*, **55**, 535–577. doi: 10.1115/1.1501080.
39. Marek, R. and Nitsche, K. (2012) *Praxis der Wärmeübertragung*, 3. Aufl//3., aktualisierte Aufl, Hanser, München.
40. Reeh, S., Music, D., Gebhardt, T., Kasprzak, M., Jäpel, T., Zaefferer, S. *et al.* (2012) Elastic properties of face-centred cubic Fe–Mn–C studied by nanoindentation and ab initio calculations. *Acta Mater.*, **60** (17), 6025–6032. doi: 10.1016/j.actamat.2012.07.038.
41. Dieter, G.E., Kuhn, H.A., and Semiatin, S.L. (Hg.) (2003) *Handbook of Workability and Process Design*, ASM International, Materials Park, OH.

42. Pöhlandt, K. (1989) *Materials Testing for the Metal Forming Industry*, 1 Band, Springer-Verlag, Berlin.
43. Henke, T., Hirt, G., and Bambach, M. (2013) Randwertermittlung für die robuste Auslegung von Schmiedeprozessen. *Werkstattstechnik online*, **103** (10), 783–788.
44. Kopp, R. and Philipp, F.-D. (1992) Physical parameters and boundary conditions for the numerical simulation of hot forming processes. *Steel Res. Int.*, **63** (9), 392–398.
45. Tadmor, E. (2012) A review of numerical methods for nonlinear partial differential equations. *Bull. Am. Math. Soc.*, **49** (4), 507–554.
46. Wriggers, P. (2008) *Nonlinear Finite Element Methods*, Springer Science & Business Media.

2.4 Heat Treatment

Martin Hunkel

2.4.1 Introduction into Heat Treatment

Heat treatment in general involves all processes using a temperature change with the aim of altering physical properties of a part. Specific properties are needed for the next process step, for example, softening after cold forming, or as resulting part property, for example, hardening. Regarding metals, in most cases, the physical properties of interest are the mechanical properties: the softening or hardening of the material. The underlying physical effects are phase transformations and precipitations as well as recovery, recrystallization, and reduction of dislocations. There are also some special applications for other properties such as optimizing the magnetism by precipitation. The physical effects are the same in different processes such as the cooling after hot forming, the quenching during press hardening, or within the heat-affected zone during welding and during a heat treatment.

Typically, during heat treatment, temperature, mechanical behavior, and microstructure (grain size and/or phases) have to be considered. All of these effects are more or less strongly interrelated (see Figure 2.4.1). Most of these interrelationships are described later. For the simulation of a part, the part geometry is needed. The material data depends on temperature, on microstructure, and in some cases on the stress or on the local chemical composition. During heat treatment (thermal and thermochemical treatments), a macroscopic diffusion can be involved, for example, due to carburizing or nitriding. Then a diffusion simulation can be helpful. Typically, thermal and mechanical boundary conditions are needed. Some of the boundary conditions can be determined with the help of simulations. For example, the heat transfer coefficient between part and fluid computational fluid dynamics (CFD) is applicable. For inductive or conductive heating, the internal heat source can be determined by electromagnetic simulations.

In most cases heat treatment is carried out in higher temperatures. At higher temperatures, normally, effects occur faster, for example, due to higher diffusion

Handbook of Software Solutions for ICME, First Edition. Edited by Georg J. Schmitz and Ulrich Prahl.

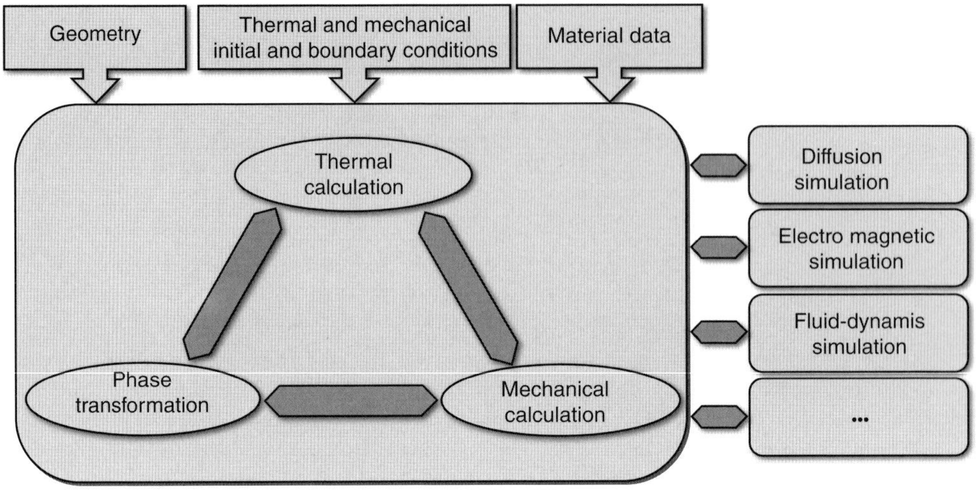

Figure 2.4.1 Interactions during a heat treatment simulation.

coefficients or dislocation mobility. Also the dissolution of precipitations or the formation of specific phases such as austenite in steels occurs only at higher temperatures. But in some cases cryogenic heat treatment is necessary. At very low temperatures, supersaturated states can be stabilized. Some transformations occur only at subzero temperatures, for example, the transformation of retained austenite to martensite.

2.4.2 Heat Transfer in and out of a Part

2.4.2.1 Thermal Conduction

The heating and cooling lead to a temperature gradient within the part, and therefore the thermal flux has to be considered. The thermal flux j in a part is given by Fourier's law:

$$j = -\lambda \nabla T. \tag{2.4.1}$$

In general, the thermal conductivity λ depends on the temperature T. The temperature rate is given by

$$\rho c_{\mathrm{p}} \dot{T} = \nabla j + \dot{Q}. \tag{2.4.2}$$

Density ρ and specific heat capacity c_{p} depend also on the temperature. The energy rate $\dot{Q}$ is given by internal heat sources or sinks, for example, due to deformation heat, transformation heat, or electromagnetic fluxes.

2.4.2.2
Heat Transfer

In general, the parts are heated up to a specific temperature, are held at this temperature for a certain time, and finally cooled down afterward. A special case is the cryogenic treatment, where the parts are first cooled down, for example, in liquid nitrogen, and are held and heated up. Often, the whole part should be heat treated, but there are special treatments where only some regions of the part should be heat treated, for example, surface hardening. Different physical effects can be used for heating and cooling. The heating up can be done by internal heat sources as well as by an ambient medium, while for cooling only an ambient medium is used. Regarding an ambient medium the heat transfer flux j from the part to the medium or vice versa is described by a heat transfer coefficient α and the difference between ambient temperature T_A and surface temperature of the part T_S:

$$j = \alpha(T_A - T_S). \tag{2.4.3}$$

Exemplarily, the heat transfer coefficient is given in Figure 2.4.2 for some media depending on the surface temperature of the part. For gases the heat transfer coefficient is assumed to be constant. Salt baths are used at higher temperatures. The high heat transfer coefficient in the salt bath is followed by the low heat transfer coefficient for the air cooling. The heat treatment coefficient for vaporizing liquids (quenching oil, still cold water) is described in Section 2.4.2.5.

2.4.2.3
Thermal Radiation

In transparent mediums and vacuum, the thermal radiation is always present. Therefore parts can be heated or cooled in vacuum to prevent oxidation or similar effects. The thermal radiation depends on the angle between the source of the radiation and the surface as well as on the frequency of the radiation, but these

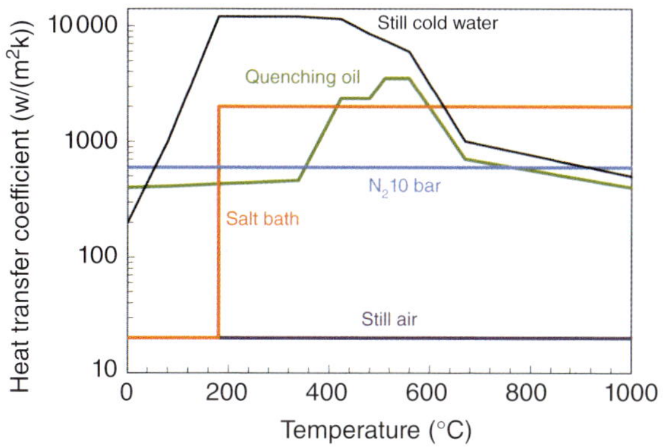

Figure 2.4.2 Heat transfer coefficient of different media.

effects are neglected in general. The main influencing factor is the surface state given by the emissivity ε. The emissivity ε is in the range between 0 and 1.

The radiative heat transfer can be calculated with the help of the Stefan–Boltzmann law and is given by

$$j = \varepsilon k_{SB}(T_A^4 - T_S^4) \quad (2.4.4)$$

using the Stefan–Boltzmann constant $k_{SB} = 5.67 \times 10^{-8}\,W/m^2\,K^4$. From the Stefan–Boltzmann law, a heat transfer coefficient can be determined by using Eq. (2.4.3):

$$\alpha = \varepsilon k_{SB}(T_A^3 + T_A^2 T_S + T_A T_S^2 + T_S^3). \quad (2.4.5)$$

The emissivity ε can change during a heat treatment, for example, due to oxidation. Regarding steel parts, for machined parts the emissivity is about 0.1, and for oxidized parts the emissivity is about 0.8.

2.4.2.4 Convective Heat Transfer

Pure convection takes place in gases and nonvaporizing liquids such as salt melts. The heat transfer coefficient depends mainly on the (local) velocity and the pressure of the medium. Often the heat transfer coefficient can be assumed to be constant. The convective heat transfer in gases or salt melts are used for heating and cooling. Typical heat transfer coefficients are $20\,W/m^2\,K$ for still air, about $400–1500\,W/m^2\,K$ for high-pressure gas quenching, and $1000–4000\,W/m^2\,K$ for liquid salts.

2.4.2.5 Cooling in Vaporizing Liquids

Very often parts are cooled in vaporizing liquids, where the part has a temperature significantly above the boiling temperature of the liquid. Then there are three stages with totally different heat transfer coefficients. Directly after moving the part into the liquid, the part is surrounded by a gas film. This gas film has an isolating effect and the heat transfer coefficient is low. The part cools down slowly. After some time the liquid starts bubbling beginning at corners, at edges, or at contamination at the surface of the part (Leidenfrost point). During bubbling the heat transfer coefficient is very high. At low part temperatures the heat transfer is convective with a low heat transfer coefficient. Altogether the heat transfer coefficient depends on liquid, liquid temperature, liquid velocity, part shape, and part surface temperature. For simplicity, only the part surface temperature is considered for a specific liquid condition (see Figure 2.4.2).

Using water sprays, the heat transfer coefficient can be much more than $20\,000\,W/m^2\,K$ depending on the water droplet size, velocity, and density. The high heat transfer coefficient has its origin in the instantaneous evaporation of the droplets.

2.4.2.6
Solid–Solid Heat Transfer

The solid–solid heat transfer, for example, between a part and a forming tool, depends on the surface qualities, the temperature, and the contact pressure between the solids. Exemplarily, the heat transfer is between 500 and 5000 $W/m^2\,K$ for press hardening [1].

2.4.2.7
Electromagnetic Heating

Very high heating rates can be achieved by electromagnetic heating, that is, conductive or inductive heating. The heating occurs directly in the material by electric currents. For inductive heating, the electric current will be induced by the electromagnetic field. In the case of inductive heating, the outer layer of a specimen is subjected to a higher electrical current density leading to its higher-temperature field compared to the specimen's core (skin effect). The heating rates for conductive or inductive heating can be much higher than 1000°C/s. Due to the high heating rates and the skin effect, inductive heating can be used for a localized (near-surface and defined region) heat treatment.

Electromagnetic heating can be considered with the help of an electromagnetic simulation. Transient FE methods can be used for the electromagnetic simulations, but there exist very efficient boundary element methods (BEMs) or steady-state FE methods. Very often it is more effective to use a weakly coupled electromagnetic heat treatment simulation due to different timescales (e.g., inductive, μs; thermal, ms).

2.4.3
Microstructure

An aim during heat treatment is the generation of a specific microstructure after the heat treatment. For example, this can be a martensitic microstructure in steels, fine-sized precipitates in Al alloys, or generally a fine-grained microstructure. Typically, the microstructure depends strongly on the chemical composition. Besides the microstructure variations, the temperature can also change the specific physical properties such as hardness, electric resistivity, or strain. The dependency on the properties is used for detecting the microstructure change, for example, by dilatometry (strain) or differential scanning calorimetry (specific heat capacity). As an example the measured length change of dilatometer specimens of low-alloy steels are shown in Figure 2.4.3. The chemical composition besides the carbon content, the austenitizing condition, and the cooling rate are the same for each experiment. For these experiments, bainite and martensite are formed from austenite. The lower the carbon content, the higher bainite start

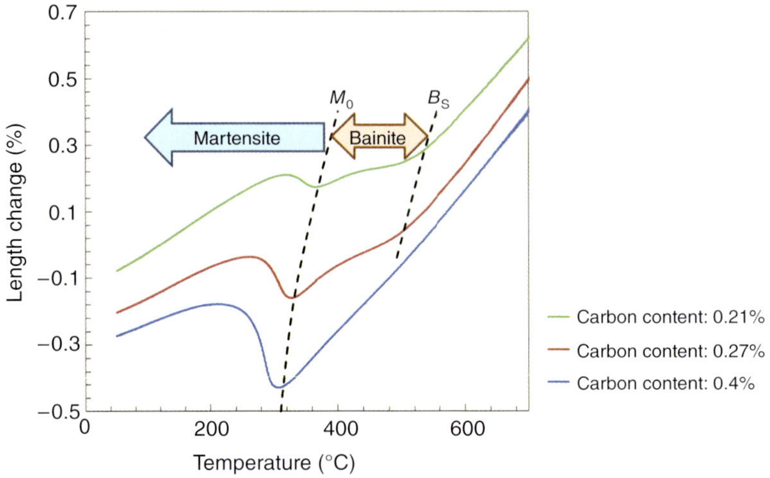

Figure 2.4.3 Measured length change during identical quenching of a low-alloy steel with different carbon contents.

temperature B_S and martensite start temperature M_S, and the more bainite and less martensite are formed.

2.4.3.1
Phase Transformations and Precipitations

While material engineers distinguish very often between solid–solid phase transformation (the whole amount can be transformed) and precipitations (only a small amount can be transformed), from the physical point of view, there is no difference. Structural phase transformations in materials science are associated with the change in lattice structure, for example, for steel the transformation of austenite to ferrite is a transformation of an fcc structure to a bcc structure and/or a change in the chemical composition. Besides the structural phase transformations, magnetic phase transitions or order–disorder phase transitions are of interest. These phase transitions are not discussed further. Most phase transformations are diffusive controlled. The diffusion leads to a time dependency of these transformations. But there are also diffusionless transformations. The most important diffusionless transformation is the austenite to martensite transformation in steels or shape-memory alloys.

The time dependency of the phase content P during isothermal diffusive phase transformations is often described by the Johnson–Mehl–Avrami equation

$$P = P_0(1 - e^{-k(T)t^{n(T)}}) \tag{2.4.6}$$

with the (sometimes temperature-dependent) maximum phase content P_0, a temperature-dependent kinetic factor k, and a temperature-dependent shape factor n. The kinetic factor k depends strongly on nucleation rate and diffusion coefficient, while the shape factor n depends on the shape (e.g., spherical or needlelike) and the nucleation rate related to the growth rate [2]. In the case of the dissolution of, for example, precipitations, the Johnson–Mehl–Avrami equation is written as

$$P = P_0 e^{-k(T)t^{n(T)}}. \tag{2.4.7}$$

For continuous temperature changes, in general, the additivity rule is not given, and a simple transfer is not possible. There exist some concepts using, for example, Scheil's sum or a correction function for calculating phase transformation during continuous temperature change. The quench factor analysis is often applied for the modeling of precipitations in aluminum, titanium, and other lightweight alloys [3].

The martensite transformation occurs only if the temperature is below the martensite start temperature M_S. Then the martensite content P_M is described by the Koistinen–Marburger equation [4]

$$P_M = P_{M0}(1 - e^{-k(T)(M_S-T)}). \tag{2.4.8}$$

Besides this athermal martensite transformation, there exists also an isothermal martensite formation (especially in high-alloy Ni-rich steels), which can be described by an autocatalytic equation [5].

2.4.3.2 Recrystallization and Grain Growth

During cold forming (i.e., forming below the recrystallization temperature), a significant amount of dislocations is formed, and the material is strongly hardened. During heat treatment above the recrystallization temperature, the material is recrystallized. The recrystallization has two effects. First the hardening is reduced and further cold forming can be applied. Second, after recrystallization of high-deformed material (minimum 20% deformation degree), a fine-grained structure occurs. For cold-formed materials, static recrystallization occurs only during heat treatment.

In principle, the recrystallization starts if the temperature is higher than the recrystallization temperature. During recrystallization, new grains with a low dislocation density are formed. Due to the low dislocation density, the material can be further formed. The size of the newly formed grains depends mainly on the degree of deformation and the temperature. The higher the degree of deformation and the higher the temperature is, the smaller the grains are. The higher the temperature, the smaller the freshly formed grains are. After recrystallization normal grain growth occurs. Precipitations can reduce the grain growth due to Zener pinning. During heat treatment these precipitations can dissolve. Then abnormal grain growth occurs [6].

2.4.4
Mechanical Behavior during Heat Treatment

Mostly, during heat treatment, plastic strains are small, that is, plastic strain is rarely above 2%. Additionally there is a thermal and transformation strain up to about 2%. Thermal and transformation strain itself is nonplastic, but a gradient in these strains leads to strain-inducing stresses. Therefore small strain approximation can be used during heat treatment simulations. Even if the strains during heat treatment are small, they result in significant residual stresses and distortion of the parts. The stress–strain curves have to be determined carefully also in metastable temperature regions. It has to be carefully checked that no significant transformation occurs during the mechanical test.

The strain consists of different terms:

$$\varepsilon = \varepsilon_{\mathrm{el}} + \varepsilon_{\mathrm{pl}} + \varepsilon_{\mathrm{cr}} + \varepsilon_{\mathrm{th}} + \varepsilon_{\mathrm{pt}} + \varepsilon_{\mathrm{tp}} + \cdots \tag{2.4.9}$$

with elastic strain $\boldsymbol{\varepsilon}_{\mathrm{el}}$, plastic strain $\boldsymbol{\varepsilon}_{\mathrm{pl}}$, creep strain $\boldsymbol{\varepsilon}_{\mathrm{cr}}$, thermal strain $\boldsymbol{\varepsilon}_{\mathrm{th}}$, phase transformation strain $\boldsymbol{\varepsilon}_{\mathrm{pt}}$, transformation plasticity $\boldsymbol{\varepsilon}_{\mathrm{tp}}$, and maybe some more terms. Elastic, plastic, and creep strain have their usual meaning and are not discussed further.

2.4.4.1
Thermal and Transformation Strain

Thermal and phase transformation strain is due to the density change during heat treatment and can be calculated by Jablonka *et al.* [7]:

$$\varepsilon_{\mathrm{th}} + \varepsilon_{\mathrm{pt}} = \sqrt[3]{\frac{\varrho(T_0, P_0)}{\varrho(T, P)}} - 1. \tag{2.4.10}$$

The strain is always referred to an initial state given by the initial temperature T_0 and the initial phase contents P_0.

2.4.4.2
Transformation Plasticity

During heat treatment very often, external or internal loads are applied. External loads are, for example, clamping or the self-weight of the part. Internal stresses can be due to thermal or phase transformation gradients. These stresses lead to an additional plasticity effect called transformation plasticity during phase transformations even in the elastic range of both initial and final phases:

$$\varepsilon_{\mathrm{tp}} = Kf(P)\sigma, \tag{2.4.11}$$

where $\boldsymbol{\varepsilon}_{\mathrm{tp}}$ is the transformation plasticity strain tensor, $\boldsymbol{\sigma}$ is the stress deviator tensor, K is the transformation plasticity constant, and f is the phase-dependent saturation function (with $f(0) = 0$ and $f(1) = 1$). Experimentally, the phase transformation is determined by the volume change, and the transformation plasticity

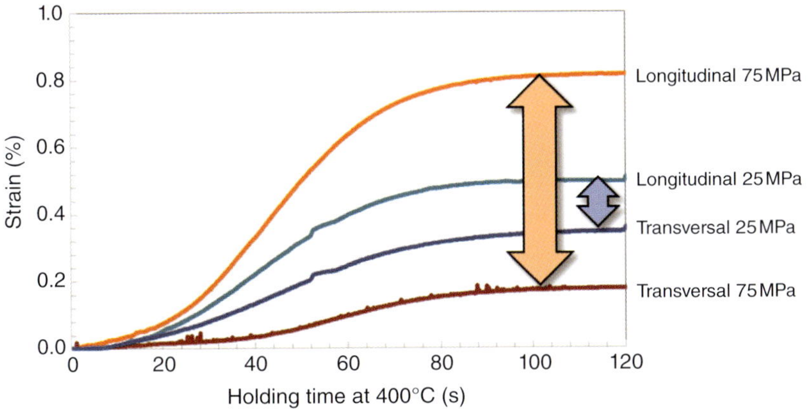

Figure 2.4.4 Measured longitudinal and transversal strain during isothermal bainite formation (100Cr6; 400°C) for two different applied stresses in longitudinal direction.

can be calculated by the strain deviator. The linearity in Eq. (2.4.11) holds only for small stresses. If the stress is higher than half of the yield strength of the weaker phase, transformation plasticity becomes nonlinear. For steels, the transformation plasticity constant is in the range of 5×10^{-5} MPa^{-1} to 5×10^{-4} MPa^{-1}. Figure 2.4.4 shows the measured longitudinal and transversal strain during isothermal bainite formation of 100Cr6 at 400°C for two different applied stresses in longitudinal direction. The higher the strain difference is, the higher the applied stress is.

There are two different physical mechanisms explaining the transformation plasticity. The Greenwood–Johnson effect [8] explains the transformation plasticity by an additional local stress incorporated with the transformation strain. Therefore the superposition of the external or internal stress with the local stress leads to an additional plastic strain. The Magee effect [9] is a directed transformation especially of the martensite correlated with the stress. This directed transformation leads to an oriented transformation strain.

Similar effects occur due to inner directed stresses during transformation of a banded microstructure (e.g., due to segregations) or layers. During transformations, directed inner stresses lead to inner transformation plasticity [10]. As a result, the macroscopic transformation strain has a preferred direction.

2.4.5
Thermochemical Treatment

Thermochemical treatments are a desired modification of the chemical composition in the near-surface region (in contrast to unwanted oxidation or decarburizing) by absorption of elements at the surface and diffusion of these elements within the parts at different temperature ranges. Depending on treatment temperature and diffusion elements, the depth of the modified layer can be from micrometers to millimeters. In general, interstitial elements such as carbon or nitrogen are used due to their significantly higher diffusion coefficients than those of substitutional elements. During the diffusion step, a layer at the surface or precipitates within the part can be formed.

2.4.5.1
Carburizing and Carbonitriding

A very popular thermochemical treatment is the case hardening of steel components. The parts are heated up to a proper austenitizing temperature followed by carburizing (addition of carbon). The carburized components are then transferred to an appropriate hardening temperature and soaked for a certain amount of time followed by quenching. In the case of gas carburizing, the carbon flux j_C across the surface is described by

$$j_C = \beta(c_P - c_S) \tag{2.4.12}$$

using carbon transfer coefficient β, gas carbon potential c_P, and the carbon concentration within the part at the surface c_S. The carbon flux within the part is described by Fick's first law:

$$j_C = -D\nabla c \tag{2.4.13}$$

with the temperature- and concentration-dependent diffusion coefficient D. If the carbon content becomes too high during carburizing, carbides can be formed. The carbon content has a significant influence on the phase transformations during cooling or quenching down to room temperature. The formation of the diffusive phase transformations such as ferrite, pearlite, or bainite is shifted to much longer times due to the enhanced hardenability. The martensite start temperature is dropped to lower temperatures (compare Figure 2.4.3). Figure 2.4.5 shows measured carbon and hardness profiles after a standard case hardening. The higher the carbon, the higher the hardness due to higher martensite content, which in turn leads to higher martensite hardness. The carbon content in the carburized layer is sufficiently high; thus, in the case of quenching at the surface, martensite and retained austenite are present. However, in the noncarburized core, there can be bainitic–ferritic microstructure. A related heat treatment is the carbonitriding. The carbon and nitrogen are added in the near-surface region. Nitrogen has similar effects on the transformations than carbon, but in general the diffusion coefficient is lower. There is a strong thermodynamic interaction between carbon and nitrogen within steels [11]. Therefore, the simple Fick's first law (Eq. 2.4.13) no longer holds.

2.4.5.2
Nitriding and Nitrocarburizing

Other thermochemical treatments applied on steel and other metals such as titanium are nitriding and nitrocarburizing [12]. These thermochemical treatments

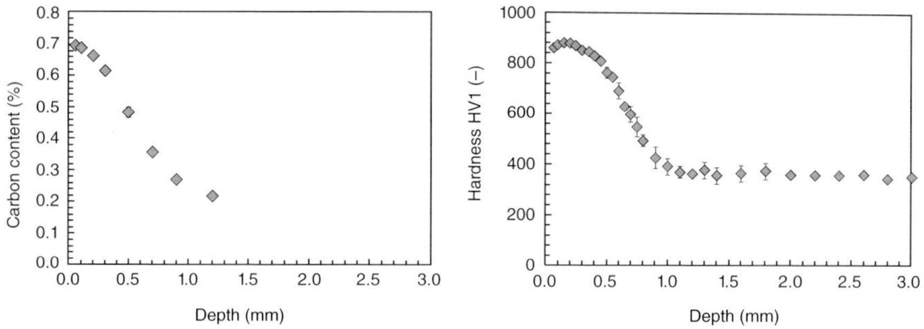

Figure 2.4.5 Measured carbon and hardness profiles for a standard case hardening process.

are applied at temperatures where the initial (room temperature) microstructure is still stable (for steels: around 500–550°C). The advantage is the reduced distortion during cooling due to the missing phase transformation. The disadvantage is the much longer diffusion time. First, the initial microstructure is enriched with nitrogen (and carbon). For longer nitriding times and higher nitrogen amounts, hard compound layers can be formed. On nitriding low-alloy steels, first γ'-layer and then ε-layer will be formed [13]. The layer growth is parabolic in most cases.

2.4.6 Heat Treatment Simulation

2.4.6.1 Specific Remarks on Heat Treatment Simulations

Usually, heat treatment simulations are done by the FE method. In most cases, the small strain approximation can be used for heat treatment simulations. Unfortunately, heat treatment simulations coupled with phase transformations need a lot of model parameters. The transformation kinetics have to be determined accurately. Also the effects related to phase transformations such as transformation plasticity are needed. All physical and mechanical parameters such as temperature-, phase-, and if necessary stress-dependent parameters have to be determined.

During heat treatment, especially during quenching, high temperature gradients can occur. The temperature gradients can lead to high stress gradients. There can be also phase gradients due to the temperature gradients or the thermochemical treatment. When meshing a part, these gradients have to be considered. Additionally, time steps have to be small enough so that the phase transformation increment is not too large.

During heat treatment, there is very often only a negligible influence of stresses and strains on temperature and transformations. To reduce the simulation duration, thermal/transformation and mechanical simulations can be simulated separately. First, the thermal/transformation simulation is executed, and then the mechanical simulation is executed. Most commercial simulation software support the transfer of temperature. Specialized software also support the transfer of phase contents. For the same reason, diffusion simulation is separated from thermal/transformation simulation.

2.4.6.2
Specialized Software

Categories: 2410 (thermal, mechanical, phase transformation) and 2420 (thermal, mechanical)

SW cat	Name of code	Website of software	Type of code (academic/ commercial)	Short description
2410	SysWELD	www.esi-group.com/ software-services/ virtual-manufacturing/ welding-assembly	Commercial	FEM analysis and CAE Main application: welding and heat treatment
2410	DEFORM-HT	www.deform.com/ products/deform-ht/	Commercial	FEM analysis and CAE Main application: forming and heat treatment
2410	LS-DYNA	www.dynamore.de/ de/produkte/dyna	Commercial	FEM analysis and CAE Main application: forming, welding, and heat treatment of sheets
2410	Simufact	www.simufact.de/en	Commercial	FEM analysis and CAE Main application: forming
2410	DANTE	www.dante-solutions .com/	Commercial	FEM analysis and CAE Main application: heat treatment
2410	FORGE	www.transvalor.com/ en/cmspages/forge-nxt.32.html	Commercial	FEM analysis and CAE Main application: forming/general use
2420	Abaqus	www.3ds.com/ products-services/ simulia/products/ abaqus	Commercial	FEM analysis and CAE General use

SW cat	Name of code	Website of software	Type of code (academic/ commercial)	Short description
2420	Marc	www.mscsoftware .com/product/marc	Commercial	FEM analysis and CAE Main application: forming
2420	ANSYS	www.ansys.com	Commercial	FEM analysis and CAE General use
2420	COMSOL	www.comsol.com	Commercial	FEM analysis and CAE General use

References

1. Merklein, M., Lechler, J., and Stoehr, T. (2009) Investigations on the thermal behavior of ultra high strength boron manganese steels within hot stamping. *Int. J. Mater. Form.*, **2**, 259–262.
2. Burke, J. (1965) *The Kinetics of Phase Transformations in Metals*, Pergamon Press, Oxford.
3. Staley, J.T. (1987) Quench factor analysis if aluminum alloys. *Mater. Sci. Technol.*, **3**, 923–935.
4. Koistinen, D.P. and Marburger, R.E. (1959) A general equation prescribing the extent of the austenite-martensite transformation in pure iron carbon alloys and plain carbon steels. *Acta Metall.*, **7**, 50–60.
5. Magee, C.L. (1971) The kinetics of martensite formation in small particles. *Metall. Trans.*, **2**, 2419–2430.
6. Hillert, M. (1965) On the theory of normal and abnormal grain growth. *Acta Metall.*, **13**, 227–238.
7. Jablonka, A., Harste, K., and Schwerdtfeger, K. (1991) Thermomechanical properties of iron and iron-carbon alloys: density and thermal contraction. *Steel Res.*, **62**, 24–33.
8. Greenwood, G.W. and Johnson, R.H. (1965) The deformation of metals under small stresses during phase transformations. *Proc. R. Soc. A: Math. Phys. Eng. Sci.*, **A283**, 403–422.
9. Magee, C.L. (1967) Transformation kinetics, microplasticity and aging of martensite in iron 21 Ni. PhD thesis. Carnegie Melton University, Pittsburg, PA.
10. Hunkel, M. (2009) Anisotropic transformation strain and transformation plasticity: two corresponding effects. *Materialwiss. u. Werkstofftech.*, **40** (5-6), 466–472.
11. Slycke, J. and Ericsson, T. (1981) A study of reactions occurring during the carbonitriding process. *J. Heat. Treat.*, **2**, 3–19 and 97–112.
12. Schaaf, P. (2002) Laser nitriding of metals. *Prog. Mater Sci.*, **47**, 1–161.
13. Hosmani, S.S., Schacherl, R.E., and Mittemeijer, E.J. (2008) Compound layer formation on iron-based alloys upon nitriding phase constitution and pore formation. *HTM, Haerterei Tech. Mit.*, **63**, 139–146.

2.5
Joining Processes

Ulrike Beyer, Gerson Meschut, Stephan Horstmann, and Ralph Bernhardt

2.5.1
Introduction

"THE WHOLE IS MORE THAN THE SUM OF ITS PARTS"

ARISTOTLE (384–322 AD)

Of course, it seems grotesque at first glance to apply this more than 2300-year-old thesis to the manufacturing process of joining. But if Aristotle's basic thought is transferred to the special conditions of joining technology, then significant amount of the primary goals of current research can be inferred. Materials should be joined together in such a way that the result intensifies the advantages of the original materials and compensates for their disadvantages. What is sought is a whole that exceeds the sum of its single parts [1].

Currently these ambitions are being driven by lightweight design and its tendency to use multimaterial designs. The technology of hybrid material designs the usage of materials of differing qualities – demands that joining techniques be thoroughly reoriented, since the previously used, mostly thermal procedures are no longer adequate in terms of process stability and cost. For this reason, adhesive bonding and mechanical joining have become key technologies (Figure 2.5.1) and drivers in the further development of lightweight construction.

However, established thermal joining techniques are also being steadily developed and offer innovative solutions for joining, even heat-sensitive materials, with great process stability.

Multimaterial design plays an important role in joining technology and is closely connected to aerospace and automotive construction where the primary goal is to reduce fuel consumption and the amount of pollutant emissions created by reversing the weight spiral. This leads to a wide spectrum of different materials being used in automotive body in white construction that demand the appropriate joining technique. As an example of the diversity of joining techniques used in hybrid design, consider the aluminum structure of the Audi A8.

A supporting structure, the so-called Audi Space Frame is created out of extruded sections and multifunctional casted components, together with sheet

Handbook of Software Solutions for ICME, First Edition. Edited by Georg J. Schmitz and Ulrich Prahl.

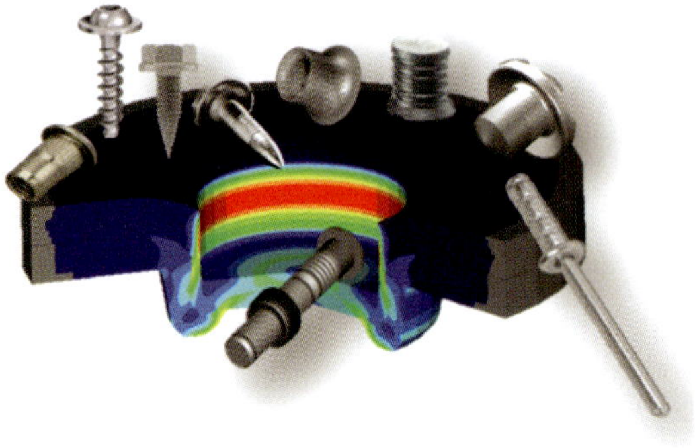

Figure 2.5.1 Overview of mechanical joining technologies. (Reprinted from [2] with kind permission from Wiley.)

metal components. Extensive sheet metal parts as shear areas or as plating and fixtures are then added. To join these different components, different techniques are employed. For example, 2200 self-pierce rivets, 200 punch rivets, and 650 flow-drill screws are used, in addition to 26 m of welding seam, 20 m of laser welding seam, 62 m of adhesive seam, and 200 resistance welding points [3]. This illustrates not only the high number of joining areas but also the necessity to employ processes that are based on different physical joining mechanisms.

2.5.2
Basics and Definitions

2.5.2.1
Definition

The word "join" has its roots in western Germanic languages and was/is used in the sense of connecting and fitting. In modern technical language, "joining" is also used as the name for a group of manufacturing processes [4]. DIN 8580 [5] (DIN being the German Industrial Norm used internationally) defines "joining" as the connection of different components of geometrically determined forms, or of components with formless or connecting elements, where the connections between the components are locally created or increased.

2.5.2.2
Joint

The result of a joining procedure is a joint. It is the point of connection between the components of a product. As a rule, the spatial degree of freedom to move, for one or more components, is either temporarily or permanently reduced at the point of connection. The properties of the joint are largely determined by the way in which the components are meant to lock together. The different locking processes are differentiated as follows:

- Positive locking: Mechanical contact between the components that are to be joined.
- Force closures: A connection is established by the interaction of forces between the components that are to be joined. The transmission of force can occur either through direct mechanical contact or without any contact (e.g., by magnetic fields).
- Adhesive bonding: The interaction of forces between the atoms and molecules in the materials the components that are to be joined are made of; chemical or physical processes are the basis for the formation of the connection.

These different ways of locking components either are applied by themselves or can be combined [4].

2.5.2.3
Process

According to [6], the process of "joining" is further divided into groups of assembling, filling, fastening, pressing in, joining by casting, joining by forming, joining by welding, joining by soldering, adhesive bonding, and textile joining. Examples and the characterization of these groups in respect to their connectivity and solubility are shown in Table 2.5.1.

Which type of joining or which hybrid solution is used depends on the following aspects:

- Whether the type of joint required for an application is temporary or permanent.
- Whether similar or dissimilar materials are to be joined. Metallurgical incompatibility can be disastrous for the performance of the joints.
- The physical and chemical metallurgical properties of materials to be joined.
- The requirements of the joint under special conditions, for example, temperature, corrosion, environment, and reliability.
- The type and nature of loading conditions (static and/or dynamic loading under tension, shear, compression, bending, etc.).
- The economy and cost effectiveness is one of the most important factors that influences the selection of the jointing solution selected when manufacturing and engineering components.

The multitude of joining techniques and their variants on the one hand, and the different fields of application/the demands on the joint on the other, result in a broad range of appropriate joining technique and process parameters to investigate. In the past, this used to be the starting point for a sequential, cost-intensive iteration process that often involved many practical experiments. Such experiments are being increasingly virtually realized by numerical simulation.

Table 2.5.1 Groups of procedures [6].

ON[a]	Name	Kind of connectivity	Detachability	Example
4.1	Assembling	Positive-locking fit Spring force Gravity	Without damage to components	Layering Telescoping Mounting
4.2	Filling	Bond	Without damage to components	Covering Soaking Impregnating
4.3.	Fastening and pressing in	Force closure	Usually without damage to components	Screwing Clamping Nailing
4.4	Joining by casting	Positive locking fit caused by casting	Usually without damage to components	Casting Embedding Galvanizing
4.5	Joining by forming	Positive locking fit caused by forming	Usually only with damage to components	Twisting Flanging Clinching Riveting
4.6	Joining by welding	Connection of materials	Only with damage	Press welding Fusion welding
4.7	Joining by soldering	Connection of materials	Only with damage	Soft soldering Hard soldering
4.8	Adhesive bonding	Adhesion	Usually only with damage to components	Wet bonding Contact bonding Active bonding
4.9	Textile joining	Positive locking fit and/or force closure of textile fibers	Without damage to components	Knitted fabrics Woven fabrics

a) Ordinal number.

2.5.2.4 Joining Simulation

Whereas 20 years ago only a few companies used a calculation method based on finite elements (finite element method (FEM)) for the simulation of numerically relatively simple joining techniques [7], modeling process variables are becoming more and more complex. Not only can techniques that involve cutting or high-speed effects be simulated today but also hybrid techniques using adhesives or materials such as CFK.

The prerequisite for simulation models is to create simulations that consider not only the process parameters but also the geometry of the tools and components that are brought into contact during the joining process. Furthermore, information regarding the mechanical and thermal properties of the materials used is also required.

FEM is routinely used to:

- Identify parameters and their multifactorial interdependence
- Visualize the material flow during joining and/or the heat-induced distortion of the component during thermal joining
- Efficiently adjust tools and/or processes for new joining tasks
- Systematically optimize the joining processes to reach the process's target [8]

Since there is such a huge variety of different joining processes, the authors focus on two selected joining technologies, namely, welding and mechanical joining. Both processes are typically representatives of processes where either the microstructure of at least one component undergoes drastic changes or the initial microstructure plays an important role in the process output. This is discussed with respect to chained processes and their interaction.

2.5.3 Welding

The European standard EN 14610 defines welding as the permanent joining of components by melting or by heat introduction and an additional force (pressure) on the components. No other joining method allows such a high-strength, tight, and space-saving connection. Welding is optimal for the joining of different component groups. More than 100 different welding processes have been classified depending on the type of heat source and applied pressure. Besides metallic materials, thermoplastics and glass can also be welded. Cladding can be done by welding to generate solid and high-strength surfaces.

The design of a welding construction is usually based on a detailed plan of the weld safety (construction), the weld ability (material choice), and the welding possibilities (manufacturing). The choice of the optimal process conditions from one involved department can have a great influence on other departments, especially regarding the distortions and subsequent proofs the other departments have to do.

Distortions due to welding are the biggest economic problem related to welding next to a reduction in strength. Welding distortions occur due to (i) the shrinkage of the melt pool and (ii) as a result of the plastically upset surrounding material.

2.5.3.1 Products and Industries

Finite element calculations are suited to model elastic–plastic material behavior, which in turn qualifies it for the simulation of welding structures. The melt pool flow and the resulting seam configuration are not the outcome of welding structure simulations, but are considered as input parameters.

There are also finite volume-based process simulations that are capable of calculating the melting pool flow. The main objective of the process simulation is to predict the parameters to be adjusted on the welding equipment. As of

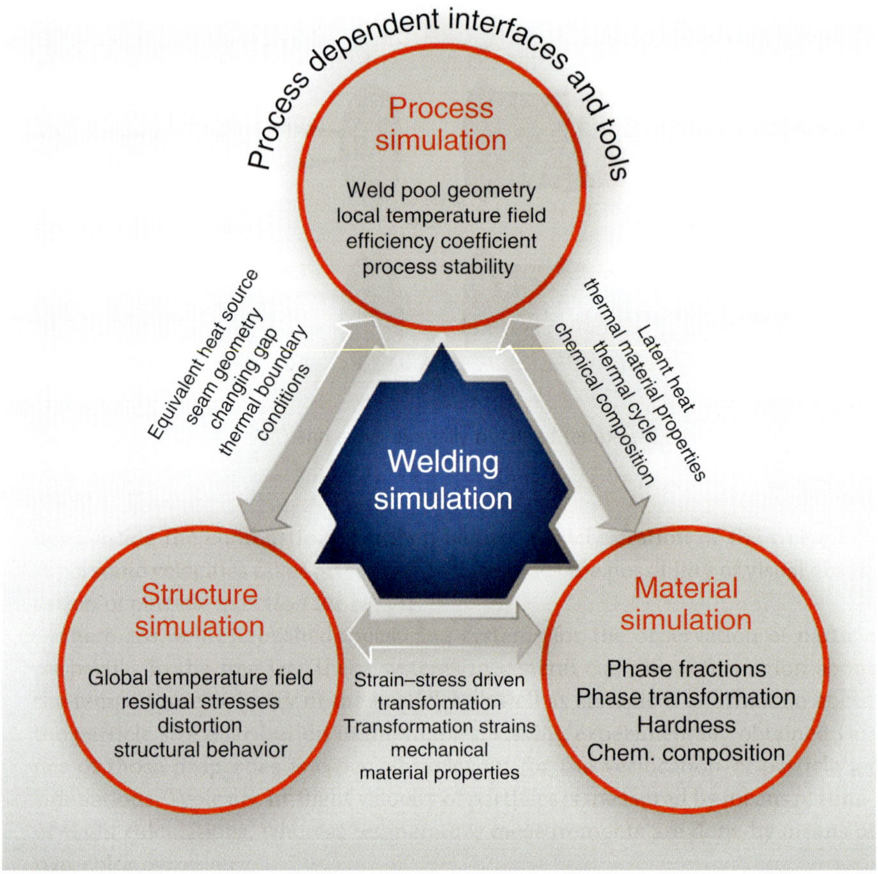

Figure 2.5.2 Different areas of welding simulation (according to [9]) and their coupling.

today, welding simulation is not yet widely spread in the industry. The solution itself exists for more than 30 years, but even with modern hardware it is possible to a certain extent to get reasonable computational performances. It is vital for process simulations to have high resolution, which results in more complex and time-consuming calculations. So, these kinds of simulation are studied separately (see Figure 2.5.2).

The structural welding simulation calculates the consequences of the heat effect from the welding process. Therefore, the exact reproduction of process-specific details for the various processes (e.g., ND:YAG laser, focusing position, submerged branch, and so on) is not critical. It is more important to model the heat flow that comes out of the melt pool isosurface via mathematical resources ("equivalent heat sources").

These models are then ideally calibrated according to the weld pool dimensions that can be observed in a micrograph.

The "disadvantages" of the "equivalent heat sources" that have to be calibrated are put into perspective:

- They are to be used in the state-of-the-art welding structure simulations.
- The question "Which welding methods are supported by the software?" becomes secondary.

All the usual methods can be reproduced with the heat sources.

With friction welding and spot welding one should be careful as there are special influences on the process that should be mapped appropriately.

The coupling of a process simulation that by itself cannot calculate deformations, residual stress, and so on is an important topic and will be implemented. This is currently a "work in progress."

The aim of structural welding simulations is predominantly to predict welding distortions. Through the standard implementation of predefined materials and material models, one can calculate the structural components and material states as well as the resulting local material (property) conditions. The phase transformation models consider the transformation strains, transformation plasticity, and martensitic and bainitic hardening.

Distortions in the welding itself are not a problem while manufacturing when the distortions in the welding exert a negative influence on a specific characteristic of the component. Otherwise these distortions have the potential to cause a costly problem. The technical requirements related to the dimensional accuracy and tolerances of the welded components are directly related to the function of the components themselves.

Table 2.5.2 gives an overview of the negatively influenced characteristics – pain points – which have been investigated using Simufact.welding.

2.5.3.2
Technical Solution Based on Structural Welding Simulation

Once the technical requirements are defined, the needed mesh quality and adequate postprocessing can be defined. One major pain caused by welding distortion is the challenge thereafter to assemble the welded components. As this problem is found in all industry sectors, the distortion calculation has to focus on the distance between the points or on the rotation of the surfaces to be connected. When gap tolerances are the focus of the welding simulation (e.g., for chassis elements of a car, see Figure 2.5.3), the results have to be transferable in order to be used in the reference positioning system (RPS). Buckles on the surface of a component can have a negative influence on the reflected light of the coated component. This is where high-resolution rendering can help.

Table 2.5.2 Pain points caused by welding distortion evaluated by Simufact.

Pain points	Automotive	Transportation	Aerospace	Consumer electronics	Industrial	Power	Special machinery/heavy vehicles	Healthcare	Civil engineering	Shipbuilding	…
Assembly	X	X	X	X	X	X	X	X	X	X	X
Optical properties curvature	X	X		X			X			(X)	
Optical properties gap tolerances	X	X	X	X			X	X	X		
Wear due to geometrical imperfections	Power train										
Tightness	Exhaust systems		X		X	X		X			
Fuel consumption/efficiency	X	X	X		X	X	X			X	
Acoustic properties	Due to ticktack					Due to imbalance		Due to imbalance			
Stability	X								X		
Correction										Water on deck	
Electrical contact and conductivity to prevent a local heating	X			X							
…											

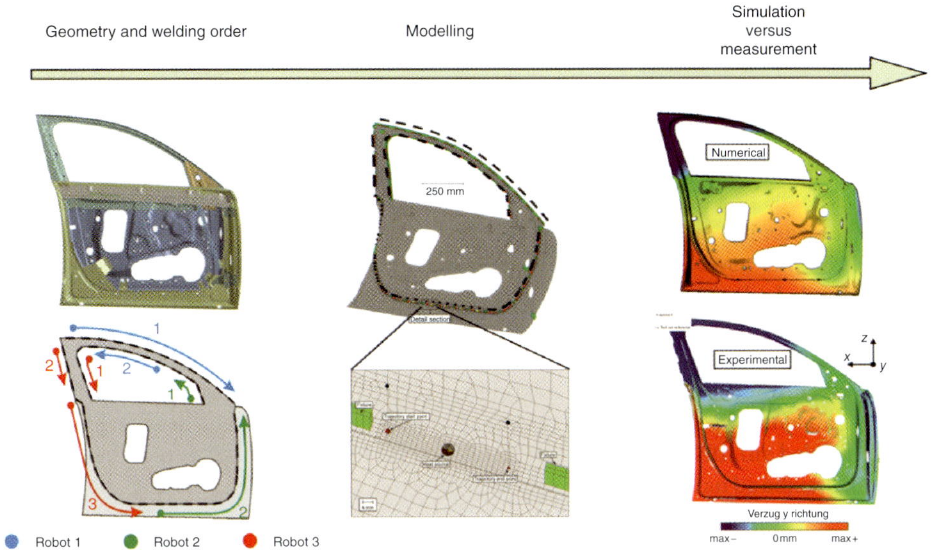

Figure 2.5.3 Using welding simulations to optimize welding direction and to keep distortions related to gaps at a minimum. (Reprinted from [10] with kind permission from Wiley.)

The calculation of distortions is neither a solution nor an improvement. However, it can help in finding useful counteractions in order to minimize the negative influence of welding distortion by virtually trying out clamping concepts, welding order, and so on beforehand.

Welding simulations can be used at all stages of the development process. In an early design phase, it can be used, for example, to compare different welding processes and strategies with regard to distortions before any investments for welding equipment have been made (see Figure 2.5.4).

When designing a welded structure, three groups/positions are involved. Depending on the company size, the three different tasks are done by one person or by a bigger team with up to three different departments:

- The design (the most CAE-experienced calculation engineers are in this department)
- The welding engineer or material specialist;
- The process engineer in the manufacturing department.

The design engineer is responsible for checking most of the calculations that have to be done: extreme loads, fatigue strength, crash behavior, and the usability. Normally, the design engineer neglects distortions due to welding.

The material specialist is responsible for hot and cold cracks, brittle failure, and the formability of the material. If they change the process parameters to get a usable microstructure, this can affect the distortions that subsequently occur.

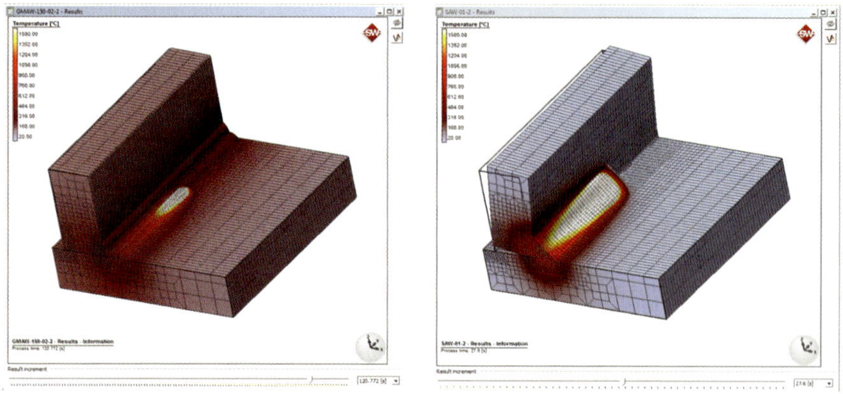

(a) (b)

Figure 2.5.4 Comparing the distortions of a one layer under powder weld with a preheated five-layer MAG weld. (a) One layer submerged arc welding without pre-heating, (b) five-layer gas metal arc welding joint with 150°C pre-heating.

The manufacturing team is responsible for the successful implementation of the process. This team generally consists of highly experienced process engineers who, however, have little experience with finite element calculations and are not generally able to generate simulation models with the special requirements for welding.

2.5.4 Joining by Forming

2.5.4.1 Technological Overview

Forming techniques are used in all industrial areas, as, on the one hand, they offer a wide spectrum of applications, a high efficiency, and process stability and, on the other hand, they display good mechanical properties both for long- and short-term use. Furthermore, the excellent utility for composite construction must be emphasized. The growing trend toward lightweight design in many areas makes this an increasingly important criterion.

Joining by forming creates a joint through positive locking and force locking and by permanently forming the materials and/or the components and the elastic spring back. For this process the materials that are to be formed must have certain ductility. Because of the great diversity of materials and the great number of joining tasks, a variety of joining techniques have been developed [12]. They mainly differ in respect to the following criteria:

- Joining with or without auxiliary elements
- Joining with or without pilot holes
- Necessity for one- or two-sided accessibility
- Forming of the auxiliary element or of the main components

In the following, a selection of joining techniques and the respective processes is discussed in greater detail.

2.5.4.1.1 Blind Riveting

For blind riveting, an auxiliary element (the blind rivet) is inserted in a pilot hole in the main component. While the mandrel is pulled, the mouthpiece exerts counterpressure on the die head. This causes the mandrel to slip into the rivet body and the rivet body changes form (step 2). During further forming, the rivet body is pressed against the walls of the pilot holes, and the components being joined are pressed against each other (step 3). When a predetermined tensile force is reached, the mandrel pulls away at a predetermined breaking point, and a positive-locking and force-locking joint is created (step 4) [11]. Since only the auxiliary element is formed, this technique is, in principle, applicable to all adhered materials used in sectional and/or tubular construction because of the one-sided accessibility (Figure 2.5.5).

Besides the advantages of this techniques, the additional effort of creating the pilot holes, the precise positioning of the components with pilot holes, and the auxiliary element must be taken into account [13].

2.5.4.1.2 Self-Piercing Riveting

In contrast to blind riveting, self-piercing riveting techniques do not require pilot holes. However, they do require accessibility from two sides. Generally, the two techniques of semitubular self-piercing riveting and solid self-piercing riveting can be differentiated. In both processes, the components that are to be joined are

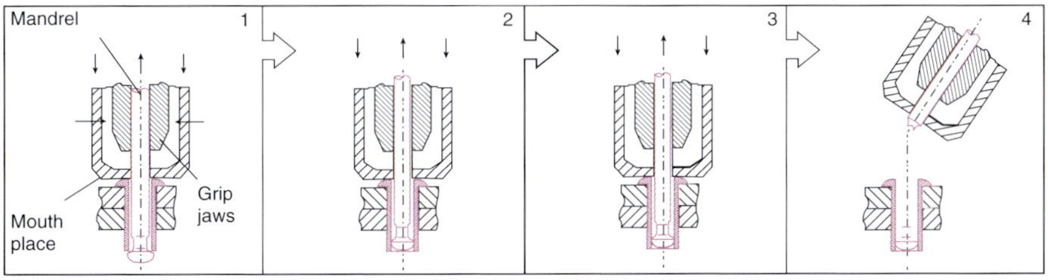

Figure 2.5.5 Process steps of blind riveting [11].

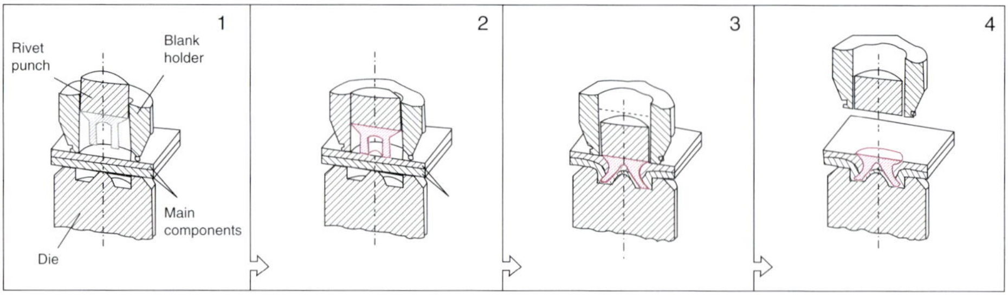

Figure 2.5.6 Process sequence of self-piercing riveting with semitubular rivet [11].

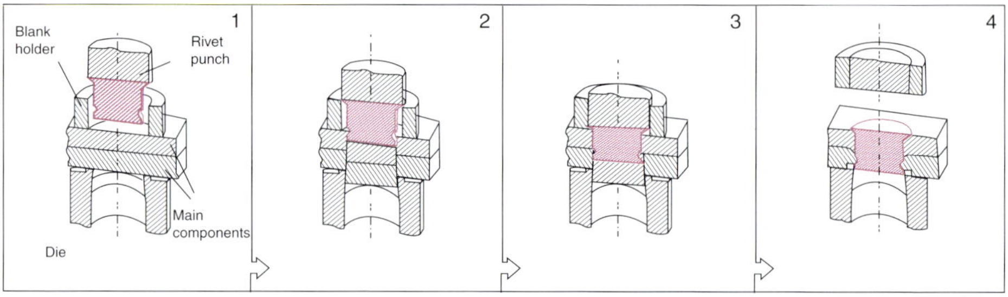

Figure 2.5.7 Process sequence of self-piercing riveting with a solid rivet [11].

positioned on the die and fixed with a blank holder (step 1). For self-piecing riveting, the rivet is then punched by the riveting punch, penetrates into the riveting punch-sided component, and usually punches through it. The rivet then penetrates the other component (die-sided), is spread apart within the component, and is formed within the die (steps 2 and 3). After reaching the maximum joining force or the maximum displacement, the tools are retracted (step 4) [11]. The material properties of the components that are to be joined, and the properties of the auxiliary elements, must be attuned for this technique to work and to allow the rivet to punch through the upper component and spread out within the lower [14]. This restricts where this technique can be applied. However, a broad spectrum of materials can still be joined in this way because of the variety of rivets available on the market and their respective hardnesses and geometries (Figure 2.5.6).

In contrast to semitubular self-piercing riveting, the rivet is not plastically formed during solid self-piercing riveting, but punches through both components. Only the die-sided component must have certain ductility so that it can fill out the shank groove of the rivet (steps 2 and 3) [11]. A disadvantage of this technique is the necessity to remove the punched-out material (Figure 2.5.7).

2.5.4.1.3 **Clinching**

Clinching, also known as toxing, requires a die too. But in contrast to other techniques, it does not require an auxiliary element. The components are positioned

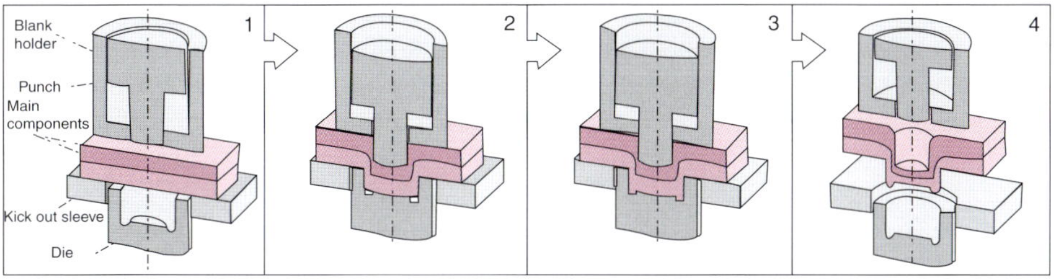

Figure 2.5.8 Joining process for one-step clinching, without cutting and with an undivided die [15].

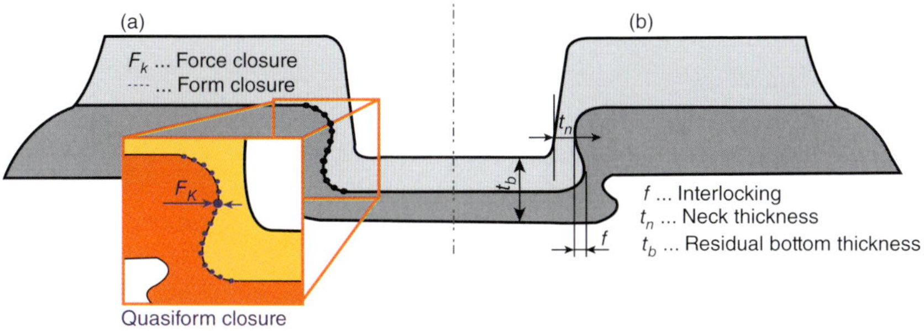

Figure 2.5.9 (a) Active principle and (b) parameters of clinching.

on the die and fixed with a blank holder. The components are then directly formed by the stamp within the die (Figure 2.5.8).

In clinching, a joint is created by forming (compressing) and possibly separating (cutting), which leads to mechanical grouting with or without the cutting stage. Apart from this differentiation, clinching techniques can be differentiated according to the kinematics of the joining process, that is, one-step and multi-step process. In general, a trend that can be observed over the past few years is a move toward the use of one-step clinching processes. The advantages of this technique include primarily the lower costs for machinery and tools. Although the process usually involves several stages of processing nowadays, the American name "clinching" has remained. Joining elements come with a great diversity of shapes (round, oval, bar shaped, star shaped) that are mainly defined by cutting although round point joining without cutting is becoming increasingly prevalent. The simpler manufacturing of rotational symmetrical tools, as well as the reduction notch effects on the joining points, favors clinching processes [1].

In line with the nature of clinching, the transmission of forces relies on the quasiform closure of the joined components. This quasiform closure is based on a combination of traction and positive locking (F_k), the relationship of which varies according to the kind of the components and the size of the indentation or undercut (Figure 2.5.9).

The form closure of the two components, the so-called undercut (f), is created during the forming of the joint by round point clinching, when the materials formed by the stamp can spread outside the level of the sheet metal. The material "flows" into the direction of least resistance to the forming, in the direction of the stamp. The joining point extruded from the level of the sheet metal causes a transmission of force and momentum both parallel and perpendicular to the metal sheet level. The proportion of traction that is created during joining, by bracing the materials, causes zero backlash in the joint and contributes to the transmission of force when the component is formed. Combining both kinds of locking offers the opportunity to combine the advantages of form fitting (transmission of strong forces) with those of force fitting (avoiding backlash by elastic bracing). Therefore, a highly resilient component, against static and dynamic stresses, can be achieved that guarantees a satisfactory amount of holding force [1].

2.5.4.2 Virtual Joining Factory

Over the past few years, engineering based on simulation methods has established itself in many areas of production technology. Efficient tools and software applications for the development and planning of products, machines, and systems have been developed that allow virtual analysis especially of critical areas, for example, microstructural analysis, forming processes, or joining applications. However, these often only offer isolated solutions, which only cover one particular aspect and are not complete calculation tools. This isolation is often caused by data transfer between simulation programs. There are hardly any automatic links between software tools, as they frequently come from different providers. Data formats have to be regularly converted first and then transferred manually. Therefore, interdependencies can only be taken into consideration to a small extent and with high resource expenditure. This affects the quality of the numerical results and hence the development and optimization process. Consequently, the necessity of a process chain simulation, which covers all production steps similar to a virtual factory, becomes obvious. This automatic simulation should make it possible to consider the numerical results from the microstructure calculations to the manufacturing of the composite tools successively and thus contribute to having a nearly fault-free perfected process for the real factory.

2.5.4.2.1 Example: Self-Piercing Riveting

2.5.4.2.1.1 Necessity

In conventional simulations, only individual partial aspects of the joining process could be examined. The examinations were carried out individually and were isolated from up- or downstream production steps. Due to this, the reciprocal impact of adjoining processes was not taken into consideration. In order to achieve a higher degree of accuracy within the simulation processes, it is necessary to consider numerically viewed production steps as input parameters for further analysis throughout the whole process. This allows a virtual joining factory, marked by a complete process chain simulation, to be realized.

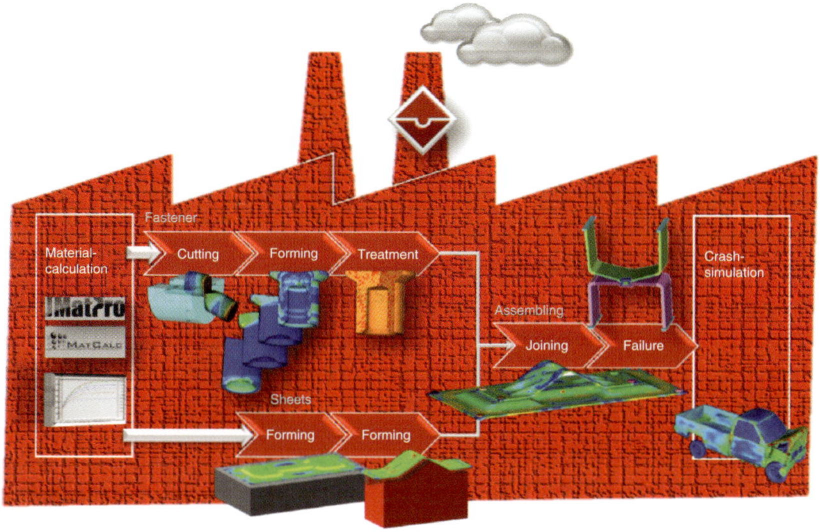

Figure 2.5.10 Virtual joining factory for simulating manufacturing composite materials.

2.5.4.2.1.2 Overview

To be able to model a virtual joining factory for the production of material composites (Figure 2.5.10) consistently, it is necessary to integrate information on microstructure simulation consequently in the forming simulation.

At present the thermokinetic software MatCalc is used for this coupling. This software is based on physical modeling and therefore makes it possible to gain a better understanding of the microstructural processes of the material. Because of this, the direct influence of the process parameters can be described and predicted. The microstructural properties are needed as input quantities for the numerical calculation of different process stages and are taken into consideration (via interfaces) in the forming and joining simulation. Figure 2.5.10 shows the entire virtual process chain for numerical consideration regarding the main production steps for producing a material composite. The numeric calculations of the joining element (rivet) production on one hand and of the sheet metal forming on the other are the input variables for the joining simulation and a static and dynamic failure analysis. Hence, there are unique possibilities for a cross-process simulation in terms of a virtual joining factory [8].

2.5.4.2.1.3 Process Steps

The main individual partial processes of the rivet production are emphasized with regard to the interaction of the results in the following descriptions.

Cutting The numerical analysis starts with the separation of the necessary wire segment. The respective numerical model produced in Simufact.forming and the required material data for the cold forming of steel 34Cr4 are derived from its own material database. Taking this data into account, the amount of smooth cuts and breakages, solidifications, and deformations, among other things, are calculated. The simulation results create the foundation for the following numerical analyses.

Forming In the same simulation solution, the wire billet input data for the numerical simulation of the following cold forming is calculated. The multistage production of self-piercing rivets is simulated together with the tool geometries that have to be implemented. The resulting data for the individual process steps function as input data for the subsequent processes. This is carried out automatically and therefore has the best preconditions for the effective consideration of variations in order to identify the optimal production solution.

Heat Treatment As in the real production process, heat treatment follows in the virtual process evaluation. There is a special module used within Simufact.forming for this. Therefore, heating, cooling, and quenching of the heat treatment process and the associated microstructural transformation and property changes of the material can be simulated. For quenching and the resulting targeted adjustments to the heat treatment, the descriptions of the microstructural development and the resulting mechanical characteristics (e.g., yield and tensile strength) are dependent on the process properties. This is currently described by experimental data provided in the form of a material file, which can be transferred via a newly developed interface in Simufact.forming. Therefore, it is possible to calculate a complete heat treatment process and to use the results for the subsequent joining processes.

Joining The cold-formed self-piercing rivet, with its special strength properties due to heat treatment, forms the starting component for the joining process. Furthermore, the materials that will be joined with the self-piercing rivet have to be initially defined for the forming and cutting processes. The required information for the mechanical–thermal description of the steel sheet (DC04), which is arranged on the punch side, is available in its own database within the program and is used for this purpose. The mechanical properties of the aluminum alloy (calculated with MatCalc) are based on the chemical composition and the alloy's microstructure. This is then exported as a material file. This file, shown in the Simufact material database after it is imported, can then be selected for the purpose of modeling the forming processes.

An exemplary flow curve simulation for an aluminum alloy for temperature and velocity dependencies is shown in Figure 2.5.11. The results of this simulation are based on the one-parameter model (Sherstnev–Kozeschnik) described in [16].

Furthermore, the geometries of the joining tools (punch, blank holder, die) and a definition of the press are necessary. The joining process is modeled using calculated and experimental obtained input parameters. It is also possible to simulate

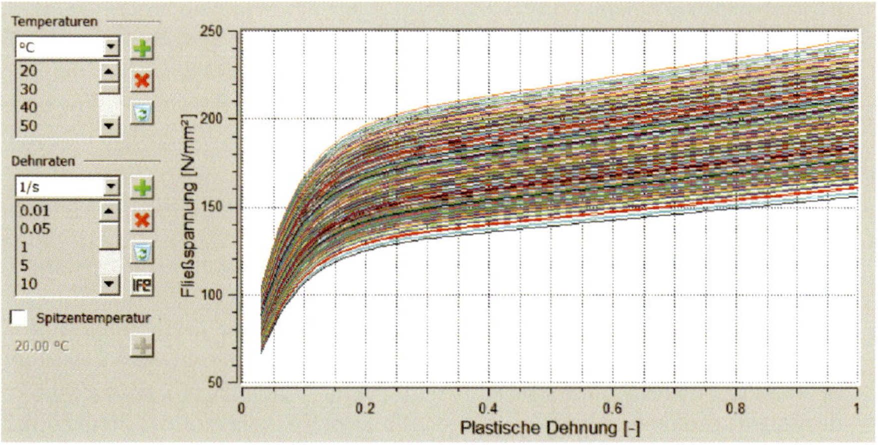

Figure 2.5.11 Yield stress of aluminum alloy depending on temperature and velocity.

Figure 2.5.12 Analysis of behavior under a load with shear tension test.

a multitude of variations in order to identify different results, for example, the robustness of the process.

Failure Analysis With the virtual joining connection, examinations regarding the determination of joint strength or failure behavior are possible in the next step. For this, the simulation result of the complete joining connection is used as a basis. The test setup is realized, simulated, and analyzed, together with the respective tools, which, for example, are necessary for shear tension (Figure 2.5.12) and cross-tension test.

From "cutting," to "forming," to "failure analysis," calculated simulation results were successively included as input parameters in the subsequent numerical processes. Because of that, the virtual results could be used throughout the process. The achieved optimization of the simulation results and the details for integration of microstructural properties will be the subject matter of the lecture.

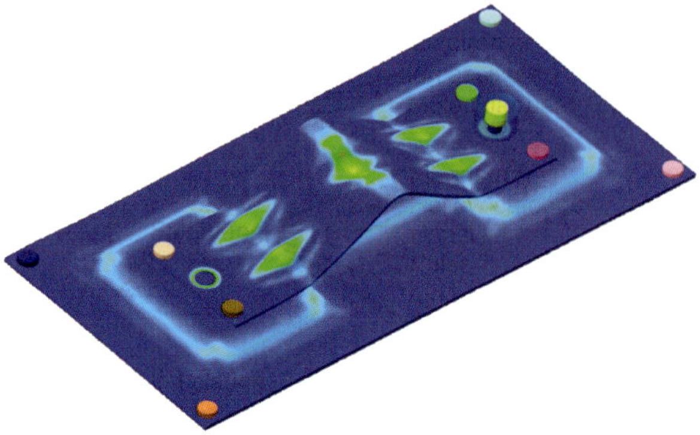

Figure 2.5.13 Multipoint considerations.

2.5.4.2.2 Conclusion

A numerical process chain is shown with the production of a material composite through self-piercing rivets, which is characterized by the consequent transfer of the calculated simulation results between the individual process steps and with an integration of microstructural properties. First qualitative optimizations of predictive accuracy could be achieved here. An intensification of these tendencies results from the examination of joining chains (Figure 2.5.13). In this case the effects exponentiate and allow substantiated statements regarding material behavior in between the individual joining points. In order to intensify the comparability between virtual and experimental results further, the aim should be to incorporate further processes, which are up- or downstream of forming or joining (e.g., primal forms), in the numerical examination.

2.5.5 Software for Joining Processes

The following table lists a number of established and academic software tools being used to simulate joining processes. The categorization has been selected as follows:

2500 General-purpose tools being frequently applied to welding and joining
2510 Thermal joining processes
2520 Mechanical joining processes
2530 Friction stir joining processes
2540 Academic codes/consultancy
2550 Practical welding trainers

Welding often involves phase transformations and microstructure evolution both in the weld and also in the heat-affected zone. For respective simulation tools, refer to Chapter 3. Further thermodynamic data as depicted in Chapter 4 are frequently needed as input for welding simulations.

Software category	Name	Website	Type	Short description/typical applications/phenomena being treated
2510	**SysWELD**	https://www.esi-group.com/resources/welding-simulation-suite-folder	Commercial	SysWELD is part of ESI's "welding suite," a single virtual welding manufacturing software; modules are available for distortion engineering, weld quality, welding, and assembly
2510	**SimWeld**	https://www.isf.rwth-aachen.de/index.php?id=35&L=1	Academic/commercial	SimWeld® allows the reliable simulation of the welding process for many geometries and materials with variable parameters. SimWeld® specifies the welding process via three submodels for (i) the arc, (ii) the heat transfer inside the workpiece, and (iii) the 3D surface of the molten pool
2510	**LaserWeld3D**	http://www.nld.rwth-aachen.de/go/id/gzaq/lidx/1	Academic	Calculates, for example, the capillary shape of a laser beam during laser welding
2510	**BeamSIM**	https://www.isf.rwth-aachen.de/index.php?id=35&L=1	Academic/commercial	BeamSIM® is used for the simulation of laser and electron beam welding processes. BeamSIM® provides the molten pool geometry and the heat distribution and opens the same application fields for laser beam welding as SimWeld® does for arc welding
2510	**Sorpas3D**	http://www.swantec.com/	Commercial	Sorpas provides tools to model resistance welding including the variants such as spot welding, projection welding, seam welding, butt welding, and microwelding

(*continued overleaf*)

Software category	Name	Website	Type	Short description/typical applications/phenomena being treated
2510	**Simufact. welding**	http://www.simufact.de/en/solutions/sol_weld.html	Commercial	Simufact.welding is a finite element-based software with state-of-the-art solver technology allowing for elastic–plastic material behavior to be modeled. It realistically predicts, for example, distortions and residual stresses occurring during welding. Simufact.welding considers microstructural properties and phase transformations in the heat-affected zone
2520 2510	**Simufact. forming**	www.simufact.de www.simufact.com	Commercial	Simufact.forming provides a number of modules tackling a variety forming processes (see also Chapter 2.3). Its mechanical joining module provides tools for mechanical joining processes such as clinching and blind and punch riveting. With the release of version 13.2, all kinds of electric resistance welding processes ("pressure welding") are supported
2510/2530	**VirFac_welding**	http://virfac.com/index-2.html	Commercial	Virfac® predicts the distortions encountered by the workpiece due to manufacturing. It also predicts the level of residual stresses and allows to assess the in-service fatigue life of the manufactured part as well as the metallurgy and quality of the manufactured component. In the area of welding processes, Virfac® allows to simulate, for example, the friction stir spot welding process

2550	**Arc+ Welding Simulator**	http://www.123arc.com/	Commercial	A training tool that powers lifelike virtual welding practice. Easy-to-use and realistic simulator
2540	**N.N.**	http://www.ipk.fraunhofer.de/en/divisions/joining-and-coating-technology/fields-of-expertise/welding-simulation/	Academic	Fraunhofer IPK holds a pool of in-depth simulation expertise and of virtual welding processes and provides respective consultancy
2540	**N.N.**	http://www.twi-global.com/capabilities/integrity-management/finite-element-analysis/simulation-of-joining-processes/	Commercial	TWI is an independent research and technology organization with expertise in materials joining and engineering processes as applied in industry
2520	**DEFORM**	http://www.deform.com/	Commercial	General forming software (see also Chapter 2.3) often finding application in mechanical joining
2500	**MSC Marc**	http://www.mscsoftware.com/product/marc	Commercial	General-purpose software that can be used for all kinds of welding processes, including friction stir welding. Electromagnetic effects can be modeled. Mechanical joining is also available. Strong in joining of composites
2500	**ANSYS**	http://www.ansys.com/de_de	Commercial	Has been used for friction stir welding (see [17])

(*continued overleaf*)

Software category	Name	Website	Type	Short description/typical applications/phenomena being treated
2500	**OpenFOAM**	http://openfoam.com/	Open source	OpenFOAM has an extensive range of features to solve anything from complex fluid flows involving chemical reactions, turbulence, and heat transfer to solid dynamics and electromagnetics. It has been applied to joining processes in several cases
2500	**Abaqus**	http://www.3ds.com/products-services/simulia/products/abaqus/	Commercial	General-purpose FE code often used for thermomechanics of weld joints and heat-affected zones
2530	**Virtua RFW**	http://www.uni-magdeburg.de/ifme/l-dynamik/quellen/schmicker_naumenko_strackeljan_-_a_holistic_approach_on_the_simulation_of_rotary_friction_welding.pdf	Academic	Rotational friction welding code; operated in MATLAB (see also [17])
2500	**FEMFAT**	http://www.femfat.com/	Commercial	Life cycle assessment of weld joints (see also [17])
2510	**WeldWare**	http://www.lttz.de/fileadmin/user_upload/Rostock/Software/WeldWare-Prospekt_SLV_M-V_engl.pdf	Commercial	WeldWare serves the calculation of the heat management before welding, the calculation of microstructure and mechanical properties in the HAZ, the determination of necessary preheating temperatures for real components, and the supply of data for FEM

References

1. Beyer, U. (2012) Multi-Material-Fügen Mittels Flach-Clinch-Technologie. Dissertation. TechnischeUniversität Chemnitz, ISBN: 978-3-942267-39-7.
2. Beyer, U. (2015) Simulation manufacturing – consistent virtual considering in car body engineering. Automotive Engineering Congress 2015, Nürnberg, June 9–10, 2015.
3. Bangel, M. and Müller, S. (2014) Trends in der Fügetechnik durch neue Karosseriekonzepte. 4. Fügetechnisches Gemeinschaftskolloquium 2014, Dresden, December 9–10, 2014, ISBN: 978-3-86776-448-3.
4. Matthes, K.-J. and Riedel, F. (Hrsg.) (2003) Fügetechnik. Fachbuchverlag Leipzig im Carl Hanser Verlag GmbH & Co. KG, pp. 252–264. ISBN: 3-446-22133-6
5. Deutsche Industrie Norm DIN 8580 (2003) *Fertigungsverfahren: Begriffe, Einteilung*, Deutsche Industrie Norm, September 2003.
6. Deutsche Industrie Norm DIN 8593-0 (2003) *Fertigungsverfahren Fügen, Teil 0: Allgemeines – Einordnung, Unter-teilung, Begriffe*, Deutsche Industrie Norm, September 2003.
7. Klasfauseweh, U. (1994) Beitrag zur Simulation nicht-schneidender Durchsetzfügevorgänge. Dissertation. Universität-Gesamthochschule Paderborn.
8. Beyer, U., Neubauer, I., Kreyca, J., and Kozeschnik, E. (2015) Virtual Joining Factory – Integration of microstructure evolution in the manufacturing process chain simulation. Material Forming ESAFORM 2015, Tagungsband, pp. 1325–1330, ISBN 13 978-3-03835-471-0.
9. Radaj, D. (2002) *Eigenspannungen und Verzug beim Schweißen – Rechen- und Messverfahren*, DVS-Verlag Düsseldorf.
10. Thater, R., Wiethop, P., and Rethmeier, M. (2015) Welding simulation in car body construction. *Laser Tech. J.*, **12** (2), 33–37.
11. Hahn, O. and Klemens, U. (1996) *Fügen durch Umformen, Nieten und Durchsetzfügen – Innovative Verbindungsverfahren für die Praxis. Studiengesellschaft Stahlanwendung e. V.* Dokumentation 707, 1. Aufl., Verlag und Vertriebsgesellschaft Stahlanwendung GmbH, Düsseldorf. ISBN: 3-930621-56-8
12. Deutsche Industrie Norm DIN 8593-5 (2003) *Fertigungsverfahren Fügen, Teil 5: Fügen durch Umformen – Einordnung, Unterteilung, Begriffe*, Deutsche Industrie Norm, September 2003.
13. Meschut, G. and Bobbert, M. (2013) Effiziente Fügetechnologien für Mischbauweisen im Fahrzeugbau. 13. Round Table Simulation in der Massivumformung, Heidelberg, April 9–12, 2013, Tagungsband.
14. Merkblatt, N.N. (2005) *DVS®/EFB 3410: Stanznieten – Überblick. Merkblatt des DVS/EFB-Gemeinschaftsausschusses*, DVS-Verlag, Düsseldorf.
15. Hahn, O., Tan, Y., and Voigt, H.-M. (2005) Maschinelles Lernen zur Vorhersage der Tragfähigkeit von Clinchverbindungen, Nr. 1, https://mb.uni-paderborn.de/en/lwf/publikationen/veroeffentlichungen/2000-2009/ (accessed 19 July 2016).
16. Sherstnev, P., Lang, P., and Kozeschnik, E. (2012) Treatment of simultaneous deformation and solid-state precipitation in thermo-kinetic calculations. European Congress on Computational Methods in Applied Sciences and Engineering (ECCOMAS 2012), Vienna.
17. Hildebrand, J., Loohse, T., Sakkiettibutra, J., and Brand, M. (eds) (2014) *Simulationsforum 2013- Schweisstechnik und Wärmebehandlung*, Förderverein Wärmebehandlung und Schweißen e.V., Weimar. ISBN: 978-3-00-045902-3.

2.6 Thick Coating Formation Processes

Kirsten Bobzin, Mehmet Öte, Thomas Frederik Linke, and Ilkin Alkhasli

2.6.1 Overview

Coating processes are conventionally divided into deposition of thin films (thickness less than 20 μm) and deposition of thick films (thickness more than 30 μm). The most common thin-film deposition techniques are chemical vapor deposition (CVD) and physical vapor deposition (PVD), which require reduced pressure environments. Thick-film deposition methods include, among others, electrochemical plating, (laser) cladding, explosion welding, and thermal spraying. Thick films are utilized when the coating thickness can influence the coating performance, as in thermal barrier coatings (TBCs) or wear- and corrosion-resistant coatings. They can also be used in order to restore the original dimensions of worn parts.

Due to the different physical phenomena and a wide range of time and length scales involved in different thick coating formation approaches, it is practically impossible to generalize them under a single physical model and describe within the scope of this chapter. There are certain physical aspects common to some of the coating processes, which allow similarities in their modeling. Thermal spraying simulation alone is a challenging endeavor that requires multiphysics modeling and multiscale resolution. This work focuses on the modeling and simulation of coating formation through thermal spraying processes, which is intended to serve as an exemplary approach for modeling other thick coating processes with similar underlying physics. In the following, brief information about the common thick coating processes is given, and their simulation methods are described.

Handbook of Software Solutions for ICME, First Edition. Edited by Georg J. Schmitz and Ulrich Prahl.

2.6.2
Typical Applications of Coating Simulations

Simulations in general are used either as a substitute to experiments that are linked to high costs and are difficult to conduct or complement experiments and theoretical knowledge about the topic of interest by shedding light onto previously poorly investigated areas. Coating simulations in particular can be divided into simulations of coating formation processes and the simulative determination of mechanical properties of coatings.

Process-oriented simulations of thick coating generation are as diverse and interdisciplinary as the available variations of coating processes themselves. Relevant physical phenomena include wide range of continuum mechanics, plasma, and laser physics as well as chemistry. An incomplete list of phenomena include compressible and turbulent flows; particle-laden multiphase flows; magnetohydrodynamics (MHD); heat radiation and laser irradiation; elastic, plastic, and elastic–plastic solid mechanics; melting and solidification of metals and ceramics; melt-pool dynamics; chemistry of explosive reactions; and electrochemical processes. Simulation of relevant coating formation process helps gain insight into the underlying phenomenon and allows more control over the overall process. The main motivation of process-oriented simulations is to gain more control over the process, in other words to optimize it in terms of lowering the costs by minimizing wasted feedstock and energy and increasing the coating quality where possible [1].

However diverse the coating formation processes might be, the approaches for the determination of mechanical properties of obtained coatings are similar, differing from each other with the level of complexity of the coating microstructure and their porosity. Coatings whose properties shall be determined are obtained either from a coating formation simulation or modeled separately as a layer/multilayer of feedstock material of uniform/varying thickness over the substrate surface. The system of substrate-coating is then subjected to a relevant boundary condition in order to virtually test and mimic the test that would be done on the actual coating. Young's modulus can be determined by applying a tensile stress on the surface of the coating and measuring the resulting strain. By applying temperature gradient on the opposite surfaces of the coating, the thermal diffusivity and the thermal conductance can be determined [2].

Thick coatings obtained by thermal spraying have in general complex microstructure, and mechanical properties of final coatings differ from the properties of the original feedstock material to a significant extent. A common practice for dealing with such microstructural morphologies is to employ mathematical homogenization methods. A homogenized microstructure of a coating represents a uniform layer of a hypothetical material, which has the identical mechanical properties as the complex morphology of the original coating. Coatings obtained from electroplating and cladding processes are mostly very dense and do not require homogenization [3a, 4].

2.6.3 Phenomena Occurring During Coating Formation

One of the common processes of thick coatings is electroplating. It is mainly used for wear/corrosion protection or aesthetic purposes and also to give an additional thickness to undersized objects. It utilizes a process called electrodeposition, in which the substrate acts as a cathode and the feedstock material acts as an anode. Both materials are bathed in an electrolytic solution that contains metal salts and other ions. By applying a direct electric current to the electrodes, the atoms on the anode, which acts as a donor material, are oxidized and dissolved in the solution. These atoms are then "plated," that is, accumulated on the cathode surface, thus creating the coating on the substrate.

One of the common problems that can occur during the electroplating process is the uneven distribution of the plated material on the substrate surface. The plating process is continued until the minimum required coating thickness is achieved all over the surface. This introduces additional costs not only due to the added material costs but also because of an extra postprocessing in the areas with excess plated material. Electroplating simulations are helpful to design optimal electroplating racks and cells that enable better plating uniformity on the substrate surface [1]. Apart from the thickness distribution of the coating material, pattern created by the dissolution on the anode surface, average current densities, operating electrode potential, and the temperature results can be obtained from the electroplating simulations using commercially available multiphysics simulation tools such as ANSYS and Comsol [5]. "Level set method" (LSM) can be utilized in electroplating simulations in order to precisely define the layers of plated material on the substrate surface [6].

Another common thick coating deposition process is cladding. This process is used for bonding of metals. It also allows bonding between two dissimilar metals. It usually involves the application of forced deformation, high pressure, or heat in order to achieve ductility of the clad material to ensure diffusion on the interface of the two metals. Roll welding is one of the variations of cladding where two or more layers of metal sheets are passed through rolls and pressed together. The applied pressure must be high enough to deform and fuse the metal layers, thus causing bonding between them. Another variation of cladding is explosive welding. In explosive welding, a layer of chemical explosive is applied on the surface of the outermost metal sheet. Enormous impulse caused by the detonation induces a sudden plasticization between the metal interfaces that enables bonding. One advantage of explosive welding is that due to the high impulse of the explosion, oxidation and porosity within the bonded materials are minimized.

The common way of simulating the roll and the explosive welding processes is the implementation of the finite element method (FEM) and taking nonlinearities into account. The employed simulation tool (Abaqus, ANSYS) for such computations is expected to have adequate material models (elastic–plastic stiffness, anisotropy, kinematic hardening), contact model with allowance of friction algorithm, and heat conduction model [7]. During the roll welding simulations, the boundary conditions are set in such a way that the substrate and the cladded materials are gradually pressed toward each other, resulting in high temperatures at the interface of both materials. Heating effects at the interface are achieved with the help of powerful contact and nonlinear friction models. Followed by the cooling phase, modeling of the plasticization at the interface is the most important aspect of roll welding simulation [8].

In explosive welding simulations, it is important to consider the rapid collision of the clad metal sheet on the base metal. The impact pressure is typically in the order of 100,000 bar. Because of the shock induced from the impact of the plates, the interfaces of the materials liquefy and a jet flow consisting of the liquid mixture of the two metals is created. The correct modeling of this jet, as well as the contact angle between the plates, is necessary for a realistic simulation of the welding of two metal plates [9]. In addition to FEM simulations, atomistic simulations have been also performed in order to resolve physical processes of explosive welding. This method has the advantage of allowing the observation of the nanocrystalline morphology of both the coating and the substrate after annealing [10].

Perhaps the most important variant of the cladding process is the laser cladding (according to DIN EN 657, it belongs to the group of thermal spraying processes, which is discussed later). It is a process for the deposition of feedstock material on the substrate surface by melting it with a laser. The melted feedstock is mixed with the local melt pool on the substrate, thus creating an excellent bonding upon solidification. Laser cladding is utilized in coating applications as well as in near-net shape manufacturing processes.

Simulation of the laser cladding process is commonly done with the help of commercial computational fluid dynamics (CFD) software with the consideration of heat generation through the laser radiation, liquid melt-pool formation, and the injection of the particle stream into the melt pool. The liquid melt pool can be modeled with the "volume of fluid" (VoF) method [11] or "level set model" (LSM) [12] that defines the borders between the liquid- and solid-phase metals as well as the liquid and the surrounding gas. Heat transfer between the melt pool and the solid particles, the melt pool and the substrate surface, as well as the solidification and the cooling of the melt are the key points for correct simulation of laser-cladded coatings [13, 14].

Simulation of thermal spraying exhibits similarities to laser cladding simulations with respect to the common physics of the process during the heat and impulse transfer to the feedstock materials. In contrast to the laser cladding, the substrate material is not melted in case of thermal spraying. The main bonding mechanism in thermal spraying is mechanical clamping, whereupon in case of laser cladding the bonding is realized through metallurgical bonding. Thermal spraying differs from other types of coating processes due to its ability to treat almost every material and the fact that the feedstock is not necessarily heated to molten phase. Furthermore, thermal spraying in general can produce microstructure morphologies according to functional requirements of the coating. This report focuses on thermal spraying as an example of a thick coating deposition technique. The methods that are described in the following sections are mostly also applicable to related technologies, except perhaps electroplating simulations. Nonlinear material models, liquidation and solidification of metals, and high gradients of heating and cooling phases are common phenomena for roll welding, explosive welding, and laser cladding, which also overlap with the important simulated physical processes of thermal spraying. Particle-laden jet flow is a further touching point between laser cladding and thermal spraying that can be simulated using CFD of multiphase flows.

Coatings are structurally complex material systems. Their characteristic properties are essentially influenced by the feedstock, the parameters of the coating process, as well as the microstructure, which is a result of both mentioned factors. A large number of degrees of freedom exist during the application of functional coatings (e.g., thermal barrier or wear protection). This is the reason for the absolute necessity of the fundamental grasp of the physics involved in order to specifically influence the characteristic properties based on knowledge. The modeling and simulation play an increasingly important role in the description of the correlation between process parameters and, for example, thermally sprayed microstructures. In spite of the fact that early researches considerably increased the level of comprehension of the coating processes, the developed models are still unable to fully establish the complex relations between the input parameters and the coating parameters.

This report first considers the subprocesses of thermal spraying, and presents approaches for modeling and simulation thereof in the following section. In the same section, the modeling and simulation approaches for the determination of effective coating properties, which in turn are results of the coating microstructure, as well as in-service coating performance are discussed. The approaches for validation of the simulation results are discussed in Section 2.6.6.6 followed by conclusion and references.

2.6.4
Basic Equations to Model the Phenomena

As was discussed earlier, processes for thick coating formation are numerous and of different physical nature. Furthermore, modeling of a single variant of thick coating formation processes is most of the time complicated due to the multiphysical phenomena involved in the process. Therefore, instead of presenting an equation that is supposed to completely model the coating formation process, it would be more meaningful to talk about a set of fundamental equations that describe the variety of physical phenomena generally occurring in coating formation processes.

Basic equations of continuum mechanics can be divided into equations of solid mechanics and fluid mechanics. For the cases when applied stresses do not exceed yield strength, materials can be modeled as elastic, and equations of linear elasticity such as Hooke's law can be used. More often involved loads do not fulfill the linear elasticity assumption, and effects of viscoelasticity, plasticity, and thermoelasticity have to be included in the model. This is particularly true for processes as particle impact on the solid substrate in thermal spraying, particle impact on the melt pool in laser cladding, and fusing of metals upon contact in roll and explosive welding processes. The principal equations describing the fluid flow are the Navier–Stokes equations, when coupled with Maxwell's equations these equations are used to model MHD phenomena. Navier–Stokes can be used to describe a variety of fluid-mechanical phenomena such as particle-laden multiphase flows of particle jet, droplet and melt-pool dynamics in thermal spraying and laser cladding processes, liquid metal jet behavior in the interface of substrate, and the coating layer materials in explosive welding simulations. MHD equations, on the other hand, describe electromagnetic processes occurring inside the plasma generator in Air Plasma Spraying (APS) process. For different physical processes occurring in a given coating process, more specific models are used.

2.6.5 Initial and Boundary Conditions

Due to various physical phenomena involved in a given thick coating formation process, simulation of the process is divided into separate subprocesses. Modeling of the subprocesses involves fundamentally different physical phenomena and can be best described in isolation from the rest of the process. Hence, interfaces of interconnected boundary conditions are necessary in order to couple these subprocesses among each other. In this context, results of a subsequent subprocess simulation play a role of input parameters for the consequent subprocess simulation, whereas the process parameters and the obtained final coating are considered global boundary conditions. Modeling and simulation of different thick coating formation processes require boundary conditions of separate physical nature and complexity. Describing all of them individually would go beyond the scope of this chapter.

2.6.6 Process Modeling on the Example of Thermal Spraying

Because of the distinct physical processes and enormous differences of the length scales within the coating process, it is nearly impossible to simulate such a process with a single model. Therefore, the coating process is divided into several subprocesses starting with the heat generation (e.g., plasma, combustion, electric arc) and ending with the coating formation, each of which has to be modeled separately. Furthermore, the effective coating properties are of great importance.

2.6.6.1
Heat Generation in Combustion Chamber/Plasma Generator

The heat generation, as an important part of the many coating processes, was already subject to various numerical studies with the goal of, on the one hand, a better understanding of the processes in the generator/chamber and, on the other hand, an optimization of the heat generation with respect to the coating properties. Among the thermal spray processes, the high-velocity oxy fuel spraying (HVOF) and the air plasma spraying (APS) are the most widespread and hold the biggest market share [15]. Both processes include huge amounts of energy conversion, whereby they differ in the ratio of the kinetic (E_{kin}) and the thermal energy (E_{therm}) of the spray jet and the particles. Typically, it is valid for HVOF and APS that $E_{therm}(\text{HVOF}) < E_{therm}(\text{APS})$ and $E_{kin}(\text{HVOF}) > E_{kin}(\text{APS})$. The modern APS systems (multielectrode systems), however, can be adjusted to have particle properties similar to that of HVOF. Another common form of thermal spraying is wire arc spraying. Two feedstock wires under application of DC current are fed into the spray gun, melted by the heat that is created from the electric arc. Liquid metal is then accelerated by the operating gas from the spray gun and deposited on the substrate. Wire arc spraying results generally in lower particle velocities than APS and HVOF, but has significantly higher deposition rates than both of the processes.

In the HVOF process, there is not just one fluid inside the combustion chamber, but usually rather a mixture of gases and fuels (gaseous or liquid), which in turn react with each other. In order to model and simulate the processes that are typical to HVOF, CFD methods are most often utilized, which basically solve the Navier–Stokes equations numerically. As a rule, the implemented models have to be able to adequately simulate time-dependent, multiphase (gaseous/liquid/solid), compressible, and turbulent flows with terms of combustion reaction. Another important aspect for the combustion models is, for example, the mathematical description of the liquid-phase atomization in the combustion chamber [16].

The working principle of plasma spraying is based on the conversion of a gas or a mixture of gases into the plasma phase and the release of a large amount of thermal energy as a result of the recombination of the plasma. The plasma phase of the gas is achieved through the application of an electric current between the electrodes in the plasma generator, hence resulting in a highly dynamic electric arc. Figure 2.6.1 shows the setup of a conventional one-cathode plasma generator with a single rod-shaped cathode and a nozzle serving as an anode.

The most precise way to describe plasma is the implementation of kinetic plasma simulation. This results in a distribution function that describes the velocity of each particle (electrons, heavy ions) at a given location and given time. The underlying equation is the Boltzmann equation, as well as the Vlasov equation, that takes into account long-range Coulomb interaction and the Fokker–Planck equation that approximates the collision terms. Despite the fact that the kinetic simulation gives much detailed description of the plasma, its use is limited due to the high computational demand imposed [18].

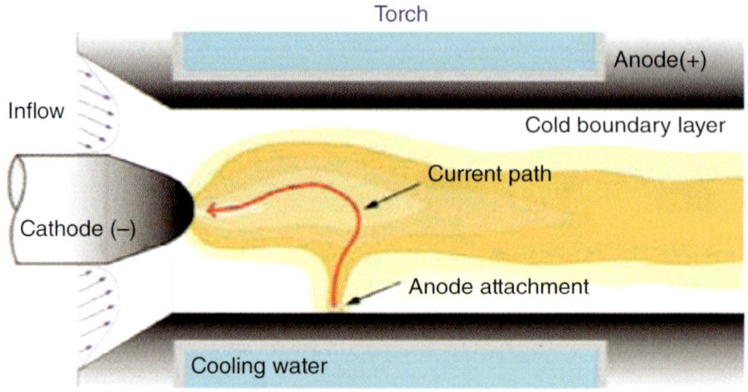

Figure 2.6.1 Schematic depiction of a one-cathode plasma generator. (Trelles *et al.* [17], reproduced with the permission of Springer.)

Another method to simulate plasma is the fluid description, which needs significantly less computational effort and is, therefore, widely used in practice. The fluid model deals with macroscopic quantities of the plasma flow, such as mean density, mean velocity, and mean energy, by solving fluid equations. The fluid equations that are derived by taking velocity moments of the Boltzmann and the Vlasov equations give a relatively precise description of the plasma behavior. However, precision of the fluid description of plasma is limited by the requirement to prescribe the velocity distribution before the simulation [18].

Interactions between hydrodynamic and electromagnetic fields lead to an oscillation of the electric arc along the nozzle wall in one-cathode plasma guns. In order to accommodate the electromagnetic fields and their effect on the overall flow, the CFD methods are used most frequently to model the plasma generation and take into account the unsteady and turbulent plasma flow, which in this case is described with the Maxwell system of equations. With this kind of coupling of fluid mechanics with electromagnetics, which is called MHD (magnetohydrodynamics), the velocity and temperature fields inside the plasma generator can be calculated with relatively high precision [3a, 17, 19, 20].

Unlike one-cathode plasma generators, where oscillation of the electric arc along the nozzle requires time-dependent simulations, in three-cathode plasma guns, the location of the electric arc inside the torch does not change significantly. This allows steady-state plasma simulations to be conducted, enabling a closer observation of the shape of the electric arc inside the torch (Figure 2.6.2). Temperature and velocity distribution of the plasma flow at the nozzle is an important parameter that influences particle conditions during the particle injection and can be obtained from steady-state plasma simulations [3b].

The mathematical modeling of coating processes enables simulations using general-purpose commercial CFD software products (ANSYS, Abaqus), as well as open-source CFD software (OpenFOAM). Because of very high gas temperatures inside the torch, real gas properties of the supplied gas should be modeled additionally [21]. In particular, important here is a high computational power

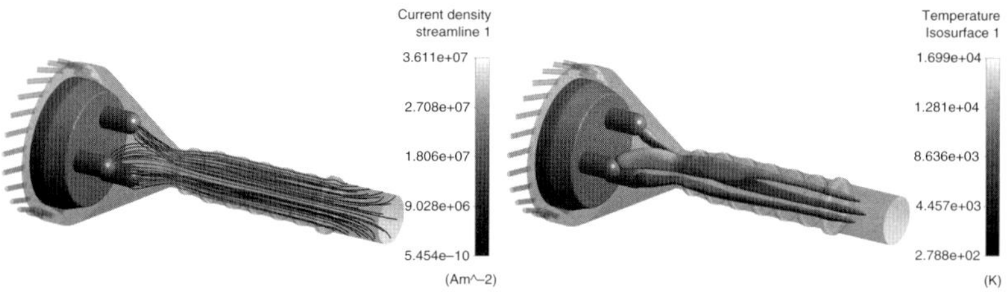

Figure 2.6.2 Electrical current paths between the electrodes (a), the core of the electrical arc with the temperature over 13.250 K (b) [19].

that has to be available for simulations in order to achieve adequately resolved numerical results.

2.6.6.2
Free Jet and Particle Jet

By separation of the thermal spraying into subprocesses, the modeling and the iterative solution of respective systems of equations are considerably simplified. The incremental modeling, however, makes sense only if the injection of the feedstock occurs outside the combustion chamber or, respectively, the plasma generator and/or there is a suitable interface for the coupling of the submodels available through the geometrical boundary conditions. Another reason for the separation of the models is the difference of the turbulence scales of each subprocess.

Simulation of this subprocess requires a correct modeling of temperature-dependent material properties of feedstock material, because of rapid heating of feedstock particles during the injection into the hot stream. Material properties such as density, viscosity, heat capacity, and thermal conductivity should be represented in forms of functions depending on the temperature and the phase of the powder particle. For determination of the required properties, values referenced in the literature can be used. If no material reference values are given in the literature, these properties are determined experimentally; for example, heat capacity of a material can be determined by using differential scanning calorimetry (DSC).

It is typical for the jet produced during the HVOF process to exhibit shock diamonds as a result of the successive expansion and compression of the gas, as it decelerates from supersonic to subsonic regime (Figure 2.6.3). In general, using the HVOF systems, depending on the process parameters, gas velocities up to 2000 m/s and gas temperatures up to maximum 3000 K can be attained. Gas temperatures in plasma spraying can reach considerably higher values (up to 30,000 K locally) than in HVOF. In APS, despite the enormous energy released at the outlet of the plasma generator, the maximum achieved plasma-jet velocities are typically

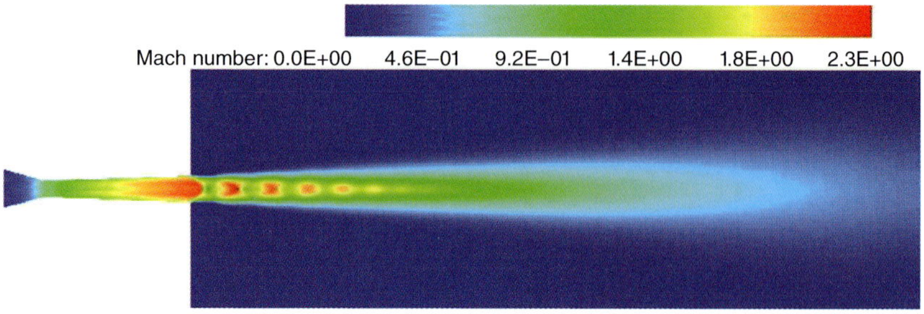

Figure 2.6.3 Simulation result of an HVOF spray jet. (Reproduced with the permission of Panagiotis Christofides, University of California.)

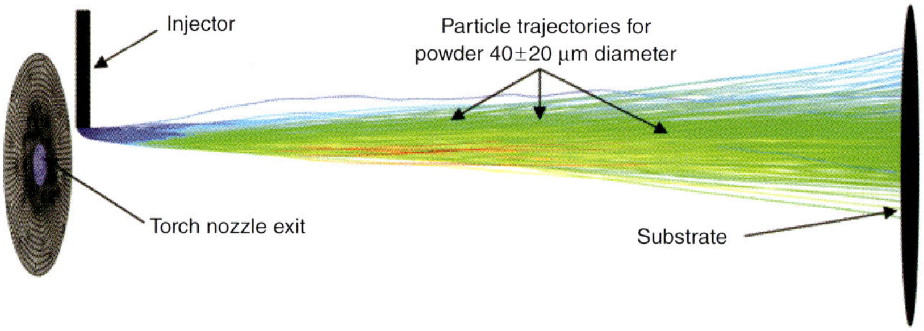

Figure 2.6.4 Steady-state simulation of pathlines of particle injection into the free jet [23].

limited to 2300 m/s. Methods of CFD simulations show very good results describing the free jet of both processes [22].

For the estimation of the particle properties in the free jet, the Euler–Lagrange approach is usually employed, where a force balance for each particle (the discrete phase in the continuum jet) is established along a moving coordinate system. The interaction between the free jet (Eulerian description) and the particles (Lagrangian description) can be described either in a one-way-coupled manner where only the jet flow influences the particles and the particles have no influence on the jet flow, or in a two-way-coupled manner where the reciprocal influence of both phases is incorporated. The particle properties can be calculated either in a time-dependent or in a steady-state manner and the three-dimensional particle pathlines (trajectories) in the spray jet can be analyzed for both cases (Figure 2.6.4).

The variations of the particle velocity and temperature values during the flight phase in the spray jet can be obtained simultaneously for each time period up until the collision with the substrate surface (Figure 2.6.5). Obtained results exhibit particle size dependency of the particle velocity and the temperature variation; for example, finer particles have higher velocity gradients and react faster to flow field disturbances.

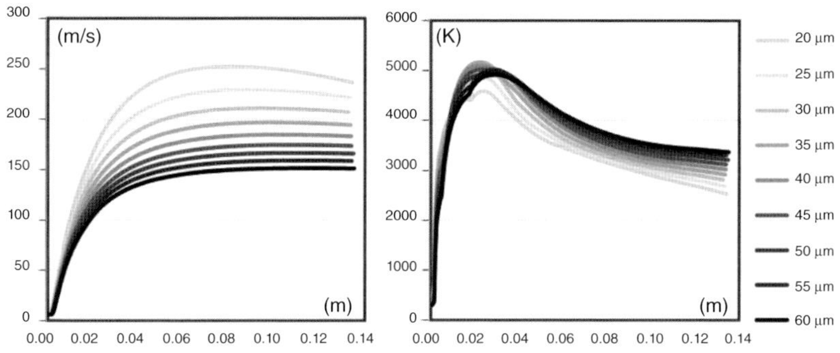

Figure 2.6.5 Calculated particle velocities and temperatures [23].

2.6.6.3
Particle Impact and Coating Growth on the Substrate

As an approach to simulate solid particle impact on the substrate, simulations using FEM approach can be indicated. This method generally allows the examination of internal stresses inside the individual particle as well as the stresses inside the substrate material due to the impact [24]. In case of molten and semi-molten particles, it is convenient to use CFD-Method. The determined in-flight particle properties can be used directly as an inlet condition for simulating the microstructural growth of the coating. The formation of the microstructure is reproduced through the simulation of the successive impact and the deformation of individual particles while rapidly cooling and solidifying on the substrate or the coating surface (Figure 2.6.6). In order to integrate this complex process in a multiphase CFD model, a method of VoF is additionally implemented in order to be able to trace the borders between the gaseous jet environment and the solid/liquid particle material. These simulation methods, however, do not incorporate the effects such as thermally induced residual stresses. The calculated microstructures help to understand the complex process of particle impact, albeit no physical quantities with respect to residual stresses can be derived from it. In order to determine residual stresses in microstructure resulting from the impact of multiple particles, FEM method is coupled with CFD simulations. At this point, the results of the CFD calculations are used as input parameters of FEM simulations. The calculations using these methods are related to a very high computational cost and, therefore, feasible to execute only on a small area of the substrate [23]. Validation of particle impact and coating growth simulations is possible through visual comparison of the calculated and real microstructure e.g., by means of scanning electron microscope (SEM) as well as through the comparison of the porosity values.

2.6.6.4
Homogenization Methods Based on Finite Elements

The most common numerical homogenization methods used to determine the effective properties of the coatings are based on the finite element analysis (FEA).

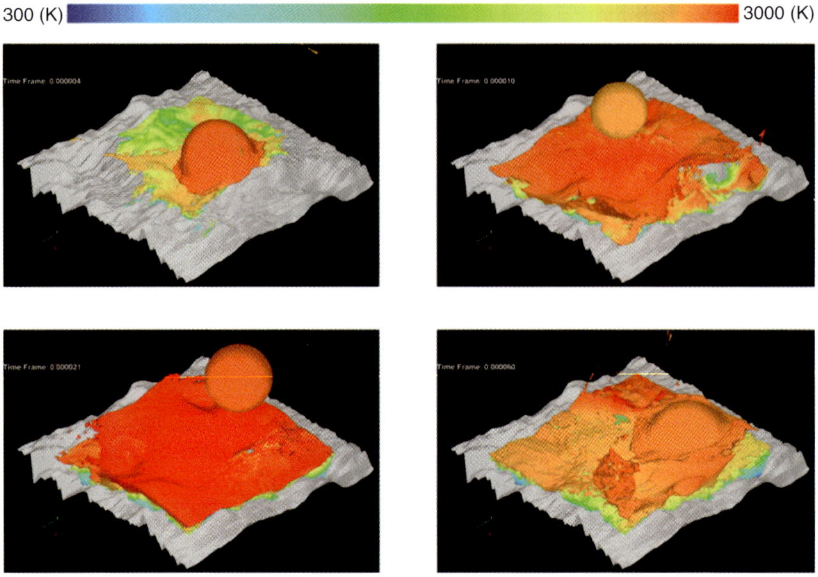

Figure 2.6.6 Simulation of the coating formation on a rough substrate surface using thermal spraying (substrate area: $120 \times 120\,\mu m^2$) [23].

The spatially distributed stress and strain behavior of the representative volume element (RVE) (see section 2.6.6.6) under a constant external load, which later can be used to calculate the effective properties, can be determined with the help of FEM. The discretization of the RVE with sufficiently small finite elements small enough to provide an adequate resolution for a realistic approximation of the microstructure is the basis for the FEA. For this purpose, first the cross-section pictures of the coating, which are usually obtained by means of optical microscopy or X-ray microscopy, are edited and meshed. For this purpose, the open-source program OOF2 developed by the US National Institute of Standards and Technology (NIST) is often used [25, 26]. It is also common, though, that the required steps, such as the reduction of the color palette to the number of the identifiable phases or adapting the mesh in accordance with the phase borders on the cross-section pictures, are executed using a commercial image editing program or macros developed by authors.

In the next step, the boundary conditions of the FEA will be defined in accordance with the tensile test setup. Young's modulus is then calculated from the stress–strain distribution. In order to calculate the thermal diffusivity, different static temperature values are prescribed as boundary conditions on the surface of the coating and the interface of the coating and the substrate. Adiabatic boundary conditions are applied to the other boundaries. The resulting heat flux is used to calculate the thermal diffusivity value [3a, 4, 27].

2.6.6.5
Modeling and Simulation of In-Service Coating Behavior

In general, coatings exhibit mechanical and thermal properties that are different from those of the substrate material. Hence, additional problems related to stress concentrations and stress peaks in the coatings arise from these differences. Coating failures usually occur at the cracks that are developed due to internal stresses in the coating. Here, locations with undesirable high stress concentrations such as defects, pores, or inhomogeneities act as a crack source. Therefore, the focus of interest of the research dealing with the modeling and simulation in the field of coating behavior is the stress distribution in the coating microstructure. State of stress of the coatings in use depends on the stresses, which result from thermal or mechanical loads, as well as from residual stresses, which are induced during the coating process itself. It is common to use FEM in order to calculate the stress distribution in the coatings under thermal and mechanical loads and estimate the crack propagation rates as well as predict possible coating delamination. Using coupled heat transfer analysis with structural analysis, the thermal expansion rates and resulting coating delamination can be determined [2, 28].

2.6.6.6
Validation of Results

Experimental validation of thermal spray simulations can be done according to the corresponding subprocesses, namely, heat generation, particle stream, impact, and coating formation. In addition, analysis of effective coating properties is a common practice for the validations of simulative results.

Experimental observation of the processes during heat generation in APS or HVOF is a great challenge due to the fact that the plasma generation and combustion occur inside the enclosure of the plasma gun in APS and the combustion chamber for HVOF. This significantly impedes visual access to heat generation processes (unless a transparent plasma torch walls are used [29]), and highly turbulent flow with enormous temperatures constitutes a highly hostile environment for sensory intrusion. One way of experimental validation of numerical results of a plasma gun is the application of computer tomography, which can capture the temperature distribution of the electric arc at the torch exit (Figure 2.6.7 [19]). In the case of wire arc spraying, however, heat generation through creation of an electric arc between two feedstock wires happens outside the spray gun, making it accessible for visual observation. Because of high-intensity radiation of the electrical arc, special optic filters and fast shutter speeds should be used in order to be able to capture the processes of droplet formation and atomization of the feedstock. Employment of optical filters plays a role also in APS where particles are injected into the plasma plume outside the plasma gun nozzle. Because of the supersonic nature of the jet exiting the HVOF nozzle, particle injection is

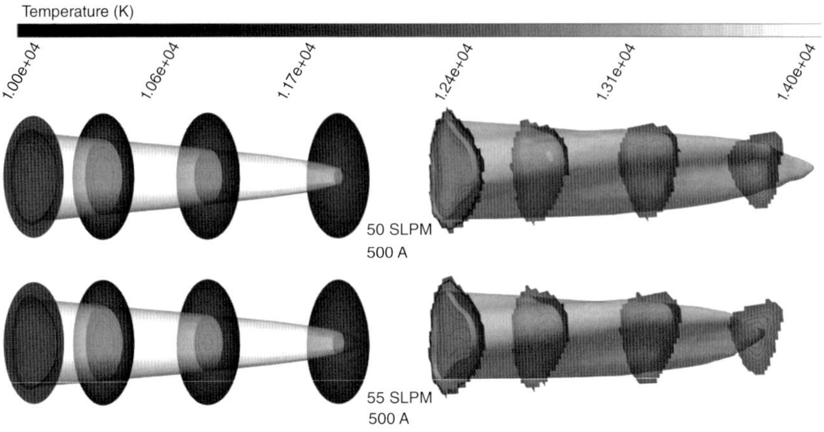

Figure 2.6.7 Simulated (left) and experimentally obtained temperature contours of plasma jet [19].

done inside the combustion chamber, before the acceleration of the free jet to supersonic velocities takes place. This fact excludes the possibility of visual observation of particle injection for HVOF.

There are well-established measuring systems for the observation of particle properties in the free jet. These systems in general can give information about the temperature, velocity of the particles as well as statistical information about the particle size distribution in the free jet. Having experimentally obtained values of those properties is especially relevant for the verification of particle jet simulations. Typically, in-flight velocity of particles is measured by means of time-of-flight calculations, whereas temperature measurements are done by means of two-color pyrometry.

Phenomena during impact and coating formation can be captured by means of photo or high-speed video cameras. Light microscopes and X-ray microscopes are helpful for post-factum analysis of splats, in other words solidified particle shapes after impact. In order to capture phenomena during the splat formation such as the moment of impact, deformation of the particle, solidification, and shrinkage during cooling, modern high-speed video cameras are required. Although such videos give no information about the temperature of the particle, the velocity can be deduced from the particle displacement prior to the impact between the time frames. Special measures should be taken in order to be able to capture the individual particles during the impact on the substrate surface. High-speed video filming brings a huge insight into the particle impact and plays a significant role in the validation and verification of particle impact simulations. Nevertheless, it is not as widely used due to its cost and complexity. Apart from experimental validation of individual processes, there exist various methods for the determination of effective mechanical properties of completely formed coatings; they are discussed in the following.

The effective elastic properties of coatings can be determined through different experimental methods such as the uniaxial tensile stress, the four-point bending test, the indentation test (micro- and nanoindentation), and the ultrasound test. The tensile test and the four-point bending test involve measurements of the elastic deformation of the specimen and linking them to Young's modulus and the flexural modulus using the force and momentum equilibrium equations. In the indentation test, a Vickers or Berkovich diamond is pressed into the surface of the specimen. The acting forces are recorded continuously and together with the resulting penetration depth, used to determine the indentation hardness. The ultrasound test is a nondestructive testing method where the acoustic emissions emanating from the vibrating specimen are recorded and used in connection with the material properties to calculate Young's modulus of the material.

Because of its versatility, practicality, and simplicity of use, the microindentation test is often employed in the literature in order to measure the effective indentation modulus, which is comparable with Young's modulus, of the coatings and to verify the simulation results. In order to determine the thermal diffusivity coefficient of the coatings, the laser flash method is widely used. In this method, the front side of the specimen is heated with a laser impulse, and the temperature values are measured on the back side by an infrared sensor, yielding the thermal diffusivity value. The actual thermal conductivity of the material can be determined from the connection between the thermal capacity, density, and the thermal diffusivity [30].

A number of methods for the characterization of the microstructural features of coatings exist, such as mercury intrusion porosimetry (MIP), helium pycnometry, electrochemical tests, optical image analysis, or small-angle neutron scattering (SANS). While the former methods mainly give information about the volume fraction of the pores/gaps in the microstructure, the latter – SANS and optical image analysis – are also able to capture qualitative aspects. As an example, the SANS can be utilized to differentiate spherical pores, interlamellar pores, and cracks and to estimate their orientation [31]. Image analysis techniques are usually based on metallographic examinations of polished cross sections, which are obtained by light microscopy or scanning electron microscopy (SEM). One of the main advantages of using the image analysis for microstructure characterization is the fact that the cross-section examinations give detailed information about the 2D spatial extent and the distribution of the pores as well as their shape [32].

Another method to identify the microstructural features of coatings is the X-ray microtomography (XMT), which can produce a three-dimensional image of the microstructure by measuring the mitigation in X-ray magnitude along different directions in the specimen. On the one hand, the XMT method is simpler and faster in application in comparison to other methods. On the other hand, this method is more expensive and is constrained by the specimen size and resolution. Moreover, materials with similar damping properties cannot be distinguished from each other [32].

It is particularly important for coatings consisting of more than two phases to be able to differentiate as clearly as possible between the individual phases of the

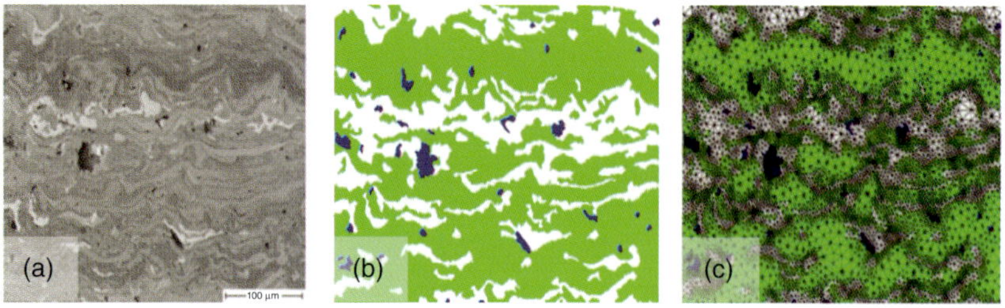

Figure 2.6.8 (a) REM cross section of the coating, (b) edited cross-section picture, and (c) FE-mesh of the coating. (Reproduced with the permission of permission of TU Dortmund.)

coating, since the numerical homogenization methods require spatially resolved material properties. The correct attribution of material properties is possible only by knowing the exact positions of the individual coating phases. In Figure 2.6.8, it is evident that the computations have to be done with severe simplifications because the individual phases are highly dispersed [33].

Furthermore, the simplification of 3D structures with 2D models is another possible factor that limits the quality of the prediction of the effective coating properties [34, 35]. It is an active field of study in the literature because of its promising perspective of improving the predictions of the coating properties. Since the experimental methods are still not capable of capturing three-dimensional structures with all the details, the researches focus on the generation of the 3D model of the coating microstructure out of 2D cross-sectional images of the coating. Bobzin *et al.* have shown in their works that the results generated using 3D models are superior to those generated using 2D models with respect to the computational accuracy [36].

2.6.7
Conclusion

In conclusion, it has to be noted that the deposition of thick coatings involves several processes as well as different disciplines. Their common trait is the generation of a coating layer larger than 30 μm, which then can be analyzed using well-established analytical and measurement tools in order to determine their functional properties. Simulations of the processes themselves that lead to the coating formation, on the other hand, require a wide range of simulative tools, such as electrochemical kinetics to plasma physics, and obviously include continuum mechanics. As it seems practically impossible to describe all physical processes involved in detail, let alone to discuss the approaches for their simulation, it was decided to concentrate on the simulation techniques of thermal spraying processes. Simulation techniques that were described for the example of thermal spraying can be applied partially to other thick coating deposition processes as well. Nonlinearities such as contact and elastic-plastic material behavior, as well as the phenomena such as rapid heating, cooling and phase changes are some of the challenges that are common to the simulation of most of the coating processes.

As was already mentioned, simulations of coating formation processes involve multiphysical models. The main difficulty, however, comes from the multiscale nature of such simulations. In other words, simulation of physical events in the spatial range of nanometers (fine particles and features of the coating morphology) to centimeters (length of particle flight path in the free stream and effective macroproperties of the coatings), which require time resolution in the range of nanoseconds to seconds, demands an enormous computational power. This factor that limits such simulations can be tackled to some extent with the help of parallel computing methods.

Despite the challenges associated with coating formation simulations, they have proven to be very useful in process understanding as well as in process control and design. Specifically carried out simulations can spare expensive experiments that are usually run multiple times in order to achieve desired result. With the help of the insight gained from the simulation results, process parameters can be adjusted in order to achieve more efficiency and minimize material waste.

2.6.8 Software Tools

Some software tools typically being used in thick-film coating simulations are depicted in the following table:

SW cat	Name of code	Website of software	Website of provider	Type of code (academic/ commercial) installations	Short description/ typical applications/ phenomena being treated
2.6	ANSYS	http://www.ansys.com	http://www.ansys.com	Commercial	General-purpose simulation software/fluid mechanics, solid mechanics, multiphysics
2.6	Comsol Multiphysics	http://www.comsol.com	http://www.comsol.com	Commercial	General-purpose multiphysics simulation tool/electrical, mechanical, fluid, chemical

SW cat	Name of code	Website of software	Website of provider	Type of code (academic/ commercial) installations	Short description/ typical applications/ phenomena being treated
2.6	Flow-3D	http://www.flow3d.com/	http://www.flow3d.com/	Commercial	Computational fluid dynamics software specialized on VoF method
2.6	OpenFOAM	http://www.openfoam.com/	http://www.openfoam.com/	Open source	Open-source computational fluid dynamics software package
2.6	SimCoat	http://simulent.com	http://simulent.com	Commercial/ dedicated	SimCoat is a probability-based algorithm using Monte Carlo technique to determine the random droplet impact and coating of various surfaces
2.6	SimPlasma	http://simulent.com	http://simulent.com	Commercial/ dedicated	SimPlasma simultaneously solves the equations for the transport of mass, momentum, and energy, along with Maxwell's equations
2.6	SimSpray	http://simulent.com	http://simulent.com	Commercial/ dedicated	SimSpray is a three-dimensional algorithm that models the spray dynamics in a transient, turbulent, chemically reactive flow
2.6	SimDrop	http://simulent.com	http://simulent.com	Commercial/ dedicated	SimDrop is a volume of fluid (VoF)-based free surface code, which allows for the simulation of single or multiple droplet dynamics with phase change

References

1. Druesne, F. (2001) A new 3D electroplating simulation and design tool. Presented at SURFIN 2001 by Roger Mouton.
2. Toparli, M., Sen, F., Culha, O., and Celik, E. (2007) Thermal stress analysis of HVOF sprayed WC-Co/NiAl multilayer coatings on stainless steel substrate using finite element methods. *J. Mater. Process. Technol.*, **190** (1-3), 26–32.
3. (a) Bobzin, K., Kopp, N., Warda, T., Schäfer, M., and Öte, M. (2013) A numerical investigation: influence of the operating gas on the flow characteristics of a three-cathode air plasma spraying system. Thermal Spray 2013: Proceedings of the International Thermal Spray Conference, Busan, South Korea.; (b) Bobzin, K. and Öte, M. (2016) Modeling multi-arc spraying systems *J. Therm. Spray Technol.* doi: 10.1007/s11666-016-0407-7.
4. Nickel, R. (2007) *Dissertation: Die beschichtungsprozessabhängige Materialmodellierungvon EB-PVD und APS-Wärmedämmschichten*, IOT, Aachen.
5. Lima, F., Mescheder, U., and Reinecke, H. (2012) Simulation of current density for electroplating on silicon using a hull cell. Proceedings of the 2012 COMSOL Conference in Milan.
6. Carr, N.Z. (2002) Simulation and estimation of copper electroplating using level set methods and image processing techniques. Mechanical Engineering dissertation. Texas Tech University.
7. Bujik, A. and Schafstall, H. (2011) Process simulation of closed profiles by chaining of forming and welding simulations, using Simufact.forming & Simufact.welding. Colsolda, XXXVII Congresso Nacional de Soldagem.
8. Feng, Z. and Champliaud, H. (2012) Modeling and simulation of mechano-welding process for cylinders. Proceedings of ASME 2012 International Design Engineering Technical Conferences & Computers and Information in Engineering Conference, August 12–15, 2012.
9. Tabbataee, M. and Mahmoudi, J. (2008) Finite element simulation of explosive welding. 49th Scandinavian Conference on Simulation and Modeling.
10. Saresoja, O., Kuronen, A., and Nordlung, K. (2012) Atomistic simulation of the explosion welding process. *Adv. Eng. Mater.*, **14** (4), 265–268.
11. Passandideh-Fard, M., Esmaeelpanah, J., and Najmi, K. (2009) Simulation of laser cladding process using enthalpy model integrated into VOF method. 17th Annual International Conference on Mechanical Engineering.
12. Qi, H., Mazumder, J., and Ki, H. (2006) Numerical simulation of heat transfer and fluid flow in coaxial laser cladding process for direct metal deposition. *J. Appl. Phys.*, **100** (2), 024903.
13. Zhao, H., Zhang, H., Xu, C., and Yang, X. (2009) Temperature and stress fields of multi-track laser cladding. *Trans. Nonferrous Met. Soc. China*, **19**, s495–s501.
14. Schulz, W., Nießen, M., Wollschläger, J., Hansen, U., Pittne, A., and Rethmeier, M. (2014) Schnelle Numerische Methoden für die Effiziente Temperaturfeldberechnung in Bauteilnahen Geometrien und Mehrlagenschweißungen (Fast numerical methods for efficient temperature field computation in near-net shape components and multilayer welds). *Schweißen Schneiden*, **66** (1–2), 34–37.
15. Hanneforth, P. (2006) The global thermal spray industry – 100 years of success: so what's next? in *Advanced Materials & Processes* (ed. R. Knight), ASM International, Materials Park, OH, Mai.
16. Bobzin, K., Schläfer, T., and Schäfer, M. (2011) Optimierungeines HVOF-prozessesdurch simulationstechniken (Optimization of an HVOF process through simulation techniques). *Therm. Spray Bull.*, **4** (2), 159–164.
17. Trelles, J., Chazelas, C., Vardelle, A., and Heberlein, J. (2009) Arc plasma torch

modeling. *J. Therm. Spray Technol.*, **18** (5–6), 728–752.
18. Kim, H.C., Iza, F., Yang, S.S., Radmilovic-Radjenovic, M., and Lee, J.K. (2005) Particle and fluid simulations of low-temperature plasma discharges: benchmarks and kinetic effects. *J. Phys. D: Appl. Phys.*, **38** (19), R283–R301.
19. Bobzin, K., Bagcivan, N., and Petkovic, I. (2011) Numerical and experimental determination of plasma temperature during air plasma spraying with a multiple cathodes torch. *J. Mater. Process. Technol.*, **211** (10), 1620–1628.
20. Fauchais, P. (2004) Topical review – Understanding plasma spraying. *J. Phys. D: Appl. Phys.*, **37** (9), R86–R108.
21. Murphy, A.B. (2000) Transport coefficients of hydrogen and argon-hydrogen plasmas. *Plasma Chem. Plasma Process.*, **20** (3), 279–297.
22. Jazi, H.S. (2012) *Advanced Plasma Spray Applications*, InTech.
23. Bobzin, K., Bagcivan, N., Parkot, D., Kashko, T., Petković, I., Beck, T., Bialas, M., Bednarz, P., Singheiser, L., Hallstedt, B., Nemna, S., and Schneider, J.M. (2009) Modelling of coating process, phase changes and damage of plasma sprayed thermal barrier coatings on Ni-base superalloys. *Adv. Eng. Mater.*, **12** (3), 110–126.
24. Yokoyama, K., Watanabe, M., Kuroda, S., Gotoh, Y., Schmidt, T., and Gärtner, F. (2006) Simulation of solid particle impact behavior for spray processes. *Mater. Trans.*, **47** (7), 1697–1702.
25. Langer, S.A., Fuller, E.R., and Carter, W.C. (2001) Oof: an image-based finite-element analysis of material microstructures. *Comput. Sci. Eng.*, **3** (3), 15–23.
26. Reid, A.C., Langer, S.A., Lua, R.C., Coffman, V.R., Haan, S., and Garca, R.E. (2008) Image-based finite element mesh construction for material microstructures. *Comput. Mater. Sci.*, **43** (4), 989–999.
27. Bolot, R., Seicheoine, J.L., Qiao, J.H., and Coddet, C. (2011) Predicting the thermal conductivity of AlSi/polyester abradable coatings: effects of the numerical method. *J. Therm. Spray Technol.*, **20** (1–2), 39–47.
28. Zhang, X.C., Xu, B.S., Tu, S.T., Xuan, F.Z., Wang, H.D., and Wu, Y.X. (2009) Fatigue resistance and failure mechanisms of plasma-sprayed CrC-NiCr cermet coatings in rolling contact. *Int. J. Fatiue*, **31**, 906–915.
29. Gruber, J., Hlina, J., and Sonsky, J. (2008) On possibilities to characterize turbulence in plasma torches by correlation dimensions. Plasma Science, ICOPS.
30. Kaßner, H. (2009) Theoretische und Experimentelle Untersuchungen zum Plasmaspritzen mit Nanoskaligen Suspensionen (Theoretical and experimental analysis of plasma spraying with nanoscale suspensions). Dissertation Fakultät für Maschinenbau der Ruht-Universität Bochum.
31. Deshpande, S., Kulkarni, A., Sampath, S., and Herman, H. (2004) Application of image analysis for characterization of porosity in thermal spray coatings and correlation with small angle neutron scattering. *Surf. Coat. Technol.*, **187** (1), 6–16.
32. Wiederkehr, T., Klusemann, B., Gies, D., Müller, H., and Svendsen, B. (2010) An image morphing method for 3D reconstruction and FE-analysis of pore networks in thermal spray coatings. *Comput. Mater. Sci.*, **47**, 881–889.
33. Klusemann, B., Hortig, C., and Svendsen, B. (2009) Modellierung und Simulation des Heterogenen Material verhaltens in Thermisch Gespritzten Schichten, 3. Öffentliches Kolloqium SFB-708, TU Dortmund.
34. Bobzin, K., Nickel, R., Parkot, D., and Kashko, T. (2007) Microstructure dependency of the material properties: simulation approaches and calculation methods. *Steel Res.*, **78**, 10–11.
35. Bobzin, K., Bagcivan, N., Parkot, D., Kashko, T., Laschet, G., and Scheele, J. (2009) Influence of the definition of the representative volume element on the effective thermoelastic properties of thermal barrier coatings with random

microstructure. *J. Therm. Spray Technol.*, **18** (5-6), 988–995.

36. Bobzin, K., Kopp, N., Warda, T., and Öte, M. (2012) Determination of the effective properties of thermal spray coatings using 2D and 3D models. *J. Therm. Spray Technol.*, **21** (6), 1269–1277.

2.7
Thin-Film Deposition Processes

Andreas Pflug, Michael Siemers, Thomas Melzig, Martin Keunecke, Lothar Schäfer, and Günter Bräuer

2.7.1
Introduction

Coatings with film thicknesses in the order of several micrometers or less are considered as *thin films*. They are applied for refinement and functionalization of devices in a large variety of technologies. Applications include hard coatings on tools, machine parts, and in automotive technology, energy-saving coatings for large area architectural glass, functional coatings in micro- and nanotechnology, photovoltaics and solar thermal devices, coatings for optical high-precision devices, as well as coatings with defined biological functionality.

Coating materials have to be selected to fulfill the requirement of products. In terms of intrinsic film properties, this may include electrical, magnetic, or optical performances such as high transmittance, magnetoresistivity, or high conductivity as well as mechanical properties such as wear and scratch resistance. In terms of chemical properties, this may include good diffusion barrier functionality and environmental resistivity, integrated hydrophilicity, or hydrophobicity, for example, for self-cleaning.

While computational material modeling has already been proven successful in identifying new candidate materials for specific technical purposes, it cannot be neglected in the case of thin films that the conditions during their deposition process also have a strong impact on their intrinsic properties. In a work of Thornton [1], the sputter deposition conditions of metallic layers with thicknesses of several micrometers are categorized into a *structure-zone model* with respect to the ratio T_s/T_m between substrate and melting point temperature and the total pressure of the sputtering gas argon. For T_s/T_m ratios below 0.1, that is, low mobility of adatoms, layer growth occurs in the form of polycrystalline columns separated by voids, whereof the growth direction is mainly determined by the flux direction of the coating species. For higher T_s/T_m ratios, the growth mode is influenced by an increased surface mobility of adatoms, and the voids between the columns become filled or even grain boundary migration occurs, leading to increased crystallite

Handbook of Software Solutions for ICME, First Edition. Edited by Georg J. Schmitz and Ulrich Prahl.

sizes and smooth surfaces. By increasing the total pressure, the energy of sputtered atoms is reduced due to scattering with the background gas and, therefore, the different growth mode zones are shifted toward higher substrate temperatures.

The situation becomes more versatile for materials where different crystalline phases can arise as a result of different growth conditions. For stoichiometric TiO_2 layers used in optical interference coatings, two different crystalline phases, namely, rutile and anatase, as well as an amorphous structure are observed. For sputter-deposited and evaporated TiO_2, it is shown that the phase composition after deposition and after postdeposition anneal strongly depends on process conditions such as total pressure, reactive gas fraction, and magnetic field configuration [2]. Furthermore, it is shown that a wide range of optical refractive indices between 2.0 and 2.7 can be realized for TiO_2 via different deposition methods and as a function of precursor energy and degree of ionization [3]. In summary, the energy and angular distribution of adatoms and other species impinging onto the substrate during film growth is an important determining factor with respect to morphology and phase composition of the resulting film material.

Besides the aforementioned process conditions during film growth, the quality and performance of thin films also depend on the history of previously coated layers and processing steps. Acting as so-called seed layers, previously coated layers can enable a growth of the actual layer in a certain preferred crystalline texture. In the field of low-emissive coatings for architectural glazing, it is shown that thin Ag layers exhibit a (1 1 1) crystalline texture and significantly enhanced conductivity when they are coated onto crystalline ZnO (0 0 2) layers [4]. Another factor can be the *surface polarity* of the layer beneath the growing film. Dielectric ZnO layers often consist of (002) oriented crystallites with vertically aligned *c*-axis, whereof the upper surface is either oxygen or zinc terminated. By turning the crystallites upside down, this surface property can be inverted. Subsequent crystallites preferably grow in the same orientation as the ZnO crystallites from the layer beneath. The surface polarity has strong impact on surface properties, for example, on the chemical etching speed or on elastic properties [5]. For thin-film solar cells based on a-Si:H/μc-Si:H, sputtered ZnO:Al layers are used as transparent conductive front electrode. In order to enhance light scattering for improved light conversion efficiency, these layers are chemically etched in diluted HCl. In this context, ZnO:Al layers fabricated with radio-frequency (RF) sputtering exhibit an improved etching texture compared with conventional DC sputtering, which is due to a modified surface polarization pattern. For efficient upscaling of the process, it is shown that the etching behavior of a thin ZnO:Al seed layer prepared by RF sputtering can be transferred to a thick ZnO:Al layer fabricated in a subsequent DC sputtering process [6, 7].

The aforementioned examples demonstrate that in thin-film deposition the intrinsic film properties result from a complex interaction between the possible atomic bond structures, the deposition process conditions as well as the condition of the substrate, the previous layer structure, and the subsequent processes. In order to meet the general trend in device production toward higher throughput, higher functional integration, and higher precision, an improved understanding of these interacting phenomena is essential. While many investigations in thin-film technology are based on phenomenological, empirical approaches, computational materials modeling in combination with process and device modeling will put more focus on knowledge-based development of novel advanced thin-film materials.

2.7.2
Overview of Thin-Film Deposition Methods

This section gives an impression about the various technical issues and applications in thin-film deposition illustrated with representative examples. A comprehensive description of all deposition techniques is beyond the scope of this chapter. An overview of vacuum-based deposition technology is given in [8], while basic principles and applications of wet-chemical-based electroplating processes are given in [9]. An overview focusing on sputter deposition but also with an introduction on other physical vapor deposition (PVD) and chemical vapor deposition (CVD) processes can be found in [10, 11].

Thin-film deposition is based on the exposure of a *precursor* to a substrate, subsequent chemical or physical reactions between precursor and substrate surface, resulting in a growing film and formation of specific structures. During this process, the substrate surface can be additionally modified by energetic supply causing changes in film structure, partial film removal, or implantation of energized species. The precursor transport to the substrate is realized either in gaseous or liquid media or via ballistic propagation in vacuum. In some cases, the final film structure is achieved in a postdeposition treatment step such as thermal annealing or light exposure. This causes chemical conversion of the deposited material via oxidation or diffusion processes. Further possible postdeposition treatment steps are treatment with UV light or chemical etching.

With respect to the precursor generation and the character of film-forming reactions, thin-film deposition can be divided into *chemical* and *physical* methods. In chemical deposition methods, the film-forming mechanism is based on precursors undergoing chemical reactions within a deposition reactor, thereby synthesizing the film-forming species. The chemical reaction can either take place in the volume of the reactor or directly at the reactor–substrate interface. In contrast, in physical deposition methods, the film material is supplied into the reactor and transported onto the substrate without essential change of its chemical structure.

A survey on various physical and chemical thin-film deposition methods is sketched in Figure 2.7.1. While the majority of deposition methods work in gaseous media under atmospheric or low-pressure conditions based on vacuum technology, there are also a few processes based on reactions and transport within liquids. Gaseous processes can be divided into PVD and CVD, while there is no PVD in liquid processes as they are essentially based on chemical or electrochemical reactions.

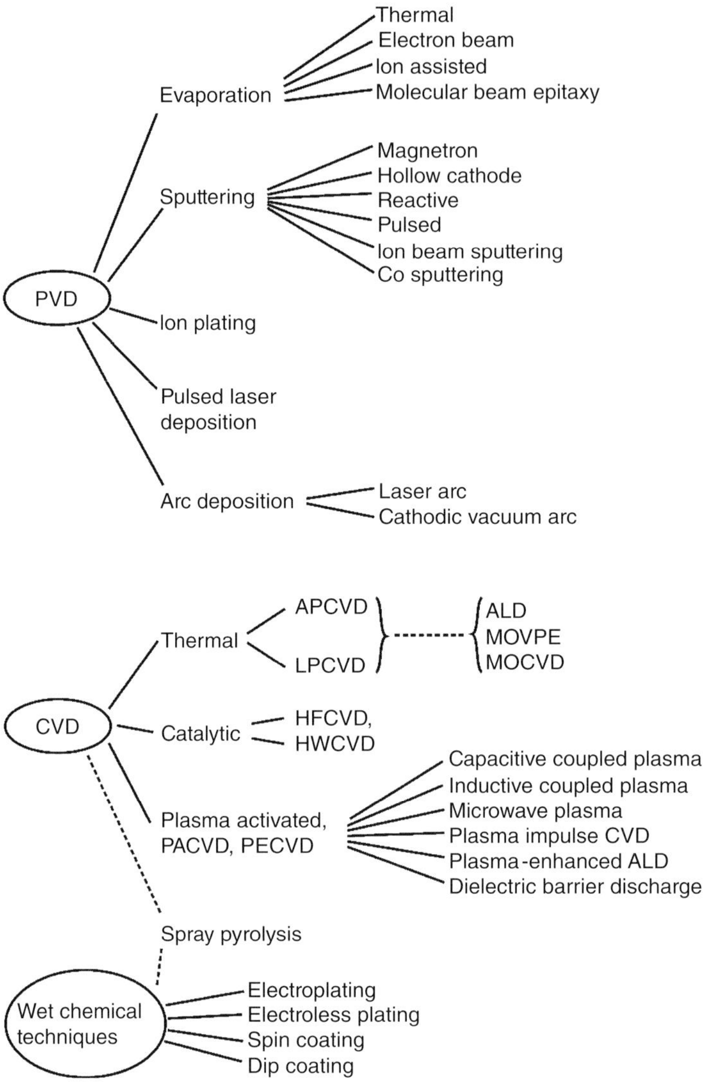

Figure 2.7.1 Mind map on thin-film deposition techniques.

In PVD, the film material is transported from a source toward the substrate while being chemically essentially unchanged. The material emission from the source is accomplished by supplying energy. In the case of evaporation processes, the energy supply is thermal either by heating a crucible containing the source material or by local evaporation of the source with a scanning electron beam, via an arc discharge or by other means. An evacuated vessel enables the evaporated material being transported toward the substrate. Optionally, additional reactive gas can be supplied under low-pressure conditions in order to form compound materials. In order to improve the density and hardness of the evaporated films, the substrate surface can be additionally treated by ion bombardment during evaporation. A special variation of evaporation is *molecular beam epitaxy,* where material is evaporated in a so-called effusion cell whereof the source design in combination with high-vacuum conditions allows the transport of evaporated material toward the heated substrate without scattering and contamination with residual background gas. Under such conditions, an epitaxial growth of essentially monocrystalline layers can be realized.

Alternatively, it is possible to release source atoms or molecules by *sputtering,* that is, by bombarding a source material referred to as *sputtering target* with high-energetic ions. This is either accomplished via a magnetically enhanced plasma discharge with the target on negative potential (magnetron sputtering) or via external ion sources (ion beam sputtering). In most cases, inert gases such as Ar are used for generating sputtering ions; however, there are exceptions such as self-sputtering with ionized sputtered metal. A high-energetic ion impact on a target results in a collision cascade, whereof most energy is dissipated into the target bulk material, while a certain fraction of the kinetic energy is being reflected back toward the surface thereby releasing atoms or molecules.

In general, the various deposition methods represent different compromises between ambiguous demands, namely, high deposition rate on large substrate areas and high throughput on the one hand while on the other hand high film quality, high reproducibility, and precise control of deposition conditions are required for smart products with increasing complexity and functional integration. An important parameter with respect to film structure formation is the energy distribution function of film-forming species. At energies well above 1 eV, the surface mobility of film-forming species is significantly enhanced enabling a rearrangement of the growing film with respect to energy minimization. As a result, films with less porosity, smoother surfaces, and more regular crystalline structure can be realized.

In PVD, the energy distribution of adatoms or film growing species vastly differs among the respective deposition techniques. As shown in the graph of Figure 2.7.2a, thermally evaporated species have a Maxwellian energy distribution according to the evaporation temperature, where most species have energies below 100 meV. In contrast, sputtered atoms have a nonthermal energy distribution resulting from the collision cascade in the target material [12], which can be approximated via $f(E) = E/(E + U_b)^3$ as a function of the binding energy U_b of the target material [13, 14]. Processes such as pulsed laser deposition, where a focused laser beam locally generates highly ionized plasma at a target surface allow for even higher ion energies of ejected material [15]. High energies due to high degree of ionization of ejected precursor material can be further obtained in arc discharge processes as summarized in [16].

On the way between source and substrate, the energy distribution of emitted species may considerably change due to scattering with the background gas, which is also referred to as *thermalization.* An example is shown in the graph of Figure 2.7.2b, where the energy distribution function of sputtered Ti atoms in

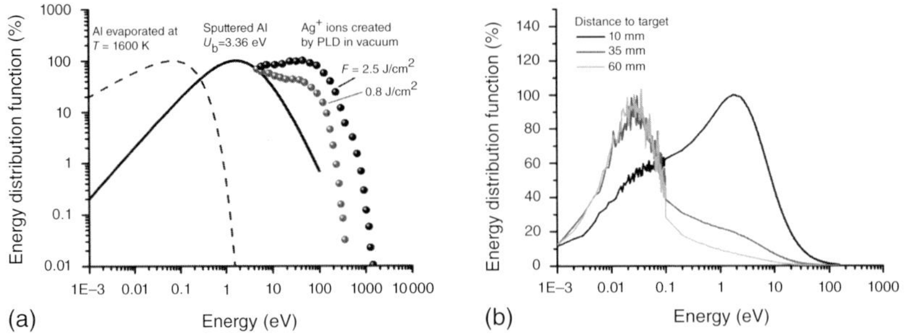

Figure 2.7.2 Characteristic energy distribution functions of film-forming species generated in PVD processes (a): Thermal evaporation of Al, sputtering of Al, and pulsed laser deposition of Ag (data of the PLD process according to [15]). Graph in (b) shows the DSMC simulated energy distribution of sputtered Ti atoms scattered in Ar at 0.32 Pa and different distances from the target.

the presence of an Ar background pressure of 0.32 Pa is shown at three different distances from the target. The data are generated by direct simulation Monte Carlo (DSMC) method, including scattering cross sections between Ti and Ar. At a distance of 60 mm, the high-energy tail of the distribution function is almost completely lost and the energy distribution function is mainly composed of thermalized particles, forming a Maxwellian peak. In contrast, in processes such as ion beam sputtering, where the background pressure is about one order of magnitude lower, the high-energetic tail of the energy distribution function can be maintained over distances of several centimeters [17].

Besides emitting the film-forming material with high energy from the source, additional energy supply for film growth can also arise from a potential difference between reactor volume and the substrate. This can be accomplished by pulsed, mid- or high-frequency plasma excitation, leading to an increase in the plasma potential versus floating substrate potential as shown by plasma simulation in [18] and by Faraday cup measurement in [19]. In ion plating processes, the substrate is being put on negative potential in order to attract positively charged precursor or gas ions, thereby enabling growth energies in the order of 100 eV and above [20].

For energies of film-forming species at the substrate above several 10 eV, implantation and buildup of intrinsic film stress, resputtering of the growing film, and defect formation may occur, which can be beneficial for hard coatings but may be detrimental in cases where a low defect level is important. An energy impact in the order of 10 eV but below the sputtering threshold of the respective coating material has been identified as mostly beneficial for dense and defect-free coatings in the field of electrooptical functional coatings.

The impact of the energy of adatoms on the film morphology is illustrated in Figure 2.7.3 as adapted from [21], where cross-section scanning electron microscopy images of TiAlN coatings fabricated by DC and pulsed magnetron sputtering are compared. In the case of the pulse sputtering, a frequency of 50 kHz and a rectangular signal with 50% duty cycle is used. While the DC

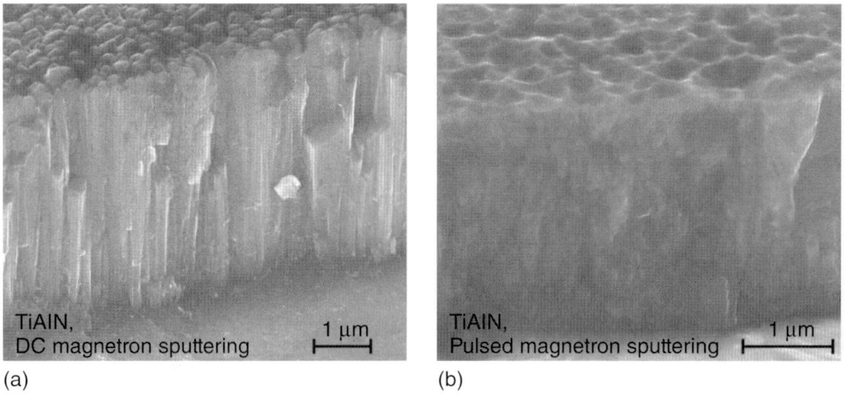

Figure 2.7.3 Cross-section scanning electron microscopy images of TiAlN coating deposited by DC magnetron sputtering (a) and pulsed sputtering at 50 kHz (b); for further details, see [21].

sputtered film clearly exhibits columnar growth and high surface roughness, the film resulting from the higher energetic pulse sputtering mode shows a more homogeneous structure and a comparably smooth surface texture.

A further aspect in PVD is that some techniques such as arc or pulsed laser deposition or so-called high-power impulse magnetron sputtering (HiPIMS) yield a high degree of ionization of film-forming species. As charged particles follow the local gradients of the electric potential, this enables conformal coating of three-dimensional structures such as deep trenches, which is of importance, for example, in microtechnology [22].

In CVD, the film material originates from chemical reactions between the substrates or the reactor walls and the precursor gases injected into a deposition reactor either under atmospheric pressure (APCVD, atmospheric pressure physical vapor deposition) or low-pressure conditions (LPCVD, low-pressure physical vapor deposition). In CVD processes, the energy for excitation and/or dissociation of precursor gas molecules is supplied either by the temperature of the substrate or via additional excitation of the gas phase. In most cases, the excitation of the gas phase is achieved by the use of plasma. These processes are referred to as plasma-activated chemical vapor deposition or plasma-enhanced chemical vapor deposition (PACVD, PECVD). The chemical film formation reaction usually takes place on the heated substrate surface and other walls of the reactor. Compound formation reactions within the gas volume are possible at higher pressure but are in most cases unwanted as this often causes enhanced dust formation. A special case of thermally activated CVD is *hot-filament* CVD [23] or *hot-wire* CVD (HFCVD, HWCVD), where an array of wires with high melting point, mostly consisting of tungsten, is electrically heated to approximately 2000°C, enabling the thermal decomposition of precursors at the wire surface. Depending on the materials, the wires can additionally act as catalysts in decomposing the precursor gases. Therefore, HFCVD or HWCVD processes are also referred to as *catalytic CVD* [24] processes. They are applied to produce either polycrystalline diamond films from CH_4 precursors or silicon-based coatings from SiH_4 precursors [25]. Due to the formation of radicals such as monoatomic hydrogen or oxygen, etching processes can also take place at the substrate or the growing film, respectively. The desired phase composition and film structure often results from concurrent deposition and etching processes.

In PECVD or PACVD reactors, the two main different concepts of plasma ignition involve either capacitive or inductive coupling of the electric power into the reactor. In capacitive coupling, the power is introduced via DC or pulsed glow discharges from biased electrodes. For coatings on large area, this is often accomplished within the so-called parallel-plate reactors, where the glow discharge occurs in a slot between two planar electrodes. For parallel-plate reactors, at the Gaseous Electronics Conference in 1994 [26], the so-called GEC reference cell is proposed as reference geometric setup, enabling different research groups in plasma diagnostics and modeling to compare and benchmark their results. There has been extensive research in upscaling large area parallel-plate RF-driven PECVD reactors in thin-film photovoltaics for the deposition of amorphous or microcrystalline Si layers [27]. It turned out that moving from RF excitation to higher frequency is advantageous for increasing the achievable power density and gas conversion efficiency. Operating such very high-frequency (VHF) capacitive plasma discharges on large area is not trivial; numerous high-frequency-related mechanisms such as standing waves, the skin effect, and so on impair the plasma uniformity; and advanced electrode concepts in combination are required to circumvent these issues, as shown, for example, in [28]. In inductively coupled plasma discharges, high plasma densities with good homogeneity and low substrate damage can be achieved, and the aforementioned drawbacks can be circumvented [29].

Typical film materials in CVD are organically modified oxides or nitrides of silicon and other metals acting as scratch protection, diffusion barriers, antireflective coatings, or electric passivation. Via metal–organic precursors, a large variety of metals and their compounds can be realized in CVD, which is then referred to as *metal organic chemical vapor deposition* (MOCVD). Under certain conditions such as high gas purity and high substrate temperature and depending on the material system even epitaxial film growth can be realized in CVD within the so-called *metal organic vapor phase epitaxy* (MOVPE) processes. MOVPE allows faster production of epitaxial layers compared with MBE and is, therefore, of high relevance in semiconductor industry.

With hydrocarbon precursors such as CH_4 or C_2H_2, the so-called diamond-like carbon coatings for mechanical protection and wear reduction are fabricated, which is a large business field within mechanical engineering. In HFCVD processes with hydrocarbon precursor, it is also possible to deposit diamond coatings from CH_4 precursor on an industrial scale [25].

A special case of a CVD process based on surface reactions is atomic layer deposition (ALD or PEALD if the reaction is plasma activated), which is based on self-saturating reactions between the substrate surface and at least two different gases, which are alternately exposed to the heated substrate. In an idealized description, in each reaction cycle, either a saturation of the surface with bond centers or a single atomic layer is grown. This allows conformal deposition of coatings with high film thickness precision and very low defect density such as pinholes. A comprehensive overview on ALD processes, their modeling, and applications is given in [30].

In wet-chemical processes, the film formation is essentially of chemical nature, while in electroplating, crystalline growth can be additionally triggered by electric current between the solution and the substrate, which is referred to as *electrocrystallization*. Conformal coatings on complex 3D substrates can be realized in electroplating by tailoring the flow pattern in the liquid solution, as the precursor transport is mainly governed by the flow and local electric field distribution [31]. In the case of self-saturating reactions, for example, in alternating adsorption of anionic and cationic electrolytes thereby reversing the surface charge, nanolaminated films can be conformally coated with high precision [32]. With a similar method, it is possible to produce nanocomposite coatings, that is, nanoparticles embedded into a metallic matrix as demonstrated for oil-containing nanocontainers in [33]. A hybrid technique between CVD and wet-chemical processing is spray pyrolysis, where a liquid precursor is transformed into droplets, for example, by ultrasonic excitation and the droplets are directed toward a heated substrate, where subsequently an atmospheric pressure CVD process takes place.

The examples shown above demonstrate a broad variety of different technical methods and resulting products within thin-film deposition technology. While the process dynamics and mechanisms vastly differ among the different reactor concepts, the resulting structure formation of the growing film always depends on the *detailed growth conditions* at the substrate in terms of thermal energy, particle fluxes, and their energy and angular distributions. Thus, extracting the characteristic detailed growth conditions from reactor scale – either by *in situ* diagnostics or by reactor-scale modeling – yields quantitative criteria for comparing the different thin-film deposition methods in terms of resulting film quality and performance.

2.7.3
Modeling of Thin-Film Deposition as a Multiscale Problem

Thin-film deposition involves the interaction of manifold mechanisms acting on different size and length scales. Consequently, a comprehensive simulation model of all aspects in thin-film deposition would be a multiscale modeling cascade. It would consist of several submodels focused on different mechanisms and acting in vastly different time and length scales. As sketched in Figure 2.7.4, the multiscale cascade can be divided into three major categories, namely, the modeling of (i) intrinsic material properties, (ii) reactor-scale transport dynamics, and (iii) heuristic global process models.

Within this scheme, reactor-scale transport simulation plays a central role. It delivers information that can be easily used for validation and optimization

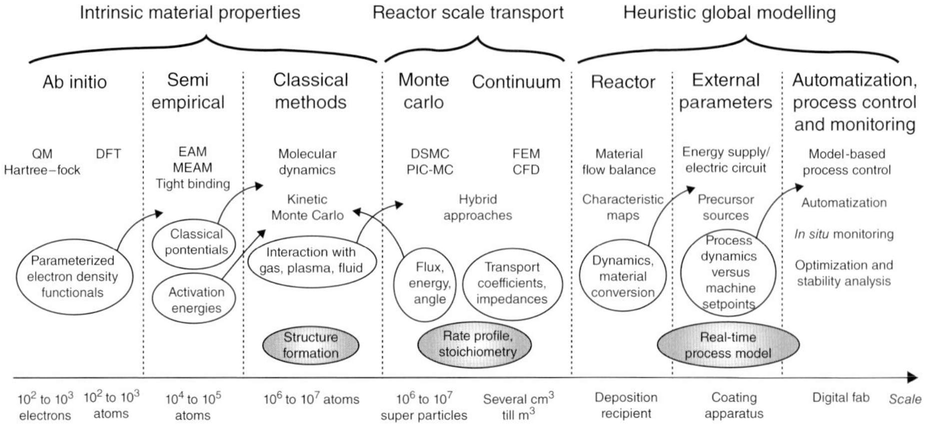

Figure 2.7.4 Different levels in modeling thin-film deposition, their scales, and interaction (adapted from [34]).

of experimental setups. Among these information, the deposition rate and stoichiometry profile of the growing film are most important, but also the stoichiometry profile, temperature distribution, electromagnetic field, and so on are well accessible and highly relevant in the experimental context.

For the creation of global heuristic models, which are preferably real-time capable, many unknown parameters can be deduced by averaging dynamic parameters of the deposition reactor such as flow conductances, time constants, flow profiles, material conversion rates, or electric impedances. Furthermore, reactor-scale modeling delivers the *detailed growth conditions* at the substrate consisting of particle fluxes as well as their energy and angular distribution functions. This information is relevant for atomistic models on intrinsic structure formation of the growing layer. Besides being input for atomistic growth modeling, the detailed growth conditions can also be used to identify correlations with experimental findings, which exist independently from machine set points of the respective reactor. This simplifies upscaling of an established lab-scale process toward industrial production lines under the boundary condition that the same growth conditions should be maintained.

The main task in reactor-scale simulation is modeling the transport of the precursor and reactive gases or fluids within the reactor geometry. Depending on the flow characteristics, which is continuous, molecular, or transitional, different numerical approaches are appropriate. As sketched in Figure 2.7.5, the flow characteristics can be quantified via the Knudsen number, which is the ratio between mean free path of molecules and characteristic geometric dimensions such as tube diameters or source–substrate distance.

For fluids or gases at atmospheric pressure, typical mean free path values are in the range of 10^{-5} to 10^{-4} m resulting in a low Knudsen number unless microcavity reactors are considered. In such systems, the individual molecules are essentially *locally* interacting via elastic collisions; hence, the flow can be approximated as

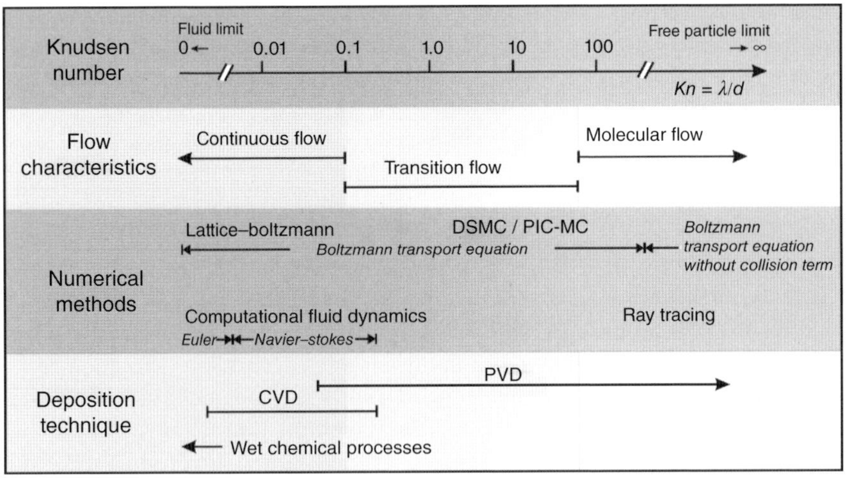

Figure 2.7.5 Complementary approaches in reactor-scale transport simulation and their validity with respect to the flow characteristics (adapted from [35]).

continuous medium and numerically represented within computational fluid dynamics (CFD). In the context of thin-film deposition, CFD is typically used for describing CVD and wet-chemical deposition processes.

In contrast, at low pressure the molecules are only weakly interacting and individually propagating through the reactor. Under such conditions, *nonlocal* effects arising from collisions between particles and the reactor wall geometry become significant. These effects are considered in the more general kinetic gas theory governed by the Boltzmann transport equation. Feasible approaches of solving the Boltzmann transport equation are statistical, particle-based algorithms such as the DSMC method [35] for neutral transport simulation. For simulation of gas discharges, the transport simulation is coupled with electromagnetic field solvers, which is referred to as particle-in-cell Monte Carlo method with collision term (PIC-MC or PIC-MCC; see, e.g., [36]). In some cases, hybrid approaches are used, where, for example, the background gas is treated as continuum while fast, diluted species such as electrons or fast ions are treated on particle basis. Particle-based simulation algorithms play an important role in describing thin-film deposition technology, where many processes fall within the categories of PVD or low-pressure CVD, while most thick-layer deposition techniques are sufficiently described within CFD simulation techniques.

Particle-based Boltzmann equation solvers are principally numerically valid at any pressure; however, at increasing pressure the practical feasibility is limited by the increasing amount of collisions, which requires a finer discretization of the simulation space and time. While, for example, DSMC gas flow simulations at low pressures as in PVD sputter processes are feasible even for reactor volumes of several cubic meters [37], the upper practical limit of this method for typical geometric setups is in the order of 0.1 mbar. For PIC-MC/PIC-MCC plasma simulation with self-consistent electric field computation, additional numerical and statistical constraints are introduced. One of the most severe restrictions is the necessity to properly resolve field gradients and the shielding effect of space charges. This yields an upper limit of the cell size of about three times of the local Debye length; otherwise, the discretization errors lead to artificial "numerical heating" of the simulated plasma [38]. For plasma densities in the order of $10^{17}/m^3s$, as are typical conditions in PVD sputter discharges, the resulting cell size limit is below 100 μm, which in the case of realistic three-dimensional geometries imposes a severe challenge in terms of required memory and CPU time. As a compromise, plasma simulations at reduced power density or in reduced 2D geometric models are feasible and nevertheless yield instructive results; however, relevant mechanisms in 3D plasma dynamics might get lost in lower dimensional models [39]. Advanced optimization techniques such as adaptive hierarchical grids and dynamic load balancing as well as advanced computing hardware such as graphical processing units (GPUs) would further lift the performance limits. Currently, three-dimensional PIC-MC plasma simulation at high power density cannot yet be considered as state of the art.

The aforementioned particle-based solving methods for the Boltzmann equation are based on discretization of the density distribution; an alternative is discretizing the velocity distribution, which is referred to as *lattice Boltzmann method* [40]. This method is especially being used in the description of multiphase flow in complex geometries such as porous media or the transport of macroparticles within background flow fields.

In the case of very low pressure, that is, molecular flow conditions, there is no significant interaction between gaseous species. This enables using simplified algorithms, where each particle is treated independently and in parallel. The deposition profile of low-density evaporation processes can, for example, be treated by simple ray tracing of the trajectory of evaporated particles, unless the particle–particle collisions at higher evaporation pressure become important.

While reactor-scale modeling gives valuable insight into the detailed process dynamics, in many cases, it is too slow for performing multidimensional optimization studies or model-based *advanced process control* schemes. Depending on the model complexity, a single model evaluation can take multiple hours or even days. Parametric optimization studies usually involve a large number of model evaluations for slightly modified input parameters, and model-based *advanced process control* (APC) scenarios need response in real time. To solve that discrepancy, global modeling of deposition processes comes into play. Such models treat a whole deposition reactor as black box, whereof externally measurable parameters such as energy and material consumption, current–voltage characteristics, average film thickness and stoichiometry, and so on are represented within a set of simplified balance equations or characteristic maps.

As an example, the detailed deposition profile of a single evaporation point source can be determined experimentally or by reactor-scale modeling in a first step. In a second step, this information can be exploited for homogeneity optimization in a simplified global model of a multisource arrangement, as, for example, demonstrated for a large area fabrication process of copper–indium–gallium selenide (CIGS) solar cells [41]. In another example [42], a wet-chemical alternating ionic layer deposition process of CuS thin films is analyzed by response surface methodology. This approach just approximates the relation between machine parameters and deposition results by a multidimensional polynomial expression, which is trained against an experimental data set by parameter fitting. In the actual example, the response surface approximates the electronic band gap of the deposited CuS layers as a function of pH value, dipping time, and number of dipping cycles. In [43], the deposition flux onto the substrate within a PVD sputter coater for optical coatings is obtained by DSMC simulation in a first step. In a second step, this information is used for optimization of uniformity masks, whereby the impact of the mask geometry on the deposition profile is obtained from the DSMC computed flux profile with a simple integration method, thereby enabling fast optimization of the mask geometry. A prominent global model in PVD is the model of Berg and Nyberg [44] for reactive sputtering, which is based on a binary reaction model between metallic deposition and gettering of reactive gas within an atomic monolayer of the inner chamber surfaces including target and substrate. Based solely on global balance equation, this model qualitatively explains the hysteresis behavior, which is often observed in reactive sputtering process dynamics. For better quantitative agreement with experimental findings, additional physical mechanisms such as reactive ion implantation into the sputtering target, knock-on implantation and redeposition are included in extensions of this model [45, 46]. With such models, realistic reactive sputtering characteristic process curves can be obtained within a few seconds.

From the aforementioned cases, it is clear that in the case of global heuristic models, a model evaluation only takes fractions of a second on a single CPU. This enables using such models within a real-time capable process control and data mining framework as sketched in Figure 2.7.6. The goal of such framework is providing systematic data logging of *in situ* process diagnostic data along with virtual process data predicted by a global model. A deposition reactor is usually controlled by visualization software with recipe control, which is communicating with the SPS device of the machine. By including a logging database and an interface translating recipe control parameters into a parameter case for the global model, it is possible to establish an online model-based process surveillance and quality control, which can be active even in industrial deposition lines. Deviations between model prediction and actual experimental data would be revealed quickly and automatically and would indicate either a potential improvement of the intrinsic global model parameters or a temporary hardware problem within the deposition reactor. This would enable continuous improvement of the involved simulation models and internal model parameter on the one hand. On the other hand, an effective, model-based quality control of deposition processes can be realized.

While reactor-scale modeling enables to predict overall film uniformity, stoichiometry, energy, and material consumption, they cannot predict the intrinsic

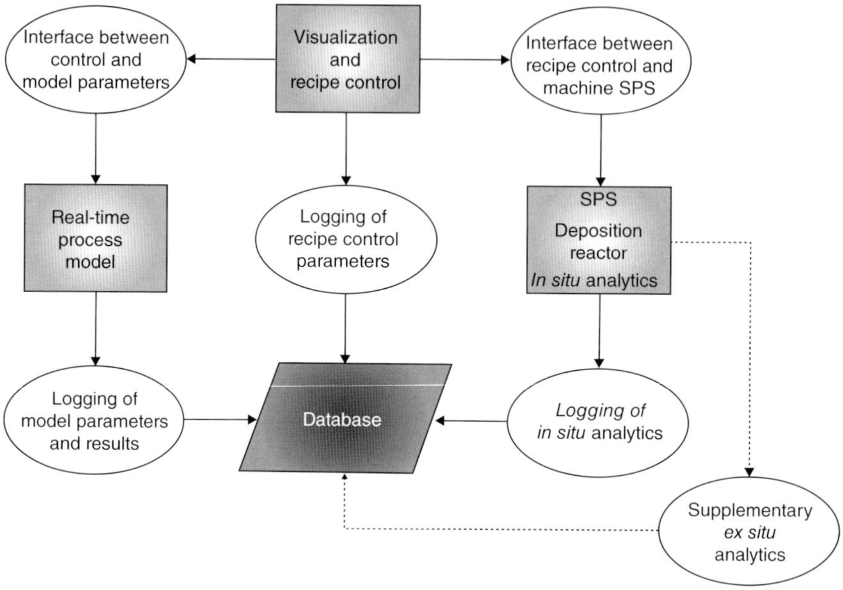

Figure 2.7.6 Concept of model-based data mining of thin-film deposition reactors.

properties of the growing films. Furthermore, many parameters in reactor-scale modeling such as chemical reaction rates, cross sections, and particle emission rates from surface reactions need to be determined. This is where atomistic simulation models of film growth and intrinsic material properties come into play.

As sketched in Figure 2.7.4, atomistic simulation methods can be divided into *ab initio*, quantum chemical methods, semiempirical methods, and classical methods, where parameterized interaction potentials need to be previously determined by the former methods. Atomistic film growth modeling can benefit from reactor-scale modeling, as it provides information about the detailed growth conditions in terms of particle fluxes, energy, and angular distribution functions. If these information are extracted at different substrate locations, multiple atomistic film growth simulations can be run in parallel resulting in a profile of the interesting intrinsic film properties.

Conversely, reactor-scale modeling can benefit from atomistic simulations as they can provide intrinsic model parameters about chemical reactions taking place in the reactor volume or at surfaces. Examples for such parameters are collision cross sections characteristic for gas collisions or plasma–gas interactions. Such parameters can be empirically obtained in beam scattering or swarm analysis experiments [47]; however, in many cases, the data obtained by literature research are incomplete or ambiguous. First tools based on quantum chemistry *ab initio* simulation for computation of reaction rates and cross sections for molecular systems of limited size are nowadays available.

In summary, the integration of atomistic simulation models in thin-film deposition modeling fulfills the following goals:

- Via atomistic film growth simulation the structure and intrinsic properties of films can be predicted, while reactor-scale modeling merely predicts the overall film uniformity and stoichiometry.
- Postevaluation plug-ins allow the determination of macroscopic parameters resulting from the arrangement of a large atomistic ensemble. This includes the band structure, mechanical properties, X-ray diffractometry, and so on, thereby enabling a link between atomistic simulations and experimentally accessible data.
- The interaction between reactor volume and walls can be modeled on atomistic scale in order to reveal reaction parameters of the surface chemistry such as sputtering yield or sticking coefficients.
- The detailed reaction chemistry can be modeled *ab initio* in order to specify reaction rates. The same holds for cross sections of plasma chemical reactions such as electron impact ionization.

If atomistic and reactor-scale simulation models are combined, a virtual coater capable of predicting dynamic process variables as well as intrinsic film properties will be achieved. An example for such concept is given in Figure 2.7.7: In an ion beam sputtering reactor, a sputtering target is bombarded by high-energetic particles originating from an ion source. Sputtered material is emitted from the target and propagates through the vessel together with reactive gas. Due to the

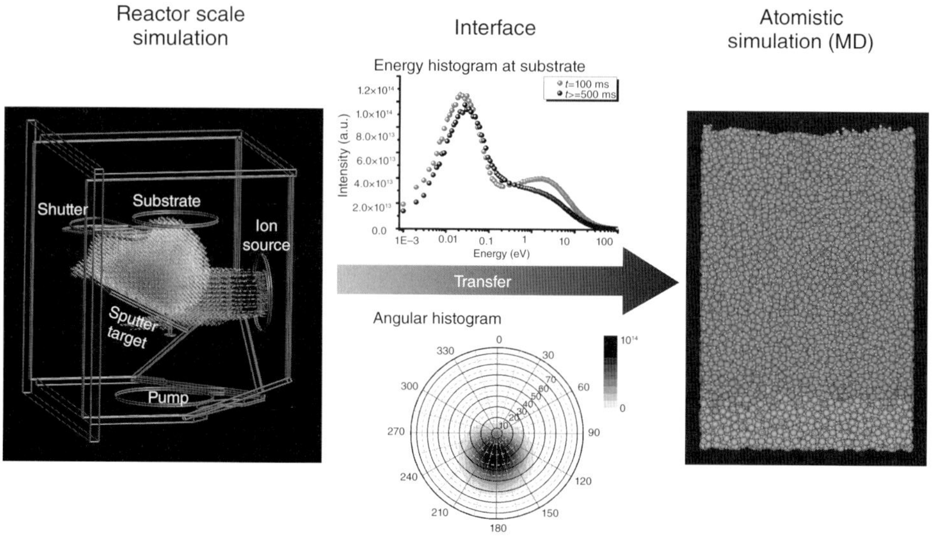

Figure 2.7.7 Example of a combined reactor-scale and atomistic simulation of thin-film growth (adapted from [17]).

rather low pressure in ion beam sputtering, an appropriate modeling method for the particle transport is DSMC. By detailed sampling of the particle flux at the substrate surface, the energy and angular distribution functions can be obtained. Subsequently, atomistic film growth simulations are performed whereof the direction and energy of arriving atoms is randomly generated according to these distribution functions. The atomistic ensemble can be further analyzed in terms of bond configuration and band structure in order to predict, for example, the optical properties of the growing film. Further details on this approach are given in [17]. Within such *virtual coater* concept, the growth conditions can be extracted at multiple positions of interest and accordant atomistic simulation models can be run in parallel. Resulting from this, a multiscale simulation scheme capable of predicting deposition rate, stoichiometry, as well as the structure and quality of deposited films can be realized.

2.7.4 Software Codes

Due to the multiscale nature of thin-film deposition methods, there are a large variety of simulation code categories spanning from global simulation models, reactor-scale models, atomistic film growth, electromagnetic solvers and associated geometric modeling, and postprocessing tools. An overview of the categories by category number is given in the following; the software codes are summarized in Table 2.7.1.

2730 Generalized multiphysics software frameworks
2740 Global modeling of deposition reactors
2741 Global modeling via characteristic maps
2750 Reactor-scale modeling
2751 Transport simulation using computational fluid dynamics (CFD)
2752 Particle-based transport simulation of neutrals
2753 Particle-based transport simulation of charged particles within electromagnetic field
2754 Hybrid particle-based and CFD methods
2755 Alternate methods involving the Boltzmann equation
2756 Electromagnetic field simulation
2757 Detailed gas phase and surface chemistry in CFD
2758 Software for pre- and postprocessing of reactor-scale transport models
2760 Modeling of thin-film growth dynamics
2761 Molecular dynamics (MD)
2762 Kinetic Monte Carlo (kMC)
2763 *Ab initio* QM simulation via Hartree–Fock or DFT
2764 Ballistic computation of collision cascades in sputtering processes

Table 2.7.1 Survey on software codes being used in thin-film deposition.

SW cat	Name of code	Website of software	Website of provider (if different than software website)	Type (academic/ commercial)	Short description
2730	COMSOL	https://www.comsol.de/		C	COMSOL is a commercial multipurpose, multiphysics software framework. In the context of thin-film deposition simulation, it is often being used as FEM-based CFD solver, for example, [48]. By coupling of the CFD solver with electromagnetic field simulation, it is also possible to describe plasma-based deposition processes in COMSOL, for example, [49]
2730	OpenFOAM	http://www.openfoam.com/	http://www.esi-group.com/	A	OpenFOAM is an open-source toolbox, which has been acquired and is further developed by the ESI group. It consists of a large variety of solvers for computational fluid dynamics, structural mechanics, electromagnetic field computation, and also particle-based Boltzmann equation solvers. A comparison of CFD and DSMC computation of the gas flow in a PVD reactor using OpenFOAM solvers is given in [50]. OpenFOAM is distributed open source under GPL v 3
2740	RSD2013	http://www.draft.ugent.be/		A	Heuristic global model of reactive magnetron sputtering, which is based on Berg's model but includes extensions such as reactive ion implantation, current–voltage characteristics, redeposition of sputtered material and can be fitted to experimental process characteristics. A binary version can be downloaded for free. Published results generated with RSD2013 should refer to [45]

(*continued overleaf*)

Table 2.7.1 (*continued*)

SW cat	Name of code	Website of software	Website of provider (if different than software website)	Type (academic/ commercial)	Short description
2741	Design Expert	http://www.statease.com/dx9.html		C	Software implementing the response surface method for the representation of relations between process parameters and quantitative results. Besides experimental optimization via design of experiments (DoE), this can also be used as global modeling representation of deposition processes [42]
2751	CFD ACE+	https://www.esi-group.com/software-services/virtual-environment/cfd-multiphysics/ace-suite/cfd-ace	https://www.esi-group.com/	C	CFD ACE+ is a CFD tool for gas flow and fluid simulation and can be used for modeling of various deposition technologies. Examples for modeling hot-wire CVD, rapid thermal CVD, and spray CVD deposition processes are given in [51, 52]
2751	ANSYS FLUENT	http://www.ansys.com/Products/Simulation+Technology/Fluid+Dynamics/Fluid+Dynamics+Products/ANSYS+Fluent		C	ANSYS FLUENT and ANSYS CFX are two independently developed CFD codes, which are now included into the ANSYS software suite. Both suites implement the CFD equations via finite volume method; CFX uses vertex-based numerics while FLUENT uses a cell-centered approach. A special feature of FLUENT is its extensibility by user-defined subroutines, which enables to extend CFD computation by additional models such as gas-wall interaction or detailed chemical reaction schemes and so on in a flexible way
2751	ANSYS CFX	http://www.ansys.com/Products/Simulation+Technology/Fluid+Dynamics/Fluid+Dynamics+Products/ANSYS+CFX		C	

2752	SIMBAD	http://www.doe.carleton.ca/~tjs/simbad.html	http://www.reactiondesign.com/	C	SIMBAD is a collection of simulation codes used for the description of thin-film growth on three-dimensional structures. While reactor-scale particle transport is realized by ray tracing, the film growth is treated by molecular dynamics according to the angular and energy distribution of arriving species [53]. While the software is originally developed at Carleton University, it is nowadays distributed by Reaction Design, which is part of ANSYS
2752	SIMTRA	http://www.draft.ugent.be/		A	SIMTRA is a binary collision Monte Carlo solver for tracking the trajectories of sputtered particles in configurable reactor geometries and for predicting deposition uniformity profile, energy, and angular distribution of deposited particles [54]. It is developed at University of Ghent. A binary version can be downloaded for free. Published results generated with SIMTRA should refer to [54]
2752	DSMC/DS1V, DS2V, DS3V	http://www.gab.com.au/		A	1D, 2D, and 3D implementation of the DSMC method by G. A. Bird [35]. Executables of the original DSMC Fortran-based codes by G. A. Bird can be downloaded for free together with sample input files. The website further contains documentation about the quantum kinetic model for dissociation and recombination of thermalized molecules as well as the related software "QKrates"
2752	dsmcFoam	http://www.physimmasters.com/wordpress/?page_id=187	http://www.physimmasters.com/wordpress/	A	DSMC software based on the algorithm of G. A. Bird and implemented within the OpenFOAM simulation framework
2752	ICARUS	http://www.sandia.gov/~sjplimp/dsmc.html	http://www.sandia.gov/	A	Open-source 2D DSMC code developed at Sandia National Lab. ICARUS is the predecessor of the parallel 3D code SPARTA

(*continued overleaf*)

Table 2.7.1 (*continued*)

SW cat	Name of code	Website of software	Website of provider (if different than software website)	Type (academic/ commercial)	Short description
2752	MONACO	http://ngpdlab.engin.umich.edu/introduction/monaco	http://ngpdlab.engin.umich.edu/	A	Parallel DSMC implementation for 2D, axial symmetric, and 3D models developed within the Department of Aerospace Engineering of University of Michigan. A copy of the code can be obtained by contacting Prof. Boyd; however, the usage of the code is restricted to within the United States.
2752	PI-DSMC	http://www.pi-dsmc.com/software.html		C	Parallel DSMC package based on the programs DS2V and DS3V by Martin Rose of the PI-DSMC company. Internal model parameters can be viewed and modified during runtime via a DLL interface
2752	SMILE	http://lnf.nsu.ru/en/smile.html	http://lnf.nsu.ru/	C	SMILE (Statistical Modeling in Low-Density Environment) is a parallel DSMC implementation developed at the Laboratory of Non-Equilibrium Flows at Novosibirsk State University
2752	SPARTA	http://sparta.sandia.gov/	http://www.sandia.gov/	A	Open-source DSMC code developed at Sandia National Lab. It is based on the former ICARUS code. It uses a hierarchical Cartesian grid for the DSMC volume cells and a triangulated surface mesh for the reactor geometry. DSMC volume cells may intersect with surface element via a hybrid grid approach
2752/2753	DSMC/PIC-MC	http://www.ist.fraunhofer.de/en/services/simulation.html	http://www.ist.fraunhofer.de	C	DSMC and PIC-MC method developed at Fraunhofer IST. The software uses OpenMPI parallelization and finite element reactor geometries generated by GMSH. DSMC volume cells may intersect with surface element via a hybrid grid approach. Plasma chemical reactions and plasma

					wall interaction can be specified in detail on script level including multiphase reactive wall models based on dynamic surface coverages. A Boundary Element magnetic field solver as well as a parallel distributed Poisson equation solver are included
2753	VSIM (VORPAL)	https://www.txcorp.com/vsim		C	VSIM is a particle motion code combined with a finite difference time domain electromagnetic code suited for plasma simulation. The electromagnetic package includes electrostatic as well as electrodynamic solvers. VSIM is the successor of the former simulation package VORPAL
2754	HPEM, nonPDPsim		http://www.quantemol.com/	C	HPEM stands for "Hybrid Plasma Equipment Model" and is developed at the Computational Optical and Discharge Physics Group (CODPG; Prof. Mark Kushner) at University of Illinois. It is dedicated to low-pressure plasma discharges. The plasma discharge can be capacitively and inductively coupled, also a module for microwave excitated ECR plasma discharges is included. The background gas is modeled via a fluid dynamics approach. HPEM is nowadays being commercially distributed by the Quantemol company. nonPDPsim is a former version including computational fluid dynamics and a Poisson equation solver restricted to electrodynamics
2754	Opera	http://operafea.com/		C	Opera is a collection of electromagnetic field solvers in connection with a FEM-based Maxwell solver for the steady-state distribution of charged particles within electrostatic and magnetostatic fields. With respect to thin-film deposition, it is used for modeling PVD sputtering sources but there are also applications in other technical fields

(*continued overleaf*)

Table 2.7.1 *(continued)*

SW cat	Name of code	Website of software	Website of provider (if different than software website)	Type (academic/ commercial)	Short description
2754	Pegasus	http://www.psinc.co.jp/en/products/index.html		C	Pegasus is a software suite for computing gas flow and plasma dynamics in plasma materials processing and vacuum technology. It comprises particle-based solvers such as DSMC and PIC-MCC, a CFD module for distribution of neutral species and a plasma hybrid module, where electrons are treated by Monte Carlo method and heavy species are treated in a continuum approach. A module for computing magnetic fields is further included. The software suite is available as commercial license
2754	Plasimo	https://plasimo.phys.tue.nl/	https://www.tue.nl/	C	Plasimo is a software collection of various flow solvers and electromagnetic modules dedicated for the description of low-pressure plasma sources. Solvers for electrostatics, inductive coupled, and microwave plasma discharges in combination with various CFD solvers are available. It is being developed and maintained at Eindhoven University of Technology
2754	Quantemol-VT	http://www.quantemol.com/products/quantemol-vt/	http://www.quantemol.com/	C	Quantemol-VT is an expert system for modeling technical plasma discharges. It is based on the "Hybrid Plasma Equipment Model" (HPEM) code developed by Prof. Mark Kushner from University of Michigan. The algorithm is dedicated for nonequilibrium, low-pressure ($p < 1$ Torr = 133 Pa) plasma discharges. Quantemol-VT comes with a user interface, data visualization and analysis as well as job and batch management tools

2754	Starfish	https://www.particleincell.com/starfish/	https://www.particleincell.com/	A/C	Starfish is a simulation package for fluid dynamics, gas kinetics, and plasma physics including both fluid dynamics and DSMC solvers. An electrostatic Particle-in-Cell method is included. The "Starfish Light" version is Java based and can be downloaded for free. Results should make reference to [55]. The full version is being used internally at Particle in Cell Consulting LLC and is not available on a regular basis
2754	USim	https://www.txcorp.com/usim	https://www.txcorp.com/	C	USim is a plasma modeling framework for the description of charged fluids. Structured and unstructured grids can be used within the package. Additional optional modules enable simulation of magnetohydrodynamics (MHD), two-fluid plasmas, and Navier–Stokes equations
2755	BOLSIG+	http://www.bolsig.laplace.univ-tlse.fr/		A	BOLSIG+ is a free program for numerical solution of the Boltzmann Equation in a zero-dimensional model of electrons in weakly ionized gases. These conditions are representative for Swarm analysis experiments of electron scattering in gases. Furthermore, electron transport coefficients for fluid models can be obtained, as described more in detail in [56]. A binary version can be used for free; published results should make proper reference of the software
2755	OpenLB	http://optilb.com/openlb/		A	C++ library implementing the lattice Boltzmann method. The library is released as open source under GPL v2. It is optimized for shared and distributed memory parallelization

(*continued overleaf*)

Table 2.7.1 *(continued)*

SW cat	Name of code	Website of software	Website of provider (if different than software website)	Type (academic/ commercial)	Short description
2756	CST Microwave Studio	https://www.cst.com/Products/CSTMWS	https://www.cst.com/	C	3D Electromagnetic simulation of high-frequency components, which can be used to simulate high-frequency field distribution in RF capacitive discharges, inductive coupled discharges, or microwave discharges
2756	GetDP	http://www.geuz.org/getdp		A	GetDP ("General environment for the treatment of Discrete Problems") is an open-source finite element solver, where partial differential equations can be formulated on script level. It allows for using mixed elements in the FEM simulation space representation. Besides many other purposes, it can be used for solving static and dynamic electromagnetic problems. It interoperates with GMSH, which can be used for mesh generation and visualization. It is released as open source under GPL v2
2757	CHEMKIN	http://www.reactiondesign.com/products/chemkin/		C	CHEMKIN is a software suite for modeling and simulating gas phase and surface chemistry that is used for the conceptual development of combustion systems in cars, trucks, jet engines, boilers, and equipment used in materials processing applications. With respect to thin-film deposition processes, it can be used in connection with CVD reactor-scale simulation. CHEMKIN-CFD can be used as plug-in linked to CFD software such as ANSYS Fluent

2757	DETCHEM	http://www.detchem.com/		A/C	DETCHEM is a software collection for providing detailed gas phase and surface chemistry reactions in connection with CFD simulation models. An example, where DETCHEM is included into a FLUENT CFD simulation of an oxidation reactor is given in [57]. The software collection also includes a few simplified flow models of specific geometries. Some of the tools can be obtained for free for academic purposes
2757	QKrates	http://www.gab.com.au/qkrates.html		A	Software for computing dissociation and recombination rates of thermalized molecules according to the "Quantum Kinetic Chemistry Model" of G. A. Bird [58]
2757	Quantemol-P	http://www.quantemol.com/products/quantemol-p/	http://www.quantemol.com/	C	Quantemol-P is an expert system for generating plasma chemical reaction data for high-pressure plasmas ($p > 0.1$ Torr = 13.3 Pa). It is based on the 0D global kinetics code GLOBAL_KIN developed by Prof. Mark Kushner from Michigan University and embeds it within a graphical user interface with data visualization and analysis capabilities. Quantemol-P can deliver plasma chemical reaction data to the hybrid plasma simulation tool Quantemol-VP within a coupled simulation
2758	GMSH	http://www.geuz.org/gmsh/		A	GMSH is a 3D finite element grid generator with a built-in CAD engine and postprocessor. It can be used for meshing of geometric models as well as postprocessing and visualization of simulation results. It is being distributed under the GPL v2 open-source license. Besides the online documentation, a comprehensive description of GMSH can be found in [59]

(continued overleaf)

Table 2.7.1 (*continued*)

SW cat	Name of code	Website of software	Website of provider (if different than software website)	Type (academic/commercial)	Short description
2761	NAMD	http://www.ks.uiuc.edu/Research/namd/	http://illinois.edu/	A/C	NAMD is a parallel MD code provided by the Theoretical and Computational Biophysics group at University of Illinois [60]. It is specialized in the MD simulation of large biomolecules. There is a free, nonexclusive license for noncommercial purposes available
2761	GROMACS	http://www.gromacs.org		A	GROMACS is a parallel classical molecular dynamics application [61]. It has a good parallel scalability on a high number of CPU cores and is specially designed for biomolecules; however, it can also be used for other types of MD simulation. It is licensed under the GNU Lesser general public license (LGPL) version 2.1
2761	Amber	http://ambermd.org/		C	Amber is a software collection around MD simulation comprising an MD kernel, tools for the preparation of simulation cases and quantitative analysis of results as well as a database of classical MD force fields of biomolecules. Industrial and academic licenses can be obtained. A detailed overview on the software package is given in [62]
2761	CHARMM	http://www.charmm.org/	http://accelrys.com/products/collaborative-science/biovia-discovery-studio/simulations.html	A/C	CHARMM stands for "Chemistry at HARvard Macromolecular Mechanics and is a MD simulation program with a primary focus on the study of biological molecules but can be applied on any kind of species. It provides a variety of tools for postprocessing and sampling. A noncommercial free license can be obtained on request. A commercial version is provided by Biovia; see provider link

2761	GROMOS	http://www.gromos.net/		A	GROMOS stands for Groningen Molecular Simulation computer program package, whereof the development started in 1980 [63]. It is focused on biomolecular simulation but can be also used for other types of molecules. It is released as open source but with a commercial license on a yearly basis
2762	NASCAM	https://www.unamur.be/sciences/physique/pmr/telechargement/logiciels/nascam	http://www.unamur.be	A	NASCAM is a kinetic Monte Carlo code developed at University of Namur. With appropriate activation energies for atom–surface interaction such as diffusion, hopping, evaporation, and coalescence, film growth and thermal annealing can be simulated (see, e.g., [64, 65]). An arbitrary surface texture can be chosen as starting condition and various substrate movement schemes during deposition can be considered
2762	SPPARKS	http://spparks.sandia.gov/		A	SPPARKS is an open-source code for the kinetic Monte Carlo (kMC), rejection kinetic Monte Carlo (rKMC), and Metropolis Monte Carlo (MMC) methods. It is provided under LGPL license. A detailed description can be found in [66]
2763	CPMD	http://cpmd.org/		A/C	The CPMD code is a parallelized plane wave/pseudopotential implementation of density functional theory, particularly designed for *ab initio* molecular dynamics. A noncommercial personal license can be obtained upon registration

(*continued overleaf*)

Table 2.7.1 (*continued*)

SW cat	Name of code	Website of software	Website of provider (if different than software website)	Type (academic/ commercial)	Short description
2763	Dalton, LSDalton	http://www.daltonprogram.org		A/C	Dalton and LSDalton are a software site for computing electronic structure of large molecules via *ab initio* methods and with DFT. A comprehensive description is given in [67]. Free noncommercial licenses are available upon signing a personal license contract
2763	GAUSSIAN	http://gaussian.com/		C	GAUSSIAN is a software package for *ab initio* and DFT simulations for prediction of the electronic structure of molecules
2763	Quantemol-N	http://www.quantemol.com/products/quantemol-n/	http://www.quantemol.com/	C	Quantemol-N computes scattering cross sections between electrons and molecules based on a quantum chemical formalism called R-matrix method [68]. Electron scattering cross sections are essential for performing subsequent reactor-scale plasma simulations, for example, by PIC-MC method
2763	VASP	http://www.vasp.at	http://www.materialsdesign.com/	C	VASP stands for Vienna *ab initio* Simulation Package. It is a software for material modeling on atomic scale via Hartree–Fock approximation, Kohn–Sham equations, or density functional theories. A commercial version of VASP is integrated in the MedeA software environment of Materials Design. A few tools for postprocessing and visualization can be obtained for free

2764	SRIM	http://www.srim.org/	A	Monte Carlo computation of the collision cascades in target materials upon ion impact. The software allows to set up target arrangements consisting of multiple layers of different atom species or species mixtures, which can be bombarded with ions at specified energies. It computes the ion implantation profile, sputter yields, energy loss, and recoiled ions. The theory behind is described in [69]
2764	TRIDYN	https://www.hzdr.de/db/Cms?pOid=21578&pNid=0	C	TRIDYN is a binary collision cascade calculator for description of sputtering processes similar to SRIM. In addition to SRIM, TRIDYN takes into account dynamic changes in the surface composition due to preferential sputtering of single elements. As a result, a surface concentration profile is formed until steady-state conditions are reached [70].

References

1. Thornton, J.A. (1974) Influence of apparatus geometry and deposition conditions on the structure and topography of thick sputtered coatings. *J. Vac. Sci. Technol.*, **11**, 666–670.
2. Löbl, P., Huppertz, M., and Mergel, D. (1994) Nucleation and growth in TiO_2 films prepared by sputtering and evaporation. *Thin Solid Films*, **251**, 72–79.
3. Vergöhl, M. and Bruns, S. (2010) Properties of TiO_2 films deposited by bipolar reactive HiPIMS. *Proceedings of the 8th International Conference on Coatings on Glass and Plastics, ICCG 2010*, pp. 301–305.
4. Arbab, M. (2001) The base layer effect on the d.c. conductivity and structure of direct current magnetron sputtered thin films of silver. *Thin Solid Films*, **381**, 15–21.
5. Hickernell, F.S. (1985) The microstructural properties of sputtered zinc oxide SAW transducers. *Rev. Phys. Appl.*, **20**, 319–324.
6. Sittinger, V., Dewald, W., Szyszka, B., Säuberlich, F., and Stannowski, B. (2010) New approach towards an optimized light trapping morphology of Al-doped ZnO films. Proceedings of the 8th ICCG, pp. 269–272.
7. Szyszka, B., Dewald, W., Gurram, S.K., Pflug, A., Schulz, C., Siemers, M., Sittinger, V., and Ulrich, S. (2012) Recent developments in the field of transparent conductive oxide films for spectral selective coatings, electronics and photovoltaics. *Curr. Appl. Phys.*, **12**, S2–S11.
8. Mattox, D.M. (2003) *The Foundations of Vacuum Coating Processes*, William Andrew Publishing.
9. Kanani, N. (2004) *Electroplating: Basic Principles, Processes and Practice*, Elsevier. ISBN: 978-1-85617-451-0
10. Wasa, K., Kanno, I., and Kotera, H. (2012) *Handbook of Sputter Deposition Technology*, 2nd edn, Elsevier. ISBN: 978-1-4377-3483-6
11. Rees, W.S. (1996) *CVD of Nonmetals*, VCH Publishers. ISBN: 3-527-29295-0
12. Sigmund, P. (1969) Theory of sputtering. I. Sputtering yield of amorphous and polycrystalline targets. *Phys. Rev.*, **184** (2), 383–416.
13. Garrison, B.J. (1989) Energy and angular distributions of atoms sputtered from polycrystalline surfaces: Thompson and beyond. *Nucl. Instrum. Methods Phys. Res., Sect. B*, **40** (41), 313–316.
14. Thompson, M.W. II, (1968) The energy spectrum of ejected atoms during the high energy sputtering of gold. *Philos. Mag.*, **18** (152), 377–414.
15. Hansen, T.N. and Schou, J. (1998) Angle-resolved energy distributions of laser ablated silver ions in vacuum. *Appl. Phys. Lett.*, **72** (15), 1829–1831.
16. Anders, A. (2012) The evolution of ion charge states in cathodic vacuum arc plasmas: a review. *Plasma Sources Sci. Technol.*, **21**, 035014.
17. Turowski, M., Jupé, M., Melzig, T., Moskovkin, P., Daniel, A., Pflug, A., Lucas, S., and Ristau, D. (2015) Practice-oriented optical thin film growth simulation via multiple scale approach. *Thin Solid Films*, **592**, 240–247. doi: 10.1016/j.tsf.2015.04.015
18. Pflug, A., Siemers, M., Schwanke, C., Kurnia, B.F., Sittinger, V., and Szyszka, B. (2011) Simulation of plasma potential and ion energies in magnetron sputtering. *Mater. Technol.*, **26** (1), 10–14.
19. Jäger, S., Szyszka, B., Szczyrbowski, J., and Bräuer, G. (1998) Comparison of transparent conductive oxide thin films prepared by a.c. and d.c. reactive magnetron sputtering. *Surf. Coat. Technol.*, **98**, 1304–1314.
20. Teer, D.G. (1977) Adhesion of ion plated films and energies of deposition. *J. Adhes.*, **8** (4), 289–300.
21. Keunecke, M., Stein, C., Bewilogua, K., Koelker, W., Kassel, D., and van den Berg, H. (2010) Modified TiAlN coatings prepared by d.c. pulsed magnetron sputtering. *Surf. Coat. Technol.*, **205**, 1273–1278.
22. Zhu, Z., Gao, X., Kubart, T., Zhang, Z.-B., and Wu, D. (2012) Conformal Ni-silicide formation over three-dimensional

device structures. *Appl. Phys. Lett.*, **101**, 053508.

23. Matsumoto, S., Sato, Y., Tsutsumi, M., and Setaka, N. (1982) Growth of diamond particles from methane-hydrogen gas. *J. Mater. Sci.*, **17**, 3106–3112.
24. Matsumura, H. (1986) Catalytic chemical vapor deposition (CTL-CVD) method producing high quality hydrogenated amorphous silicon. *Jpn. J. Appl. Phys.*, **25**, L949–L951.
25. Schäfer, L., Höfer, M., and Kröger, R. (2006) The versatility of hot-filament activated chemical vapor deposition. *Thin Solid Films*, **515**, 1017–1024.
26. Hargis, P.J., Greenberg, K.E., Miller, P.A. *et al.* (1994) The Gaseous Electronics Conference radio-frequency reference cell: a defined parallel-plate radio-frequency system for experimental and theoretical studies of plasma-processing discharges. *Rev. Sci. Instrum.*, **65** (1), 140–154.
27. Rech, B., Repmann, T., van den Donker, M.N., Berginski, M., Kilper, T., Hüpkes, J., Calnan, S., Stiebig, H., and Wieder, S. (2006) Challenges in microcrystalline silicon based solar cell technology. *Thin Solid Films*, **511–512**, 548–555.
28. Takatsuka, H., Noda, M., Yonekura, Y., Takeuchi, Y., and Yamauchi, Y. (2004) Development of high efficiency large area silicon thin film modules using VHF-PECVD. *Sol. Energy*, **77**, 951–960.
29. Ge, J., Tang, M., Wong, J., Zhang, Z., Dippell, T., Doerr, M., Hohn, O., Huber, M., Wohlfart, P., Aberle, A.G., and Mueller, T. (2014) Excellent silicon surface passivation achieved by industrial inductively coupled plasma deposited hydrogenated intrinsic amorphous silicon suboxide. *Int. J. Photoenergy*, **2014**, 752967 (12 pp.).
30. Hwang, C.S. (2014) *Atomic Layer Deposition for Semiconductors*, Springer. ISBN: 978-1-4614-8053-2
31. Paatsch, W., Mollath, G., and Ban, A. (2010) Anwendungsorientierte Simulation zur Planung und Produktion maßgeschneiderter, elektrolytisch erzeugter Oberflächen (AnSim), Project Report, http://publica.fraunhofer.de/dokumente/N-238856.html (accessed 5 May 2016).
32. Decher, G., Hong, J.D., and Schmitt, J. (1992) Buildup of ultrathin multilayer films by a self-assembly process: III. Consecutively alternating adsorption of anionic and cationic polyelectrolytes on charged surfaces. *Thin Solid Films*, **210** (211), 831–835.
33. Moustafa, E.M., Dietz, A., and Hochsattel, T. (2013) Manufacturing of nickel/nanocontainer composite coatings. *Surf. Coat. Technol.*, **216**, 93–99.
34. Roths, K., Gochermann, J., Albinsky, K., Bonhoff, K., Borchers, C., Frenzel, R., Kölbl, S., Mock, U., Ondratschek, D., Patyk, A., Pflug, A., Plagemann, P., Schorb, A., Schmidt, B., Speith, S., Stenzel, V., Szyszka, B., Uhlmann, P., Voit, B., and Weissenberger-Eibl, M. (2006) *Forschungsagenda Oberfläche: Analyse des Innovations- und Nachhaltigkeitspotenzials im Bereich der Oberflächenbehandlung*, DFO Service. ISBN: 10 3834912301
35. Bird, G.A. (1994) *Molecular Gas Dynamics and the Direct Simulation of Gas Flows*, Oxford Engineering Science Series, vol. **42**, Oxford University Press, 458 pp.
36. Birdsall, C.K. and Langdon, A.B. (2005) *Plasma Physics via Computer Simulation*, Taylor & Francis, 479 pp.
37. Pflug, A., Siemers, M., and Szyszka, B. (2006) Parallel DSMC gasflow simulation of an in-line coater for reactive sputtering. *Lect. Notes Comput. Sci.*, **4192**, 383–390.
38. Langdon, A.B. (1970) Effects of the spatial grid in simulation plasmas. *J. Comput. Phys.*, **6**, 247–267.
39. Pflug, A., Siemers, M., Melzig, T., Schäfer, L., and Bräuer, G. (2014) Simulation of linear magnetron discharges in 2D and 3D. *Surf. Coat. Technol.*, **260**, 411–416.
40. Huang, H., Sukop, M.C., and Lu, X.-Y. (2015) *Multiphase Lattice Boltzmann Methods: Theory and Application*, Wiley-Blackwell.
41. Patrin, J., Breshahan, R., and Miller, D.L. (2008) Thin film deposition system optimization using flux profile modelling. Proceedings of the 23rd European Photovoltaic Solar Energy Conference, pp. 2607–2609.

42. Yücel, Y., Yücel, E., and Gökhan, D. (2015) Optimization and modelling of preparation conditions of CuS thin films deposited by successive ionic layer adsorption and reaction (SILAR) method using response surface methodology. *J. Mater. Sci. - Mater. Electron.*, **26**, 4105–4112.
43. Pflug, A., Siemers, M., Melzig, T., Rademacher, D., Zickenrott, T., and Vergöhl, M. (2014) Numerical optimization of baffles for sputtering optical precision filters. *Surf. Coat. Technol.*, **241**, 45–49.
44. Berg, S. and Nyberg, T. (2005) Fundamental understanding and modeling of reactive sputtering processes. *Thin Solid Films*, **476**, 215–230.
45. Strijckmans, K. and Depla, D. (2014) A time-dependent model for reactive sputter deposition. *J. Phys. D: Appl. Phys.*, **47**, 235302 (13 pp.).
46. Strijckmans, K., Leroy, W.P., De Gryse, R., and Depla, D. (2012) Modeling reactive magnetron sputtering: fixing the parameter set. *Surf. Coat. Technol.*, **206**, 3666–3675.
47. Phelps, A.V. (1991) Cross sections and swarm coefficients for nitrogen ions and neutrals in N_2 and argon ions and neutrals in Ar for energies from 0.1 eV to 10 keV. *J. Phys. Chem. Ref. Data*, **20** (3), 557–573.
48. Veneroni, A., Omarini, F., Moscatelli, D., Masi, M., Leone, S., Mauceri, M., Pistone, G., and Abbondanza, G. (2005) Modeling of epitaxial silicon carbide deposition. *J. Cryst. Growth*, **275**, e295–e300.
49. Umeozor, C.J. (2013) Computer simulation of the methane plasma enhanced deposition of carbon film. Master thesis. University of Missouri, http://hdl.handle.net/10355/43030 (accessed 5 May 2016).
50. Bobzin, K., Brinkmann, R.P., Mussenbrock, T., Bagcivan, N., Brugnara, R.H., Schäfer, M., and Trieschmann, J. (2013) Continuum and kinetic simulations of the neutral gas flow in an industrial physical vapor deposition reactor. *Surf. Coat. Technol.*, **237**, 176–181.
51. Bouteville, A. (2005) Numerical simulation applied to chemical vapour deposition process. Rapid thermal CVD and spray CVD. *J. Optoelectron. Adv. Mater.*, **7** (2), 599–606.
52. Zhou, J. and Wolden, C.A. (2003) Modeling and measurement of thin film deposition in a one-dimensional hot-wire CVD system. *Thin Solid Films*, **430**, 28–32.
53. Smy, T., Dew, S.K., and Joshi, R.V. (2001) Efficient modeling of thin film deposition for low sticking using a three-dimensional microstructural simulator. *J. Vac. Sci. Technol., A*, **19**, 251–261.
54. van Aeken, K., Mahieu, S., and Depla, D. (2008) The metal flux from a rotating cylindrical magnetron: a Monte Carlo simulation. *J. Phys. D: Appl. Phys.*, **41**, 205307 (6 pp.).
55. Brieda, L. and Keidar, M. (2012) Development of the starfish plasma simulation code and update on multiscale modeling of hall thrusters. Proceedings of the 48th AIAA Joint Propulsion Conference.
56. Hagelaar, G.J.M. and Pitchford, L.C. (2005) Solving the Boltzmann equation to obtain electron transport coefficients and rate coefficients for fluid models. *Plasma Sources Sci. Technol.*, **14**, 722–733.
57. Deutschmann, O., Schmidt, L.D., and Warnatz, J. (1999) Simulation of reactive flow in a partial oxidation reactor with detailed gas phase and surface chemistry models, in *Scientific Computing in Chemical Engineering II* (eds F. Keil, W. Mackens, H. Vofl, and J. Werther), Springer, pp. 368–375.
58. Bird, G.A. (2011) The Q-K model for gas-phase chemical reaction rates. *Phys. Fluids*, **23**, 106101 (13 pp.).
59. Geuzaine, C. and Remacle, J.-F. (2009) Gmsh: a three-dimensional finite element mesh generator with built-in pre- and post-processing facilities. *Int. J. Numer. Methods Eng.*, **79**, 1309–1331.
60. Phillips, J.C., Braun, R., Wang, W., Gumbart, J., Tajkhorshid, E., Chipot, C., Skeel, R.D., Kalé, L., and Schulten, K. (2005) Scalable molecular dynamics with NAMD. *J. Comput. Chem.*, **26** (16), 1781–1802.
61. Berendsen, H.J.C., van der Spoel, D., and van Drunen, R. (1995) GROMACS: a message-passing parallel molecular

dynamics implementation. *Comput. Phys. Commun.*, **91**, 43–56.

62. Salomon-Ferrer, R., Case, D.A., and Walker, R.C. (2012) An overview of the Amber biomolecular simulation package. *Wiley Interdiscip. Rev. Comput. Mol. Sci.*, **3**, 198–210.
63. Christensen, M., Hünenberger, P.E., Bakowies, D., Baron, R., Bürgi, R., Geerke, D.P., Heinz, T.N., Kastenholz, M.A., Kräutler, V., Oostenbrink, C., Peter, C., Trzesniak, D., and van Gunsteren, W.F. (2005) The GROMOS software for biomolecular simulation GROMOS05. *J. Comput. Chem.*, **26** (16), 1719–1751.
64. Godinho, V., Moskovkin, P., Álvarez, R., Caballero-Hernández, J., Schierholz, R., Bera, B., Demarche, J., Palmero, A., Fernández, A., and Lucas, S. (2014) On the formation of the porous structure in nanostructured a-Si coatings deposited by dc magnetron sputtering at oblique angles. *Nanotechnology*, **25**, 355705 (11 pp.).
65. Lucas, S. and Moskovkin, P. (2010) Simulation at high temperature of atomic deposition, islands coalescence, Ostwald and inverse Ostwald ripening with a general simple kinetic Monte Carlo code. *Thin Solid Films*, **518**, 5355–5361.
66. Plimpton, S., Battaile, C., Chandross, M., Holm, L., Thompson, A., Tikare, V., Wagner, G., Webb, E., Zhou, X., Garcia Cardona, C., and Slepoy, A. (2009) Sandia Report SAND2009-6226, http://spparks.sandia.gov/pdf/sand09.pdf (accessed 5 May 2016).
67. Aidas, K., Angeli, C., Bak, K.L., Bakken, V., Bast, R., Boman, L., Christiansen, O., Cimiraglia, R., Coriani, S., Dahle, P., Dalskov, E.K., Ekström, U., Enevoldsen, T., Eriksen, J.J., Ettenhuber, P., Fernández, B., Ferrighi, L., Fliegl, H., Frediani, L., Hald, K., Halkier, A., Hättig, C., Heiberg, H., Helgaker, T., Hennum, A.C., Hettema, H., Hjertenæs, E., Høst, S., Høyvik, I.-M., Iozzi, M.F., Jansik, B., Jensen, H.J.A., Jonsson, D., Jørgensen, P., Kauczor, J., Kirpekar, S., Kjærgaard, T., Klopper, W., Knecht, S., Kobayashi, R., Koch, H., Kongsted, J., Krapp, A., Kristensen, K., Ligabue, A., Lutnæs, O.B., Melo, J.I., Mikkelsen, K.V., Myhre, R.H., Neiss, C., Nielsen, C.B., Norman, P., Olsen, J., Olsen, J.M.H., Osted, A., Packer, M.J., Pawlowski, F., Pedersen, T.B., Provasi, P.F., Reine, S., Rinkevicius, Z., Ruden, T.A., Ruud, K., Rybkin, V., Salek, P., Samson, C.C.M., Sánchez de Merás, A., Saue, T., Sauer, S.P.A., Schimmelpfennig, B., Sneskov, K., Steindal, A.H., Sylvester-Hvid, K.O., Taylor, P.R., Teale, A.M., Tellgren, E.I., Tew, D.P., Thorvaldsen, A.J., Thøgersen, L., Vahtras, O., Watson, M.A., Wilson, D.J.D., Ziolkowski, M., and Ågren, H. (2014) The Dalton quantum chemistry program system. *WIREs Comput. Mol. Sci.*, **4**, 269–284. doi: 10.1002/wcms.1172
68. Tennyson, J., Brown, D.B., Munro, J.J., Rozum, I., Varambhia, H.N., and Vinci, N. (2007) Quantemol-N: an expert system for performing electron molecule collision calculation using the R-matrix method. *J. Phys. Conf. Ser.*, **86**, 012001 (12 pp.).
69. Behrisch, R. and Eckstein, W. (2007) *Sputtering by Particle Bombardment*, Springer-Verlag. ISBN: 987-3-540-44500-5
70. Möller, W., Eckstein, W., and Biersack, J.P. (1988) TRIDYN – Binary collision simulation of atomic collisions and dynamic composition changes in solids. *Comput. Phys. Commun.*, **51**, 355–368.

2.8 Machining

André Teixeira, Markus Krömer, and Roland Müller

This chapter provides an overview of common machining processes and approaches for assessing significant process values. These approaches include analytic–geometric simulation models and FEM models for calculating relevant values such as undeformed chip geometry, resulting workpiece geometry, stresses, temperatures, machining process dynamics, and surface integrity properties.

2.8.1 Introduction to Machining Processes

Machining is the name given to the controlled processes that mechanically remove small parts of material from a raw workpiece to achieve a desired geometry. This fundamental process for geometrically defined cutting-edge tools can be easily understood in an orthogonal view being represented in Figure 2.8.1, as well as the primary, secondary, and tertiary shear zones, regions where physical and, occasionally, chemical phenomena occur. These zones are the primal reason for the cutting resistance, responding to the process parameters and the interaction tool–workpiece with opposite forces to the cutting movement.

For processes with undefined cutting-edge geometry tools, the cutting logic is, in part, the same (Figure 2.8.2). As the tool consists of a combination of abrasive grains, bonding material, and pores, it can be compared with a tool with a vast amount of small cutting edges generating consequently a great number of small chips. These processes are mainly used in operations that require a high surface finish quality and geometrical precision.

The generated forces in machining processes are inherent in the mechanical resistance of the workpiece material when acting against the separation of its atomic planes, summed up to the frictional effect between the pair tool–workpiece and the sheared planes. These forces are represented in Figure 2.8.3, illustrated in a turning process.

Handbook of Software Solutions for ICME, First Edition. Edited by Georg J. Schmitz and Ulrich Prahl.

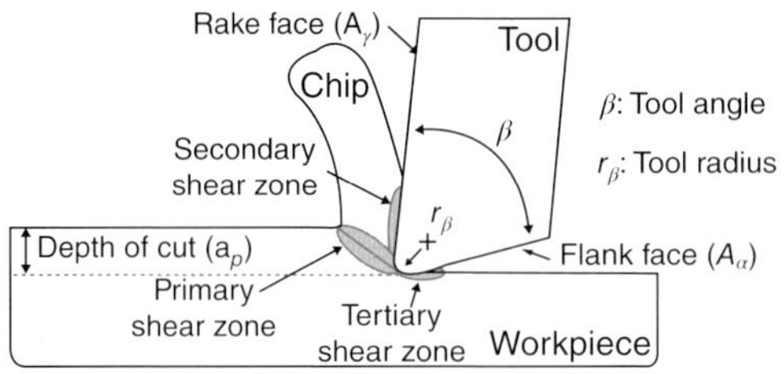

Figure 2.8.1 Representation of an orthogonal cutting process with defined tool edge.

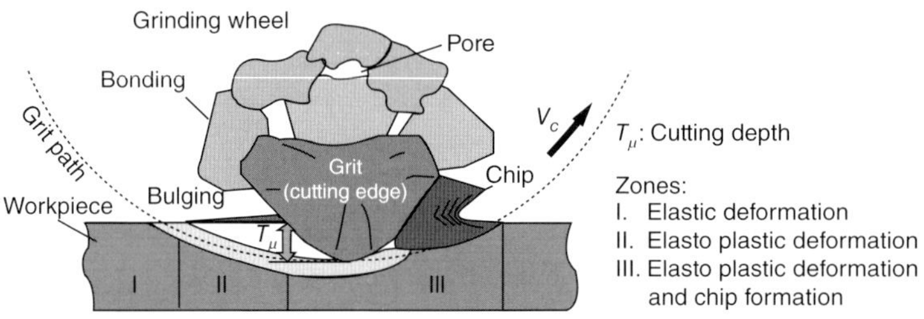

Figure 2.8.2 Representation of an orthogonal cutting process with undefined tool edge.

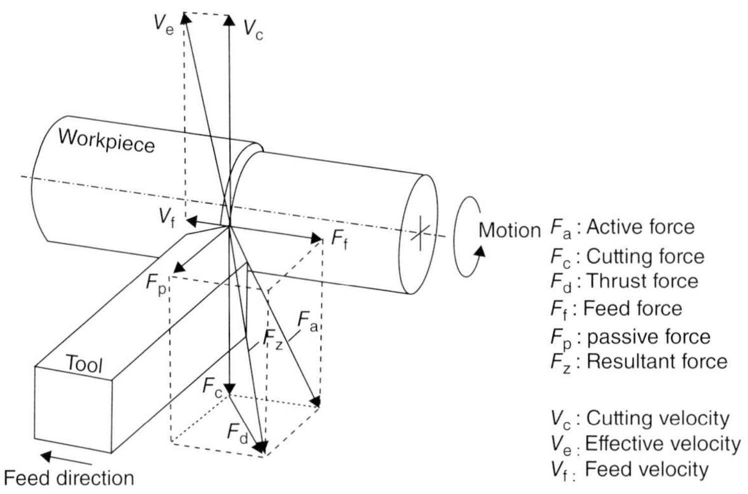

Figure 2.8.3 Cutting forces in a turning process.

Moreover, during the cutting, due to the friction in the three shear zones, considerably high heat loads are generated, increasing the system temperatures up to 1000°C (Figure 2.8.4). The control of the temperatures is extremely important in order to optimize the tool lifetime and avoid undesired property modifications in the workpiece surface integrity, such as crystalline phases, grain size distribution, and residual stress. In undefined tool edge processes, the heat generation must

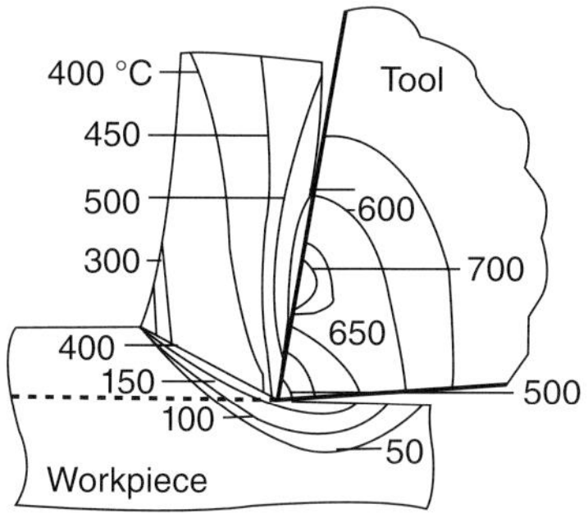

Figure 2.8.4 Temperature distribution in an orthogonal cutting process.

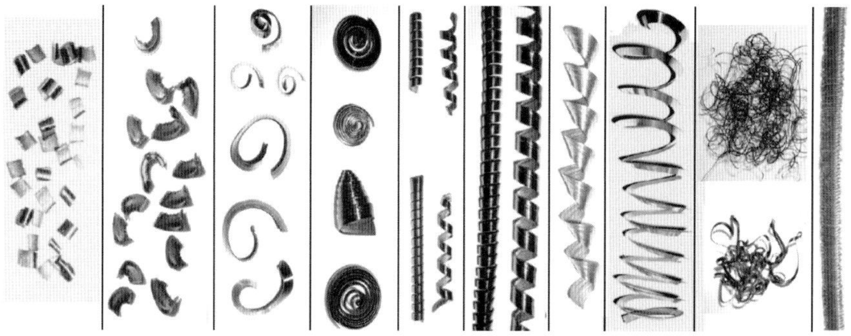

Figure 2.8.5 More favorable to less favorable chip geometries (from left to right) [1].

also be carefully analyzed; otherwise, when achieving a very high temperature, the bonding might be damaged.

The cut chip geometry (continuous, lamellar, segmented, or discontinuous), as can be seen in Figure 2.8.5, varies according to the machining process, cutting parameters, tool geometry, and material machinability. Its consideration in projects has mainly economic reasons, that is, to avoid tool/surface damages from the chip returning to the cutting system and to prevent production idle times stopping the machine. In general, small chips are considered favorable due to the improved evacuation from the cutting region. To augment the removal process and to promote the heat dispersion, a coolant fluid jet aiming at the cutting zone can be used, depending on the tool and the workpiece materials and the amount of generated heat.

There are two distinct cutting scenarios:

Roughing: A large quantity of material is removed in a short period of time without high surface quality requirements, in order to prepare the workpiece geometry for a following manufacturing process.

Finishing: It is characterized by a small material removal rate, having the focus on the dimensional tolerances and surface roughness.

By a combination of material properties and cutting velocity, summed up with other minor cutting parameters, a generation of built-up edge on the tool rake face

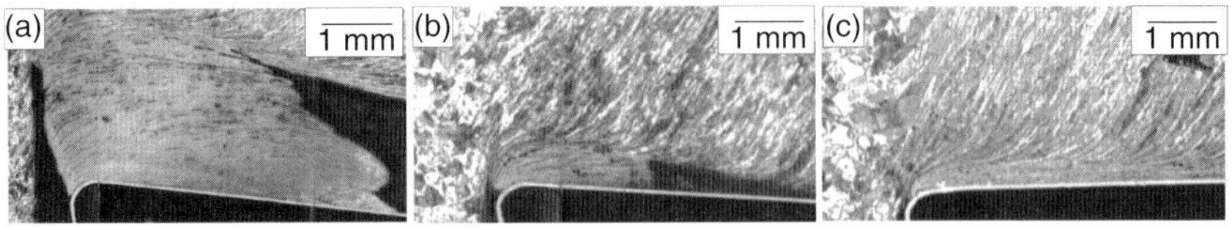

Figure 2.8.6 Built-up edge formation with different cutting velocities: (a) 20 m/min, (b) 30 m/min, and (c) 40 m/min.

during machining can come about (Figure 2.8.6). This has negative consequences for the surface quality values and reduces the tool lifetime due to excessive forces and temperatures. Thereby, since the surface integrity is not relevant in roughing, this phenomenon is admissible, but it must be avoided in finishing.

Manufacturing processes involving defined edge machining can be classified according to their main cutting movement: rotational (e.g., turning, milling, drilling, and sawing) or translational (e.g., planning, shaping, broaching, rasping, and scraping). This division is established based on the cutting concept of each process and groups them basically for studying purposes, promoting a better understanding of similar processes and facilitating mathematical modeling and simulations.

Due to the flexibility and economic viability, three manufacturing processes cover the major requirements of machining projects in the industry:

1) Turning: Removes the material from a rotating workpiece producing cylindrical final forms (Figure 2.8.7a). Variations can occur with different tool geometries and with the addition of one more degree of freedom in the tool holder in specific cases.
2) Milling: A cutting tool positioned in a spindle, with a determined number of cutting edges, advances sideway or diving to the workpiece producing final geometries with a wide variety, particularly in five-axis machines (Figure 2.8.7b).
3) Drilling: With the single intention of making holes, a tool is positioned in a spindle vertically advancing through the surface of the workpiece (Figure 2.8.7c).

There are also distinctions between processes with undefined tool edge geometry processes (e.g., grinding, honing, or tape finishing). The most common process in industries and scientific studies is grinding, varying the tool positioning in relation to the workpiece (Figure 2.8.8).

In particular, in the last few years, with significant increase of the research publication quantity around the world, machining processes are more than ever being improved due to the integration of technologies increasing the range of manufacturing possibilities; for example, a laser can be used for the application in an area just before the cut to soften the material, decreasing the required cutting

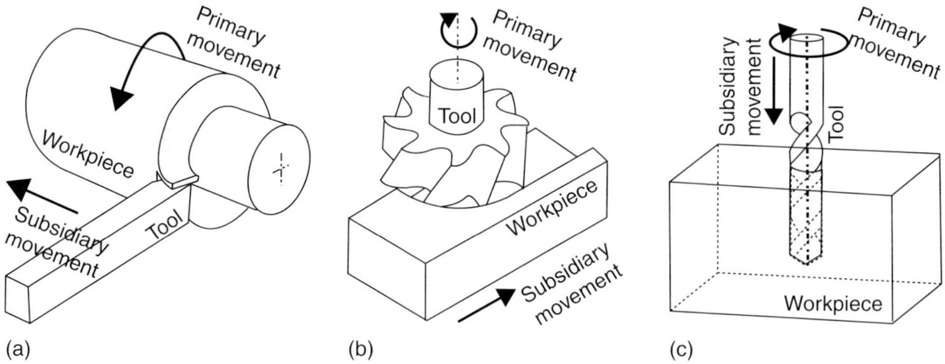

Figure 2.8.7 Most common machining processes in the industry: (a) turning, (b) milling, and (c) drilling.

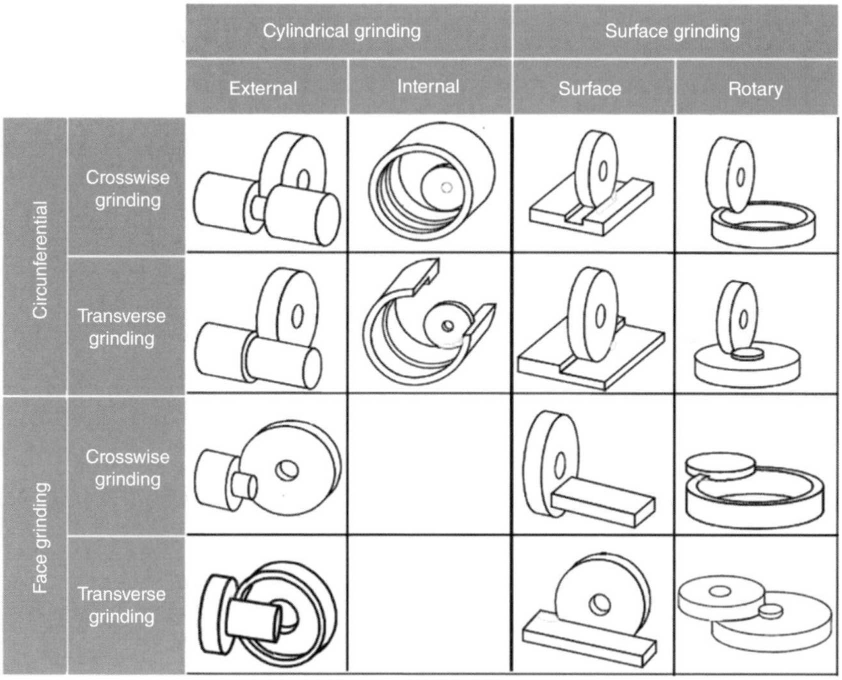

Figure 2.8.8 Different grinding techniques [2].

forces and increasing the tool lifetime – even advanced ceramic materials can be cut. Cryogenic cooling uses liquid nitrogen (at −196°C), which is applied directly to the cutting edges to provide benefits such as significant tool lifetime increase, reduction of surface distortion, white layer, and burr formation. Besides that, the process speed is increased, and it is considered environmentally friendly. Water jet cutting uses high-pressure water with carbides to cut workpieces in a pregeometry. The final geometry is achieved by a regular machining process. It has benefits such as an improvement of the process speed and no heat-affected zones, and it can also be considered as environmentally friendly.

2.8.2
General Aspects of Machining Simulations

State of the art in the modeling of machining processes is the simulation of processes with a defined cutting edge. There are two main approaches for the development of simulation models for cutting processes. On the one hand, there are analytic–geometrical models for simulating the machining process, and on the other hand, numerical simulations based on finite element method (FEM) models exist.

2.8.2.1
Analytic–Geometric Simulation Models

The analytic–geometric simulation models are able to calculate the resulting workpiece geometry and the geometry of the undeformed chips. These results can then be analyzed and later used to develop characteristic values. A prominent example for simulation programs based on this approach is the computer-aided manufacturing (CAM) system. These software products are used for a fast simulation and evaluation of machining processes in five-axis milling as well as turning machines. The aim of such a simulation is a collision check as well as the generation and optimization of the machine code to describe the axis movements. Common CAM software products include Siemens NX CAM, Open Mind hyperMILL, and CosconProfiCAM among others. All of these systems can simulate $2\frac{1}{2}$-, 3-, and 5-axis milling processes and calculate the resulting workpiece geometry. CAM software often uses geometrical models such as voxels, dexels, or surfaces. A voxel represents a value on a regular grid in three-dimensional space (Figure 2.8.9). Voxel is a portmanteau for "volume" and "pixel." The accuracy of voxel models depends on the size of the voxels, resulting in a minimal accuracy of one voxel in each direction in space. For describing machining processes, which generate a surface, dexel or surface models are common. Dexel is a portmanteau of "depth" and "pixel." The minimal accuracy of a dexel model is defined in two dimensions by the grid structure. The third dimension is represented by length

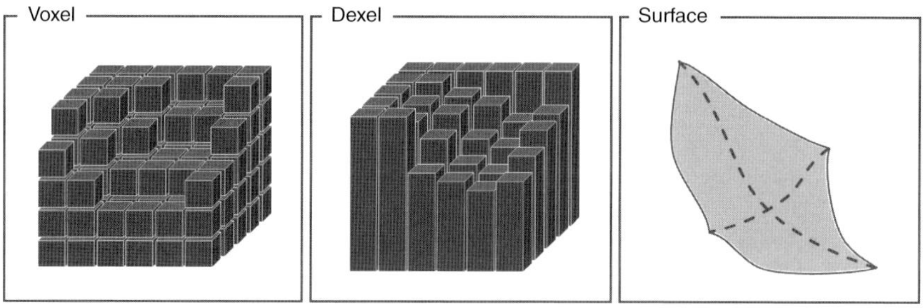

Figure 2.8.9 Voxel, dexel, and surface models.

or depth of the dexel, which results in a much higher resolution in this dimension. Besides dexel models, surfaces can be described with the help of polynomials. These surface models or sheet bodies have the highest accuracy of all mentioned models, but the virtual machining requires calculating the intersections of these polynomials, which complicates the simulation calculation.

In the academic sector, often geometrical penetration calculations based on planar intersections are used. An example for simulating the machining of cylindrical gears by the continuous gear hobbing process is SPARTApro, which is currently developed at the WZL of the RWTH Aachen University. The software uses planar planes of workpiece and tool and positions these according to the given kinematics. Later, the splines that represent the outline of workpiece and tool are intersected, and the resulting workpiece geometry as well as the chip geometry is determined. Figure 2.8.10 shows an exemplary approach for developing a geometrical penetration calculation. First, the input data such as workpiece and tool geometry has to be prepared. Therefore, the workpiece and tool will be discretized in the required accuracy. Second, the penetration calculation can be performed. The tool is positioned according to the kinematics, and the intersection of tool and workpiece is calculated. Next, the new workpiece and the resulting chip geometry are computed. As the final step in the penetration calculation, a repositioning of both tool and workpiece is performed, and the subsequent cut is simulated. The results of a geometrical simulation are the geometries of all undeformed chips occurring in the machining process. These can be analyzed and characteristic values such as maximum and mean chip thickness, cutting length, and specific volumes can be calculated.

All mentioned simulations based on analytic–geometrical models can be used to analyze, optimize, or design a machining process. With the characteristic values described in the last pattern, it is also possible to describe and examine the cause of tool wear, but a profound investigation of tool wear development is not achievable because important effects such as deformation of chips, workpiece, and tool as well as temperature at the cutting edge and stresses are neglected. To simulate stresses and temperature occurring in the machining process, FEM simulations are used.

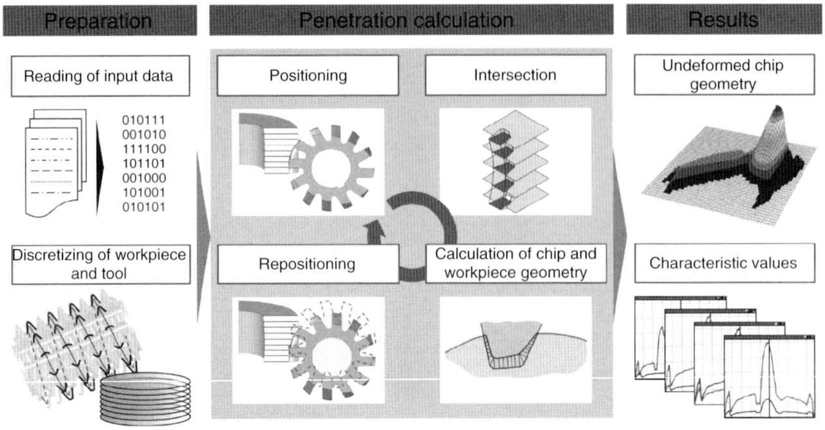

Figure 2.8.10 Process of a geometrical simulation based on planar intersections. (WZL of RWTH Aachen University.)

2.8.2.2
Finite Element Method Simulation Models

If the stresses, temperatures, or dynamics of machining process have to be calculated, the FEM is the preferred simulation method. To avoid mistakes in developing simulation models, it is helpful to know the general concept of FEM. Each finite element analysis (FEA) consists of five steps:

1) Discretization of the continuum
2) Selection of the interpolation functions
3) Defining the characteristics of the elements
4) Composition of the elemental equations
5) Solving the defined equation system

While discretizing the continuum, for example, workpiece and tool, the type, size, number, and distribution of the finite elements are defined. Often the interpolation functions are selected in the same step. The functions are used to describe the behavior of the variables within the finite elements. Later, the characteristics of the influence quantities are defined with the help of equations. These elemental equations are then combined to equation systems that are solved in the last step.

Different approaches for discretizing the continuums are used. The two most common ones are the approaches according to Lagrange as well as to Euler (also see Chapter 7). While the finite elements in the approach of Lagrange are deformed with the material, the approach of Euler uses stationary elements. Because of these differences, in most machining simulations the approach according to Lagrange is used, and for stationary processes such as flow simulations, the

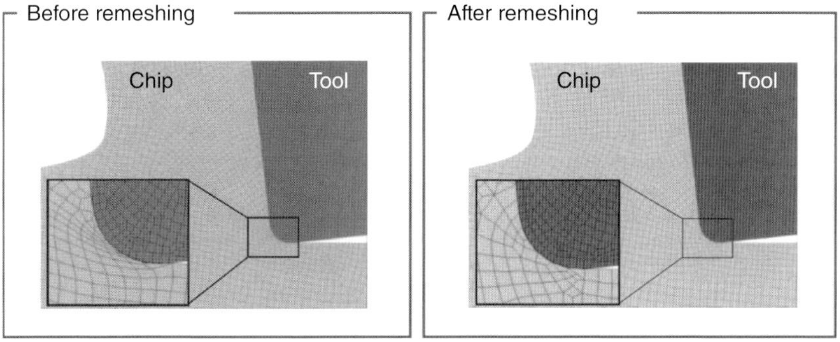

Figure 2.8.11 Necessary remeshing of FE mesh grid after critical deformation.

common method is to use stationary elements. In addition, it is also possible to use a combination of both approaches. Commercial software solutions for the FE machining simulation are, for example, DEFORM or Abaqus FEA. Besides that, open-source software for FEA such as CAELinux exists.

The simulation of cutting process in an FEA can be performed in two- and three-dimensional models. In both models, a separation criterion for the workpiece material has to be defined. In the most commonly used FE approach for machining, the mentioned approach according to Lagrange, the criterion can be geometrical or physical. In the geometrical criterion, the separation is performed as soon as the distance between the tool cutting edge and an FE node is below a given limit. In the case of the physical criterion, a separation of chip and workpiece will take place if a predefined equivalent stress is exceeded. In both ways, it is necessary to define the limit for the criterion as well as the line or, in the case of a three-dimensional model, the plane in which the separation of the material will be performed. This line or plane of action is determined by the kinematics of the machining process. In addition, it is possible to renounce any separation in the mesh grid. In this case, a continuous remeshing of the grid is necessary. After each time step in the FEA, the deformation of the grid is analyzed, and if a predefined limit of deformation is exceeded, a remeshing algorithm is executed. Figure 2.8.11 shows the two-dimensional grid of a cutting simulation before and after remeshing. As it is depicted, after remeshing the elements of the FEM are nearly equal in size and orientation, whereas before remeshing they were stretched and orientated in one direction. During remeshing, it is important that the outline of workpiece and tool be represented correctly by the nodes of the new grid. An easy technique to fulfill this requirement is to use the outline nodes of the old mesh as the outline in the remeshing algorithm as well. Also, it is important that the variables of the FEA be properly interpolated to new nodes; therefore, the predefined interpolation functions are used.

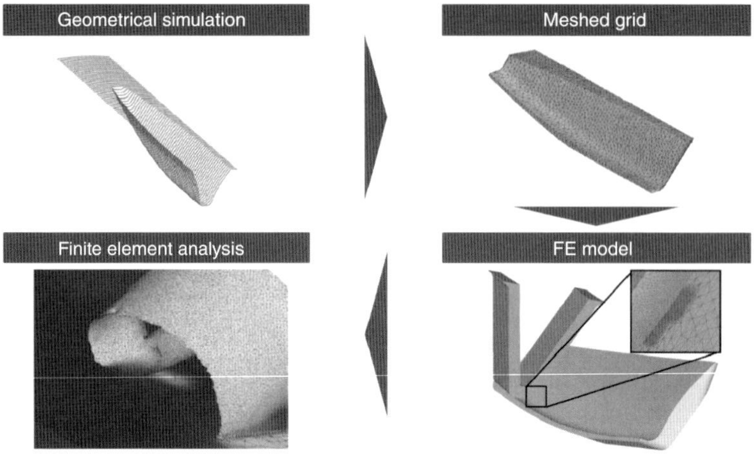

Figure 2.8.12 Combination of geometrical penetration calculation and FEA.

2.8.3
Combination of Analytic–Geometric Simulation Models and FEM Simulation Models

For simulating complex machining processes, a combination of a fast geometrical penetration calculation and a more complex but more precise FEA can be used. To combine these two simulation methods, at first the geometry of the workpiece and the tool at a defined time has to be calculated by the geometrical calculation. To do so, a specific time in the machining process is defined and the geometry and positioning of tool and workpiece is computed. Also, it is important to obtain the kinematics at this time of the process, which describes the path of action in the FEA. Then, the resulting geometry of workpiece and tool must be converted into a meshed grid, which can be used by the FE software. Figure 2.8.12 shows in the first image the geometry of a gap of a spur gear, which is calculated in the software SPARTApro. As it is shown, the gap is described by points that are defined in parallel planes. In the FEA, the chip formation on one flank should be analyzed in detail. In order to do so, half of the gap is meshed (second image in Figure 2.8.4). Then, the tool is positioned to the workpiece (image 3) and simulation of a chip formation is performed (image 4). This can be repeated for specific cutting positions, which are determined to be important by the geometrical simulation. Despite the need of designing two different simulations, the overall simulation time can be reduced drastically. While a geometrical simulation takes seconds to minutes, an FE simulation of a chip formation can take weeks. Because of that, it is not practical to analyze the full machining process in an FEA.

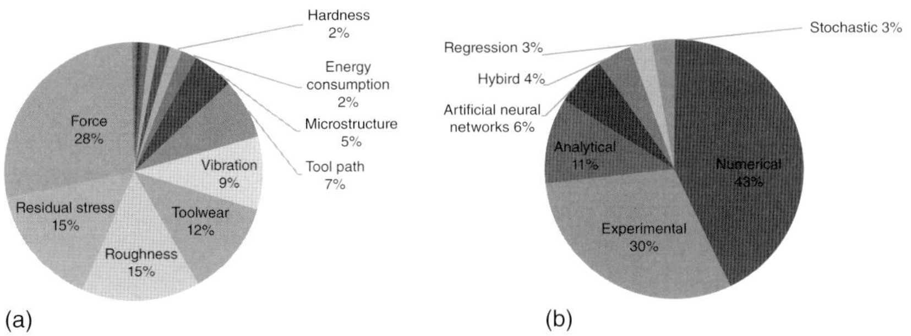

Figure 2.8.13 Approximated percentage of articles about machining simulation: (a) Machining model main topics and their (b) machining model methods.

2.8.4 Simulation of Surface Integrity Modifications

In the beginning of the scientific study of machining processes, modeling of surface integrity modifications in machining was not possible. This was due to economic factors and limited computational capabilities. Nowadays, with the computational resources significantly increased, there is a great diversity of models with different predictions of properties, such as force, residual stress, roughness, hardness, tool wear, vibration, more adequate tool path to minimize forces or to reduce operation time, recrystallization and grain size redistribution, built-up edge and burr formation, tool geometry optimization, cooling strategy and efficiency, and effects of the combination of machining processes with other techniques as laser or cryogenic (Figure 2.8.13a). The methods to model machining processes are diverse: numerical, experimental, analytical, artificial neural networks, stochastic, regression, and, more recently, hybrid models, combining more than two methods (Figure 2.8.13b).

Depending on the simulated property and the method used, a special attention can be given to the material behavior. Most of the time, analytical and experimental models are not as complex as numerical models. Between these models, the FEM is highlighted as very powerful and robust to solve complex simulations, being able to use advanced material models. The Johnson–Cook model (Eq. 2.8.1) is actually the most used material model in machining, and very often it is adapted with empirical or experimental observations to improve the precision of the behavior description. The model equation describes the material behavior dividing it into three parts: plastic behavior, strain rate sensitivity, and thermal softening. Their coefficients A, B, C, n, and m can be acquired by a specific test called Split–Hopkinson pressure bar. Anisotropic materials, as composites, are still not well described, needing complex adaptations or specific material models:

$$\delta = (A + B\varepsilon^n)\left(1 + C\ln\frac{\dot{\varepsilon}}{\dot{\varepsilon}_0}\right)\left(1 - \left(\frac{T - T_1}{T_m - T_f}\right)^m\right) \quad (2.8.1)$$

Depending on the chosen machining simulation, the influences on the surface integrity of the workpiece can vary. Simulations involving the tool wear and tool lifetime most of the time ignore the effects on the workpiece, as the focus is on the tool. On the contrary, simulations exist where the focus is not on the workpiece surface integrity, such as force, uncut chip geometry, vibration, ideal tool path, and electrical energy consumption of the machine.

The surface integrity study in machining can be basically divided into two parts: simulations and the phenomena that affect the surface roughness and/or the subsurface microstructural properties. Depending on the model approach, the surface integrity can be taken into consideration or not, for example, while one force model is purely analytical and considers primarily the uncut chip geometry, another force model can consider the material phases and the effects of the mechanical loads on the workpiece surface integrity.

The first machining models focused on the solving of problems involving processes forces, avoiding premature tool failures, damage of the workpiece, and also of the machine. In a second development stage, the surface roughness started to be taken into consideration, as it has direct influence on the final product quality and performance in properties, such as wear resistance and wear rate, friction coefficient, lubrication, fatigue strength, and corrosion resistance. Surface roughness simulations can be divided in two groups – exclusively geometrical simulations, considering the tool radius (r) and the cutting feed (f), as shown in Eq. (2.8.2), and those with microconsiderations from tribological phenomena, as adhesion and abrasion, mainly:

$$h = \frac{f^2}{8r} \quad (2.8.2)$$

As the phenomena behind the variations of the microconsiderations of surface roughness are not well described in the scientific literature, experimental models are very common. Regressions and artificial neural network are used that require a lot of experimental testing. Simulations of the contact between the tribological pair of tool and workpiece and the resulting wear phenomena are very complex and require the use of FEM or molecular dynamics approaches.

Recently, with the fast augmentation of computer process capabilities in the last 20 years, complex simulations considering the material behavior in cutting were improved. In particular, FEM simulations in this area are highlighted due to their performance, but they also bring the disadvantage of long computation times. Modification phenomena of subsurface properties occur, in general, up to 1000 μm below the workpiece surface and are affected by the thermomechanical loads of the cutting process (Figure 2.8.14). Some examples of the simulation possibilities comprise the following:

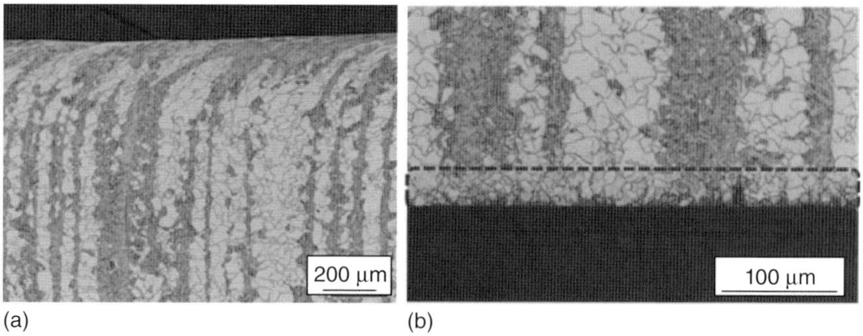

Figure 2.8.14 (a) Microstructure deformed by machining effect; (b) reduction of grains' size and distribution change due to recrystallization effect.

1) Phase transformation: generates completely new phases on a certain depth range.
2) Phase gradients change: variations of the existing phase proportions, but no new phase is generated.
3) Grain size distribution: the surface crystals are "cut in half" and deformed, reducing the size, also when achieving a very high temperature, due to recrystallization, a fast nucleation and growth of the grains occurs – as the nucleation and growing time of the grains are too small, there is a tendency to have smaller grains near the surface.
4) Oxidation: a very thin oxide layer can be generated due to contact with the atmosphere or cutting fluid.

Models are rarely so complex that they take into consideration all of the material properties. The best example of a machining simulation that requires complex models, connecting machining effects and surface properties, is a residual stress simulation. Complex phenomena are highly dependent on the workpiece process chain, and, consequently, closely connected to the material's history and its properties, such as phases, phase distributions, grain size, presence of carbides, or defects such as inclusions and microcracks. Therefore, residual stress has an influence in many other properties, such as wear, friction, corrosion resistance, fatigue strength, coating integrity, and geometric deformation. The detailed simulation of machining processes thus provides an impressive example for the potential of ICME-type approaches.

In the recent years, a tendency in mapping the processes based on the thermomechanical loads generated in the cutting is observed (Figure 2.8.15).

A combination of the material properties and the cutting parameters is responsible for the process identity: forces and temperatures from tool and workpiece surface interaction. Based on the tool geometry and contact conditions, the intensity and geometry of the thermomechanical loads can be designed. These loads can be applied on a workpiece surface substituting conventional cutting simulations, having no chip formation, significantly reducing calculation time.

Figure 2.8.15 Turning simulation showing the application of thermomechanical loads in four passes.

This point summed with the confidence in the results can be considered the biggest barrier for machining simulation being massively adopted in industries.

2.8.5 Summary

Machining is still one of the industrial fields with more potential to be technologically advanced regarding ICME. In general, most of those studies involving simulations have a small comprehension lack between the mechanical processes and the materials science. The simplification of the models is one statement to be surpassed in the future with the advance of more efficient simulation techniques and computer science development. A complete simulation of real processes englobing different models or linking them is necessary for a macrocomprehension of the final properties of a manufactured product. A milling simulation to find the best tool path with the optimal residual stress condition is one example of how the interaction of models is needed, mixing force, residual stress, and machine models. Having in mind the development of the last 20 years and tracing a projection for the near future, machining simulation has a huge potential to be explored inside ICME approach, with extremely significant industrial and scientific benefits that can be accomplished for the improvement of production.

2.8.6 Simulation Tools for Machining Processes

Categories: 2810 (general solver), 2820 (general machining processes), 2830 (machining parameters optimization), and 2840 (software solutions for gearing components)

SW cat	Name of code	Website of software	Type of code (academic/commercial)	Short description
2810	ABAQUS	www.3ds.com/products-services/simulia/products/abaqus	Commercial	FEM analysis and CAE
2810	Deform	www.deform.com	Commercial	Analyzes the 3D flow of complex metal forming
2810	ANSYS	www.ansys.com	Commercial	Multiphysics simulations, highlighted in fluid dynamics
2810	COMSOL Multiphysics	www.comsol.com	Commercial	General-purpose software platform, based on advanced numerical methods, for modeling and simulating physics-based problems
2810	LS-DYNA	www.lstc.com	Commercial	General-purpose finite element program capable of simulating complex real-world problems
2810	Marc	www.mscsoftware.com/product/marc	Commercial	Nonlinear finite element analysis solution to simulate the response of products
2820	AdvantEdge	www.thirdwavesys.com/advantedge	Commercial	Optimized FEM tool of metal cutting
2830	NX CAM Software	www.plm.automation.siemens.com/en_us/products/nx/for-manufacturing/cam	Commercial	Solutions for machine tool programming, postprocessing, and machining simulation
2830	hyperMILL	www.openmind-tech.com/en/products/hyperMILL-cam-software	Commercial	Development of machining strategies for simple and complex geometries

SW cat	Name of code	Website of software	Type of code (academic/commercial)	Short description
2830	CosconProfiCAM VM	www.coscom.de/produkte	Commercial	Improvement of machining parameters
2840	GearGenerator	www.wzl.rwth-aachen.de	Commercial	Continuous generation grinding of gears
2840	KegelSpan	www.wzl.rwth-aachen.de	Commercial	Manufacturing process for bevel gears
2840	Kraftkoppelelement	www.wzl.rwth-aachen.de	Commercial	Three-dimensional tooth contact analysis
2840	SPARTApro	www.wzl.rwth-aachen.de	Commercial	Designing of gear hobbing processes
2840	FE-Stirnradkette (STIRAK)	www.wzl.rwth-aachen.de	Commercial	Gear design based on simulated mechanical loads
2840	WZL Gear Toolbox	www.wzl.rwth-aachen.de	Commercial	Graphical user interface for the gear manufacturing simulation
2840	ZaKo3D	www.wzl.rwth-aachen.de	Commercial	Analysis of complex kinematics and geometries for special transmissions and gears
2840	Kisssoft	www.kisssoft.ch	Commercial	Design for machine components such as gears, shafts, and bearings
2840	Load Distribution Program – Windows LDP4	gearlab1.eng.ohio-state.edu	Commercial	Computes the load distribution, root and contact stresses, and so on
2840	Run Many Cases Program – RMC	gearlab1.eng.ohio-state.edu	Commercial	Macrodesign parameters (number of teeth, pressure angle, helix angle, etc.)

References

1. Klocke, F. (2011) *Manufacturing Processes 1*, Springer-Verlag, Berlin.

2. Deutsches Institut für Normung (ed.) (2003) DIN 8589: Fertigungsverfahren Spanen, Beuth Verlag, Berlin.

2.9
Fatigue Modeling: From Microstructure to Component Scale

Mohamed Sharaf and Sebastian Münstermann

2.9.1
Influence Factors on Component Fatigue Limit

Including corrosive assistance, fatigue is considered the leading cause of component failures [1]. From an engineering perspective, this has been revealed by recent statistics on long-term statistics of structural and mechanical failure cases [2]. As a process, however, fatigue is understood as a series of four successive stages. From a micromechanical perspective, these are cyclic crack incubation, short crack propagation, stable long cyclic crack propagation, and finally unstable cyclic crack growth or final macroscopic failure [1, 3–7]. Carefully distinguishing between crack nucleation and propagation from crack growth in later stages is a key aspect when tackling fatigue modeling problems.

Fatigue limit is not the critical stress for crack initiation [8]. It is rather the threshold stress for crack nonpropagation for a given number of cycles. The investigations of the fatigue limit using experimental approaches always exhibits large scatter bands. The latter fact results in difficulties in the precise determination of the fatigue limit of steel. The dispersion observed in experimental data is attributed to the abnormality of the behavior of cyclic microcrack initiation and propagation in polycrystalline microstructures in comparison to that of physically small and long cyclic cracks [5, 9]. Microcracks do initiate and propagate under load amplitudes corresponding to the component's fatigue limit after a relatively low number of cycles but are often stopped by microstructural barriers such as grain boundaries [8, 10–15]. This process lasts for several million cycles and thereby occupies a significant portion of a component's lifetime. For ductile materials, it becomes obvious that this crack stagnation phenomenon plays a crucial role concerning material endurance, so that engineering enhancement measures should be oriented toward microcrack behavior.

Handbook of Software Solutions for ICME, First Edition. Edited by Georg J. Schmitz and Ulrich Prahl.

Table 2.9.1 Influencing factors on component fatigue limit.

Material effects	Structural effects	External effects
• Ultimate tensile strength • Maximum effective defect size • Microstructure morphology (material notch factor) • Residual stresses	• Geometrical notch factor • Size factor • Surface roughness	• Loading notch factor • Stress ratio • Load triaxiality factor • Load history/profile • Surface removal • Temperature effect • Corrosion effect

Table 2.9.1 classifies the factors affecting the fatigue limit of steel components into material-dependent, geometry-dependent, and external factors. Many studies considered the influences of these factors as well as their interdependencies. In particular, based on two quantitative works [5, 8], it can be supposed that all factors are interdependent and that their effects on the fatigue behavior are significantly different with respect to the anticipated component application. For example, when structural notches are present in the component, the material microstructure and defects have no critical effect on the fatigue limit, whereas for cases where the component geometry and surface roughness are accurately controlled, the microstructure and defect effects are surely of high significance. Since material-dependent influences on the fatigue limit are usually the control instruments in steelmaking, these are the ones handled here.

A very extensive review of a large number of works on steel fatigue performed by many authors was conducted in the last decade by Murakami [8], who quantitatively approached the subject of material-related effects on fatigue behavior. His work was motivated by the dispersion of experimental outcomes of fatigue lifetime with respect to material hardness and extreme-value inclusion sizes, as shown in Figure 2.9.1. According to his conclusions, the fatigue limit is mainly dependent on the intrinsic microstructure strength represented in terms of Vickers hardness H_V and the maximum nonmetallic inclusion size. In fact, Figure 2.9.1 forms a graphic representation of the relations of the upper bound and the lower bound fatigue limits developed by Murakami in megapascal:

$$\sigma_{wU} \simeq 1.6\, H_V$$

and

$$\sigma_{wL} \simeq \frac{1.41(H_V + 120)}{(\sqrt{\text{area}_{max}})^{1/6}}.$$

The square root of the extreme value of the effective projected area of a nonmetallic inclusion onto a plane perpendicular to the maximum principal stress direction is represented in the latter equation by $\sqrt{\text{area}_{max}}$ in micrometer.

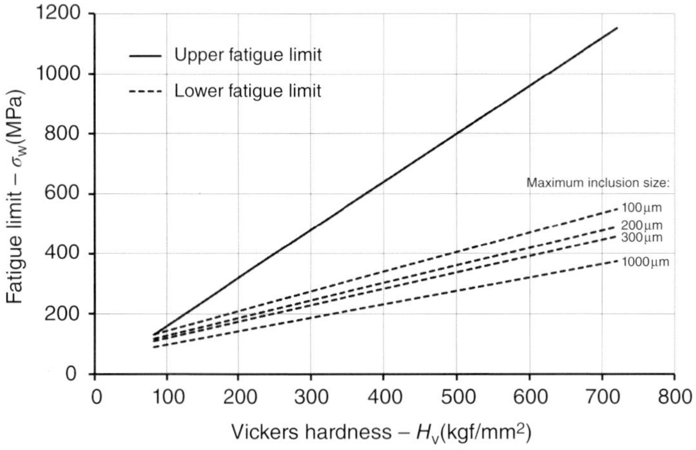

Figure 2.9.1 Effect of average steel hardness on the fatigue limit for different extreme-value nonmetallic inclusion sizes according to calculation formulae by Murakami [8].

These simple but rather handy prediction formulae show that the increasing scatter with increasing strength is due to the higher sensitivity of the microcrack nonpropagation stress to small defects. However, as some material scientists and engineers involved in the development of new materials may believe, one drawback of Murakami's approach is neglecting the effect of microstructure morphology from a metallurgical point of view. As a matter of fact, several other authors [8, 16] considered the intrinsic microstructure strength to be the main microstructure-related factor affecting the fatigue limit and that tailoring microstructures by any metallurgical processes improves the fatigue resistance only through material hardness. Although this might be true, the interdependence between those neglected morphological aspects and the shape or size of nonmetallic inclusions becomes out of consideration too. However, this interdependence may have a large potential to affect the microstructure resistance to cyclic microcracking, and it may become more significant for high-strength steels and/or at high number of cycles to failure ($N_f > 10^7$). Variability in the morphology of the microstructure may therefore also be a cause of scatter in fatigue experiments.

2.9.2 Micromechanics as a Modeling Approach

To date, there exist numerous extensive works on the mechanism of fatigue of metallic materials on the microscopic scale (e.g., Mughrabi *et al.* [17–23]). Nevertheless, these studies are qualitative and are still too limited to deliver handy, quantitative solutions of practical engineering problems [3, 8]. They, however, definitely present indispensible fundaments that support the general understanding and therefore modeling of cyclic microdeformation and the mechanisms of microcrack incubation and initial propagation. As it is the case for the works of Socie *et al.* [24–27] and Sun *et al.* [28], there is a range of approaches for multiaxial fatigue loading applied at the grain scale and having an intimate potential connection to fine scale driving forces for microcrack nucleation and early growth. When one takes the fundamental outcome of these works as well as the remarkable development in computational material mechanics implementations of crystal plasticity into account, it appears viable to conduct mesoscopic computations to quantify the influence of microstructure from the perspective of a frequency distribution of driving forces within crystals of a polycrystalline ensemble. These computations can then help fill the gap between the fatigue-related bodies of work tackling dislocation substructure under cyclic loading, slip offsets, and progressive crack nucleation and another separate but parallel research stream applying smooth specimen testing for the prediction of fatigue lifetime of components under different loading conditions.

By constructing, verifying, and applying a modeling framework that is able to quantify the effects of microstructure and inclusion content of steel on its fatigue behavior, the comparison of responses of ranges of steel microstructures in parametric studies becomes achievable. Accomplishing this goal can contribute in many industrial application in relevance to fatigue. The targeted correlation of steel fatigue and microstructure can guide the continuous improvement of steel production processes toward better fatigue strength. It can also support material selection for certain engineering applications. The method presented in the following sections, in combination with other fatigue life computation methods, may help improve the overall prediction accuracy, with integration of the role of steel microstructure.

2.9.3
Numerical Representation of Microstructure

A new method for the construction of customizable and statistically representative volume elements (RVEs) has very recently been developed. These are digital instantiations of virtual or existing microstructures for application in finite element computations.

The mathematical basis of the RVE generator algorithm is provided by the multiplicatively weighted Voronoi tessellation [29]. The main idea here is to create statistical distributions of seed weights w_i and then correlate these distributions to those of the microstructure grain size for each constituted microstructural phase. For example, for the structural steel S355 J2, which is a ferritic–pearlitic steel, two seed sets are created: one for the ferrite phase and one for the pearlite colonies. After fitting the log-normal distribution function to the grain size distributions characterized, the seed positions p_i for the Voronoi algorithm are determined. These positions are carefully determined prior to applying the weighted Voronoi tessellation algorithm. In the phase fraction allocated in the RVE space, the number of grains to be generated for that phase is calculated by

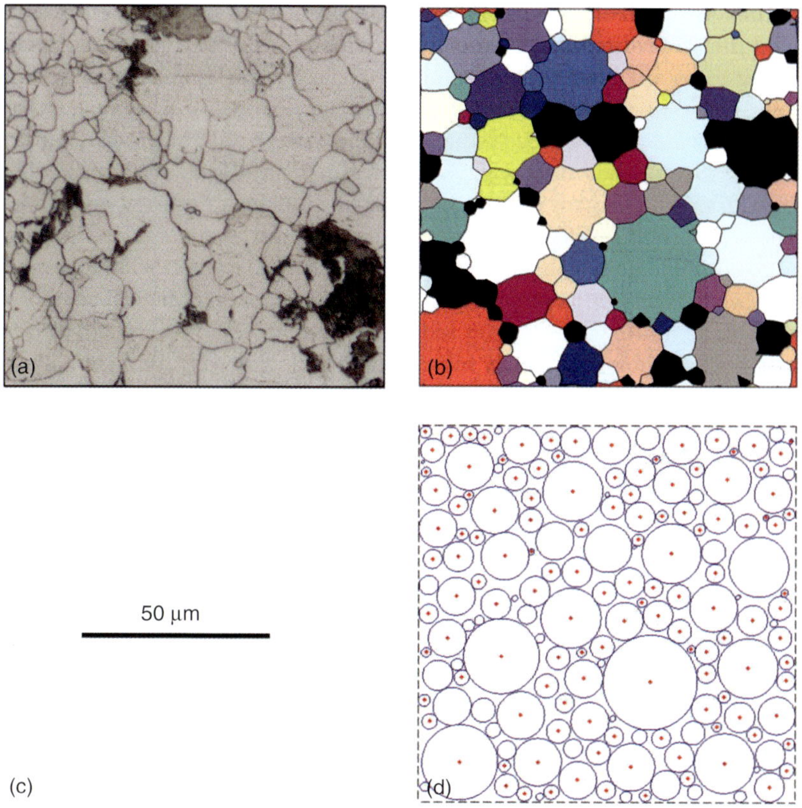

Figure 2.9.2 (a) LOM micrograph of S355 J2. (b) Created RVE of $100 \times 100\,\mu m^2$ size; black regions represent pearlite colonies and colored regions represent ferrite grains with various crystallographic orientations. (d) RSA output arrangement of virtual circles. The scale (c) applies to all three panels [31].

correlation of the area distribution of a list of virtual circles to the desired grain size distribution of the phase. The random sequential adsorption (RSA) algorithm [30] is applied to place the circles of all phases into the predefined representative area. Sequentially and in random order, the algorithm locates the circles without overlapping. The centers of the circles together with their diameters as their assigned weights are input to the tessellation algorithm to calculate the geometry of the grain boundaries and enforce periodic structure of the created RVE. Figure 2.9.2 presents a created RVE together with the corresponding RSA output.

2.9.4
Cyclic Elastoplasticity of Crystals and Microsubstructures

An empirical law can be implemented to model the deformation behavior of pearlite substructures. By assuming that only the ferrite lamellae between the cementite layers can deform plastically and that the resulting deformation is isotropic, the elastoplastic response of pearlite can be described in the form of a Young's modulus and a Poisson's ratio in addition to a traditional flow curve.

Mainly based on the work of Allain and Bouaziz [32], the flow behavior of pearlite seems to be suitably predicted by the following phenomenological formula:

$$\sigma^{\mathrm{Pear}} = \sigma_0^{\mathrm{Fer}} + \frac{\mu M b}{s} + \frac{K}{g}\left[1 - \exp\left(-\frac{g\varepsilon^{\mathrm{Pear}}}{2}\right)\right],$$

where the pearlite flow stress σ^{Pear} is given in terms of the plastic strain of pearlite $\varepsilon^{\mathrm{Pear}}$. μ is the shear modulus of ferrite ($\mu = 80$ GPa), M evaluates the mean Taylor factor ($M = 3$), and b represents the Burgers' vector of perfect lattice dislocation ($b = 0.25$ nm). s is the interlamellar spacing of the pearlite colonies. K and g are empirical parameters. Based on the experiments performed on fully pearlitic steels with different interlamellar spacings by Dollar *et al.* [33], they are estimated to, respectively, amount to 38 GPa and 70. σ_0^{Fer} is the critical lattice friction stress in the ferrite layers of pearlite. Considering the recent outcomes of an extensive European project on local deformation modeling of high-strength steels [34], the elements in solid solution contribute to the flow stress of pearlite so that the critical friction stress can be expressed in terms of weight percent contents:

$$\sigma_0^{\mathrm{Fer}}(\mathrm{MPa}) = 77 + 80(\%\mathrm{Mn}) + 750(\%\mathrm{P}) + 60(\%\mathrm{Si}) + 80(\%\mathrm{Cu}) + 45(\%\mathrm{Ni}) + 60(\%\mathrm{Cr}) + 11(\%\mathrm{Mo}) + 5000N_{\mathrm{ss}}.$$

Several scanning electron microscope (SEM) images are produced on material samples of S355 J2 (see Figure 2.9.3). The true average lamellar spacing s is evaluated in this case to amount to 0.2 μm. This value is substituted into the pearlite flow curve calculation formula to describe the average local deformation behavior of the pearlite colonies in the RVEs produced.

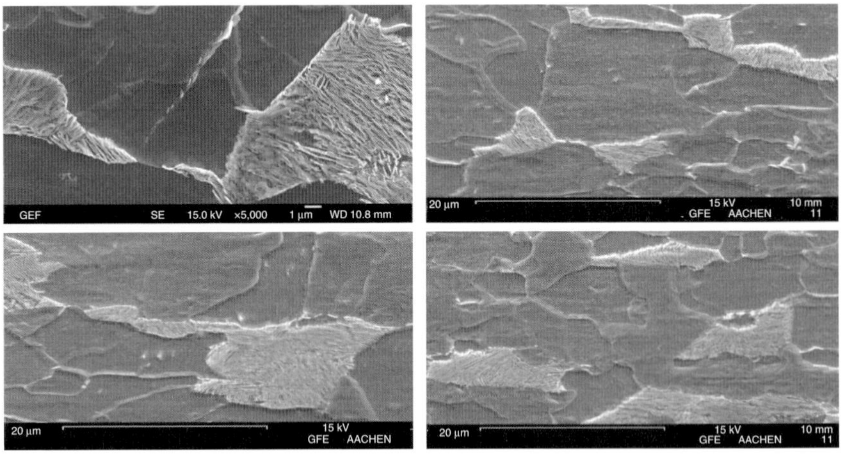

Figure 2.9.3 SEM images utilized in measuring the interlamellar spacings of pearlite in S355 J2 [31].

The main driving force of short cyclic crack initiation and early growth in structural steels is slip irreversibility [23, 24]. A mechanism-based material model of ferrite is hence indispensible. A dislocation slip crystal plasticity model is suitable for such cases. A thorough review of available crystal plasticity finite element (CPFE) models has recently been given by Raabe *et al.* [35]; also see Chapter 3. The formulations of the CPFE constitutive equations can accordingly be modified to incorporate the effects of grain boundary strengthening and kinematic hardening. This way, the increase of stress around the grain boundary, which hinders the onset of slip, and the back stresses, which mitigate the stress increase around the grain boundary and increase the stress toward the core of the grain, can be taken into consideration when computing the stress response in cyclic deformation simulations with RVEs. The governing constitutive CPFE equations can be summarized in the following form:

$$\dot{\gamma}^{\alpha} = \dot{\gamma}_0 \left| \frac{\tau^{\alpha} - \chi^{\alpha}}{\tau_c^{\alpha}} \right|^m \tau^{\alpha} - \chi^{\alpha},$$

$$\tau_c^{\alpha} = \tau_0 + \int_0^t q_{\alpha\beta} h_0 \left(1 - \frac{\tau_c^{\beta}}{\tau_s} \right)^a |\dot{\gamma}^{\beta}| \mathrm{d}t,$$

$$\tau_0 = \tau_{\mathrm{av}} + \frac{k_y}{\sqrt{d}},$$

and

$$\dot{\chi}^{\alpha} = C\dot{\gamma}^{\alpha} - D|\dot{\gamma}^{\alpha}|\chi^{\alpha},$$

where the terms are explained in Table 2.9.2.

Table 2.9.2 Nomenclature of terms used in the crystal plasticity constitutive equations.

Term	Description
$\dot{\gamma}^{\alpha}$	Shear rate of a slip system α
$\dot{\gamma}_0$	Material parameter: reference shear rate
τ^{α}	Resolved shear stress on slip system α
χ^{α}	Kinematic back stress on slip system α
$\tau_{\mathrm{c}}^{\alpha}$	Critical resolved shear stress (slip resistance)
m	Material parameter quantifying the rate sensitivity of slip
τ_0	Initial critical resolved shear stress
$q_{\alpha\beta}$	Material parameter: anisotropic latent hardening matrix
$h_0 \tau_{\mathrm{s}}$ and a	Material parameters: hardening parameters
τ_{av}	Material parameter quantifying the lattice resistance to dislocation motion
k_y	Material parameter: strengthening coefficient
d	Average ferrite grain size
C and D	Material parameters: linear and nonlinear kinematic hardening coefficients

On the whole, the (modified) CPFE constitutive equation formulations constitute a number of material parameters. These constants can be calibrated using experimental and numerical investigations. One of the approaches applied therefore is based on nanoindentation tests. This idea is motivated by the general capability of crystal plasticity models to reproduce load–penetration depth curves and imprint topologies in a directly comparable manner to experimental results and therefore enables the estimation of true material properties by inverse fitting methods, as discussed by Hartmaier *et al.* [36]. In the example presented here, a $15 \times 10 \times 1\ \text{mm}^3$ specimen is wire-cut and firmly embedded in a resin, such that the two flat, parallel surfaces are exposed to air. The top surface is electropolished to minimize the surface roughness. An electron backscatter diffraction (EBSD) measurement is performed on the latter surface to obtain the orientations of the grains in the region of interest. The pearlitic regions are excluded from the analysis by setting a threshold image quality value for filtering. A number of ferrite grains are then selected based on their orientations and average sizes, whereby the crystallographic orientation of (1 0 1) is mainly chosen for the simplicity of its resulting imprint pattern. A spheroconical indenter with a tip radius of 1 μm and a vertex angle of 90° is used for the nanoindentation to guarantee uniform load distribution around the tip. A maximum load of 6000 μN with a loading/unloading rate of 250 μN/s is applied for indentations including a holding time of 5 s. The surface is imaged after each indentation process using the scanning probe microscopy (SPM) technique. A preliminary comparative analysis has shown that the cube-corner indenter delivers the best image quality because it possesses the sharpest tip among common nanoindenters. The convolution effect due to probe–sample interaction can hence be minimized. A cube-corner indenter is thus used to scan each indentation area and to evaluate the topology of the whole imprint profile.

The target responses are twofold: the load–depth curve and the imprint profile. A number of iterative optimization computations can be performed until an acceptable agreement between all computed and experimental responses is reached. The effect of each material constant of the CPFE constitutive equations is thereby first estimated separately in a parameter sensitivity analysis, followed by the iterative optimization computations. For one of the performed nanoindentation experiments and corresponding computations, Figure 2.9.4 displays the best-fit achieved. The ferrite crystal plasticity constants obtained accordingly can then be used in microstructure-sensitive cyclic simulations, as is discussed in Section 2.9.6.

2.9.5
The Notion of Fatigue Indicator Parameters (FIPs)

When the constitutive equations governing the ferrite crystal plasticity are numerically implemented in a stress-update material algorithm, the definition of necessary solution-dependent variables can be conducted in a computational mechanics context. Defining the so-called fatigue indicator parameters (FIPs) is

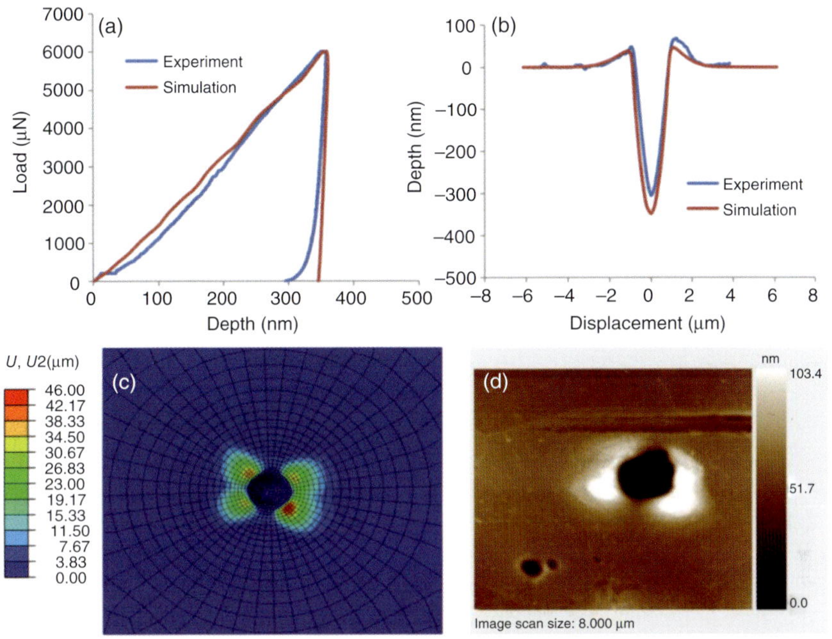

Figure 2.9.4 Reverse fitting of a nanoindentation experiment: (a) load–depth curve, (b) imprint profile, (c) computed pile-up topology, and (d) SPM micrograph of pile-up [31].

essential. The values of these represent the driving forces of microstructurally small crack (MSC) nucleation and propagation and facilitate the fatigue dependence on the microstructure. In other words, they invariably appeal to the notion of slip irreversibility in linking to fatigue damage [3].

Dunne *et al.* [37, 38] presented the accumulated dislocation slip parameter p_{acc} and used it to predict the initiation location and propagation direction of short cyclic cracks in a polycrystal under plane strain LCF and HCF conditions. Some authors [39, 40] argued that the cyclic behavior of microcracks is dependent on energy dissipation. The Fatemi–Socie parameter [24] has been shown to correlate very well fatigue cracking events at and above the grain scale. It considers the influence of the normal stress to the plane of maximum shear, which favors crack nucleation and supports its further growth by decreasing frictional forces that decelerate crack growth [27]. It is defined as follows:

$$p_{\text{FS}} = \frac{\Delta\gamma_{\max}}{2}\left(1 + k\frac{\sigma_{n,\max}}{\sigma_y}\right),$$

where $\Delta\gamma_{\max}/2$ is the maximum cyclic shear averaged over a predefined volume of the material. $\sigma_{n,\max}$ and σ_y are, respectively, the maximum stress normal to the plane of $\Delta\gamma_{\max}/2$ and the yield strength. The effect of normal stress is mediated by the constant k that controls the influence of normal stress in supporting fatigue crack incubation and early propagation.

2.9.6
Fatigue Limit as a Function of Microstructure

Classical findings have confirmed by many fatigue-related research studies that the main influencing factors to fatigue limit are nonmetallic inclusions and the intrinsic microstructure strength, which can be expressed in terms of Vickers hardness, as discussed in Section 2.9.1. Murakami's formulae are very simple and therefore useful for industrial and practical use. The upper and lower bounds of the fatigue limit can be very easily estimated just by measuring the tensile strength or the hardness of the material and the maximum inclusion size using a simple inclusion rating method. In contrast, the main assumption made here for the sake of simplicity of the equations is that the morphology of microstructure barely has any absolute influence on the fatigue behavior. Consequently, it is assumed that enhancing fatigue strength cannot be achieved by metallurgical processes or heat treatments, unless a measure of average resistance of microstructure to plastic deformation or the maximum size of inclusions present is increased. The validity of this hypothesis can be explored.

We therefore aim at inspecting whether the phase sizes of the microstructure and the inclusion shape have any direct effect or at least interact with the effect of the inclusion size on the fatigue limit. We analyze the effects of the average inclusion diameter D_{Inc}, the inclusion shape factor f_{Inc}, the average ferrite grain size D_{Ferr}, and the average pearlite colony size D_{Pear} on the fatigue limit $\Delta\sigma_{\mathrm{D}}$. This example study can hence be considered a virtual experiment that can be efficiently planned by statistical experimental design methods. The idea is therefore basically to create an effects model that relates the four influence factors to a response quantity. Figure 2.9.5 abstractly illustrates this idea.

Defining the design space to include ferrite grain and pearlite colony sizes roughly ranging from 2 to 12 μm and inclusion voids ranging from 10% to 50% of average ferrite grain size, we choose to construct a three-level-per-factor model. As shown in Figures 2.9.6 and 2.9.7, 27 ($3 \times 3 \times 3 \times 3$) possible factor-level combinations are accordingly implemented in the form of RVE realizations. The phase contents set for all RVEs are constant and amount to 85% and 15% for ferrite and pearlite, respectively.

By applying identical, cyclic, periodic boundary conditions to all RVEs, the Fatemi–Socie FIP can be calculated for each finite element. The extreme values of the calculated FIPs are then identified around the inclusion voids and correlated to the radial distance outward from inclusion for all influence factor-level

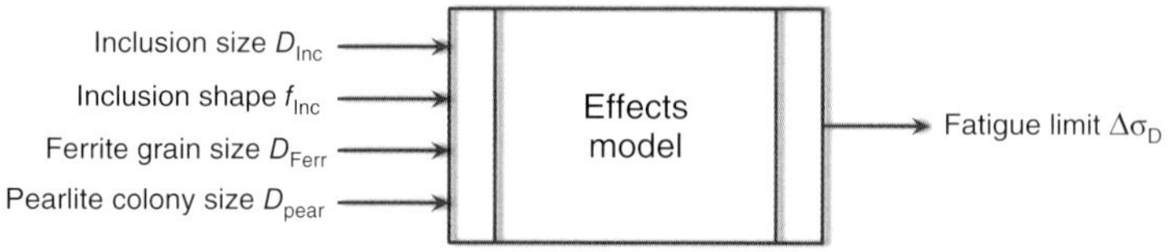

Figure 2.9.5 Virtual experiment to be designed to render the fatigue limit function of four influence factors that describe the microstructure [31].

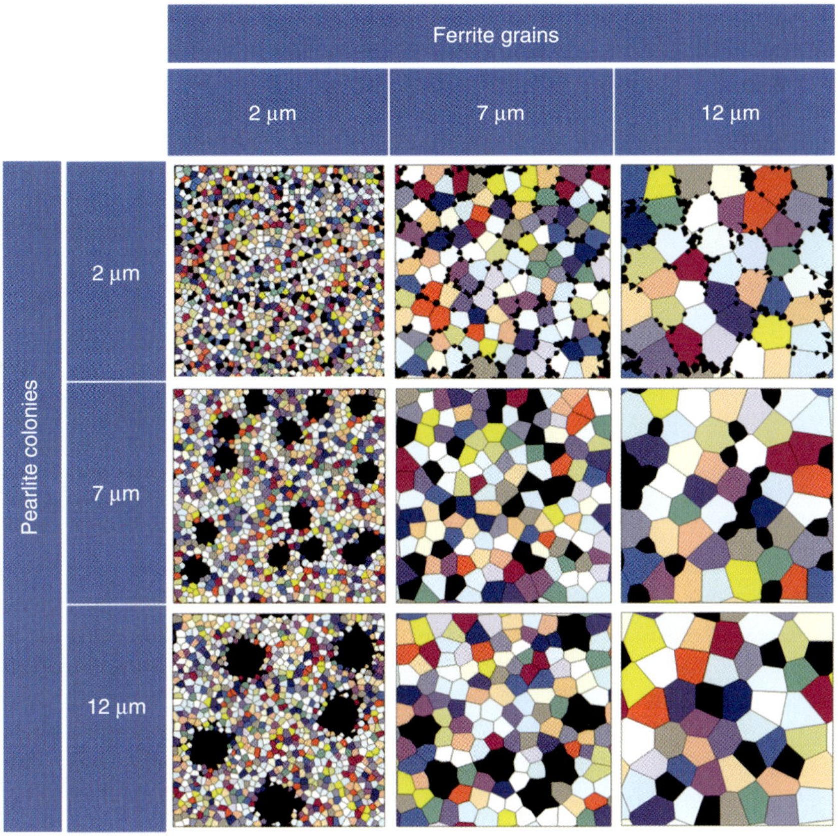

Figure 2.9.6 Factor-level combinations of ferrite and pearlite sizes. Black regions represent pearlite colonies and colored regions represent ferritic structures with different crystallographic orientations. All RVEs are of $100 \times 100\,\mu\text{m}^2$ size [31].

combinations. As the extreme-value FIP distribution $\text{EX}p_{\text{FS}}(x)$ over the distance from inclusion x appears to be almost linear on the logarithmic scale, it can be fitted to an exponential decay function, as the one employed in the work of Musinski *et al.* [41]:

$$\text{EX}p_{\text{FS}}(x) = \text{EX}p_{\text{FS}}(0)\text{e}^{\left(\frac{-\nu x}{a_{\text{EPFM}}}\right)},$$

where $\text{EX}p_{\text{FS}}(0)$ and ν are two parameters that control the linear function and a_{EPFM} is the crack length at which the crack can no longer be influenced by the microstructure and whose propagation can be handled by elastic plastic fracture mechanics approaches. In order to use the fit lines of the extreme-value FIPs to predict the propagation of MSCs within the inclusion field, $\text{EX}p_{\text{FS}}(x)$ can be employed into an MSC growth law, as the one proposed by Shenoy *et al.* [42]:

$$\left.\frac{\text{d}a}{\text{d}N}\right|_{\text{MSC}} = A_{\text{MS}}\tau_y \text{EX}p_{\text{MS}}(x)\, a,$$

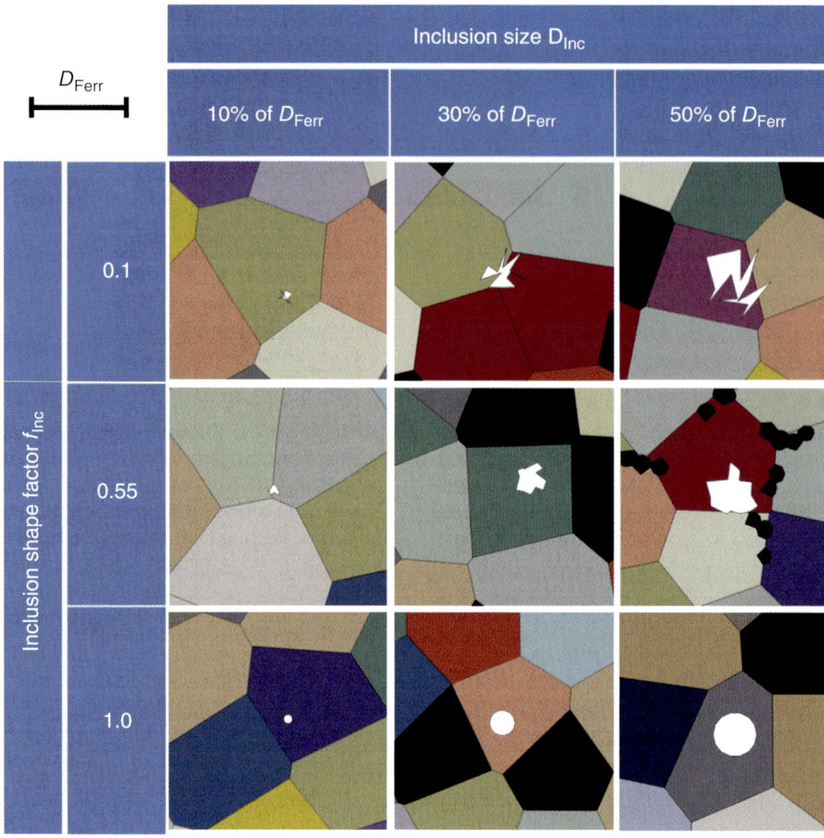

Figure 2.9.7 Factor-level combinations of inclusion size and shape. Black regions represent pearlite colonies and colored regions represent ferritic structures with different crystallographic orientations. All RVEs are of $100 \times 100\,\mu\text{m}^2$ size [31].

where a and N are the two variables: crack length and number of loading cycles, respectively. A_{MS} is a constant regulating the sensitivity of growth rate toward accumulation of microplastic deformation and τ_y is the resolved shear stress. By integration of the MSC growth law, the number of cycles to incrementally propagate a crack can be obtained. The achieved a–N relationship can be used to establish the probability of fatigue failure at any given number of cycles. Similar to the probabilistic approach adopted by Musinski *et al.* [41], the cumulative distribution function (CDF) is assumed to depend on the length of a preexisting crack a at a given number of cycles N, as in

$$\text{CDF}(N) = 1 - \mathrm{e}^{-\left(\frac{\eta}{a_{\text{EPFM}}}\int_{N_i}^{N}\frac{\mathrm{d}a}{\mathrm{d}N}\mathrm{d}N\right)},$$

where η is a normalization parameter, which can be set to amount to $-\ln(0.5)$ to normalize the probability function to 50% at $a = a_{\text{EPFM}}$.

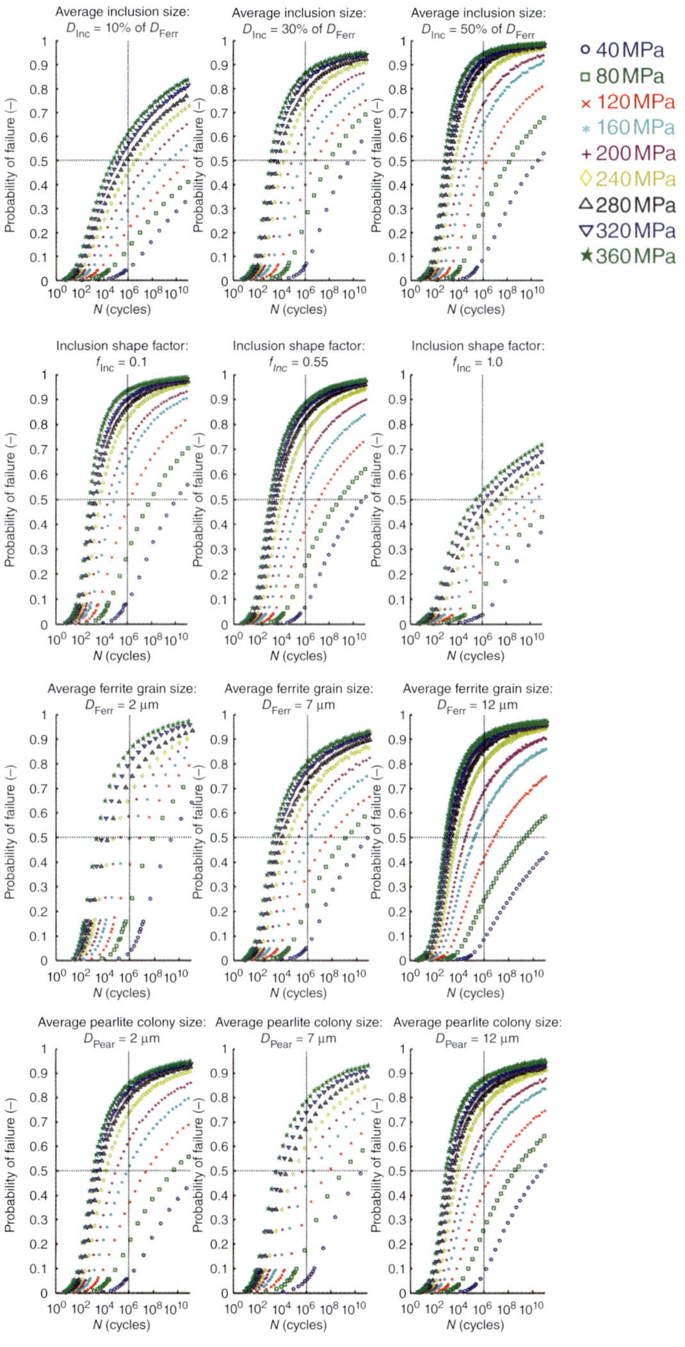

Figure 2.9.8 Probabilities of failure depending on the number of cycles for nine stress amplitudes at different levels of influence factors. (Reproduced from [31].)

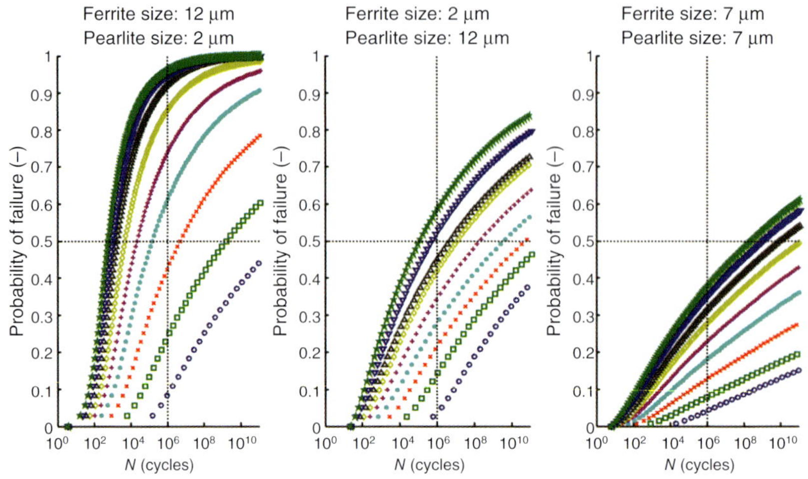

Figure 2.9.9 Probabilities of failure depending on the number of cycles for nine stress amplitudes at three-level combinations of D_{Ferr} and D_{Pear} [31]. The legend shown in Figure 2.9.8 applies.

By applying the cumulative distribution function framework, the results for nine imposed stress amplitude steps (40–360 MPa) are presented in Figure 2.9.8. The results shown are average failure probabilities over all RVE simulations realizing an individual level of each of the four influence factors: inclusion size D_{Inc}, inclusion shape factor f_{Inc}, average ferrite grain size D_{Ferr}, and average pearlite colony size D_{Pear}.

The obtained relationships between the influence factors and the probability of fatigue failure can reveal several conclusions. The extreme inclusion size in microstructure has the strongest effect on the fatigue limit as confirmed by many authors. The significance of this factor is present over most of the stress amplitude interval applied in this study. Not as pronounced as the influence of maximum inclusion size is the effect of the inclusion shape on the failure probability. One can identify that a strong effect of this factor exists only within the range from 0.55 to 1. However, it can generally be assumed that smoother inclusion surfaces have a positive influence on the fatigue limit.

At the first glance, the average ferrite and pearlite sizes do not seem to have a clear direct effect on the failure probability. In order to examine whether an interaction is present between these two factors or between them and any of the other two influence factors, Figure 2.9.9 is presented showing the failure probabilities averaged at three-level combinations of average ferrite grain size D_{Ferr} and average pearlite colony size D_{Pear}, namely, at high (12 : 2), low (2 : 12), and middle (7 : 7) values of ferrite to pearlite size ratio. Note that the phase fraction of the corresponding RVE models used is kept constant. It now becomes clear that there exists a strong interaction between the two concerned factors. It can therefore be concluded that a homogeneous distribution of ferrite and pearlite sizes enhances the resistance of microstructure to fatigue microcracking. Microstructures with fine ferrite grains along with few but coarse pearlite colonies possess a higher tendency to assist microcrack growth than do microstructures of nearly equal ferrite and pearlite sizes.

2.9.7
Software Tools for Modeling Fatigue

Fatigue modeling currently is essentially performed on the basis of classical FEM tools that are complemented by crystal plasticity routines (see tools listed in Chapter 3). The Voronoi tessellation to create synthetic microstructures depicted in this chapter can be expected to be further refined by simulated microstructures in the future (also see Chapter 3).

Chapter	Name of code	Website of software	Type of code	Short description
2.9.3 2.9.4 2.9.6	ABAQUS Unified FEA	www.3ds.com/products-services/simulia/products/abaqus/	Commercial	A software suite for finite element analysis and computer-aided engineering
2.9.3 2.9.6	MATLAB	www.mathworks.com/products/matlab/	Commercial	A multiparadigm numerical computing environment

References

1. Pokluda, J. and Sandera, P. (2010) *Micromechanisms of Fracture and Fatigue: In a Multi-Scale Context*, Engineering Materials and Processes, Springer.
2. McEvily, A.J. (2002) *Metal Failures: Mechanisms, Analysis, Prevention*, John Wiley & Sons, Inc., Hoboken, NJ.
3. McDowell, D.L. and Dunne, F. (2010) Microstructure-sensitive computational modeling of fatigue crack formation. *Int. J. Fatigue*, **32**, 1521–1542.
4. Przybyla, C., Prasannavenkatesan, R., Salajegheh, N., and McDowell, D.L. (2010) Microstructure-sensitive modeling of high cycle fatigue. *Int. J. Fatigue*, **32**, 512–525.
5. Radaj, D. and Vormwald, M. (2007) *Ermüdungsfestigkeit: Grundlagen für Ingenieure*, 3rd edn, Springer, Berlin, Heidelberg.
6. Tanaka, K. (2003) Fatigue crack propagation, in *Comprehensive Structural Integrity*, vol. **4** (eds R.O. Ritchie and Y. Murakami), Elsevier.
7. Przybyla, C.P. and McDowell, D.L. (2010) Microstructure-sensitive extreme value probabilities for high cycle fatigue of Ni-base superalloy IN100. *Int. J. Plast.*, **26**, 372–394.
8. Murakami, Y. (2002) *Metal Fatigue: Effects of Small Defects and Nonmetallic Inclusions*, Elsevier, Oxford, Boston, MA.
9. Richard, H.A. and Sander, M. (2012) *Ermüdungsrisse: Erkennen, sicher*

beurteilen, vermeiden, 2. Aufl. 2012 ed., Vieweg+Teubner Verlag, Wiesbaden.
10. Murakami, Y. (1988) Fundamental aspects of fatigue threshold of metals containing small defects, small cracks and nonmetallic inclusion – a unified quantitative evaluation and its application, in *Basic Mechanisms in Fatigue of Metals* (eds P. Lukas and J. Polak), Academia, pp. 343–350.
11. Murakami, Y. (1989) Quantitative evaluation of effects of defects and nonmetallic inclusions of fatigue strength of metals. *Tetsu To Hagane*, **75** (8), 1267–1277.
12. Murakami, Y. and Endo, T. (1981) The effects of small defects on the fatigue strength of hard steels, in *Materials, Experimentation and Design in Fatigue*, Proceedings of Fatigue (eds F. Sherrat and J.B. Sturgeon), Warwick University, pp. 431–440.
13. Murakami, Y., Kodama, S., and Konuma, S. (1988) Quantitative evaluation of effects of nonmetallic inclusions on fatigue strength of high strength steel. *Trans. Jpn. Soc. Mech. Eng. A*, **54** (500), 688–696.
14. Murakami, Y. and Shimizu, M. (1988) Effects of nonmetallic inclusions, small defects and small cracks on fatigue strength of metals. *Trans. Jpn. Soc. Mech. Eng. A*, **54** (499), 413–425.
15. Murakami, Y. and Usuki, H. (1989) Prediction of fatigue strength of high-strength steels based on statistical evaluation of inclusion size. *Trans. Jpn. Soc. Mech. Eng. A*, **55** (510), 213–221.
16. Chalant, G. and Suyitno, B.M. (1991) Effects of microstructure on low and high cycle fatigue behaviour of a micro-alloyed steel. Proceedings of the 6th International Conference of Mechanical Behavior of Materials, Kyoto, pp. 511–516.
17. Mughrabi, H. (2013) Cyclic slip irreversibility and fatigue life: a microstructure-based analysis. *Acta Mater.*, **61**, 1197–1203.
18. Mughrabi, H. (2009) Cyclic slip irreversibilities and the evolution of fatigue damage. *Metall. Mater. Trans. A*, **40**, 1257–1279.
19. Mughrabi, H., Borbely, A., Eisenmeier, G., and Höppel, H. (2002) A finite element modelling study of strain localization in the vicinity of near-surface cavities as a cause of subsurface fatigue crack initiation. *Int. J. Fract.*, **115**, 227–232.
20. Mughrabi, H. (1993) Cyclic plasticity and fatigue of metals. *J. Phys. IV France*, **03**, C7-659.
21. Mughrabi, H., Herz, K., and Stark, X. (1981) Cyclic deformation and fatigue behaviour of alpha-iron, mono-and polycrystals. *Int. J. Fract.*, **17**, 193–220.
22. Mughrabi, H. and Wüthrich, C. (1976) Asymmetry of slip and shape changes during cyclic deformation of α-iron single crystals. *Philos. Mag.*, **33**, 963–984.
23. Mughrabi, H. (1999) On the life-controlling microstructural fatigue mechanisms in ductile metals and alloys in the gigacycle regime. *Fatigue Fract. Eng. Mater. Struct.*, **22**, 633–641.
24. Fatemi, A. and Socie, D. (1988) A critical plane approach to multiaxial fatigue damage including out-of-phase loading. *Fatigue Fract. Eng. Mater. Struct.*, **11**, 149–165.
25. Hoshide, T. and Socie, D. (1988) Crack nucleation and growth modeling in biaxial fatigue. *Eng. Fract. Mech.*, **29** (3), 287–299.
26. Marquis, G.B. and Socie, D.F. (2003) Multiaxial fatigue. *Compr. Struct. Integrity*, **4**, 221–252.
27. Socie, D.F. (1993) in *Advances in Multiaxial Fatigue* (eds D.L. McDowell and R. Ellis), ASTM, Race Street, PA, pp. 7–36.
28. Sun, Z., de los Rios, E.R., and Miller, K.J. (1991) Modelling small fatigue cracks interacting with grain boundaries. *Fatigue Fract. Eng. Mater. Struct.*, **14**, 277–291.
29. Voronoi, G. (1908) Nouvelles applications des paramètrescontinus à la théorie des formesquadratiques. Deuxième mémoire. Recherches sur les parallélloèdres primitifs. *J. Reine Angew. Math.*, **134**, 198–287.
30. Tarjus, G., Schaaf, P., and Talbot, J. (1991) Random sequential addition: a distribution function approach. *J. Stat. Phys.*, **63**, 167–202.

31. Sharaf, M. (2015) The microstructure influence on fatigue life variability in structural steels. PhD thesis, RWTH Aachen University.
32. Allain, S. and Bouaziz, O. (2008) Microstructure based modeling for the mechanical behavior of ferrite–pearlite steels suitable to capture isotropic and kinematic hardening. *Mater. Sci. Eng., A*, **496**, 329–336.
33. Dollar, M., Bernstein, I.M., and Thompson, A.W. (1988) Influence of deformation substructure on flow and fracture of fully pearlitic steel. *Acta Metall.*, **36**, 311–320.
34. Buessler, P., Bouaziz, O., Lung, T., Gil Sevillano, J., Bonifaz, E., Ocana, I., Meizoso, A.M., Esnaola, J.M.M., Vrieze, J., and Kaluza, W. (2002) Modelling of Mechanical Properties and Local Deformation of Multiphase High Strength Steels. ECSC Steel RTD Final Report. CECA 7210-PR-044, Office for Official Publ. of the E.U.
35. Roters, F., Eisenlohr, P., Bieler, T.R., and Raabe, D. (2010) *Crystal Plasticity Finite Element Methods*, Wiley-VCH Verlag GmbH.
36. Engels, P.S., Begau, C., Gupta, S., Schmaling, B., Ma, A., and Hartmaier, A. (2014) in *Nanomechanical Analysis of High Performance Materials* (ed. A. Tiwari), Springer, pp. 285–322.
37. Dunne, F., Wilkinson, A., and Allen, R. (2007) Experimental and computational studies of low cycle fatigue crack nucleation in a polycrystal. *Int. J. Plast.*, **23**, 273–295.
38. Manonukul, A. and Dunne, F.P.E. (2004) High- and low-cycle fatigue crack initiation using polycrystal plasticity. *Proc. R. Soc. A Math. Phys. Eng. Sci.*, **460**, 1881–1903.
39. Korsunsky, A., Dini, D., Dunne, F., and Walsh, M. (2007) Comparative assessment of dissipated energy and other fatigue criteria. *Int. J. Fatigue*, **29**, 1990–1995.
40. Skelton, R.P., Vilhelmsen, T., and Webster, G.A. (1998) Energy criteria and cumulative damage during fatigue crack growth. *Int. J. Fatigue*, **20**, 641–649.
41. Przybyla, C.P., Musinski, W.D., Castelluccio, G.M., and McDowell, D.L. (2013) Microstructure-sensitive HCF and VHCF simulations. *Int. J. Fatigue*, **57**, 9–27.
42. Shenoy, M., Zhang, J., and McDowell, D.L. (2007) Estimating fatigue sensitivity to polycrystalline Ni-base superalloy microstructures using a computational approach. *Fatigue Fract. Eng. Mater. Struct.*, **30**, 889–904.

2.10
Corrosion and Its Context in Service Life

Daniela Zander, Daniel Höche, Johan Deconinck, and Theo Hack

2.10.1
Overview

The application of metal in commercial industry does not only require the consideration of mechanical, physical, constructive, and economical aspects, but chemical and electrochemical influences and their consequences on the material and its durability must also be accommodated for. However, corrosion and its control bring with them significant risks and challenges in various fields of engineering and applications. In order to select an appropriate material or evaluate a failure analysis, the various corrosive processes and corrosion types must be known. Additionally, corresponding testing situations must be established in order to minimize economic loss due to both seen and unforeseen corrosion and corrosion protection measures. This costly experimental approach is expected to be optimized by corrosion modeling and related predictive service life modeling.

There are many varying types of corrosion within different sectors. The oil and gas industry is often threatened by pipeline failure or tank defects. Hot gas corrosion is another important topic to be tackled. In general, environmentally assisted cracking (EAC) is a topic with high-risk aspects. Still, the most common and widely distributed type of corrosion is related to exposure to aqueous mediums containing aggressive ions. Technical impact is generated by galvanic couples, underpaint/cavitation and crevice corrosion, pitting, filiform corrosion, or, with environmental exposure, biofouling. In many cases, the material is affected by exposures coupling corrosion and wear, stress, or cycling effects, leading to erosion corrosion, stress corrosion cracking, fretting corrosion, and, at the minimum, corrosion fatigue. In nuclear reactors, radiolysis can form part of the corrosion environment [1].

Generally, corrosion can be described as surface damage due to the reaction of a material and the surrounding medium, most commonly electrolytes, hot gases, and molten salts and metals. Oftentimes, this reaction is electrochemical in nature and can lead to electrolytic metal dissolution; however, this reaction can also be chemical in nature, as in the case of high-temperature oxidation, or can be

Handbook of Software Solutions for ICME, First Edition. Edited by Georg J. Schmitz and Ulrich Prahl.

characterized by metal-physical processes such as those involved in alloying or leaching (dealloying or alloy dissolution). Regardless of the type of reaction, they always occur at the phase boundary between the material and the surrounding medium and incorporate transportation of chemical components of the material and/or medium through this interface.

Altogether, corrosion is a system property that is conveniently defined and itemized by ISO 8044 as a "physicochemical interaction between a metal and its environment that results in changes in the properties of the metal, and which may lead to significant impairment of the function of the metal, the environment, or the technical system, of which this form a part." This chapter focuses on the impact of electrochemical corrosion on various industrial fields and the state of the art of corrosion modeling at both the process and component scales.

Of the various manifestations of electrochemical corrosion in metals, *uniform corrosion* is one of the most important. It is during this form of corrosive attack that no preferential site is established and a rather even attack is instigated. However, structural weakening, through cross-sectional area loss, for example, still manifests itself. Advantageously, this form of corrosion supplies a method to predict electrolytic erosion through rates such as thickness loss in millimeter/year. This provides a means for constructive corrosion protection since appropriate safety supplements can be selected during component design. By contrast, *localized corrosion* develops at specific locations and is amplified conditionally on heterogeneous structures or concentration gradients in the electrolyte. Examples of this phenomenon include intergranular corrosion, pitting corrosion, or crevice corrosion. Corrosion damage is usually only being assessed through the utilization of a damage pattern evaluation. Crack formation due to the superimposition of a corrosive medium and mechanical loading presents another form of corrosion that can lead to hydrogen embrittlement, stress corrosion cracking, or corrosion fatigue, in part due to semibrittle crack propagation in ductile materials.

Aqueous corrosion of metals can lead to diverse corrosion processes, subcategorized into those which result in even, uniform surfaces or uneven, locally varied surfaces. The latter can be further divided depending on the presence or lack of mechanical loading. The following corrosion types are industrially relevant:

- Uniform corrosion
- Pitting corrosion
- Shallow pitting corrosion
- Concentration cell corrosion
- Galvanic corrosion
- Intergranular corrosion with irregular, local corrosion without mechanical loading
- Hydrogen embrittlement
- Stress corrosion cracking
- Fatigue corrosion with irregular, local corrosion mechanical loading

Uniform corrosion can be described using the current density, which is constant regardless of the location on the surface (cathode area = anode area), and since there is no external current, $\Sigma I_A = \Sigma I_C$; this situation is considered a *homogeneous mixed electrode*. The partial current densities for local corrosion are surface site dependent and therefore irregularly distributed, defining a *heterogeneous mixed electrode*. In this case, if $I_A = I_C$ and the cathode area ≫ anode area, then $i_A \gg i_C$ and local corrosion occurs.

Preferential attack on grain boundaries due to the extensive corrosion endurance of the inner grain area is known as intergranular corrosion. This phenomenon is triggered by the enrichment of alloying elements that promote corrosion (e.g., P or Si in steels) or the depletion of corrosion protective elements (e.g., Cr in steels) in the grain boundary region.

Electrolyte composition has an additional influence on local corrosion processes. If local concentration gradients are present in the electrolyte, especially dissolved oxygen content, the development of concentration cell corrosion increases its probability. Pitting corrosion, however, shows a stronger dependence on halide ions, as an increase in concentration can lead to the weakening of the passive layer and autocatalytic pit growth in the material.

Uneven local corrosion in collaboration with simultaneous tensile stress (stress corrosion cracking) or cyclic loading (corrosion fatigue) leads to crack initiation and subsequent propagation at extreme rates. The course of the crack can be either intergranular or transgranular, depending on the type of corrosion. This phenomenon is always associated with critical conditions that are material, electrolyte, potential, stress, and temperature relevant and, therefore, only specific combinations of susceptible materials and electrolytes are prone to this form of corrosion. An example of this type of corrosion occurs, for example, in austenitic steels in electrolytes containing Cl^- ions, where transgranular stress corrosion cracking takes place.

In particular the prediction of the reliability of components by experimental means, undergoing selective corrosion mechanism, is known to be limited and costly. Therefore, the contributions of new tools, such as corrosion modeling and predictive service life modeling, are of high interest considering their impact and potential on service life assessment.

2.10.2
Corrosion Modeling and Applications

Corrosion modeling and also the related predictive service life modeling are growing fields in computer-aided engineering. The main issues still remaining are attributed to the high complexity of corrosion predictive models. The major concerns are the multiscale nature of corrosion processes as introduced in Figure 2.10.1 and the need of many model parameters limiting the "validity window" of simulations. Thus, for many applications, experimental input and validation are crucial. The field of potential applications is huge, and the progress in industrial-oriented corrosion modeling is only at the beginning.

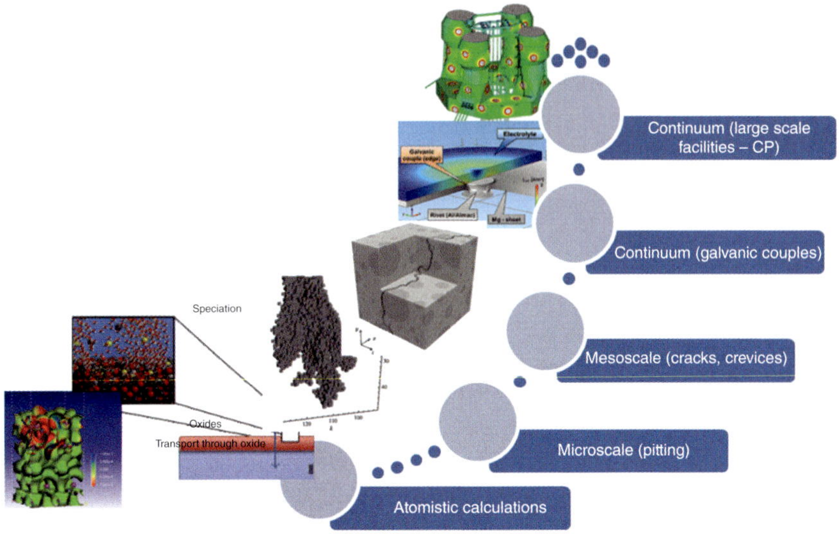

Figure 2.10.1 Scale aspects in corrosion modeling. (Reprinted from [2, 3, 4] with kind permission from Elsevier and Springer and from Dr. M. Amkreutz (IFAM).)

Damage phenomena which are related to corrosion aspects often differ between industrial sectors. Therefore, the separation according to application fields is helpful. This does not exclude the occurrence of other mentioned phenomena in the different sectors.

2.10.2.1 Phenomena Occurring during Service Life

2.10.2.1.1 Energy Sector

Energy infrastructures and all their complementary facilities such as pipelines and tanks for oil and gas production as well as concrete structures for nuclear plants or offshore wind parks are exposed to various severe conditions. Giving a full insight is not possible; however, most common issues shall be named.

For more than 50 years, steel pipelines have proven to be a safe, reliable, and economically attractive means for transporting oil and gas, although especially in offshore structures and pipelines the occurrence of cracks can cause catastrophic failures. In recent years several incidents of hydrogen-induced failure were reported in subsea oil and gas pipelines and installations in the North Sea. Hence, there is a need of establishing reliable and practical test methods and implementing robust requirements in standards and guidelines for materials under hydrogen influence in oil and gas industry.

Research activities are associated with both existing and new pipelines. For existing pipelines, managing and mitigating the threats of leaks and failures during service – due, for example, to corrosion, mechanical impact, or load cycling of defects – are a major focus of attention. For new pipelines in terms of long-distance transportation from remote gas fields, attention has focused on the application of higher-strength steels and the utilization of advanced design methods, requiring a sophisticated understanding of the behavior of pipeline materials as they approach their performance limits. Recent activities in the field of corrosion and corrosion protection have focused on stress corrosion cracking and hydrogen-induced cracking (HIC) as well as on the performance of coatings and the influence of the external environment on the lifetime of mechanically damaged pipes. One aim of these activities is to improve the methods of fit-for-purpose HIC testing and to simulate the interaction of mechanical damage and cathodic overprotection which may result in rapid failure due to hydrogen embrittlement.

The primary means of preventing corrosion of buried pipelines is a combination of a protective coating and an impressed current (*impressed current cathodic protection* (ICCP)) or sacrificial anodes (sacrificial anode cathodic protection (SACP)) to maintain cathodic protection (CP). As the pipeline age increases, the ability to maintain corrosion-free conditions is related to the resistance of the coating to aging processes and local degradation. Sixty percent of the existing buried pipelines have thermoplastic coatings, with the majority of the remainder being polyolefin coatings. The coating technology for buried pipelines has changed over the years, and nowadays almost 75% of the newly constructed pipelines are coated with three layers of polyethylene.

Sufficient CP being needed, overprotection is dangerous as the produced hydrogen causes cracking and failure particularly in stainless steel subsea pipelines and fittings. Proper design is therefore very important. Low-carbon martensitic stainless steels are an economical option to the demand for high strength and corrosion performance offshore. These steels offer sufficient corrosion resistance for sweet and mildly sour environment but will suffer corrosion under sour conditions. Supermartensitic stainless steels (SMSSs) are a new generation of the classical 13%Cr martensitic steels, lower in carbon and with additional alloying of nickel and molybdenum, offering better weldability and low-temperature toughness. The mechanical properties vary with chemical composition and tempering procedures. Due to their primary martensitic structure, the SMSS are susceptible to hydrogen embrittlement related to welding, H_2S, or CP. They are especially sensitive in the nonheat-treated condition, containing untempered martensite and unstable retained austenite. This is often the case in weld heat-affected zones, where most of the hydrogen-assisted cracking incidents occur.

Duplex and superduplex stainless steels (DSS and SDSS) have been used in offshore applications for many years. These steels offer high strength and toughness as well as excellent corrosion resistance. The steels are characterized by the two-phase microstructure of ferrite (α) and austenite (γ). To maintain the mechanical and corrosion properties, it is essential to have a dual-phase structure as close to the 50/50 percentage distribution as possible. This can often be a challenge, especially in welds where the brittle ferritic structure tends to dominate. An extra challenge in testing of HIC is the time aspect because of the low hydrogen diffusion rates in austenitic stainless steels. DSS are considered to be safe due to the duplex structure of ferrite and austenite with the ductile austenite acting as a crack stopper. At the low corrosion protection potentials applied especially in the North Sea, hydrogen production is considerable. Combined with stresses during production, hydrogen-induced stress cracking (HISC) is observed even in 25% Cr DSS. The diffusion in DSS mainly takes place in the ferrite phase. At subsea temperatures, diffusion in the austenite phase is very slow and has an insignificant influence on the effective diffusion coefficient. However, the austenite concentration of hydrogen may be high, representing local hydrogen "supply" if the duplex steel is brought up to a certain temperature, or under highly stressed/strained conditions.

Despite their general superior toughness and corrosion resistance, DSS and SDSS can suffer from hydrogen embrittlement, and hydrogen cracking can occur in the heat-affected zone after welding. Over the last years, cracking incidents related to CP and HISC have also been reported. The diffusion in the ferrite phase is much slower than in plain ferritic steels. This is an effect of an increased diffusion length in the ferrite due to the austenite islands and trapping at the austenite phase boundaries. Due to the high solubility of hydrogen in austenite, the hydrogen content can be high with a marginal reduction of the crack resistance in this phase, whereas ferrite resolves very little hydrogen but is prone to brittle cracking even at low hydrogen concentrations.

2.10.2.1.2 Construction Sector

Very similar aspects are seen in the construction sector. Corrosion design engineering is a fast-growing field which tackles corrosion issues normally related to concrete, steel in concrete, and structural integrity. Although hydrogen embrittlement may play a role, the deterioration of reinforcement bars and steel structures being in contact with salts coming from seawater, coastal conditions, or anti-icing of roads and bridges is the major problem in long-term maintenance of many concrete and steel constructions. Most influencing issues are water, sulfates, chlorides, and CO_2 ingress. Chlorides and low pH resulting from the CO_2/carbonation process initiate steel corrosion that is engraved by stress induced by corrosion products. Besides the use of coatings, CP is one of the standards in corrosion protection in the electrolyte environment [5]. Sacrificial and impressed current anodes are typically used to protect offshore metallic and reinforced concrete structures: oil rigs, windmill farms, harbor infrastructure, bridges, and so on.

2.10.2.1.3 Marine Transportation

CP is standard in ship engineering as well. Normally SACP is used but more and more IICP and mixed SACP and IICP systems are designed today. Their position and amount can be optimized in detail by simulations [6, 7]. These systems are applied to the hull and ballast tanks. Special attention is to be given to the region around the rudder where different metals are combined. Overprotection leads to hydrogen embrittlement and delamination of coatings. In the design, particular attention is to be paid to interactions of individual protected systems. Moored ships can temporarily be detrimental for connected infrastructure (platform or dock).

2.10.2.1.4 Aviation

Service life aspects for aircrafts are essential part of engineering limitations. The selection of materials and processes for the aircraft designs with good corrosion protection properties is of high importance due to its potential impact on structural integrity and operating cost in terms of inspection and maintenance efforts.

An aircraft is exposed to sometimes extreme conditions, temperature gradients, and icing. The critical corrosive exposure differs strongly depending on the position in the aircraft. Especially the lower internal surfaces of the structure are affected by a huge variety of contaminations and aggressive fluids during in-service. Efficient corrosion protection starts already during the design phase. Many of the corrosion findings can be avoided by proper material selection and adequate design solutions. The EU SiCOM project [8] was a framework to tackle this aspect by modeling in a widespread fashion.

Due to the lightweight background, aluminum-related issues such as galvanic corrosion or crevice corrosion at painted structures are impinging mechanisms. With increased use of carbon fiber-reinforced plastics (CFRP), the situation becomes even more challenging. The higher-strength aviation Al–Cu alloys (Al2xxx) are anyhow affected by pitting corrosion or with selective corrosion and require high efforts such as anodizing treatments and effectively inhibiting paints to provide satisfying long-term corrosion protection and keep the maintenance efforts in an economical range [9, 10]. Particular attention is to be paid to aircrafts that operate at sea level (carriers, coastal airports) where salts introduce increased corrosion risks.

2.10.2.1.5 Automotive

Corrosion protection in car design is a major concern. Steel protected with a metallic coating (Zn, Al–Zn–Mg), plating, E-coatings, and phosphate treatments are keywords and commonly known in surface engineering for all car vehicles. Besides classical corrosion issues related to paint scratches such as underpaint corrosion or simple red rust formation, lightweight design brings in new challenges, such as in aviation. Multimaterial mix (Fe, Al, Mg, alloys, and sometimes CFRP) is a source of highly complex corrosion processes and requires adequate protection effort. Pretreatments and the whole production chain require modification. Upcoming issues with CFRP joints are still not solved as well. Also blister corrosion is a really challenging phenomenon which needs to be predicted.

Prediction of material behavior in standard corrosion tests is still not realized but is an actual research topic [11, 12]. The following section introduces the complexity of the topic and figures out the Integrated Computational Materials Engineering (ICME) aspect of this very important issue.

2.10.2.1.6 Electronics and Equipment

The reduction of the size of electronic components in computers but more importantly in portable devices (mobiles and tablets), cars, and so on subjected to atmospheric conditions results in an increasing risk of defects [13]. This is not only due to the increasing use of multimaterial use on printed circuit boards that suffer from galvanic corrosion in the presence of very thin condensation layers. Also the potential difference (e.g., the 5 V supply voltage) that is present between circuitry and components at very short distances causes corrosion at anodes and plating at cathodes such that defects and short circuits can occur.

Corrosion design is also applied in boilers, heat exchangers, filters, mixers, valves, and many other devices. Either worst-case scenarios of galvanic corrosion provide insight into efficient application of coatings, or SACP is designed and installed.

2.10.2.2 Multidisciplinarity

The main issue of getting reliable modeling results is associated with high complexity related to interdisciplinary background of most technical case scenarios. The expectations on accuracy of end users are often very high. This heavy burden hinders industrial distribution and at least the visibility of well-working software in many sectors.

Figure 2.10.2 schematically shows the broad background to be handled for corrosion-related ICME. Besides the mentioned general aspects and knowledge on the specific application, corrosion, damage, and subsequent predictive modeling require a wide background knowledge in materials science, (electro)chemistry, physics on heat, flow and mass transfer, thermodynamics, and mathematical methods. The second nonnegligible aspect is the many different and aggressive environmental situations that need to be handled in the models. For instance, the fluid flow, air access, species concentrations, friction or cycling temperature, and humidity conditions can drastically affect the corrosion rate.

2.10.2.3 Mathematical Aspects (Basic Equations) of Corrosion Modeling

When considering corrosion modeling, one has to make a clear distinction between the ionic conducting electrolyte and the electronically conducting materials being in contact with each other and the electrolyte. In electronic conductors the potential distribution V is governed by the conservation of charge. This requires the Laplace equation be solved with a given electronic conductivity. This is very often not needed and electrodes may be considered as equipotential. In the ionic conducting medium, the situation is much more complicated. Depending on the specific ion composition and the presence of dissolved gasses, the conductivity, the chemical reactions controlling, for instance, the pH or reaction products, and the electrode reactions can be very different. This explains the huge complexity of corrosion modeling. Each case is specific and requires quantified data. There is also no unique set of equations that can deal with corrosion simulations. Nevertheless, if one considers aqueous corrosion and related technologies typically in all cases, the same basic equations are applied.

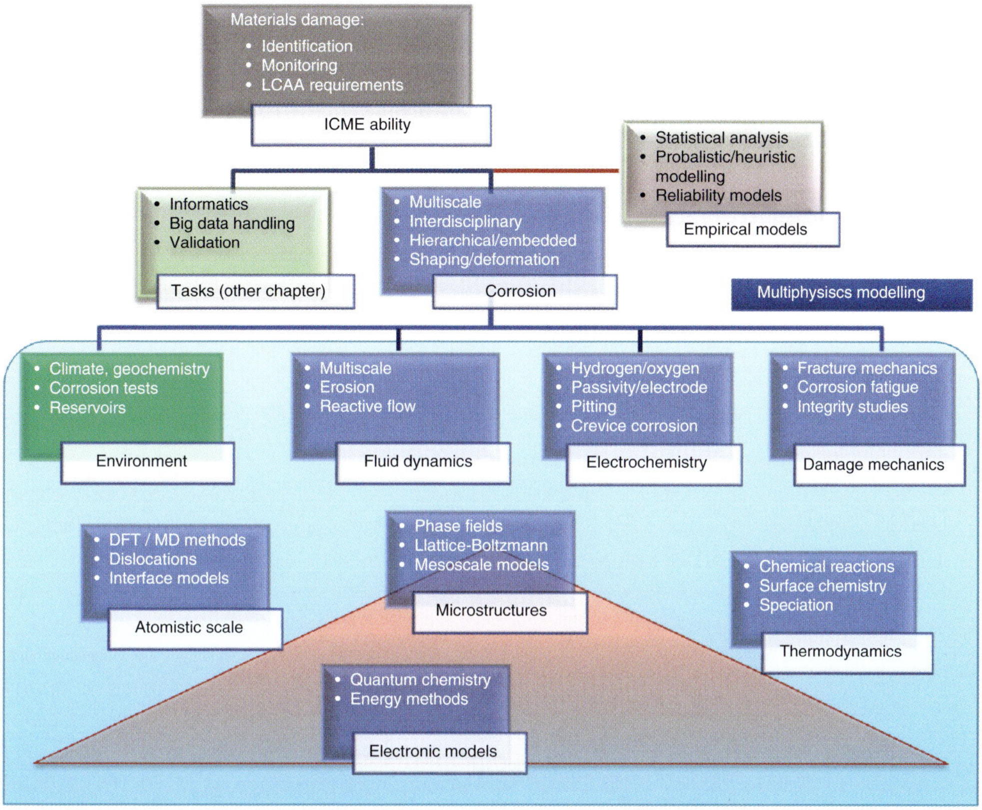

Figure 2.10.2 Integrative aspects of corrosion modeling based on approaches of C. D. Taylor. (Reprinted from [14, 15] with kind permission.)

They are derived from fundamental conservation laws (mass, momentum, energy) and Gauss's law providing the potential U in a charged medium. In most cases Gauss's law can be replaced by the electroneutrality condition. In general a partial differential equation is to be solved for each relevant species and for the potential together with proper boundary conditions. From this general approach, simplifications can be derived. When assuming no concentration differences, the potential distribution U in the electrolyte reduces to the Laplace equation with given ionic conductivity. When mass transfer is dominant, only diffusion is considered in the modeling approach. In other cases the mentioned simplifications are coupled by separating well-mixed regions from diffusion-controlled regions that are mainly seen at electrodes and occlusions. Important sets of equations are listed in Tables 2.10.1–2.10.4.

The difference in the local electrode potential V and the local electrolyte potential U is the driving force for the electrode reactions. At equilibrium, anodic and cathodic reactions are balanced, and an equilibrium potential difference U_0

Table 2.10.1 Important basic equations and quantities.

Fields of application	Equation	Description
Cathodic protection Well-mixed regions far from electrodes	$\nabla(\underbrace{-\sigma_{\text{electrolyte}}\nabla U}_{\vec{J}_{\text{tot}}}) = 0$ $\nabla(-\sigma_{\text{electrodes}}\nabla V) = 0$	Conservation of charge
Diluted solutions/ electrolyte (aqueous) exposure and dissolved species transport	$\Delta U = -\frac{F}{\varepsilon}\sum z_i c_i$ or $\sum z_i c_i = 0$ $\frac{\partial c_i}{\partial t} + \nabla\left(\underbrace{-D_i\nabla c_i - z_i u_i F c_i \nabla U_{\text{el}} + (c_i\mathbf{u})}_{N_i}\right) = R_{i,\text{tot}}$ $\vec{J}_{\text{tot}} = \sum_i \vec{J}_i = \sum_i F z_i \vec{N}_i$	Gauss's law, electroneutrality Mass balance for each species i
Diffusion-controlled corrosion (e.g., oxygen reduction)	$\frac{\partial c}{\partial t} + \nabla(-D\nabla c) = 0$ $\vec{J}_{\text{tot}} = -FzD\nabla c$	Mass balance for one controlling species

Table 2.10.2 Important basic equations and quantities.

Electrodes and electrochemical reactions	At equilibrium: $V - U = U_0 + \frac{RT}{z_i F}\ln\frac{[\text{Ox}]}{[\text{Red}]}$	Nernst equation
	At certain polarization: $V - U = b_{\text{Tafel}}\ln\left(\frac{j}{j_{\text{corr}}}\right)$	Tafel equation
	$J_{\text{ri}} = k_{\text{ox}}c_{\text{o,ox}}\text{e}^{\frac{\beta_{\text{ox}}zF}{RT}(V-U)} - k_{\text{red}}c_{o,\text{red}}\text{e}^{\frac{\beta_{\text{red}}zF}{RT}(V-U)}$	Butler–Volmer equation

Table 2.10.3 Important basic equations and quantities.

Crevices (underpaint corrosion)	First-order approximation: $L/\sqrt{\delta_0}$ and t/δ_0	Dimension scaling and time scaling of crevice corrosion evolution [16]
Pitting (steel or aluminum component failure due to crack initiation hotspot)	Initial cumulative distribution: $F(t) = 1 - \text{e}^{\left[-(t/\varepsilon)^{\nu}\right]}$ Power law of pit depth description: $d(t) = \gamma(t - t_k)^{\eta}$	Pit evolution and parameters [17, 18]

Table 2.10.4 Important basic equations and quantities.

Stress corrosion cracking–integrity failure (e.g., hydrogen embrittlement-related problems)	$K_{\text{iSCC}} \simeq \sigma\sqrt{\pi a}$	Threshold SCC–stress intensity factor at the tip [19, 20]

is observed. Out of equilibrium, different expressions for the current density are used (e.g., Tafel or Butler–Volmer equations). They are all characterized by a strong nonlinear dependency of the current density on the local potential difference $V-U$ (Table 2.10.2).

When an expression for the current density is obtained, Faraday's law is used to calculate the removal rate and finally the material removal in time. In this way different models are derived for specific applications. Some important equations are given in Table 2.10.3. The parameters in these equations are often fitted on experimental data.

In many applications functionality is in focus, and interaction between corrosion and mechanical properties is of main interest. Therefore, corrosion fatigue, hydrogen embrittlement and cracking (Table 2.10.4) endageres the general structural integrity. Crack initiation and propagation should be linked to corrosion directly. Models are very complex and require careful interpretation [21].

2.10.2.3.1 **Methods**

The mentioned partial differential equations are solved by discretizing the domain. The finite difference method (FDM) or the finite element methods (FEM) are applied. Laplacian-like problems can be solved by the boundary element method (BEM). It offers an acceptable amount of degrees of freedom even for complex shapes and geometry. Today, however more and more, also for solving the Laplace equation, domain methods have gained interest. This is due to the much improved grid generation techniques. Only in the case of large geometrical situations of buried pipelines, particular developments of the BEM remain very interesting.

At small scales, discrete methodologies based on density functional theory (DFT) become more and more useful in delivering the input for corrosion models. Also molecular dynamics (MD) simulations have become efficient tools, as shown by Taylor [14, 15]. For example, the Vienna *Ab initio* Simulation Package (VASP) [22, 23] can be applied to model the initiation of localized corrosion. Another field of application is the modeling of oxidation [24], passivation [25], or stress corrosion [26].

Other mainly discrete methods such as cellular automaton [27] or lattice models [28] have shown specific advantages at defined conditions. Such often mesoscopic approaches offer great opportunities and future perspectives for multiscale predictive modeling especially in terms of corrosion.

2.10.2.4
Model Input Data and Their Origin

It was mentioned that each corrosion case is specific and requires quantified data. A high-level characterization, action is typically required to generate reliable model parameters. Generating a corrosion database would be a tremendous, in fact impossible, work. The combination of materials, their surface treatments (coatings), and possible electrolyte solutions is endless. It also explains why corrosion modeling is not evident. Nevertheless, depending on the aimed result, there is a specific way of obtaining the input data.

When the Laplace equation is solved for, which is today the standard approach in industry, we need (i) the electrolyte conductivity, eventually as a function of composition and temperature, and (ii) the polarization behavior of the different electrodes involved. These data need to be recorded in conditions that are as close as possible to the real situation. So, validation action and experimental effort are always necessary.

When mass balance equations are involved, diffusion constants as well as flow and reaction kinetics are needed. Typically, the effort increases with smaller length scale as more details are looked at. Making a valid quantified model for a given corrosion system is a real challenge for modeling and experimentation. The broader the applicability, the more difficult this is.

Thus, big data handling and databases considering corrosion modeling demands need to be set up in the future, for example, in EU FP6, the project "Degradation modeling" developed databases (Swerea KIMAB AB) for paint delamination in automotive and building corrosion and for confined zone corrosion. Such action has to be widened and tailored.

2.10.3
Industrial Demands in ICME-Related Corrosion Modeling

Modeling and simulation tools are well established in many fields of product development and processing. Corrosion modeling has still not reached a performance level compared to, for example, structural engineering. Nevertheless, industrial end users would be interested to have a tool that is suitable to characterize their products in terms of corrosive properties and predict their long-term performance.

Relevant corrosion modeling strategies with potential for industrial implementation as an engineering tool have to consider modeling solutions at different length and timescales and should cover the full structural life of the parts and components. The holistic computational description of the corrosion performance of a structural part is at the time being by far not yet achievable. Nevertheless, there has been significant progress made in terms of corrosion and degradation modeling on different levels and areas and in different scales, and this work is continuously improving. The integration of these advances within a multilevel modeling approach and expected incremental improvements as well as immediate radical integration achievements will enable early industrial utilization. Early cooperation with industrial partners will also assure valuable engineering support to the modeling developments. The aim of modeling and simulation of corrosion and surface protection degradation is to provide more realistic and accurate methods to understand and predict the behavior of any new material systems under in-service conditions.

The corrosion behavior of metals for transport vehicles such as trains, cars, and aircrafts is a complex phenomenon governed by a huge variety of parameters. The specific material properties, a wide range of environmental conditions, interactions between the different materials, employed surface protection methods, accidental damage during manufacturing and during in-service, and quality of inspection and maintenance actions determine the corrosion protection performance. The diversity of the influencing parameters, their statistics of occurrence, and time fluctuation is quite demanding for establishing reliable characterization and testing methods and are of course also very challenging for modeling. These issues display and enforce also the huge demand for improvements for performance prediction by using modeling approaches to understand, analyze, and assess the effects, their importance, and their interactions. Individual modeling success obtained in some the different areas can be early integrated by multiscale modeling. If appropriate, missing deterministic models can be temporarily replaced by probabilistic approaches. Modeling progress may continuously be exploited to facilitate the optimization of new surface protection methods and processes for the development of new corrosion protection design principles. The application of the modeling approach should reduce the wide field of variables by analyzing and prioritizing their impact to minimize time-consuming and costly experimental testing. Thereby new opportunities will be provided at an early stage to reduce the development time and cost for future durable designs and structures.

There are a number of tools available for the different involved corrosion process steps at different modeling scales. At short length and timescales, the quantum mechanics and quantum chemistry methods are used and can give fundamental information in the understanding of occurrence of reactions. Already the exclusion of certain variables by modeling can be very useful for, for example, selection of new environmentally benign inhibiting compounds. High potential for industrial exploitation can be expected to be obtained by MD for the computation of adsorption, absorption, and diffusion, and other transport reactions can give valuable information about coating adhesion, inhibition, and inhibitor release properties and matrix compatibilities [29, 30].The state of the art of modeling at this level appears still very moderate, and current progress in corrosion engineering seems to be still very much dependent on empirical data and high-throughput experiments.

The biggest progress in modeling that can be exploited for describing the corrosion behavior has been achieved in the mesoscale modeling. This is mainly due to the big advances in localized electrochemical techniques during the last decade (e.g., scanning electrochemical microscopy (SECM), scanning droplet cell (SDC), scanning Kelvin probe (SKP), scanning vibrating electrode technique (SVET), localized electrochemical impedance spectroscopy (LEIS)) to evaluate the corrosion on heterogeneous surfaces such as intermetallic and grain boundary coating defects and monitor coating degradation behavior. The application cases of interest are intergranular corrosion, pitting corrosion behavior, crevice corrosion and paint creepage of organic coatings (filiform corrosion), or bondline corrosion phenomena.

A lot of modeling work has been performed in the area of macroscale modeling of galvanic corrosion. Several groups have successfully been working in this area, since the necessary software tools are available and almost relatively user-friendly. The tools have been successfully applied for simulation of CP in marine industry or for optimization of electrochemical deposition processes and were adapted for galvanic corrosion modeling at different scales [31].

In the field of galvanic corrosion modeling, significant progress has been achieved for the determination of corrosion protection needs and principles for new aircraft structures. The models had to be adjusted for thin electrolyte films with thicknesses below 100 μm. The challenge here was to make proper assumptions for the simplification of environmental conditions in terms of electrolyte composition representing the different areas in the aircraft and to define an appropriate experimental setup for model verification. This goal has been successfully met and the tool has been used for decision support to define corrosion protection requirements for structural design elements [8, 9]. Nevertheless, the biggest challenge for galvanic corrosion modeling is still to acquire reasonable electrochemical input data especially under atmospheric conditions. Further developments are required to address the effect of electrolyte composition changes with time on the electrochemical reactions and to consider mass transport and coating degradation phenomena to be able to more precisely predict the long-term behavior. Most corrosion findings obtained during inspections are linked with dissimilar material combination, inoperative drainage, or crevices due to degraded sealing.

On the contrary to the situation in automotive engineering, it is not feasible in aviation to perform vehicle testing in representative environments. Therefore environmental input data generation would be welcome for aircraft structures by, for example, *in situ* atmospheric monitoring to obtain more precisely the data of corrosion environmental cycles in aircraft for interior as input data for simulation. Simplification and compression of the environmental cycles may also lead to improved and more realistic accelerated corrosion test methods.

Multiscale modeling and ICME seem to be promising approaches to integrate and incrementally improve different types of models at different scale and to finally aim at service life prediction up to component level, achieving exploitable high-quality state-of-the-art software (codes).

2.10.4 Software Tool-Related Corrosion Modeling

Today corrosion software is most often very limited to well-studied corrosion systems. They are developed by research institutes to meet their own needs. Some of this software is made available (e.g., Ohio University). General applicable software is mainly found in the field of galvanic corrosion and CP. The user has to define the corrosion-specific information. Table 2.10.5 reveals examples of the availabilty of commercial and academic codes.

Table 2.10.5 Commercial and academic software to evaluate corrosion processes.

SWcat	Name of code	Website of software	Type of code (academic/ commercial) installations	Short description/typical applications/phenomena being treated
2100	Elsyca	http://www.elsyca.com/	Commercial	Elsyca is offering modeling software for evaluation and design of corrosion including galvanic corrosion and cathodic protection using sacrificial and impressed current anodes; AC interference on pipelines mitigation, surface finishing (plating and e-coating), and electrochemical manufacturing (forming and machining) Elsyca CPMaster, Elsyca Catpro, Elsyca V-PIMS, Elsyca IRIS, Elsyca CorrosionMaster, Elsyca CurveAnalyzer
2100	BEASY	http://www.beasy.com/	Commercial	BEASY offers a range of computer simulation software products to predict the effectiveness of cathodic protection systems, the risk of galvanic corrosion, corrosion-related electric and magnetic signatures, the risks associated with crack-like defects and crack growth simulation
2100	COMSOL	https://www.comsol.de/corrosion-module	Commercial	The Corrosion Module includes features, interfaces, and example models that enable a straightforward approach to the simulation of all electrochemical corrosion processes, such as galvanic, pitting, and crevice corrosion. It also allows you to design effective corrosion protection systems. This includes the simulation of impressed cathodic current protection (ICCP), sacrificial anodes, and anodic protection, where anodic current is impressed on corroding material to enforce passivation

(*continued overleaf*)

Table 2.10.5 *(continued)*

SWcat	Name of code	Website of software	Type of code (academic/ commercial) installations	Short description/typical applications/phenomena being treated
2100	InterCorr Software	http://www.intercorr.com/software	Commercial	The Predict 4.0 system is a new generation software tool that addresses one of the most significant issues in corrosion evaluation, that is, assessment and prediction of corrosion rates for steels exposed to corrosive environments. The PredictPipe 3.0 system addresses the assessment of corrosion rates in dry gas transmission pipeline systems exposed to corrosive environments due to water condensation/accumulation. The Socrates 8.0 system is a comprehensive material selection tool for oil and gas applications
2100	FREECORP/ Multicorp	http://www.corrosioncenter.ohiou.edu/software/freecorp/	Academic	FREECORP V1.0 is a simple corrosion model, strongly rooted in theory. Currently, this model is capable of predicting uniform corrosion of carbon steel at a single point in an environment containing carbon dioxide, acetic acid, oxygen, and/or hydrogen sulfide. Iron carbonate film formation, a key factor in carbon dioxide corrosion, is simulated using an empirical correlation to improve the accuracy of corrosion rate prediction. Multicorp is a corrosion prediction engine that enables the simulation of corrosion under various conditions and in various environments
2100	MiTREM	MuPhys	Academic	Muphys is an academic software toolbox based on the Finite Element Method. It is a combination of C++ code and Python. It is dedicated to develop and solve non-standard multi-physics models. The user can construct sets of coupled partial differential equations that apply in different regions. This set can be combined with other equations and be solved in time. MuPhys is applied to electrochemical problems like (atmospheric) corrosion, nucleation and growth, transport in fuel cell membranes, batteries, etc.

2100	OLI Studio: Corrosion Analyzer	http://www.olisystems.com/new-corrosionanalyzer.shtml	Commercial	Corrosion Analyzer offers several unique features which comprise OLI aqueous corrosion simulation technology: polarization curves, Pourbaix diagrams, tendency for localized corrosion
2100	Seacorr	forcetechnology.com		Cathodic protection
2100	enpICDA	http://www.enpicda.com/	Commercial	enpICDA predicts internal pipeline corrosion in flowing aqueous-bearing systems for wet natural gas, wet sour gas, wet crude or wet stabilized oil, wet gas condensate, H_2O injection, or disposal pipeline systems
2100	Predict 6.0 + other products	www.honeywellprocess.com		Carbon steel in CO_2 and H_2S environments
2100	ECE Electronic Corrosion Engineer	www.intetech.com		Estimation of corrosion rates for oil and gas
2100	List of old tools	http://www.corrosionist.com/Software_for_Corrosion_Prediction.htm	Commercial and academic	

The complexity, variety, and often multidisciplinarity of corrosion systems are the main obstacles for general applicable corrosion software. In many cases the corrosion mechanisms are even not known or studied in detail. So quantified data are missing. A combined effort on EU level could resolve that issue, at least for common applications. Related to the mentioned difficulties to get the proper system information for corrosion modeling, it is evident that software for given dedicated conditions is available.

2.10.5 Future Tasks and Limits

Within the simulation and modeling community, the challenges of realizing a well-predictive corrosion software tool are known. This type of models has become an essential part of research as shown by recent EU activities and calls. For various industrial sectors, corrosion protection engineering is assisted by simulation tools. They exhibit well-established performance such as in the CP sector. Nevertheless, CAE regarding corrosion toward virtual engineering performance, for example, in metallurgy is still in the developmental period.

The roadmap to be paced off includes "on-the-fly" coupling/modeling action, computational homogenization and linking of discrete and continuum models, database management, model input data generation, and many more. Full-chain corrosion-induced failure and damage modeling also require clear definition of test conditions being able to upscale to service life performance and even prediction. Required exposure case simulation is one of the main reasons of extraordinary complexity. Thus, future ICME developments in terms of corrosion-related predictive modeling require progress in:

- Multiphysics/multidisciplinary cross-linking based on computational models
- High-level experimental validation action and a feedback loop toward new characterization action in order to generate a database regarding corrosion/damage prediction
- Improved temporal validity and reproducibility

2.10.6 Acknowledgments

The authors thank Dr. M. Amkreutz and Dr. W. Dietzel for their kind contribution.

References

1. Vankeerberghen, M., Weyns, G., Gavrilov, S., Henshaw, J., and Deconinck, J. (2009) *J. Nucl. Mater.*, **385**, 517–526.
2. Taylor, C.D. (2012) *Int. J. Corros.*, **2012**, 204640.
3. Nilenius, F., Larsson, F., Lundgren, K., and Runesson, K. (2015) *Comput. Mech.*, **55**, 359–370.
4. Van der Weeën, P., Zimer, A.M., Pereira, E.C., Mascaro, L.H., Bruno, O.M., and

De Baets, B. (2014) *Corros. Sci.*, **82**, 133–144.
5. von Baeckmann, W., Schwenk, W., and Prinz, W. (1997) *Handbook of Cathodic Corrosion Protection*, Gulf Professional Publishing.
6. Mathiazhagan, A. (2010) *Int. J. Chem. Eng. Appl.*, **1** (3), 217–221, ISSN: 2010-0221.
7. Baete, C. (2012) *Mater. Perform.*, **51**, 18–20.
8. DeRose, J.A., Suter, T., Hack, T., and Adey, R. (2013) *Aluminium Alloy Corrosion of Aircraft Structures: Modelling and Simulation*, WIT Press, Southampton.
9. Palani, S., Hack, T., Deconinck, J., and Lohner, H. (2014) *Corros. Sci.*, **78**, 89–100.
10. Abodi, L.C., DeRose, J.A., Van Damme, S., Demeter, A., Suter, T., and Deconinck, J. (2012) *Electrochim. Acta*, **63**, 169–178.
11. Bösch, N.C., Höche, D., Mittelbach, A., and Kainer, K.U. (2014) *Euro Hybrid*, DGM, Stade.
12. Deconinck, J. and Dolgikh, O. (2013) RFCS TGS5 Workshop Future Trends and Opportunities in Finishing and Coating for the European Steel Sector, Brussels.
13. Ambat, R., Jensen, S.G., and Møller, P. (2008) *ECS Trans.*, **6**, 17–28.
14. Taylor, C.D. (2014) *Strategic Research & Innovation Position Paper 8*, DNV GL.
15. Taylor, C.D. (2015) *Corros. Eng. Sci. Technol.*, **50** (7), 490–508.
16. Song, F.M. (2012) *Corros. Sci.*, **57**, 279–287.
17. Valor, A., Caleyo, F., Alfonso, L., Rivas, D., and Hallen, J.M. (2007) *Corros. Sci.*, **49**, 559–579.
18. Ascher, H. and Feingold, H. (1984) *Repairable Systems Reliability: Modeling, Inference, Misconceptions and their Causes*, Marcel Dekker, New York.
19. Turnbull, A. (1993) *Corros. Sci.*, **34**, 921–960.
20. Gavrilov, S., Vankeerberghen, M., Nelissen, G., and Deconinck, J. (2007) *Corros. Sci.*, **49**, 980–999.
21. Dietzel, W. and Turnbull, A. (2007) Stress Corrosion Cracking. Report GKSS 2007/157, GKSS-Forschungszentrum Geesthacht GmbH, Geesthacht.
22. Kresse, G. and Furthmüller, J. (1996) *Comput. Mater. Sci.*, **6**, 15–50.
23. Bouzoubaa, A., Diawara, B., Maurice, V., Minot, C., and Marcus, P. (2009) *Corros. Sci.*, **51**, 2174–2182.
24. Costa, D., Ribeiro, T., Mercuri, F., Pacchioni, G., and Marcus, P. (2014) *Adv. Mater. Interfaces*, **1**. doi: 10.1002/admi.201300072.
25. Liu, S.-Y., Shang, J.-X., Wang, F.-H., Liu, S.-Y., Zhang, Y., and Xu, H.-B. (2009) *Phys. Rev. B*, **80**, 085414.
26. Kart, H.H., Uludogan, M., and Cagin, T. (2009) *Comput. Mater. Sci.*, **44**, 1236–1242.
27. di Caprio, D., Vautrin-Ul, C., Stafiej, J., Saunier, J., Chaussé, A., Féron, D., and Badiali, J.P. (2011) *Corros. Sci.*, **53**, 418–425.
28. Šavija, B., Luković, M., and Schlangen, E. (2014) *Cem. Concr. Res.*, **61–62**, 49–63.
29. Stampfl, C., Veronica Ganduglia-Pirovano, M., Reuter, K., and Scheffler, M. (2002) *Surf. Sci.*, **500**, 368–394.
30. Lohner, H., Poupon, C., and Beneke, M. (2010) Deutscher Luft und Raumfahrt Kongress, Hamburg.
31. Bortles, L. and Rose, A. (2011) Sustainable Surface Engineering for Aerospace and Defense Workshop, ASETSDefense, New Orleans, LA.

2.11
Recycling Processes

Klaus Hack, Markus A. Reuter, Stephan Petersen, and Sander Arnout

2.11.1
Overview

In the context of Integrated Computational Materials Engineering (ICME) recycling covers processes at the end of the lifetime (EoL) of mechanical, electronic, or other components of machines and devices. However, since the components are all made of particular materials, it is the material-related aspects that govern the field of recycling. In principle, one could consider the materials that are to be recycled as a kind of artificial raw materials. These require for their conversion into new products processes, which can be compared with those applied when treating natural raw materials. In fact, they are in almost all cases ultimately integrated into the usual production routes as an additional class of feed materials. Depending on the state in which they exist as EoL material they can either be used directly for new material production, for example, used glass in the form of cullet, used beverage can (UBC) Al or steel scrap, or indirectly, for example, zinc (oxide) dust, which is fed into a special refining process (zinc fuming) in order to generate a pure zinc oxide, which can in turn be used as feed material for the normal production route.

This chapter provides a short overview of materials that are being recycled today and the methods applied. There is a special focus on computational methods applied in the field of recycling, the foremost being methods of computational thermochemistry since mutual solubilities, phase transformations, and reactions as well as heat balances play the most important part in recycling just as they do in standard production. It should, however, be noted that this chapter serves as a key to more information (see the keyword list) rather than as a complete picture of the present state of "ICME in recycling."

Keyword short list: recycling, end of lifetime (EoL) materials, materials-centric recycling, product-centric recycling, physical separation methods, chemical separation methods, computational thermochemistry, steel, aluminum, zinc, glass, valorization of slags, heat balances, liquid and solid solubilities

Handbook of Software Solutions for ICME, First Edition. Edited by Georg J. Schmitz and Ulrich Prahl.

2.11.2
Materials-Centric versus Product-Centric Approach

Resource efficiency, if it is to be of proper value, should cover the big picture of metal production and particularly the aspect of "closing" the loop. It requires considering the whole value chain as being circular, not linear, as we are most often still accustomed to. It must be remembered that metals can be reused (eternally, speaking on the geological timescale), which makes them an important component of a resource-efficient society. This requires recovering metals from residues, scrap, and used products within the limiting constraints of both thermodynamics and economics. In short, with metallurgical knowledge and associated infrastructure, a circular economy can be realized. According to the laws of physics (e.g., Second Law of Thermodynamics), there is, however, inevitably always something left over from any process. Ultimately, economics with a strong relationship to thermodynamics and the mineralogy of the secondary material, that is, a product-centric approach, defines how much of that which is left over will end up as landfill.

The wealth of knowledge about processing complex minerals has evolved over many years. It is standard practice to recover as far as possible the minor elements that accompany the major element of a mineral to maximize profit. The same principle is applied when processing EoL goods and materials. In order to maximize resource efficiency, one must always consider all the elements in a product (much like a mineral) and quantify the destination of all these in order to obtain a true picture [1, 2].

In order to be sustainable, that is, to establish a resource-efficient circular society, we must attempt to decrease the amount of secondary resources we create. The base metal industry (copper, lead, nickel, tin, zinc) has always emphasized on extracting as much value from minerals as possible in the high-tech processing infrastructure. Now, there is a need to do the same for the increasingly complex EoL consumer products. Various solutions already exist for treating, for example, e-waste, residues, waste to energy, or slag cleaning. However, the complexity of

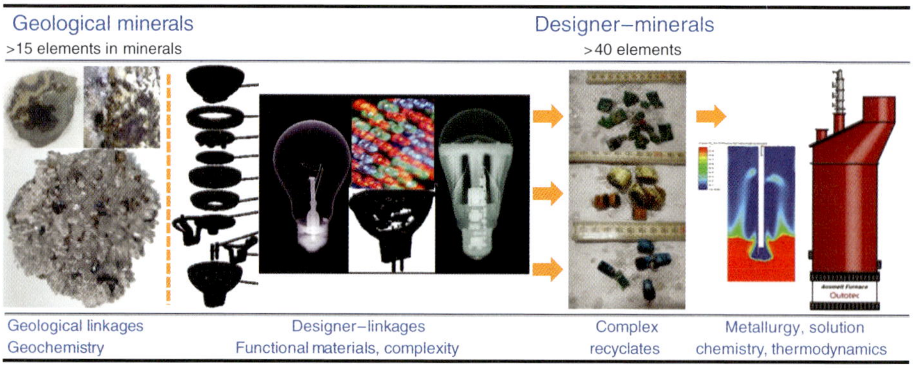

Figure 2.11.1 Complexity of minerals vis-à-vis designer minerals – the basis of product-centric recycling [1, 2].

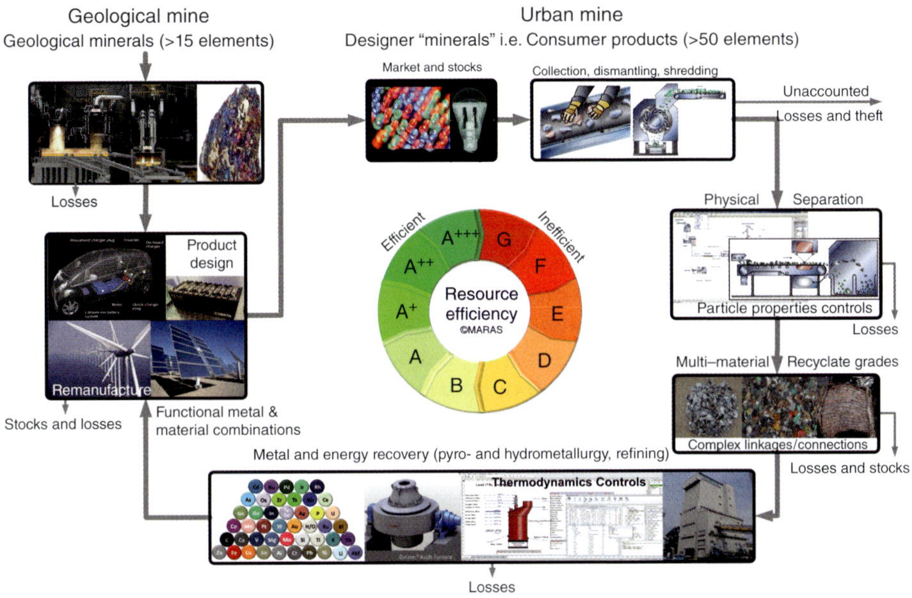

Figure 2.11.2 SIMP requires a consideration of the system and all its elements at the same time to quantify the losses. These can be related back to basic scientific methods, such as solution thermodynamics. (Reproduced from [3, 4] with the permission of Elsevier.)

EoL goods (Figure 2.11.1) makes it rather a challenging task to capture everything from these goods that has value and at the same time create the minimum amount of final waste. Doing so is one of the key drivers of a resource-efficient society but requires an understanding how the minor elements (such as precious metals – PMs, or platinum group metals – PGMs) behave in the recycling system and how the complex interactions of materials and product design affect their loss from the circular economy. At this point, system integrated metal production (SIMP) comes in, which is described in Figure 2.11.2 showing the interlinkage of reactors, product design, processing plants, losses, and so on.

2.11.3 General Phenomena: LED Lamp Recycling as an Example

As already indicated earlier, without detailed thermodynamic as well as fluid dynamic knowledge of metallurgical reactors a circular economy cannot be realized to its fullest. Furthermore, a detailed understanding of the chemical and physical complexities of recyclates and the material combinations is necessary for the understanding of the inevitable losses from the system. Figure 2.11.2 highlights a variety of these losses and the phenomena creating them. In fact, there are already tools (discussed and listed later) permitting the quantification of these losses. Indicated in Figure 2.11.2 are simulation models of sections of the system. Ultimately product design must be linked to complete system models based on solution thermodynamics, fluid dynamics, reaction kinetics, and other basic scientific methods in order to quantify the resource efficiency.

From Figure 2.11.2, it can be deduced that a deep understanding of the big seven metals (aluminum (+magnesium), copper, lead, nickel, iron, zinc, tin) and their metallurgy (always in the context of the complex Web-of-Metals (WoM) and their associated minor elements) is a key to realizing a circular economy. These key metals are used in various products and their source minerals are the suppliers of important technology elements that create key functionality in products. In addition, the capability of these elements to dissolve/process various metals in molten metallic or related aqueous solutions and the availability of an appropriate metallurgical infrastructure are crucial to recover these minor elements. Furthermore, in a product-centric system, the true value of metallurgy becomes apparent. Just considering metals in isolation can lead to erroneous conclusions and harm the sustainability of the Circular Economy system.

Due to the relative nobility of particularly lead and copper, these elements can be recycled well, if collected, and excellently dissolve and release minor elements for further recovery in suitable refining infrastructure. These elements are enablers of closed-loop recycling and fit excellently into a product-centric (PC) approach. The relative ignobility of iron (steel) and aluminum on the other hand fit into a more materials-centric (MC) open-loop recycling system as, generally, dissolved elements must be diluted in order to produce the desired alloys. This explains why in Figure 2.11.2 good product design and physical separation play key roles to minimize open-loop recycling that has an intrinsic upper recycling rate determined by the amount of dilution required when producing new metallic material.

In recent papers [1, 2], the recycling of LED lamps was estimated, starting from product design and ending in produced metal as shown in Figure 2.11.2, using process simulation tools to link the product "mineralogy" to the metallurgical and energy recovery of the metals, materials, and compounds.

A typical LED consists of numerous metals and materials, which during recycling creates numerous complex recyclates as shown in Figure 2.11.1. It is shown that from the results of the simulation linking product design to metallurgical recovery recycling rates can be calculated for the complete product already during product design. It is shown that due to complex process metallurgy the recovery and thus recycling rates of metals and thus of the complete product is low. Of importance is that in the smelting and other metallurgical technology thermochemistry is considered to ensure that metals distribute into the correct phases from which these can be recovered, or if uneconomic, stored in a benign form. It is important also to derive from these complex models easy-to-understand results such as the recycling label shown in the right of Figure 2.11.2. Fundamental of resource efficiency is thus to visualize complex thermodynamic and simulation results in a form that is understandable to everyone.

2.11.4 Methods Available

Figure 2.11.2 shows various aspects and phenomena that affect the final recovery of all elements from a product. These include the lifetime of products, that is, the residence time of materials in the market, a multitude of physical separation techniques, and subsequent metallurgical processing as discussed in detail in [1, 2, 4].This requires among others:

- A well-founded analysis of use of materials after their production in order to know which materials will return when into the recycling loop
- A detailed understanding of the metallurgical technology as well as underlying physics and thus also thermodynamics as well as kinetic and transfer processes in the system to ensure that metal recovery is maximized
- Digitalization of the system with simulation system models, thermodynamic solution and kinetic models, and so on and using these to control operating conditions. The left part of Figure 2.11.3 shows some of these techniques, that is, simulation of processes and systems. Also see Table 2.11.2 for other simulation tools that can be applied.
- Linking computer-aided design with metallurgical recovery in order to predict recyclate material associations. This permits the understanding of the destination of elements from a product as dictated by the solution thermodynamics.
- Determination of all streams with a process simulator inclusive of their mineralogy of the complete system as shown in Figure 2.11.2 in order to investigate a detailed environmental impact as indicated by the tools on right part of Figure 2.11.3.

At the heart of sustainability and resource efficiency is the optimization of a complete system as depicted in Figure 2.11.4. This figure provides some details of what can be the reality within the bottom block of Figure 2.11.2. A deep understanding of the physics in each reactor and the quantification of the flows of all materials in the complete system by suitable simulation software permits the quantification of the economic and environmental performance of the system.

Figure 2.11.3 depicts on the left-hand side a typical simulation tool and the usually extracted design parameters for a reactor or a complete plant or system as shown below the left block. Once all the flows, temperatures, compositions of all streams, and so on have been calculated, appropriate environmental analysis can be performed. This is enhanced by collecting all the design data in an appropriate format and transferring it directly to the environmental software enabling an effortless analysis.

Thus, the combination of the usual simulation data that provides the basis for Capital Expenditure (CAPEX) and Operational Expenditure (OPEX), with environmental indicators as shown in the right side of Figure 2.11.3 as well as simulation-based Exergy information, can provide a good basis to optimize complete systems as shown in Figure 2.11.4.

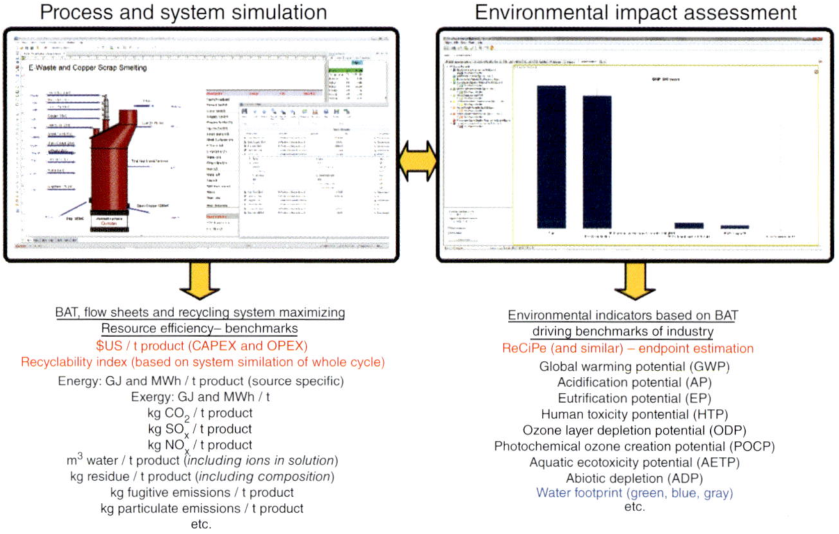

Figure 2.11.3 Simulation and the environmental footprint of the recycling system. On the left is HSC Sim8 and on the right is GaBi, which is linked via suitable file transfer format to the simulator enabling a direct environmental analysis of the process design. (Reproduced from [1, 2, 4] with the permission of Elsevier.)

Figure 2.11.4 reveals the type of processing complexity that resource efficiency considerations have to deal with. Maximizing the recovery of metals from a wide variety of materials requires a detailed understanding of the system. This can arguably be best done with rigorous modeling and simulation as has been shown in [1, 2, 4, 5].

Figure 2.11.4 shows, for example, the position in the bigger system of Top Submerged Lance (TSL) and plasma fuming technologies that are mentioned later.

Ultimately, complete system optimization hinges on the fundamentally simulation-based quantification of environmental performance. Some of the examples in this chapter show this.

2.11.5
Thermochemical Aspects of Recycling

As indicated earlier, recycling can be considered as using an additional class of materials, the recyclates, as a further input to the otherwise classical routes of metallurgical processing. In some cases, for example, for zinc recovery, additional process loops (see 2.11.7) are necessary before the material is ready to enter the classical route again. In almost all cases, there are reactive metal–slag–gas systems at elevated temperatures to be considered. Gibbs energy-based chemical thermodynamics lends itself easily to the investigation and, ultimately, simulation of these processes. The Gibbs energy with its natural variables temperature, pressure, and overall composition of the system is most suited to treat the majority

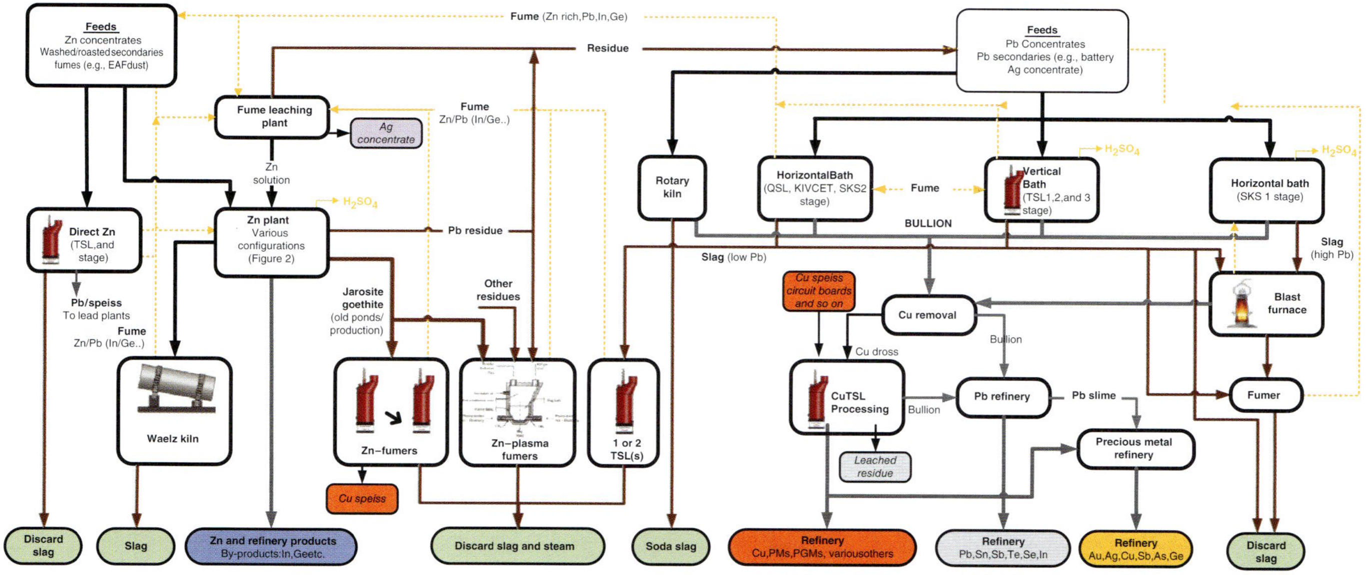

Figure 2.11.4 The various zinc, lead, and copper processing options that have to be optimized to increase resource efficiency. System optimization models are defined by the thermodynamic, kinetic, and transfer processes in each reactor and by the economical, technological, and environmental constraints of the system as a whole. (Reproduced from [5] with the permission of GDMB Verlag.)

of systems, although sometimes it may be more practical to replace pressure by its conjugate extensive property volume. Gibbs energy has yet another intrinsic advantage, which can be seen directly from the Gibbs–Helmholtz equation: $G = H - T \times S$. It contains both the enthalpy and the entropy of the system in question. Beyond obtaining complete equilibrium states by way of its minimization, on the one hand it thus enables the calculation of heat balances, that is, it covers the energetic aspects of the respective process, and on the other hand entropy balances can be derived, which are necessary for exergy analyses of the processes, that is, investigation of the quality losses of both energy and materials (see [6, 7]). Furthermore, there is now a wide range of both standard software and databases available (see Chapter 4 for more details), which enable easy use of the calculations involving equilibrium constants for given reactions, activities of constituents of solution phases with many components exhibiting nonideal chemical interactions, or two-dimensional sections through phase diagrams of multicomponent systems. In addition, there are programmer's libraries such as ChemApp [8, 9], which permit the user to apply rigorous multicomponent multiphase thermochemistry in complex environments such as process modeling and flowsheeting tools (e.g., ASPEN+, PRO/II, or gPROMS; see Table 2.11.2), computational fluid dynamics (CFD) codes, or a user's own software. In fact, there is even the tool SimuSage [10], which is provided by the producers of ChemApp and links ChemApp via a suite of both visual and nonvisual software components to the programming environment Delphi® [11], thus enabling the user to generate easily and rapidly the code of a flowsheet type of process model with a few drag-and-drop actions using graphical icons. Such models, based on the principle of interlinked local equilibria, permit to consider simultaneously the mass and enthalpy balances of the process and its individual parts in an interconsistent way. It is now even possible by way of constraint Gibbs energy minimization [12, 13, 14] to integrate proper reaction kinetic data into such models, thus permitting time to become an explicit parameter of the process models. However, it must be stated clearly that such kinetic data are badly lacking.

At this point, it is also necessary to focus on a sometimes still severe limitation of the standard Gibbs energy-based approach: the incompleteness of the thermochemical databases themselves. Although for many cases, particularly with respect to the major components of the respective chemical systems, for example, steels, Al- or Mg-based light metal alloys, slags or salts, the required Gibbs energy data are available, it still so happens that minor components are not covered by the databases. A presently pressing case is, for example, the recycling of EoL car catalysts. It is no problem to model the phase relationships of the components of the ceramic support structure on melting. However, there are presently no data that would at the same time permit the modeling of the thermochemical behavior of the PGMs, which actually are the vital chemical ingredients of the catalyst and also the most valuable ones. It must, therefore, be emphasized strongly here that a continuation of classical experimental work on the generation of basic thermochemical data such as enthalpies of formation or solution, component activities, or temperatures and heats of phase transition is vital for the improvement of future process modeling. It is hoped that the experimental work will soon find extensive support by quantum chemical methods, so-called *ab initio* calculations (e.g., [15, 16]) for the basic thermochemical properties of substances, that is, enthalpies of formation, absolute entropy, and heat capacity as function of temperature and pressure.

Figure 2.11.3 summarizes a variety of simulation techniques mentioned in Section 2.11.4 and their results. The cases shown have their roots in thermochemistry as well as plant and reactor knowledge to ensure that metal distributions to different phases are captured correctly.

Very large recycling systems in which a multitude of metals and complex particulate recyclates, derived from a large number of EoL products, are processed in reality and their simulation is indicated in Figures 2.11.2 and 2.11.3. Approaches for the simulation of reactors with complex liquid and solid solutions such as ChemApp/SimuSage can be useful to optimize such systems. However, various technological aspects may nevertheless limit these approaches due to process complexity, particularly if there are very strong temperature gradients and complex mixing relationships as well as distributed reaction rates and kinetic phenomena [17].

In each of the examples here, we briefly refer to the complexity and specifically to the detailed solution and thermochemistry detail required to complete this task. We have selected three examples, namely, aluminum, zinc as well "waste" valorization or, in other words, repurposing. The reader is referred to [4] for details on a variety of examples of recycling of metals and materials.

2.11.6 Recycling of Aluminum

The recycling of metals and various other materials has been discussed in detail in [4]. In this section, the discussion is limited to aluminum (interested reader can refer to [4]).

Aluminum metal is sourced from primary aluminum and from recycled aluminum. When in use, it is an energy and resource bank, but due to the long life of many aluminum products, and due to growing demand, this "bank" can only supply 20–25% of the current demand. The rest must be produced from primary aluminum.

When it comes to recycling, aluminum has two faces: On the one hand, there is a large cycle that deals with essentially just one product, that is, an MC cycle. This cycle pertains to UBCs. Aluminum contained in such cans can be remelted and used again for the purpose of beverage cans with only small compositional corrections. On the other hand, aluminum is part of many products, for example, cars, construction materials, or household appliances, in which it is used together with other elements. Therefore, this kind of Al must first be separated by physical means from the other accompanying materials. This is a more complex PC cycle. This complexity has been investigated and simulated in some detail. Xiao

Table 2.11.1 Al qualities as used in a weight-optimized car (Audi A8) in kilograms.

Engine, gear box, etc. (cast)		243
Car body		
Sheet	150	
Profiles	70	
Cast joints	58	278
Total		521

and Reuter [18] show the effect of impure scrap on recovery and relates this to surface tension effects within the salt slag. Zhou *et al.* [19] goes further and simulates scrap smelting and shows the effect of various aspects such as scrap size on the rotary kiln performance using CFD.

There is another point in this which is not immediately obvious. Aluminum of different quality is, for example, being used in cars, and some parts are made of wrought aluminum, for example, door panels or front and rear bonnets, while others are made of cast aluminum, for example, joints or suspension parts. When these materials enter together into a shredder plant, it is virtually impossible to separate the high-quality wrought aluminum from the lower-quality cast aluminum. Thus, a downgrading of the wrought aluminum takes place, which can only be replaced by new material. Table 2.11.1 shows the large part of wrought Al that would be downgraded by simply mixing it with the cast Al that is also being used for the case of an Al-intensive weight-optimized car (Audi A8).

In order to maintain the amount of high-quality wrought Al upon recycling, it is important to be able to predict the material behavior of recycled aluminum depending on its microchemistry. Being able to model and thus predict the influence of the locally occurring composition on, for example, hardening or softening becomes essential when more of the so-far avoided elements occur in wrought Al because of recycling. Such modeling is one of the links of the ICME chain. A brief summary of such modeling is given later.

The so-called ClaNG model (classical nucleation and growth [20, 21, 22]) that was initially developed to handle microstructural changes on heat treatment was adapted to predict the matrix composition and the particle distribution of the precipitates as well as material properties such as the electrical resistance and mechanical properties. The treatment of nucleation and growth of the precipitates has also been incorporated in this model. Figure 2.11.5 shows the general flow scheme of the algorithm.

A prerequisite for the modeling of heat treatments is a good knowledge of the solid state that is produced during solidification of the original alloy composition. For Al alloys, it was found that the so-called Scheil approach [23] for solidification can be used to obtain the fraction solid curve as a function of temperature together with the sequence of solidifying phases as well as the amount and

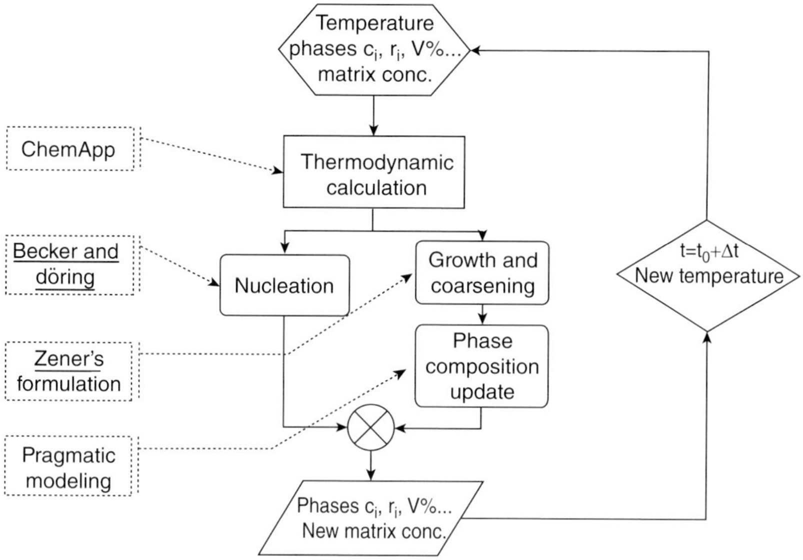

Figure 2.11.5 Flow scheme of the ClaNG model.

composition of the final eutectic particles. Such calculations can nowadays be executed in standard thermochemical software [e.g., Pandat, Thermo-Calc, FactSage; see Table 2.11.2 and Chapter 4]. The fraction solid curve for an AA5182 alloy is given as an example in Figure 2.11.6.

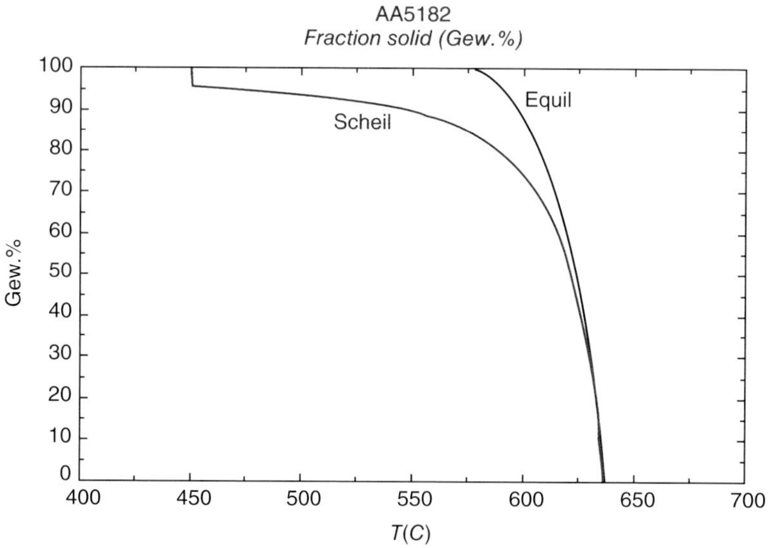

Figure 2.11.6 The fraction solid curves for AA5182, both from equilibrium and Scheil approach. Final solid at 450.23°C contains AL13FE4, FCC matrix, AlMg_beta, and Mg2Si.

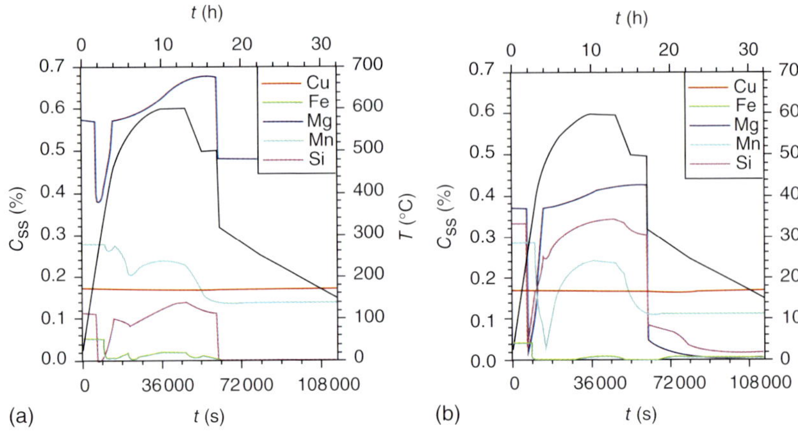

Figure 2.11.7 ClaNG simulation of the matrix solubilities by a homogenization heat treatment of an AA 3105 alloy, based on (a) pure components and on (b) recycling material with increased scrap contents

The final result of the modeling and simulation with the ClaNG model is given in Figure 2.11.7 in the form of a comparison between an AA3105 alloy based on pure component input (a) and a recycling alloy with increased scrap content (b). It becomes obvious that the recycling alloy shows considerably lower matrix solubility than the alloy made of "pure components." This indicates a higher density of secondary phases, which was confirmed by considerably higher electrical conductivity. More investigations are needed in order to understand how the differences in microchemistry affect the recrystallization behavior and thus certain anisotropic material properties upon subsequent forming operations.

2.11.7
Recycling of Zinc: Fuming

Optimizing the system shown in Figure 2.11.4 requires deep understanding of each unit operation within the metallurgical systemic context. Huda *et al.* [17] investigated the mass and heat transfer in detail within the fuming TSLs that are depicted in Figure 2.11.4 using CFD coupled to zinc fuming reaction kinetics. The example discussed in the following shows another type of zinc fumer, namely, a plasma fumer as also shown in Figure 2.11.4. This example explores important issues of this type of reactor.

Due to the environmental impact of zinc, the processing of Zn-containing residues has been of significant importance to the metallurgical industry. Examples of Zn-containing residues include zinc leach residue, a coproduct of zinc-producing processes, and electric arc furnace dust, which is a coproduct of steel recycling. In recent years, newly developed continuous zinc slag fuming processes are trying to overcome the technical and economic limitations of

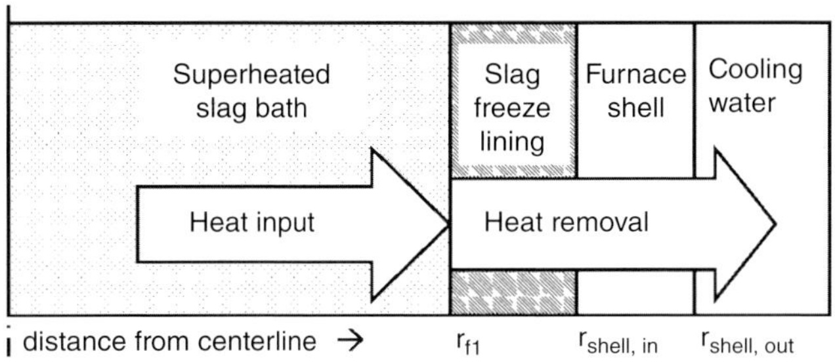

Figure 2.11.8 Schematic representation of the principal heat flows in freeze linings. (Reproduced from [24] with the permission of Springer.)

conventional batch operations. To achieve a high zinc fuming rate, the slags need to be superheated, resulting in highly aggressive slags that readily dissolve conventional refractory materials. In order to protect the water-cooled steel furnace shell of the reactor, newly developed processes are instead operated under conditions that allow the formation of a stable protective frozen slag lining separating the molten slag from the steel vessel. In order to maintain the freeze lining, heat must be continuously extracted through the reactor's steel shell.

While using a freeze lining instead of a conventional refractory lining appears to be a simple and elegant solution for the material-related challenges of the process, selecting optimal operating conditions for the process is much more difficult and dramatically changes the required approach to process design. It involves a significant number of process parameters, which need to be adjusted but which in their entirety must always be selected so as to maintain a stable freeze lining. Finding such conditions requires a mathematical model that determines the position of the "hot face," the interface between the superheated slag bath and the freeze lining (Figure 2.11.8).

It is obvious that the properties of the hot face are strongly dependent on the thermochemical properties of the slag, which are dependent on the slag composition, thus reliable freeze lining models depend on high-quality thermochemical slag data. For the new submerged plasma zinc fuming process (Figure 2.11.9), a mathematical freeze-lining model was developed, which simultaneously describes the chemical, thermal, and heat transfer properties of this process [24, 25].

The model, based on ChemApp [8, 9] and using thermochemical data from FactSage [26], considers the overall process to take place in a sequence of four reaction stages, where thermodynamic equilibrium is assumed to be attained in every stage (Figure 2.11.10).

In the preheater and the mixer stages, blast air is preheated with electrical energy and mixed with fuel. The mixed gas and the feed (residue, flux, reductant) are then introduced into the zinc fumer, where actual fuming reactions take place. The resulting liquid slag and matte phases are tapped from the furnace, whereas the gas phase moves into the postcombustor where a postcombustion reaction takes place with the secondary air. The off-gas and solid oxide particles are then removed (Figure 2.11.10).

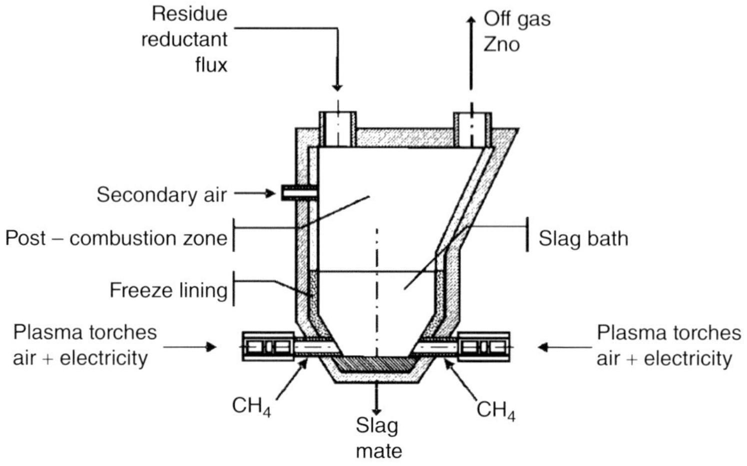

Figure 2.11.9 Schematic of the submerged plasma process for the high-temperature fuming of zinc from zinc-containing residues (see Figure 2.11.4 for its position in the bigger picture). (Reproduced from [25] with the permission of Springer.)

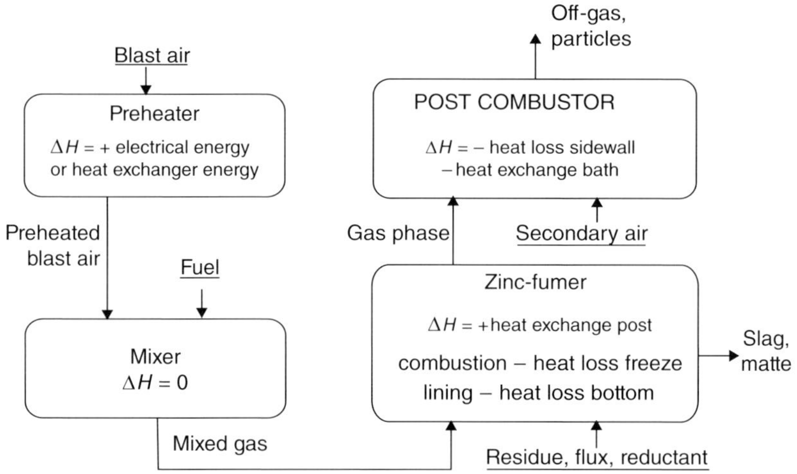

Figure 2.11.10 Flowsheet of the general zinc fuming model. (Reproduced from [24] with the permission of Springer.)

The model also takes the enthalpy differences into account, a prerequisite both for the determination of the stable freeze lining, as well as for optimizing the energy efficiency of the entire process. The energy transfer, particularly from the postcombustion to the slag bath, is not entirely predictable from models and needs to be adjusted to match production data. The model was used to investigate the potential effects of operating parameters such as feed composition, fuel/oxygen ratio, electrical power, and fluxing parameters on bath temperature, heat loss, Zn concentration and content in slag, and Cu concentration and content in matte. The use of the high-quality thermochemical data together with ChemApp has enabled complex phase equilibria calculations for multicomponent slags to be introduced to the freeze lining module, a feature that was not available in previous freeze lining models, and which is thus considered to be one of the strengths of this approach. This integrated model thus provides a tool for "slag engineering" to optimize process outcomes and maximize furnace integrity. Predictions from this model resulted in a number of important conclusions, some of which are initially counterintuitive. In contrast to conventional practice, high liquidus temperatures lead to low heat losses through the freeze lining and stable operation of the process. Limiting the heat losses through the freeze lining and the enthalpy required to maintain high bath temperatures may be done through appropriate control of the freeze lining and through adjustment of the slag chemistry.

Another technology of major industrial importance for zinc recovery in steel production is the use of Waelz kilns as shown in Figure 2.11.4. While Waelz kilns have been used for over 80 years, recent developments have resulted in significant improvements, primarily with respect to the efficiency of this energy-intensive process. However, the final slag often still contains some heavy metals in a leachable state, which makes the reuse of the slag difficult. Nevertheless, the Waelz process is considered to be the best available technology (BAT) for processing zinc/lead-containing dusts and wastes. One recent development, the SDHL process, promises a drastic reduction in energy consumption and CO_2 emissions, while still increasing the zinc yield [27]. This is achieved by reoxidizing the metallic iron fraction at the end of the kiln, thus providing additional process heat, which in turn allows for a reduction in the amount of coke supplied, and rendering the use of gas burners unnecessary. At GTT-Technologies, ChemSheet [28] was used to design an SDHL process model, which allowed for the simulation of the material and extensive property balances of the process before it was implemented on an industrial scale. The schematic flowsheet of the core part of the process is shown in Figure 2.11.11, characterizing the main reaction zones. Zinc-containing dusts or wastes are fed together with coke and flux into the top end of the kiln, while air is supplied at the bottom end. In contrast to a conventional Waelz kiln

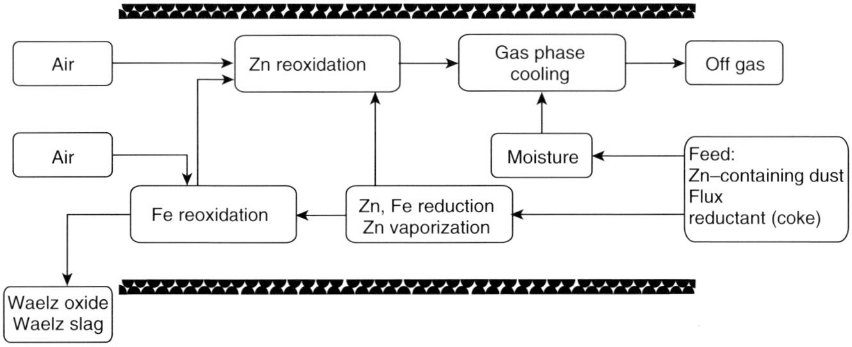

Figure 2.11.11 Flowsheet of the ChemSheet-based Zn Waelz kiln model.

process, additional process air is supplied directly to the solid charge leaving the kiln, thus resulting in the exothermic oxidation of metallic iron. This reduces the coke consumption by up to 58% compared with a basic Waelz process, while the Zn output increases by approximately 11% [27]. As long as no iron recovery is required, this new process currently constitutes the most economical option for zinc Waelz kilns, particularly since existing kilns can be retrofitted at low technical expense (Figure 2.11.11).

2.11.8
Valorization of "Wastes"

An important aspect when talking about end-of-lifetime use of materials is that not all materials can be redirected into a recycling circle. Nevertheless, it is often possible to put them to some use that is outside their normal range of application. Thus, these materials that have been dumped in the past can at least contribute to some extent to a sustainable use of materials. While wastes are untreated materials from a process that are not usable in any other way and are, therefore, dumped by the producer or required to be dumped by the law anything that is treated purposefully for a certain application and is sold or bought on the market for a price is legally considered a product. Examples for this valorization can be found in the field of metallurgical slags, particularly from steel production, but also for glass. In practice, one must emphasize that slags are by-products since no metallurgical plant is run for the purpose of producing slag.

Figure 2.11.12 shows the partitioning of steel slags into different fields of application. It is obvious that only a minor part remains as waste while almost 80% are put to some other use. One interesting aspect is the application of the slags in internal recycling cycles, which shows that recycling is directly related to the steel production itself. However, the most important part in the partitioning,

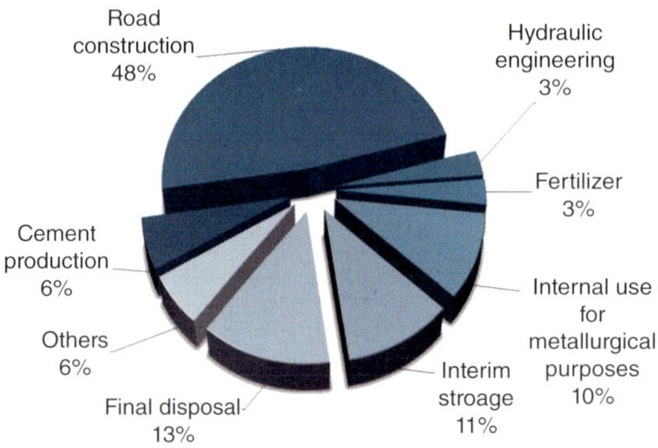

Figure 2.11.12 Distribution of usage of steel slags (22.3 mio tons 2010). (Reproduced from [29] with the permission of FEhS-Institut fur Baustoff-Forschung e.V. http://www.euroslag.com/fileadmin/_media/images/statistics/Statistics_2010_download.pdf.)

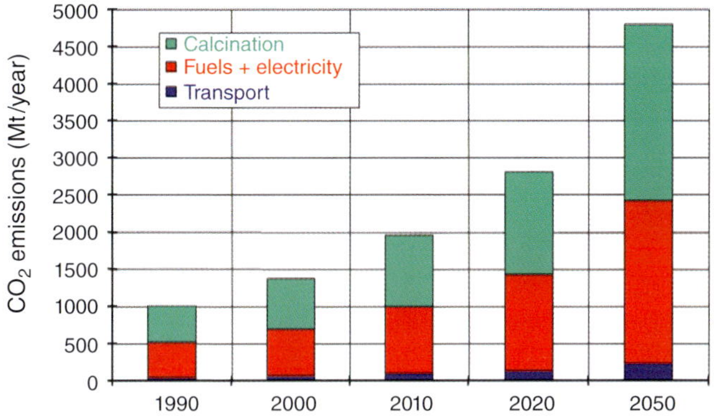

Figure 2.11.13 CO_2 output from cement industry. (Reproduced from [30] with the permission of Batelle Memorial Institute.)

not only because of its magnitude, is the use as construction material. In this role, the slags are used directly as construction materials, mainly as aggregates in road and waterway construction and as feed material to the cement industry. The main part of blast furnace slags is converted into a glassy material being used as a Portland cement clinker substitute (about two-thirds of 25.6 mio tons in 2010 [29]). In both cases, the CO_2 footprint of the cement industry is drastically reduced: in the latter by avoiding the production in the cement clinker plant completely and in the former by reducing the amount of virgin calcium carbonate. Nevertheless, the total CO_2 output of the cement industry is still increasing because of the increasing worldwide demand. The effect is depicted in Figure 2.11.13.

Data from the following countries: Austria, Belgium, Denmark, Finland, France, Deutschland, Greece, Italy, Luxembourg, Poland, Romania, Spain, Slowakia, Slowenia, Sweden, Netherlands, United Kingdom.

In addition to the very positive effect of the usage of slags with respect to the CO_2 output of cement production, there is also an equally positive effect on the energy requirements. This is depicted in Figure 2.11.14 for the normal Portland cement process in comparison to three different slag cements with increasing amounts of blast furnace slag. Up to 62% of the energy needed for the standard process can be saved. With respect to the production of concrete, this figure is still a very convincing 42% (all data courtesy of FEhS-Institut, Duisburg [31]).

Similar to the valorization of slags, there are applications of remains of the glass industry, which show that great efforts are made not to waste any material. Glass aggregates are used in a way similar to slags in construction, for example, cement applications, or as pipe bedding for around sewers, storm water, or drinking water pipes. Furthermore, recycled crushed glass is being used to produce fiber glass insulation material or ceramic sanitary ware, or in agricultural or landscape architecture, for example, as top dressing and root zone material, and even as special sand for golf bunkers. Other applications are in water filtering media or abrasives.

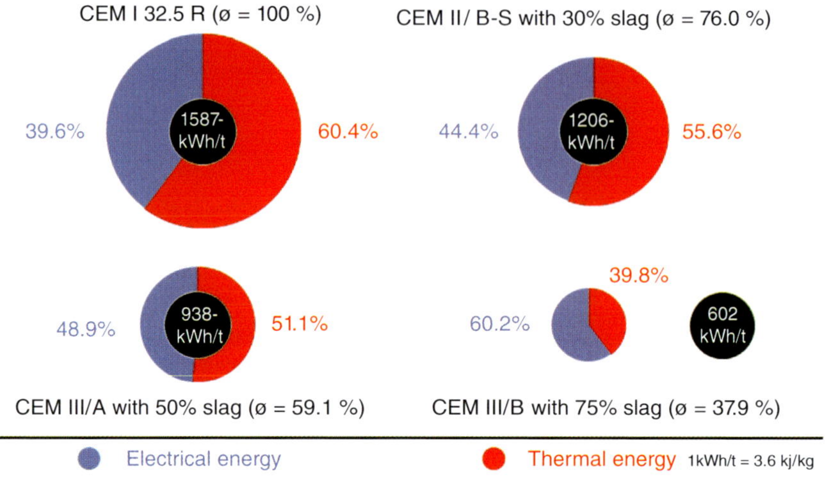

Figure 2.11.14 Influence of slag usage on the energy demand of the cement process. (Source: Reproduced from [31] with the permission of FEhS-Institut fur Baustoff-Forschung e.V. http://www.euroslag.com/fileadmin/_media/images/statistics/Statistics_2010_download.pdf.)

For further valorization of slags, the material properties are important. Depending on the application, there may be a required strength, hardness, chemical stability, or reactivity. These properties are different for different mineral phases that can be formed in slag, and therefore, slag valorization researchers have started to build a materials science approach (links processing–structure–properties) for metallurgical slags. The link between properties and structure is often only investigated on a case-by-case basis, and therefore limited to a specific number of phases.

On the other hand, the link between composition, cooling speed, and microstructure (or at least the present phases) can be investigated with more general approaches. Slags are multicomponent mixtures, and many minerals have solubilities for several elements, which means that simple phase diagrams are not sufficient to predict the stable phases at a given composition. Multiphase thermodynamic equilibrium calculations are, therefore, most suitable to understand these relations. Good results were obtained using the Scheil–Gulliver approach [23, 32], as it was shown that even for slowly cooled slags, several high-temperature phases remain up to room temperature.

Also, it was shown that the diffusion of oxygen in a solidifying slag is very limited, and oxidation can, therefore, be neglected [33].

These models focus mostly on basic slags (rich in CaO). For more acidic slags, a complexity is the formation of an amorphous phase, which cannot yet be predicted well with an equilibrium approach. Also, the full microstructure cannot be predicted only with a thermodynamic approach. To study both phenomena, attempts have been made to couple equilibrium with phase-field models. Up to now, good results have been obtained for limited numbers of components and small domains [34, 35, 36].

On the processing side, additions of SiO_2 to hot BOF steel slag have been considered to improve slag quality with respect to volume stability by reducing the free lime content. The required quantities and heat need can easily be modeled with thermodynamic process models [37, 38].

The chemical stability of slags is important for the leaching properties (environmental compatibility) as well as for binding properties, for example, through hydration, carbonation, or cementitious reactions. Again, there is a clear connection between the minerals in the slag and the behavior [39, 40, 41, 42]. This can also be modeled by dedicated low-temperature models for minerals and aqueous phase interactions, for example, Phreeqc [43].

2.11.9 Summary of Simulation Tools

Table 2.11.2 gives the reader an overview of useful software in the field of recycling. For details of the various programs, the reader is referred to the respective web pages. More details on the thermochemical software are given in Chapter 4.

Table 2.11.2 Summary of software.

Software	Producer	Description	Link
HSC 8.1	Outotec	Pure substance database, various standard application modules, process modeling tool HSC-Sim included	www.outotec.com
FactSage	GTT-Technologies and Thermfact	Various databases for metallic and nonmetallic systems including nonideal solutions, pure substance data, standard thermochemical application modules for reactions, complex equilibria and phase diagrams, and pre- and postprocessing of input and output	www.factsage.com
ChemApp	GTT-Technologies	Programmer's library for thermochemical applications, needs thermochemical data, for example, supplied via FactSage, "own data" possible	www.gtt-technologies.de
ChemSheet	GTT-Technologies and VTT	ChemApp integrated into Microsoft Excel® spreadsheet environment, has interactive GUI for the definition of input and output, needs thermochemical data, for example, supplied via FactSage, "own data" possible	www.gtt-technologies.de

Table 2.11.2 (*Continued*)

Software	Producer	Description	Link
SimuSage	GTT-Technologies	Software component library for Delphi®, enabling rapid process modeling based on the concept of interlinked local equilibria, needs thermochemical data, for example, supplied via FactSage, "own data" possible	www.gtt-technologies.de
Thermo-Calc	Thermo-Calc Software AB	See Chapter 4	www.thermocalc.se
Pandat	Computherm	See Chapter 4	www.computherm.com
Aspen Plus	AspenTech	Chemical process optimization software, has official link to ChemApp	www.aspentech.com
Fluent	ANSYS	CFD code, has official link to ChemApp	www.ansys.com
OpenFOAM	ESI-OpenCFD	Open-source CFD code, has been linked to ChemApp	www.openfoam.com
PRO/II	Schneider Electric Software	Process simulator, has official link to ChemApp	www.software.schneider-electric.com
GaBi	Thinkstep	Next-generation product sustainability solution with a powerful life cycle assessment engine	www.gabi-software.com
gPROMS	Process Systems Enterprise	Process modeling platform, has official link to ChemApp	www.psenterprise.com
CALPHAD		Phase diagrams	http://www.opencalphad.com/
MTDATA	National Physics Laboratory	Phase diagrams	http://www.npl.co.uk/science-technology/mathematics-modelling-and-simulation/mtdata/
METSIM	METSIM	Metallurgical steady-state and dynamic flowsheeting and simulation	http://www.metsim.com/
Simapro 8	Pre Consultants	Environmental assessment	http://www.pre-sustainability.com/simapro8
OpenLCA	Open LCA organization	Environmental assessment	http://www.openlca.org/
ClaNG	Institut für Metallkunde und Metallphysik RWTH Aachen	Formation, growth, and dissolution of second-phase particles in solid Al alloys	www.imm.rwth-aachen.de

References

1. Reuter, M.A., van Schaik, A., and Gediga, J. (2015) Simulation-based design for resource efficiency of metal production and recycling systems, cases: copper production and recycling, e-waste (LED lamps), nickel pig iron. *Int. J. Life Cycle Assess.*, **20**, 671–693.
2. Reuter, M.A. and van Schaik, A. (2015) Product-centric simulation-based design

for recycling: green printing of lamp recycling. *J. Sustainable Metall.*, **1** (1), 4–28.
3. EC (2014) Communication from the commission to the European Parliament, the Council, the European Economic and Social Committee and the Committee of the Regions. Towards a Circular Economy: A Zero Waste Programme for Europe, COM/2014/0398 (2.7.2014).
4. Worrell, E. and Reuter, M.A. (2014) *Handbook of Recycling*, Elsevier BV, Amsterdam, 595 pp. ISBN: 978-0-12-396459-5
5. Reuter, M.A., Matusewicz, R., and van Schaik, A. (2015) Lead, zinc and their minor elements: enablers of a circular economy. *World Metall. Erzmetall*, **68** (3), 132–146.
6. Kotas, T.J. (1995) *The Exergy Method of Thermal Plant Analysis*, Krieger Publishing Company, Malabar, FL.
7. Reuter, M.A., Heiskanen, K., Boin, U., van Schaik, A., Verhoef, E., Yang, Y., Georgalli, G. (2005) in *The Metrics of Material and Metal Ecology*, (ed. B.A. Wills), Elsevier, pp. 520–524.
8. Eriksson, G., Hack, K., and Petersen, S. (1997) ChemApp – A programmable thermodynamic calculation interface. Simulation, Modellierung, Informationssysteme, Vol. 8, pp. 47–51.
9. Petersen, S. and Hack, K. (2007) The thermochemistry library ChemApp and its applications. *Int. J. Mater. Res.*, **98** (10), 935–945.
10. Petersen, S., Hack, K., Monheim, P., and Pickartz, U. (2007) SimuSage – the component library for rapid process modeling and its applications. *Int. J. Mater. Res.*, **98** (10), 946–953.
11. Embarcadero www.embarcadero.com/delphi (accessed 5 May 2016).
12. Pajarre, R. (1991) Modelling of equilibrium and non-equilibrium systems by Gibbs energy minimisation. Master's thesis. Helsinki University of Technology, Helsinki.
13. Koukkari, P. and Pajarre, R. (2006) Calculation of constrained equilibria by Gibbs energy minimization. *Calphad*, **30**, 18–26.
14. Koukkari, P. and Pajarre, R. (2006) Introducing mechanistic kinetics to the Lagrangian Gibbs energy calculation. *Comput. Chem. Eng.*, **30**, 1189–1196.
15. Atkins, P.W. and Friedman, R. (2008) *Quanta, Matter and Change: A Molecular Approach to Physical Change*, Oxford University Press, Oxford. ISBN: 978-0-7167-6117-4
16. Levine, I. (2008) *Physical Chemistry*, 6th edn, McGraw–Hill Science. ISBN: 978-0-07-253862-5
17. Huda, N., Naser, J., Brooks, G., Reuter, M.A., and Matusewicz, R.W. (2012) Computational fluid dynamic modeling of zinc slag fuming process in top-submerged lance smelting furnace. *Metall. Mater. Trans. B*, **43B** (1), 39–55.
18. Xiao, Y. and Reuter, M.A. (2002) Recycling of different aluminium scraps. *Miner. Eng.*, **15** (IIS1), 763–970.
19. Zhou, B., Yang, Y., Reuter, M.A., and Boin, U.M.J. (2006) Modelling aluminium scrap melting in a rotary furnace. *Miner. Eng.*, **19** (3), 299–308.
20. Engler, O., Liu, Z., and Kuhnke, K. (2013) Impact of homogenization on particles in the Al-Mg-Mn alloy AA 5454 – experiment and simulation. *J. Alloys Compd.*, **560**, 111–122.
21. Engler, O., Laptyeva, G., and Wang, N. (2013) Impact of homogenization on microchemistry and recrystallization of the Al-Fe-Mn alloy AA 8006. *Mater. Charact.*, **79**, 60–75.
22. Liu, Z., Mohles, V., Engler, O., and Gottstein, G. (2014) Thermodynamics based modelling of the precipitation kinetics in commercial aluminium alloys. *Comput. Mater. Sci.*, **81**, 410–417.
23. Scheil, E. (1942) Bemerkungen zur Schichtkristallbildung. *Z. Metallkd.*, **34**, 70–72.
24. Verscheure, K., Camp, M.V., Blanpain, B., Wollants, P., Hayes, P., and Jak, E. (2007) Continuous fuming of zinc-bearing residues: part I. Model development. *Metall. Mater. Trans. B*, **38** (1), 13–20.
25. Verscheure, K., Camp, M.V., Blanpain, B., Wollants, P., Hayes, P., and Jak, E. (2007) Continuous fuming of zinc-bearing residues: part II. The

submerged-plasma zinc-fuming process. *Metall. Mater. Trans. B*, **38** (1), 21–33.
26. Bale, C.W., Bélisle, E., Chartrand, P., Decterov, S.A., Eriksson, G., Hack, K., Jung, I.-H., Kang, Y.-B., Melançon, J., Pelton, A.D., Robelin, C., and Petersen, S. (2009) FactSage thermochemical software and databases - recent developments. *Calphad*, **33** (2), 295–311.
27. Mager, K., Meurer, U., Garcia-Egocheaga, B., Goicoechea, N., Rutten, J., Saage, W., and Simonetti, F. (2000) in *Recycling of Metals and Engineered Materials* (eds D.L. Stewart Jr.,, J.C. Daley, and R.L. Stephens), John Wiley & Sons, Inc., pp. 329–344.
28. Koukkari, P., Penttilä, K., Hack, K., and Petersen, S. (2000) in *Microstructures, Mechanical Properties and Processes*, vol. **3** (ed. Y. Bréchet), Wiley-VCH Verlag GmbH, Weinheim, New York, pp. 323–330.
29. http://www.euroslag.com/fileadmin/_media/images/statistics/Statistics_2010_download.pdf (accessed 5 May 2016).
30. Batelle Memorial Institute (2002) *Toward a Sustainable Cement Industry, Substudy 8, Climate Change*, Battelle Memorial Institute, Columbus, OH.
31. Ehrenberg, A. (2002) Proceedings of the 3rd European Slag Conference, Keyworth, UK, October 2–4, 2002, pp. 1151–1166.
32. https://www.researchgate.net/publication/230262466_Slag_Solidification_Modeling_Using_the_ScheilGulliver_Assumptions (accessed 5 May 2016).
33. https://www.researchgate.net/publication/230344986_AirCooling_of_Metallurgical_Slags_Containing_Multivalent_Oxides (accessed 5 May 2016).
34. Sciencedirect http://www.sciencedirect.com/science/article/pii/S0009254111003652 (accessed 5 May 2016).
35. Sciencedirect http://www.sciencedirect.com/science/article/pii/S1359645410008372 (accessed 5 May 2016).
36. http://nele.studentenweb.org/docs/Fray_Symposium_JeroenHeulens.pdf (accessed 5 May 2016).
37. Kühn, M. and Mudersbach, D. (2004) Treatment of liquid EAF-slag from stainless steelmaking to produce environmental friendly construction materials. Proceedings of the 2nd International Conference on Process Development in Iron and Steelmaking (SCANMET II), Lulea, p. 369.
38. http://www.gbv.de/dms/tib-ub-hannover/474783510.pdf (accessed 5 May 2016).
39. http://publications.sckcen.be/dspace/handle/10038/865 (accessed 28 July 2016).
40. https://pure.ltu.se/portal/files/32438302/DOC.pdf (accessed 28 July 2016).
41. Apul, D.S., Gardner, K.H., Eighmy, T.T., Fällman, A.-M., and Comans, R.N.J. (2005) Simultaneous application of dissolution/precipitation and surface complexation/surface precipitation modelling to contaminant leaching. *Environ. Sci. Technol.*, **39**, 5736–5741.
42. Huijgen, W.J.J. and Comans, R.N.J. (2006) Carbonation of steel slag for CO2 sequestration: leaching of products and reaction mechanisms. *Environ. Sci. Technol.*, **40** (8), 2790–2796.
43. USGS http://wwwbrr.cr.usgs.gov/projects/GWC_coupled/phreeqc/ (accessed 5 May 2016).

3
Microstructure Modeling

Markus Apel, Robert Spatschek, Franz Roters, Henrik Larsson, Charles-André Gandin, Gildas Guillemot, Frigyes Podmaniczky, László Gránásy, Georg J. Schmitz, and Qing Chen

3.1
Overview and Definitions

3.1.1
What is a Microstructure and why it is Important?

Actually, "microstructure" is what you see if you observe a certain piece of material under a microscope. As an example, Figure 3.1 shows the as-cast microstructure of an Al cast alloy. The nominal average composition of this alloy is Al−7wt%Si−0.3wt%Mg plus Fe and Cu impurities. Clearly, it can be seen that the solidified metal is far from being a uniform, homogeneous, and isotropic piece of matter; instead a closer look reveals a rich variety of internal structures on different length scales. This is not only the case for metallic materials; in practice all solid materials reveal internal, more or less ordered structures on a mesoscopic scale, the length scale between the scale of atomic spacing and the scale of a few millimeters. Polymers, wood, concrete, ceramics, and others have their own characteristic microstructures.

"What you see under the microscope" seems to be a bit vague definition because it depends on the type of microscope (optical microscope OM, scanning electron microscope SEM, transmission electron microscope TEM, X-ray, etc.), on the magnification or on the sample preparation. Nevertheless, we will consider this as our starting point. In the optical micrograph, individual grains, the dendritic morphology of the fcc-Al phase, Si particles, and hydrogen pores can be seen (Figure 3.1). The different crystallographic orientations of the individual grains have been made visible by a dedicated etching technique and by using polarized light. In general, the following microstructure constituents can be derived from microscopic characterization: grains, including their crystallographic orientation, grain boundaries, phases, and phase boundaries, which all can be easily visualized by an OM. In order to identify precipitates, dislocations, or the local variation of the composition (microsegregation), higher resolutions or dedicated analytical techniques may be necessary which are offered by, for example, SEM, TEM,

Handbook of Software Solutions for ICME, First Edition. Edited by Georg J. Schmitz and Ulrich Prahl.

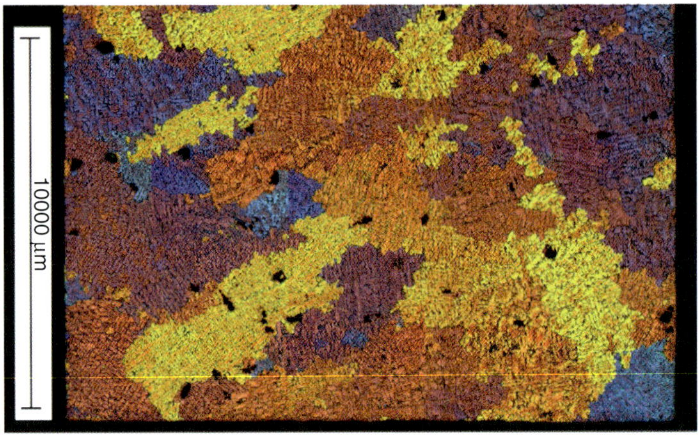

Figure 3.1 Metallographic image of an Al–7wt%Si–0.3wt%Mg as-cast alloy being color-etched to identify the different grains. The black objects are hydrogen gas pores mainly located at grain boundaries.

HRTEM, and EDX. Besides the individual constituents and their individual properties, the microstructure comprises the spatial distribution of these objects which plays a decisive role for the material properties.

There is no simple and unique definition for "microstructure." However, several important aspects are connected to the term "microstructure": (i) the individual constituents are deviations from the equilibrium state, (ii) the microstructure depends on the material history and thus is (iii) not a unique state for a given temperature, pressure as other thermodynamic state variables, and, most important from an application point of view, (iv) the microstructure determines the engineering material properties! This can be condensed into a definition given by G. Gottstein:

> Microstructure comprises all internal structural features that affect material properties.

The microstructure is therefore not necessarily confined to a mesoscopic or microscopic level.

It is important to stress that there is no direct correlation between *process* and *properties* and the engineering point of view of a process–property relation is only an indirect link. Instead, from a physical point of view, two different interdependencies exist: a *process–microstructure* and a *microstructure–property* relationship. This needs to be considered whenever physical-based property models are used. Thus, a second important statement is:

> Material microstructures are the carrier of material properties.

Microstructures – along with the properties of the phases constituting the microstructure – are the state variables eventually determining the properties.

This implies that besides microstructure models which should comprise a thorough characterization and description of the microstructure itself, accurate property models based on the microstructure are of equal importance.

Microstructures for most of the structural and functional materials are rather complex. Their comprehensive characterization is virtually impossible, but usually not necessary for material simulations in the Integrated Computational Materials Engineering (ICME) context because only those microstructural parameters which determine the targeted material property are needed, for example, for the target property "strength," a simplified property model can be used:

$$\tau_0 = \tau(\rho) + \Delta\tau(D) + \Delta\tau(c) + \Delta\tau\left(\frac{f}{r_\mathrm{p}}\right) + \cdots$$

with ρ being the dislocation density, D the grain size, c the solute concentration, f the particle volume fraction, and r_p the particle radius. In this example only ρ, D, c, and f are the relevant microstructural elements to be predicted and traced along the simulation chain. However, in order to derive these particular parameters from the microstructure simulation, it would be necessary to consider more microstructural details.

Most of the features within the microstructure typically vary across a component and lead to different properties in different parts of the component. Because the microstructure is not a state variable and the local evolution along the process history needs to be traced for different places in a component, the concept of representative volume elements (RVEs) has been introduced. An RVE representation of the microstructure is characteristic and representative for a certain part of the component, and the variation of the microstructure on the component level can be traced by a sufficient number of RVEs distributed over the component. Somehow a RVE matches everyday life metallography which is usually also restricted to the important regions in a component only. The spatial distribution of the microstructure over the component is the result of local variations of the process conditions, for example, local differences in the thermal history. With the exception of single crystalline components, there is not a single microstructure being typical for an entire component, and an important question is always how many RVEs are sufficient for a valid description of the property variation over the whole component.

Several aspects determine the size of an individual RVE to become "representative," especially in the case of nonperiodic arrangements of microstructure constituents. The guideline should be that the average properties of an RVE do not depend on the size of the RVE.

3.2
How to Describe and Store a Microstructure?

The basis for the representation of a microstructure is given by the RVE which defines a representative volume of material under consideration. A RVE can be two-dimensional (2D) or three-dimensional (3D). Whereas 3D would be always the natural choice, 2D is often a necessary approximation to reduce the computational demands.

The RVE is also the vehicle for scale bridging, that is, providing microstructure information from the natural length scale of the material microstructures to the macroscopic component and process scale. Furthermore, the RVE is the important entity for the microstructure – property relation as it should be used to determine the local material properties, averaged over the RVE for the higher-level simulations dealing with properties instead of the microstructure.

The RVE can comprise different levels of complexity. Different types of RVEs can be sorted along the type of information or microstructure constituents which are represented by an individual RVE:

Level 0 (isotropic): The RVE is considered as homogeneous and isotropic; thus no further details about the microstructure are encoded in the RVE. The RVE is rather a "property element" instead of a representation of the microstructure. In particular, its properties do not vary over the RVE. The average property is equal to the local property at each point. Anisotropy is not considered for the simplest case. Isotropic level 0 descriptions hold for liquids and gases or amorphous structures.

Level 1 (anisotropic): For crystalline solids, lattice orientation and certain crystallographic directions may lead to an anisotropic material behavior. In this case, the properties need to be formulated, for example, as tensors. Level 1 enriched by anisotropy is sufficient to represent, for example, single crystals.

Level 2 (isotropic): For multiphase materials or multiphase states, the regular case for multicomponent materials, a first approach is to consider the fractions of the individual phases being present in the RVE, but neglecting their spatial distribution. Equilibrium phase fractions can be easily determined by thermodynamic software tools as a function of the boundary conditions on the RVE. Knowledge about the phase fractions already allows for an estimate of the effective properties of the RVE, for example, by using volume averaging concepts.

Level 3 (anisotropic): Next, also the average grain sizes, grain size distributions, grain orientations, or particle size distributions (PSD) should be represented by the RVE in a statistical manner. This leads to effective anisotropic properties of the RVE.

Level 4: In a spatially resolved representation of the microstructure in the RVE, all individual objects, for example, grains or precipitates, have homogeneous properties, but they are distinguishable by their orientation.

Level 5: Full, that is, down to discretization level, spatially resolved description of the microstructural constituents, for example, phases or grains as observed under a microscope including local variations of composition or properties even within the individual microstructure element.

3.2.1
Digital Microstructures

Significant progress has been made in the last decade both in the area of 3D characterization and in the field of 3D modeling and simulation of microstructures. 3D microstructures provide the full information depth with respect to the anisotropy of the effective properties of the materials and also with respect to material processing and evolution under operational conditions. Networking scientists from the area of 3D microstructure characterization with scientists from the area of 3D microstructure modeling will be inevitable for the definition of a future common standard for the digital description of 3D microstructures. Such a standard will largely facilitate information and knowledge exchange between virtual and real worlds and also the integration of data obtained by different methods or models in each of these fields. This should also comprise the representation and analysis of experimental and simulation data by the same postprocessing tools.

Various 3D experimental microstructure characterization techniques are nowadays available which comprise either *postmortem* or *in situ* analysis of a static microstructure (e.g., serial sectioning metallography) or even capture the dynamic evolution by computer tomography.

A recent trend in experimental microstructure analysis is to combine and integrate the information gained from different techniques providing different kinds of data, for example, optical microscopy (grain sizes over large areas), EDX (element distribution on smaller scales), and EBSD (for crystallography and texture analysis). Of course, the additional integration with "virtual" data gained from simulations is highly beneficial for a seamless integrative concept. In the case of comprehensive microstructure simulations, that is, on a level 3 and above, similar algorithms for analyzing experimental and virtual microstructures can be used which also contributes to better matching models with the behavior of the real materials. The combination of synthetic, artificial microstructures with experimental microstructures is an important target for future developments.

For the generation and further handling of digital microstructures, algorithms for digitizing, meshing, and remeshing, for property and structure interpolation, or for the mapping between different representations are essential. In order to overcome computational limitations, 3D extrapolations from 2D microstructure data are beneficial.

3.3
Phenomena Affecting Microstructure Evolution

The microstructure is by definition a nonequilibrium state, thus showing in particular spatial variations in its properties, for example, local variations of the chemical composition (microsegregation pattern) or grain size variations from the surface to the bulk of a component. The nonequilibrium character implies that material microstructures can occupy an infinite number of different configurations which is the fundamental reason that the same material can show a rich variety of properties dependent on its processing or service history. Whenever the microstructure is exposed to elevated temperatures or external energy fluxes, kinetic barriers may be overcome and the microstructure will further evolve in space and time toward a more stable state. High-temperature processing is certainly the most relevant case, integrated in many fabrication processing chains, but material exposure to high temperatures is also an important aspect in material service, for example, for tubes in a power plant or for polymers used in vehicle interiors. Just imagine a situation of a car being parked close to the beach, directly in the sun on a hot summer day. In this case, not only the heat is relevant for the microstructure evolution, but additionally the direct energy transfer by absorption of photons may trigger the microstructure evolution. In summary, instability and potential further evolution are inherent aspects of microstructures.

Microstructure evolution requires the movement of atoms or structural defects, for example, the propagation of interfaces. The early stages of material processing very often comprise melting, refinement, and solidification, thus the molten liquid state. The melt is a rather well-defined homogeneous state showing no long-range structural order, only ordering at an atomic length scale. Microstructure formation upon cooling evolves by the birth of ordered structures, for example, the nucleation of solid and its further growth (see Section 3.5.3). Exceptions are amorphous solids or glasses, where the liquid-to-solid transition is a rather continuous process accompanied by steady changes in the material properties, for example, viscosity, but only minor changes of the local atomistic configuration.

The formation of many microstructure constituents like precipitates or new grains is accompanied by the formation of interfaces (e.g., phase or grain boundaries), and further growth means a movement of these interfaces. The local growth or interface velocity is determined by external fields and resulting fluxes together with local boundary conditions at the interface. In addition, an interface has its own physical properties, for example, an excess free energy (interfacial energy) or a collective kinetic behavior, described as interface mobility. Via long-range fluxes, the local interface movement is a result of external boundary conditions applied to a particular RVE. For many cases this results in a severe multiscale problem and requires special scale bridging solutions, for example, bridging between long-range transport phenomena and short-range local atomic rearrangements at phase boundaries.

In summary, a physical-based mathematical model for the microstructure evolution has to address phenomena and material properties such as interface properties, phase equilibria at interfaces, diffusion, melt flow, mechanical stress, local temperatures, temperature evolution, or electromagnetic fields in a coupled manner. Which of the previously listed phenomena are indispensable for a predictive model depends on the specific type of microstructure evolution.

3.4
Basic Equations/Models

Despite the fact that the microstructure is a nonequilibrium state, the equilibrium state of the material should be the limiting case for all microstructure evolution models because it is the natural endpoint of the microstructure evolution. Therefore, a thermodynamic material model is required as a sound basis for microstructure simulations. Idealized or simplified approximations, for example, parabolic-shaped free energy functions, ideal solution, and so on, may lead to a sufficient description of the basic phenomena and provide deeper insight into the governing mechanisms. However, a quantitative modeling of "real" materials requires more complex thermodynamic models to describe multiphase and multicomponent microstructures. The *CALculation of PHAse Diagrams* (CALPHAD) method (see Chapter 4) provides a framework to build thermodynamic databases containing comprehensive and quantitative thermodynamic information about complex alloy systems. Today, for many material and alloy systems, such thermodynamic databases are available, and it is an ongoing effort to complete and expand these databases or to develop thermodynamic models for new materials or innovative material combinations.

Beyond the thermodynamic models for homogeneous bulk phases, structural defects such as dislocations and phase or grain boundaries have to be cast into a comprehensive model description. This area is less developed in the sense of a common notation. However, different models have been developed to describe structural defects in a thermodynamic framework as well. Numerous approaches describe the dynamics of structural defects based on phenomenological descriptions. Atomistic computations increasingly improve our understanding of structural defects. Incorporation of these models into an ICME framework is an important recent activity and will especially support the design and development of new materials.

Thermodynamics is about energy and entropy and both quantities can be used to characterize the equilibrium state. Besides the equilibrium, thermodynamics also provides information about the off-equilibrium state. It can thus be applied to compute driving forces for phase transformations or chemical reactions. Thermodynamics alone provides the energy scale and does not contain any information about time and space. In order to further evolve thermodynamics into models for the microstructure evolution, two important aspects have to be added: time and space. Irreversible thermodynamics offers a fundamental physical methodology, based on energy dissipation and fluxes, which takes into account time and length scales. Following the principles of irreversible thermodynamics, the transport equations for mass or energy can be derived from the same thermodynamic functional used to describe the ground states. Historically, the transport equations have been derived in a rather phenomenological manner. Please refer to Section 3.5.4 about diffusion modeling as an example.

Eventually, the microstructure evolution can be described by a set of coupled equations, connecting the movement of interfaces in an explicit or integral form, transport phenomena like heat or mass transport, and the response on external fields (temperature, pressure, magnetic field, etc.) as the driving force for further microstructure evolution.

3.5 Models for Microstructure Evolution

3.5.1 Overview

One possibility to classify models for the dynamics of microstructure evolution is to arrange them according to their level of complexity, measured by the number and type of variables (see Table 3.1). Following this concept, one can define three major classes of microstructure models: class 1, models which describe average integral properties of the microstructure as a function of state variables; class 2, models based on a combination of integral state variables with analytical or simplified solutions for the transport equations to account for an internal length scale λ; and finally class 3, models which have a similar complexity as the microstructure itself, meaning that they provide a full spatial and time-dependent description of the microstructure.

3.5.2 Example for Integral Models

A simple and thus fast access to the microstructure evolution is through integral models (class 1), that is, models which formulate the evolution of the microstructural feature of interest. An example is the local phase fraction as a function of local state variables. Equilibrium thermodynamics is an important example for this class and provides the first access to microstructure information. Providing the state variables such as temperature, pressure, or composition, the equilibrium phase fractions can be computed as an integral measure of a microstructure characteristic in multiphase materials. Especially, for the analysis whether particular

Table 3.1 List of different microstructure evolution models

Class 1: integral models $p = f(t)$	Class 2: intermediate/kinetic models $p = f(t,\lambda)$	Class 3: fully resolved models $p = f(t,x)$
Thermodynamics + approx.	CAFE (dendritic growth)	Monte Carlo Potts model
Scheil–Gulliver	Needle crystal	Phase field/phase-field crystal
JMAK kinetics	Precipitation/ripening (LSW)	Cellular automaton
Volume averaging	Envelope dynamics	Level set
…	1D diffusion	Lattice Boltzmann
	…	…

λ is the typical length scale for transport.

phases are likely to be formed during a process, equilibrium thermodynamics provides valuable answers. The phase fraction implicitly also provides morphological information. If one length scale is known – for example, the grain size of the primary phase – the area occupied by the remaining eutectic structure can be estimated based on the constraint of complete space filling.

Instead of sticking to the global equilibrium, further constraints can be added to the equilibrium calculations in order to account for kinetic effects especially for larger RVEs where transport phenomena are decisive. One example is the Scheil–Gulliver model (SGM), introduced for the calculation of the solidification paths. In the SGM, the additional condition for the computation of phase fractions is that there is no diffusion in the solid but perfect mixing (infinite diffusion) in the liquid. Thus, the composition of the solid will change along the solidification path, and concentration gradients in the solid are kept until complete solidification. The SGM thus also provides an estimate for the microsegregation resulting from solidification. As an example, in Figure 3.2, the solid fraction for the Ni-based superalloy Inconel 713 is plotted in relation to the temperature for both equilibrium conditions and the SGM. As a consequence of the neglected diffusion in the solid, the SGM typically predicts a larger solidification interval, that is, lower solidus temperature.

A further example for integral models is the Johnson–Mehl–Avrami–Kolmogorov (JMAK) model. This approach leads to an equation for the phase fraction f as a function of time. In this regard, JMAK could be also assigned to class 2, intermediate or kinetic models; however, it does not contain a particular

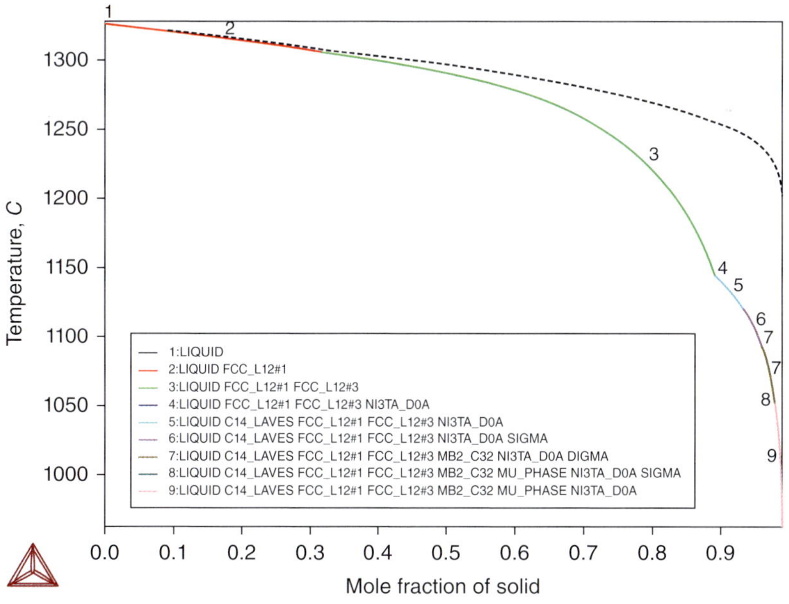

Figure 3.2 Solidification path in the Ni-based superalloy Inconel 713, equilibrium versus Scheil–Gulliver model.

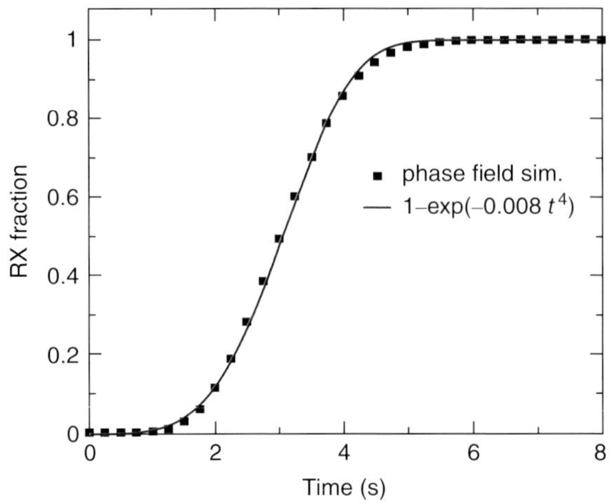

Figure 3.3 Recrystallized fraction as a function of time: comparison between a JMAK equation and a direct phase-field simulation shown in Figure 3.16.

length scale as the JMAK kinetic relates to phase fractions. However, size effects can alter the coefficients, but this is an implicit effect. The general JMAK equation is of the form

$$f = 1 - e^{-kt^n}.$$

The coefficients k and n depend on the type of reactions involved (diffusion controlled, interface controlled, etc.), the geometry (2D, 3D, dilute or dense, etc.), and nucleation conditions (e.g., constant or varying nucleation rates, etc.). Values for k and n for particular materials and process conditions are typically derived from experimental data; thus JMAK should be considered as an empirical model with limited predictive power. Figure 3.3 shows the example of the recrystallized fraction as a function of time for an aluminum alloy after deformation, fitted to the results from the phase-field (PF) simulation shown in Figure 3.16. This example demonstrates how results from computational intense PF simulations can be described by a JMAK fit which leads to the coefficients n and k. The calibrated JMAK can then be used very efficiently in large-scale component simulations and thus serves as a tool to bridge the scales between microstructure and component. A more comprehensive introduction on this topic can be found in [1].

In general, so-called criteria functions providing empirical or semiempirical relations between state variables, process time, and property can all be considered as class 1 models. Criteria functions have been developed for many different microstructural features, such as phase fraction, grain size, dendrite arm spacing, porosity, dislocation density, and many more. Criteria functions are frequently implemented in commercial software packages operating at the component scale, often being certified and validated for certain materials and conditions.

In the following, important physically based models addressing different phenomena in microstructure evolution will be introduced.

3.5.3 Nucleation Models

The crystalline freezing of a pure undercooled melt starts with nucleation, a process during which crystal-like fluctuations appear, whose size exceeds a critical value determined by the solid–liquid interface free energy (γ_{SL}) and the thermodynamic driving force of crystallization (volumetric grand potential difference between the solid and liquid, defined so that $\Delta\omega < 0$ in the undercooled state). Crystal-like fluctuations larger than the critical size tend to grow to a macroscopic size, whereas the smaller ones decay with a high probability, though there is traffic from both sides of the maximum to the other due to stochastic capturing/releasing molecules by the clusters.

3.5.3.1 Classical Approach to Nucleation

In the classical nucleation theory (CNT), a mathematically sharp solid–liquid interface is assumed, whose free energy acts at the *surface of tension*. Accordingly, the free energy of formation of the crystal-like (heterophase) fluctuations (assumed to be spherical for the sake of simplicity) is expressed as a sum of a volumetric and an interfacial term: $W_{hom} = (4\pi/3)R^3\Delta\omega + \pi R^2\gamma_{SL}$, where R is the radius of the surface of tension. Since for small sizes the positive surface term dominates, W_{hom} shows a maximum as a function of size, which yields $W_{hom}{}^* = (16\pi/3)\gamma_{SL}{}^3/\Delta\omega^2$ and $R^* = -2\gamma_{SL}/\Delta\omega$ for the free energy and radius of the critical fluctuation or *nucleus*. Since in this case, the heterophase fluctuations appear via the internal fluctuation of the homogeneous liquid, this process is known as *homogeneous nucleation*. The formation rate of critical fluctuations can be calculated within the framework of the CNT considering single molecule attachment and detachment. The respective master equations for the number of the monomers and n-mers can be written as follows:

$$\dot{N}_1 = a_2^- N_2 - a_1^+ N_1 + \sum_{n>1} (a_n^- N_n - a_{n-1}^+ N_{n-1}), \tag{3.1a}$$

$$\dot{N}_n = a_{n-1}^+ N_{n-1} + a_{n+1}^- N_{n+1} - (a_n^+ + a_n^-) N_n, \quad \text{for } n > 1, \tag{3.1b}$$

where $a_n^+ = O_n \Gamma \exp\{-(W_{n+1} - W_n)/2kT\}$ and $a_n^- = O_{n-1}\Gamma \exp\{-(W_{n-1} - W_n)/2kT\}$ are the frequencies of molecule attachment and detachment, $O_n = 4n^{2/3}$ is the number of sites to which a molecule can be attached on the cluster $\Gamma = 6D/\lambda^2$, D is the self-diffusion coefficient, and λ is the molecular jump distance, whereas W_n is the free energy of formation of an n-molecule cluster, k is the Boltzmann's constant, and T is the temperature. Solving these equations numerically, after a transient period of length $\tau \approx K\lambda^2 kT(n^*)^{2/3}/(Dv_m\Delta\omega)$, where K is a geometrical factor, n^* the number of molecules in the critical fluctuation, and v_m the molecular volume, a steady-state nucleation rate (net number of critical fluctuations formed in unit volume and unit time) of form

$$J_{SS} = \frac{1}{\left\{ \sum_{n=1} \left(a_n^+ N_{eq,n} \right)^{-1} \right\}} \approx J_0 \exp\{-W_{hom}{}^*/kT\} \tag{3.2}$$

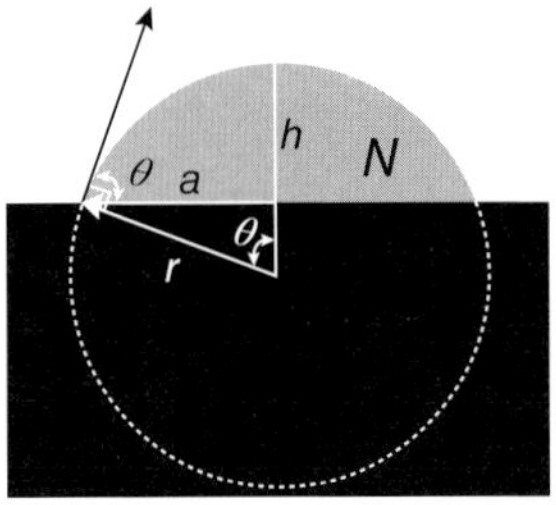

Figure 3.4 Classical "spherical cap" approach to heterogeneous nucleation on a flat surface. The white dotted line shows the contour of the homogeneous nucleus, and the gray area is the heterogeneous nucleus (N). (White, liquid; black, substrate; and gray, crystal; *a*, radius of contact surface; *h*, height of nucleus; *r*, radius of homogeneous nucleus; and *θ*, contact angle.)

is established, where $N_{eq,\,n} = \rho_0 \exp\{-W_n/kT\}$ and $J_0 = \rho_0 O_{n^*} \Gamma Z$ is the nucleation prefactor, ρ_0 is the number density of molecules, and $Z = \{|d^2 W_{hom}/dn^2|_{n^*}/(2\pi kT)\}^{1/2}$ is the Zeldovich factor that takes the dissolution of critical clusters into account. Nucleation experiments on oxide glasses indicate that the magnitude of the classical prefactor is reasonable, a conclusion also supported by molecular simulations. Extensions of the classical kinetic approach to cluster coagulation and splitting have also been explored.

In practice, the formation of the heterophase fluctuations is usually assisted by various kinds of heterogeneities present (e.g., container walls, floating particles, molecular impurities, etc.), which may reduce the height of the nucleation barrier significantly ($W_{het}^* < W_{hom}^*$), making the formation of nuclei easier. In the spherical cap model, this is captured in the following way: let us assume that the foreign particles that are much larger than the nuclei are homogeneously distributed in the undercooled liquid and have flat walls that realize a contact angle, θ, with the solid–liquid interface, related to the interfacial free energies via the Young–Laplace relationship: $\gamma_{WL} = \gamma_{WS} + \gamma_{SL}\cos(\theta)$. Here γ_{WL} and γ_{WS} are the wall–liquid and wall–solid interface free energies. It can be shown that the critical fluctuation has the shape of a spherical cap (a fraction of the homogeneous nucleus that realizes the contact angle; see Figure 3.4) and has the free energy of formation $W_{het}^* = W_{hom}^* f(\theta)$, where $f(\theta) = 1/4[2+\cos(\theta)]\{1-\cos(\theta)\}^2$. Considering the spherical cap form, the number of sites on the nucleus to which molecules can be attached is $O_l = 2\{1-\cos(\theta)\}l^{2/3}$. Taking into account that only those molecules can participate in heterogeneous nucleation that are effectively adsorbed on the surface of the heterogeneities, the steady-state nucleation rate may be expressed as $J_{SS,het} = x_A\, 1/2\{1-\cos(\vartheta)\}/f(\vartheta)^{1/2} J_0 \exp\{-W_{hom}^* f(\vartheta)/kT\}$, where x_A is the fraction of molecules adsorbed on the heterogeneities. Extensions of this approach to various geometries (spherical substrate, rough surface, nucleation in depressions, etc.) have been worked out.

3.5.3.2 Free Growth-Limited Model

A simple but rather effective approach has been put forward to describe particle-induced freezing in undercooled liquids by Greer and coworkers. Starting from the expectation that the largest particles of the smallest contact angle would initiate freezing, the foreign particles are assumed to be cylindrical of radius R and exhibit ideally wetting on the circular faces ($\theta = 0$), whereas nonwetting on the sides ($\theta = \pi$). Such idealized particles remain dormant during cooling until the

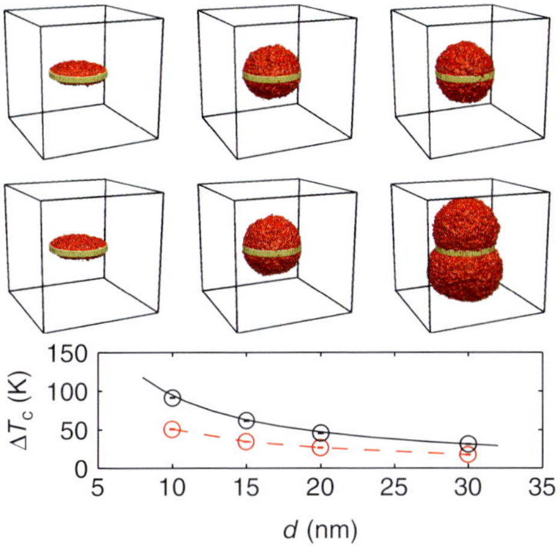

Figure 3.5 Phase-field simulations illustrating the *free growth-limited mode of particle-induced crystallization* of pure Ni. Cylindrical particles ($d = 20$ nm) with contact angles of 45° and 175° on the horizontal and vertical surfaces were used. Upper row: $\Delta T = 26\,\text{K} < \Delta T_c$, central row: $\Delta T = 27\,\text{K} > \Delta T_c$, time elapses from left to right. Bottom row: ΔT_c versus particle diameter d. Original theory, solid line; phase-field simulation. dashed line. The deviation in ΔT_c between theory and simulations is due to the thermal fluctuations considered in the latter. (Gránásy *et al.* [2]. Reproduced with the permission from American Physical Society.)

radius of the homogeneous nuclei becomes smaller than R and free growth sets in. The critical undercooling for this is $\Delta T_c \approx 2\gamma_{SL}/(\Delta s_f R)$, where Δs_f is the volumetric entropy of fusion. The approach has been extended to various shapes of the foreign particles: configurations before free growth are shown in Figure 3.5. The model proved to be highly successful in various areas of particle-induced freezing ranging from materials science to cryobiology.

3.5.3.3 **Molecular Dynamics (MD) Simulations**

The mechanical equation of motion (EOM) is solved for classical particles interacting with appropriate (e.g., hard sphere, Lennard–Jones, embedded atom) potentials. In the highly supercooled/supersaturated state, nucleation is observed in the simulations, which can be analyzed in detail. With special techniques (umbrella sampling), the nucleation barrier can be mapped out, and the solid–liquid interface free energy and other thermodynamic properties can also be determined. MD studies have shown that in the regime accessible for such simulations, the droplet model used in the CNT is rather inaccurate and (as also seen in colloid experiments) a disordered precursor may assists the formation of crystal nuclei, whereas the solid heterogeneities lead to an ordering of the adjacent liquid, offering a microscopic way to understand how the heterogeneities assist the formation of the crystalline phase.

3.5.3.4 Phase-Field Theory and Simulations

Continuum models of the van der Waals/Cahn–Hilliard/Ginzburg–Landau type have been used to describe nucleation for a long time. These models represent the earliest/simplest examples of PF-type approaches (they are often categorized also as simple classical density functional theories). More complex models have been proposed for multiphase multicomponent cases. The basic ideas are illustrated here in a simple case, where a single structural order parameter ϕ is assumed to capture the structural change between the liquid and the crystal. A simple example of such order parameter is evident if the number density of particles in the crystal is assumed to be a sum of Gaussian peaks at the molecule sites (quasiharmonic approximation). Then, the Fourier coefficients of the crystal are expressible by the first one as $\phi_n = \phi_1^{(k_n/k_1)^2}$, where ϕ_n is the nth Fourier component and k_n the length of the respective wave vector. Thus the dominant Fourier component $\phi = \phi_1$ can be used as the structural order parameter. Our starting point here is a free energy functional, expressed in terms of this order parameter and its spatial derivatives. Addressing crystal nucleation in a single component melt, we assume

$$F = \int \mathrm{d}V \left\{ \frac{\varepsilon^2}{2} |\nabla\phi|^2 + wg(\phi) + p(\phi)\Delta f \right\}, \tag{3.3}$$

where the double-well and interpolation functions are of the form $g(\phi) = 1/4\, \phi^2(1-\phi)^2$ and $p(\phi) = \phi^3(10 - 15\phi + 6\phi^2)$ and Δf is the thermodynamic driving force of crystallization. Here, $\phi = 0$ corresponds to the liquid, and $\phi = 1$ stands for the crystalline phase. Note that $g(\phi)$ and $p(\phi)$ functions that reflect the crystal structure can be derived using the Ginzburg–Landau expansion. The free energy has now three model parameters, ε^2, w, and Δf. Note that at the melting point $\Delta f = 0$. There the free energy and the thickness of the planar interface can be expressed as $\gamma_{\mathrm{SL}} = \{\varepsilon^2 w/(6\sqrt{2})\}^{1/2}$ and $d_{10-90\%} = 2\ln(9)\{2\varepsilon^2/w\}^{1/2}$, where the latter thickness corresponds to the range $0.1 \leq \phi \leq 0.9$, thus anchoring the model parameters to measurable quantities. The critical fluctuation (nucleus) represents a saddle point in the functional space; therefore, its properties can be obtained as the solution of the Euler–Lagrange equation under the boundary conditions that the unperturbed undercooled liquid exists in the far field, whereas the field gradients are zero at the center due to symmetry reasons. Considering an isotropic interface free energy (spherically symmetrical problem), the respective equation boils down to $\phi'' + \left(\frac{2}{r}\right)\phi' = (1/\varepsilon^2)\partial f/\partial\phi$, where $'$ stands for d/dr, $\phi' = 0$ at $r = 0$, and $\phi = 0$ for $r \to \infty$. Inserting back the solution into the free energy and integrating the excess free energy density relative to the far-field value (for typical solutions see Figure 3.6), one obtains the nucleation barrier $W_{\mathrm{hom}}{}^*$, which can be then plugged into the classical expression to obtain the nucleation rate. An advantage of this approach relative to the droplet model is that it considers that the solid–liquid interface is diffuse (as shown by MD simulations) and that nonbulk properties and the curvature dependence of the interface free

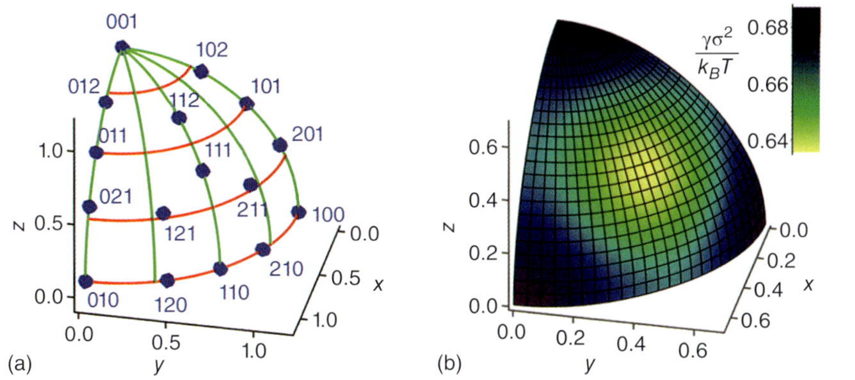

Figure 3.6 Orientation dependence of the solid–liquid interface free energy of the hard-sphere system in the fundamental measure theory. (a) Computations were performed in these directions. (b) The Wulff plot: the colors on the surface show the local values of the interface free energy. (Härtel *et al.* [3]. Reproduced with the permission from American Physical Society.)

energy are automatically handled. Remarkably, even this simple model leads to a considerable improvement in estimating the nucleation rate, especially if the structure-related $g(\phi)$ and $p(\phi)$ from the Ginzburg–Landau approach are used, as in the case of fcc–bcc phase selection in the Fe–Ni system. One can also perform *PF simulations of nucleation* via adding a noise ζ_ϕ that satisfies the fluctuation–dissipation theorem to the EOM:

$$\frac{\partial \phi}{\partial t} = -M_\phi \frac{\delta F}{\delta \phi} + \zeta_\phi \tag{3.4}$$

where $\delta F/\delta\phi$ is the functional derivative of the free energy F with respect to the PF, whereas the noise has the following correlator: $\langle \zeta(t,\mathbf{r})\zeta(t',\mathbf{r'})\rangle = 2M_\phi kT\,\delta(t\text{-}t')\,\delta(\mathbf{r}\text{-}\mathbf{r'})$. An example of such PF simulations that illustrates the *free growth-limited mechanism* mentioned earlier is shown in Figure 3.5. Here the contact angles have been set via appropriate boundary conditions at the surfaces of the substrate.

3.5.3.5 Density Functional Theory and Phase-Field Crystal Modeling

The first molecular-scale approach to crystalline freezing has been presented in 1979 by Ramakrishnan and Yussouff (RY). The local state of matter is described by the time-averaged number density of the molecules (called also particle density or singlet density) that is homogeneous in the melt and has peaks at the molecule sites in the crystal. A variety of such models have been put forward ranging from the early perturbative density functional theory (DFT) of RY (which did not improve by incorporating higher-order terms), via the more accurate nonperturbative approaches, to the most accurate of them, the fundamental measure

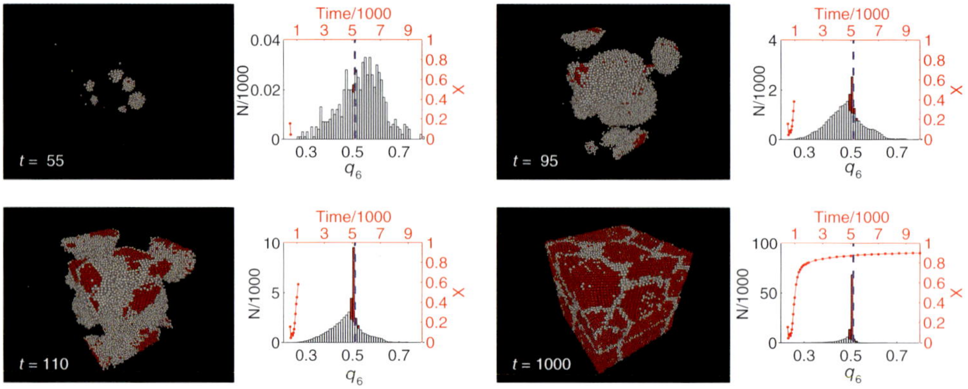

Figure 3.7 Two-step nucleation in the single-mode PFC model at reduced temperature $\varepsilon = 0.1667$. Left: snapshots of the particle density taken at dimensionless times are shown. Spheres of the diameter of the inter-particle distance centered on density peaks higher than a threshold are shown in red if the bond-order correlation functions q_4 and q_6 fall in the ranges $q_4 \in [0.02, 0.07]$ and $q_6 \in [0.48, 0,52]$ (bcc-like) and white otherwise. Right: population distribution of q_6 (histogram painted similarly) and the time dependence of the fraction X of bcc-like neighborhoods (dots and solid line). (Tóth *et al.* [4]. Reproduced with the permission from American Physical Society.)

theory. These models became gradually more accurate and more complicated. Most of them have been applied for the solid–liquid interface (see the orientation dependence in the hard-sphere system in Figure 3.6). However, nucleation has been addressed relatively rarely. A newcomer to the field is the phase-field crystal (PFC) model (for further information see Section 3.6.9), which approximates the direct correlation function in the RY model by its Fourier expansion, retaining terms up to the fourth order. Although an approximate theory, it is easy to handle numerically; simulations up to 24 million particles have been performed with relative ease. Nucleation studies in the framework of the PFC model imply the presence of an *amorphous precursor* that assists the formation of the crystalline phase (Figure 3.7), as found in experiments for colloids and in Monte Carlo (MC) simulations for the hard-sphere system. For heterogeneous nucleation, it has been shown that the *contact angle and the nucleation barrier are nonmonotonic functions of the lattice mismatch* between the substrate and the growing crystal. *Limitations of the free growth model* of particle-induced freezing have also been identified: it works for small anisotropies and large foreign particles; however, it fails for faceted crystals or for nanoscale foreign particles.

3.5.3.6 **Incorporating Nucleation into Simulations**

Another important issue is how one should incorporate nucleation into PF simulations. A straightforward solution is to add noise to the equations of motion of the PF model as explored in the case of the time-dependent Ginzburg–Landau approach. This can be regarded as another way to treat nucleation on the basis of the free energy functional. The nicety of this approach is that nucleation happens automatically. Here, the dynamics of the system comes into play. Furthermore, the addition of noise influences the position and depth of the free energy minima, the magnitude of the interfacial free energy (via the capillary waves), and so on, which requires the renormalization of the model parameters to recover the original values for such quantities. Another problem is that subnanometer resolution is needed for doing this, which restricts the simulations to small sizes (nanoscale) and short times and normally to very large undercoolings. To circumvent these problems, various techniques have been adopted. To accelerate noise-induced nucleation, one may compute the nucleus and the nucleation rate by solving the Euler–Lagrange equation and then place nuclei spatially randomly of statistics that realizes the computed nucleation rate into the simulation box, and the rest happens automatically (some of the nuclei decay, while the others grow as decided by the local fluctuations). In the case of *quantitative PF simulations* working with large interface thickness and spatial steps (a few tenth of a micron), while correcting for the associated artifacts via "antitrapping" currents, one needs to compute first the nucleation rate consistently with the free energy functional (solving the Euler–Lagrange equation(s)) and insert small particles (of size that is able to grow) randomly in space and time with the appropriate frequency. The free growth model (1) can also be implemented similarly for sub-Δx particles: assign particles of given size distribution randomly to the pixels/voxels of the simulation, compute the respective critical undercoolings considering the local temperature and composition, and compare it with the actual undercooling. Once the latter becomes larger than the critical undercooling for the given particle, flip the PF to 1 (results from such a procedure are shown in Figure 3.8). As Δx is usually much larger than the critical size, this will initiate growth. A more detailed strategy is

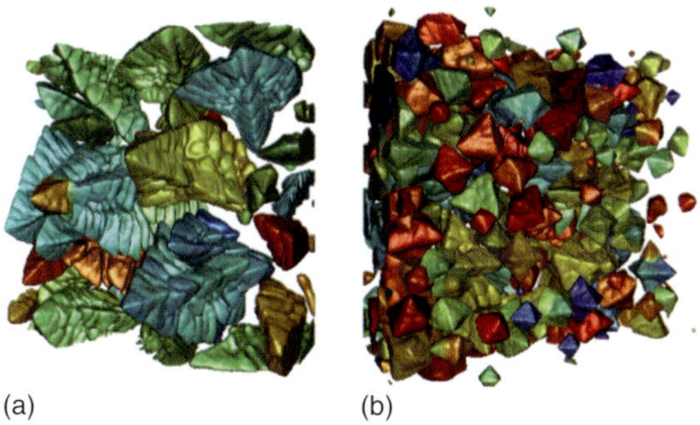

Figure 3.8 Quantitative phase-field simulation for an Al–Ti alloy relying on the free growth-limited mechanism of particle-induced freezing by Greer and coworkers. Transition from columnar dendritic growth (a) to solidification initiated by foreign particles (b). Different colors stand for different crystallographic orientations. For an animation see http://www.esa.int/spaceinvideos/Videos/2012/07/Kaleidoscopic_nanocomposite_metals.

also possible that considers growth from the size of the foreign particle to the size of the pixel/voxel.

3.5.4
Diffusion Models

From absolute reaction rate theory [5], the following expression for the vacancy-mediated flux J_k (in units mol/m^2 s) of a component k in the z direction may be derived (neglecting so-called correlation effects):

$$J_k = -M_k c_k \frac{\partial \mu_k}{\partial z}, \tag{3.5}$$

where M_k (mol m^2/J s) is the mobility, c_k (mol/m^3) the concentration, and μ_k (J/mol) the chemical potential of component k. Here we have chosen to include the local vacancy concentration in the mobility. If there are additional gradients in other potentials, such as temperature and electric potential, more terms must be added. This expression is the basis of diffusion simulations.

The flux expression (Eq. 3.5) may, if necessary, be expanded in terms of concentration gradients

$$J_k = -\sum M_k c_k \frac{\partial \mu_k}{\partial c_j} \frac{\partial c_j}{\partial z}. \tag{3.6}$$

Phenomenologically, the flux may be written as

$$J_k = -\sum D_{kj} \frac{\partial c_j}{\partial z}, \tag{3.7}$$

where D_{kj} (m^2/s) is the diffusion coefficient of component k with respect to the concentration gradient of component j. Equation 3.3 is often referred to as the Fick–Onsager law [6, 7]. With the assumption stated earlier, we may identify

$$D_{kj} = M_k c_k \frac{\partial \mu_k}{\partial c_j}, \tag{3.8}$$

where the derivative $\partial \mu_k / \partial c_j$ is often referred to as the thermodynamic factor; it contain no kinetics and can be evaluated from equilibrium thermodynamics.

3.5.4.1 Single-Phase Diffusion Problems

Diffusion process simulations in a single-phase domain are straightforward as long as the composition does not fall inside a miscibility gap; the treatment of spinodal decomposition will not be discussed here. To obtain the parabolic diffusion partial differential equation (PDE), an expression for the flux is combined with the equation of continuity

$$\frac{\partial c_k}{\partial t} = \frac{\partial}{\partial z}(-J_k). \tag{3.9}$$

Given the initial state and boundary conditions, it is then possible to solve for the evolution of the composition profiles.

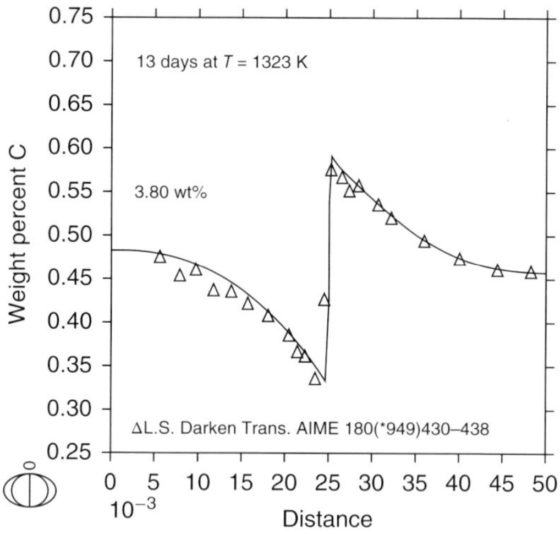

Figure 3.9 Carbon profile after heat treatment of a Fe–Si–C diffusion couple. The silicon content is 3.8 mass% in the left half of the couple and 0.05 mass% in the right half, distance in meters.

The simplicity of these simulations does not make them any less useful. Common types of simulations are, for example, homogenization heat treatments and simulations of carburization.

An example of a single-phase diffusion problem is the famous Darken [8] experiment. In this experiment two pieces of steel with similar carbon content but dissimilar silicon content were put together and heat treated at 1050 °C for 13 days. The resulting simulated carbon profile along with Darken's experimental data is shown in Figure 3.9. It can be seen that carbon has diffused against its own concentration gradient, thereby proving that it is the chemical potential gradient rather than the concentration gradient that is the driving force for diffusion. This was the purpose of the original experiment. The diffusion of iron and silicon is negligible in this case.

3.5.4.2 Moving Phase Boundary Simulations under Local Equilibrium Conditions

Assume that we have a diffusion-controlled phase transformation $\gamma \rightarrow \alpha$ where a phase α is growing into a parent phase γ. Assume that there is a sharp, zero-width moving interface between the phases. The following mass balance for the transfer between the phases must then hold [9]:

$$v^{\alpha/\gamma}(c_k^{\alpha} - c_k^{\gamma}) = J_k^{\alpha} - J_k^{\gamma} \quad k = 1, \ldots, n-1, \tag{3.10}$$

where $v^{\alpha/\gamma}$ is the velocity of the interface. The concentrations and the fluxes are to be evaluated at the interface, but on the α and γ sides, respectively. These so-called flux balance equations must in general be solved iteratively. In the case of a binary system, the composition at the interface can be read directly from a phase diagram, but for ternary and higher-order systems, the operating tie line must be found such that mass balance is fulfilled.

If the interface has a finite width, but it is still assumed that equilibrium holds locally in the interface, the following expression for the interface velocity holds [10]:

$$v^{\alpha/\gamma} = \frac{\sum_{k=1}^{n} \frac{\partial f^{\alpha}}{\partial c_k}(J_k^{i\alpha} - J_k^{i\gamma})}{\sum_{k=1}^{n} \frac{\partial f^{\alpha}}{\partial c_k}(c_k^{i\alpha} - c_k^{i\gamma})}. \tag{3.11}$$

In this expression ($J_k^{i\alpha}$-$J_k^{i\gamma}$) is the net flux of k into the finite width interface, and $c_k^{i\alpha}$ and $c_k^{i\gamma}$ are the concentrations of k in the α and γ phases inside the interface, respectively. The volume fraction of the α phase and overall concentration of k in the interface are denoted by f^{α} and c_k, respectively.

To obtain the interface velocity $v^{\alpha/\gamma}$, there are thus two alternative approaches. One can solve either the set of flux balance equations (Eq. 3.10) or the explicit expression (Eq. 3.11). It is obviously simpler to just evaluate the velocity from an explicit expression, but the drawback then is the finite width of the interface that effectively will limit the size of time steps. When solving the set of flux balance equations poses no difficulty, that approach is computationally more efficient.

When there is supersaturation in the system or when a reaction eventually gives rise to supersaturation, then it is necessary to solve a moving phase boundary problem. If the supersaturation in the system can be neglected, then a diffusion-controlled phase transformation can in many cases be simulated by the multiphase technique described in the next section.

Figure 3.10 shows the resulting chromium profile after a simulation of dissolution of cementite in austenite along with experimental data from Liu *et al.* [11]. An Fe–Cr–C steel was first equilibrated in the ferrite + cementite two-phase region at 735 °C after which it was heat treated for different times in the austenite single-phase region at 910 °C. The figure shows the chromium profile after 100 s.

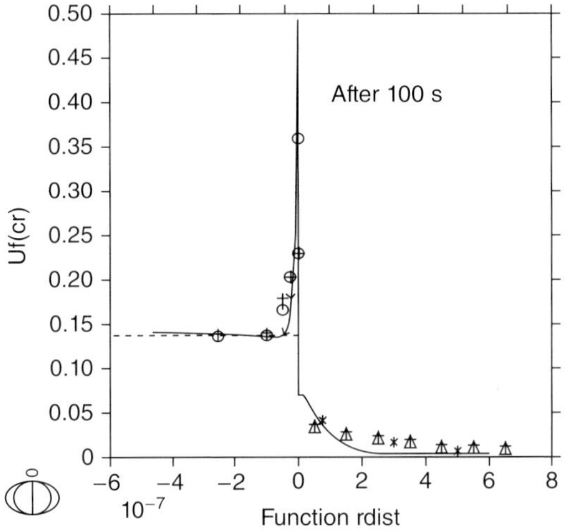

Figure 3.10 Chromium profile after 100 s dissolution treatment of cementite in austenite. From Liu *et al.*

3.5.4.3 1D Multiphase Simulations

In 1D multiphase simulation settings, diffusion is through a multiphase mixture. The interfaces between the phases are then not considered explicitly. From a simulation point of view, these types of simulations are essentially treated as a single-phase diffusion problem. It is assumed that the material is locally equilibrated; the local phase fractions and phase compositions are given by an equilibrium calculation with the local overall composition as condition. The local chemical potentials are then the same in all phases present locally, and the driving force for diffusion, that is, the gradient in chemical potential, may be evaluated in the same manner as for a single-phase problem. The local averaged, coarse-grained kinetics must be estimated from knowledge of the kinetics of the phases present. A suitable measure of the kinetics of an individual phase ϕ is the product of mobility and solubility, and it is therefore convenient to define [12]

$$\Gamma_k^{\phi} = M_k^{\phi} c_k^{\phi}. \tag{3.12}$$

The local overall kinetics Γ_k^* may then be estimated in many different ways. This can, for example, be done by means of either of the Wiener bounds (rule of mixtures and inverse rule of mixtures):

$$\Gamma_k^* = \sum_{\phi} f^{\phi} \Gamma_k^{\phi}, \tag{3.13}$$

$$\Gamma_k^* = \left[\sum_{\phi} \frac{f^{\phi}}{\Gamma_k^{\phi}} \right]^{-1}. \tag{3.14}$$

The flux is then given by

$$J_k = -\Gamma_k^* \frac{\partial \mu_k}{\partial z}. \tag{3.15}$$

The resulting carbon profile after a simulation of carburization of a Ni–25% Cr alloy is shown in Figure 3.11 along with experimental data from Bongartz *et al.* [13]. The majority phase is austenite, but during the process, M_7C_3 and M_3C_2 carbides form. Most of the carbon is tied up in these carbides.

3.5.5 Precipitation Models

Precipitation microstructure is probably the most important and most studied material microstructure, where particles of a second phase or second phases are scattered around in a solid matrix phase. These second-phase precipitates can not only impede the movement of dislocations and lead directly to the precipitation strengthening effect but also pin grain boundaries and therefore stop or slow down grain growth so that grain boundary strengthening effect can be increased. Many metallic alloys attain their high strength, ductility, and toughness through the formation of an optimal precipitation microstructure that has the right amounts of particles of different appropriate sizes, that is, the best PSD.

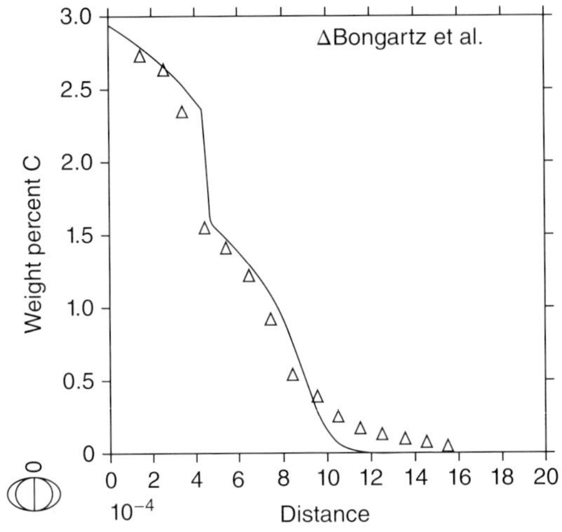

Figure 3.11 Simulated carbon profile after 1000 h carburization at 850 °C of a Ni–25% Cr alloy. (Experimental data from Bongartz *et al.* [13])

The PSD fully characterizes a precipitation microstructure and is usually represented by the distributed number density function, $F(r,t)$, a time (t)-dependent distribution function for the number of particles per unit volume (number density) over the particle size (r) dimension. By definition, once this PSD function $F(r,t)$ for a precipitate is known, the temporal evolution of the number density, the mean radius, and the volume fraction of the precipitate can be calculated by $\int_0^\infty F(r,t)\mathrm{d}r$, $\int_0^\infty rF(r,t)\mathrm{d}r$, and $\int_0^\infty 4/3(\pi r^3 F(r,t)\mathrm{d}r)$, respectively. Moreover, the interparticle spacing, a crucial parameter in predicting material strength, can also be approximated on the basis of the calculated mean radius and volume fraction.

The formation of precipitates is mostly a diffusion-controlled first-order phase transformation. Precipitates form in supersaturated solid solutions because the solubility limits of solute atoms have been surpassed. A supersaturated solid solution is a thermodynamically metastable or unstable system and should decompose into a mixture of solute-rich precipitates and the solute-depleted solid solution at the solubility limits. The decomposition needs obviously migration or diffusion of solute atoms so that their enrichment in precipitate particles becomes possible. Typically, a precipitation process in a metastable system has three distinctive or overlapping stages: nucleation, growth, and coarsening. Therefore a comprehensive precipitation model ought to treat these three stages concurrently. For an unstable system, the precipitation should proceed spontaneously through a continuous development of tiny fluctuations. This process is called spinodal decomposition and will not be covered here.

Several precipitation models exist. The most popular ones are the Langer–Schwartz (LS) model and the Kampmann–Wagner numerical (KWN) model. These computationally efficient models can treat the whole precipitation process continuously from nucleation and growth to coarsening, during which particle sizes can span from subnanometers to micrometers. Moreover, these mesoscopic models can be directly coupled with CALPHAD thermodynamics and kinetics and leave no or only a few model parameters to be adjustable for even multicomponent alloys of industrial relevance and thus possess a very high predictive capability. Other precipitation models include the atomistic kinetic Monte Carlo (AKMC) model, the cluster dynamics (CD) model, and the particle-size-grouping (PSG) model.

The KWN model is an extension of the LS model. They are all based on the continuity equation for the temporal evolution of the PSD: $\partial F(r,t)/\partial t = -\partial/\partial r[v(r,t)F(r,t)] + j(r,t)$, where $v(r,t)$ is the particle growth rate and $j(r,t)$ is the distributed nucleation rate, which is related to transient nucleation rate $J(t)$ by $J(t) = \int_{r^*}^{\infty} j(r,t)\mathrm{d}r$. This governing PDE has been solved differently in these two models under the condition of mass conservation.

By assuming a simplified PSD function and suitable analytical form for the rates of nucleation and growth, LS managed to transform the continuity equation into two coupled set of rate equations for the mean radius and number density so that their temporal evolution over the whole course of precipitation can be conveniently predicted. This model is usually implemented as a first-approximation and fast-acting model in available computational tools.

Instead of forcing a PSD function a priori, the KWN model discretizes the PSD function $F(r,t)$ into size classes and solves the continuity equation numerically by at each time step (i) following the motion of each size class due to particle growth/dissolution, (ii) adding particles to new or old size classes upon nucleation, and (iii) updating matrix concentration for mass balance. In this way, the formation and evolution of a true $F(r,t)$ and its nth moment (zeroth, number density; first, mean radius; third, volume fraction) can be obtained altogether during the simulation. The key ingredients in the KWN model are the nucleation and growth rates of particles.

Practical nucleation models follow the CNT, which gives the transient nucleation rate as $J(t) = Z\beta^* N_0 \exp(-\Delta G^*/kT)\exp(-(\tau/t))$, where Z is the Zeldovich factor, β^* is the atomic attachment rate, N_0 is the number of available nucleation sites per unit volume, ΔG^* is the Gibbs energy of formation of a critical nucleus, τ is the incubation time, k is the Boltzmann's constant, and T is the absolute temperature. The parameters Z, β^*, ΔG^*, and τ can be calculated from system (bulk and interface) thermodynamics and kinetics and can be easily adapted for heterogeneous nucleation and multicomponent systems. Especially, interfacial energy plays a critical role in calculating ΔG^*.

There are numerous growth models for binary and multicomponent systems, and within these models various approximations of the Gibbs–Thomson effect have been adopted. However, by linking to a CALPHAD thermodynamic calculation engine, precipitation simulation tools can take into account the Gibbs–Thomson effect exactly. Models that have been implemented in available simulation tools include the steady-state growth model, the thermodynamic extremal principle model, and the analytical multicomponent growth model. Without any exception, the use of these models for industrial alloys has to rely on

CALPHAD thermodynamic and kinetic databases to provide data on local phase equilibria, chemical potentials, and diffusivities.

In summary, precipitation modeling simulates precipitation microstructural evolution by capturing the temporal variation of PSD. The most efficient and widely used mesoscopic precipitation model is the KWN model in which the PSD is discretized into size classes and its governing continuity equation is solved numerically under the law of mass conservation by incorporating new particles into new or existing size classes according to CNT and tracking the growth or dissolution of each size classes on the basis of growth rate equations of different sophistications.

3.5.6
Cellular Automaton Models

The cellular automaton (CA) method has been originally developed in the 1940s by the mathematician John von Neumann in order to model macroscopic complex physical phenomena through the use of simple microscopic laws. A grid of cells is superimposed on the area of interest and the evolution of the physical processes is reproduced. For this purpose, cells are attributed with state indices and variables that evolve according to transition rules. Interactions of the cells with their neighborhood can be defined in two or three dimensions. The success of the CA methods is explained by application in numerous fields and at various scales, with the use of memory and computationally time-efficient algorithms. The CA methods have been successfully applied in metallurgy in order to simulate a wide range of physical processes including grain growth, mechanical deformation, cracking, or diffusion phenomena. In solidification, several variants have been proposed to model microstructure development during cooling. One of them couples the CA method with solution of macroscopic conservation equations using the finite element (FE) method. The interest of this so-called CAFE model is to take advantage in the resolution at each scale of the information computed at the other scale thanks to advanced coupling algorithms.

A CA grid is superimposed onto a domain defined by a FE mesh. When the two tessellations are fixed over time, each cell of the CA grid is thus precisely located in a specific element of the FE mesh. It is generally proposed to use a cell size of the order of the secondary dendrite arm spacing, λ_2, in order to correctly model grain growth competition, while the FE mesh size obeys the standard criteria for convergence of the solutions of a conservation equation problem. The state of the cells is identified as solid, mushy, or liquid during the whole computation time. The evolution of the cell states simulates the grain structure evolution and reproduces the solidification process. Cells are randomly selected as container for a potential nucleation site particle. Their number is yet imposed by the choice of a given density of the nucleation sites. In casting simulations, particles are defined in the liquid bulk with typical density $10^7 - 10^{12}$ sites per cubic meter, while on mold surfaces typical values are $10^5 - 10^8$ sites per square meter. A critical nucleation undercooling is also randomly defined for each particle while choosing a distribution law (e.g., Gaussian distribution). The growth process is initiated at a liquid cell when its temperature reaches the critical undercooling for activation of its nucleation particle. The state index of the cell is then modified to a value corresponding to a mushy state, and its solidification begins. The growth of a grain envelope associated with each mushy cell locally mimics the development of the mushy domain. The envelope geometry depends upon the crystallographic structure of the alloy and the local orientation, also chosen randomly after nucleation proceeded (e.g., attribution of a set of Euler angles). This envelope connects the tips of the dendrite arms which develop inside the cell. Its volume represents the mushy domain associated with the cell. The grain structure propagation is modeled by progressive overlapping of the centers of the neighboring liquid cells. Each cell captured by a growing envelope is then considered as a new mushy cell that propagates its own specific part of the mushy zone as part of the global grain envelope. In the following examples, envelopes preferentially develop along the $\langle 1\,0\;0 \rangle$ crystallographic directions of cubic structures, as this is generally accepted as a good representation of the preferred growth directions of dendrites. A dendrite growth kinetics is thus applied to compute the spatial evolution of the $\langle 1\,0\;0 \rangle$ vertices of the local envelope and the development of the mushy domain in each cell. The kinetic accounts for the effect of the local undercooling, the fluid flow intensity, and its direction with respect to each of the $\langle 1\,0\;0 \rangle$ directions.

Thus, while rules are relatively simple for the CA computations, coupling with the macroscopic scale solutions for momentum conservation equations is more complicated to implement. At the microscopic scale, the temperature, composition, and fluid flow velocity are interpolated from the FE mesh to the CA cells. They are used for computation of the local rules of evolution of the structure and phase composition. But in turn, evolution of the structure is fed back to the FE solutions, providing with the mush fraction developed due to the integration of the nucleation and growth algorithms using the CA rules.

Figure 3.12 presents a CAFE simulation for the solidification of a Sn−3wt%Pb alloy in a rectangular cavity. A temperature difference of 30 °C is first settled between the left-hand-side and the right-hand-side small vertical surfaces of the cavity (gray cooling curves), all other surfaces being kept adiabatic. This heat flow configuration creates a natural thermal convection in the melt still visible in (a) of Figure 3.12. The two vertical controlled temperature surfaces are then cooled down at −0.03 °C /s. One can note that temperature records for thermocouples located at position labeled L24 and L27 (green and blue curves, respectively) are superimposed up to 2000 s, thus not reflecting the 30 °C temperature difference between the two surfaces. This is retrieved by simulation and reflects the effect of the convective transport of energy that only maintains a horizontal temperature gradient close to the two surfaces while a vertical temperature gradient is created in toward the center of the cavity. The columnar grain structure initially nucleates on the colder right-hand-side surface and progressively develops. Due to solute enrichment in lead, clockwise current is further enhanced in the liquid domain. The simulation shows the influence of the fluid flow regime on the grain structure which develops upward. It also shows how the presence of the structure

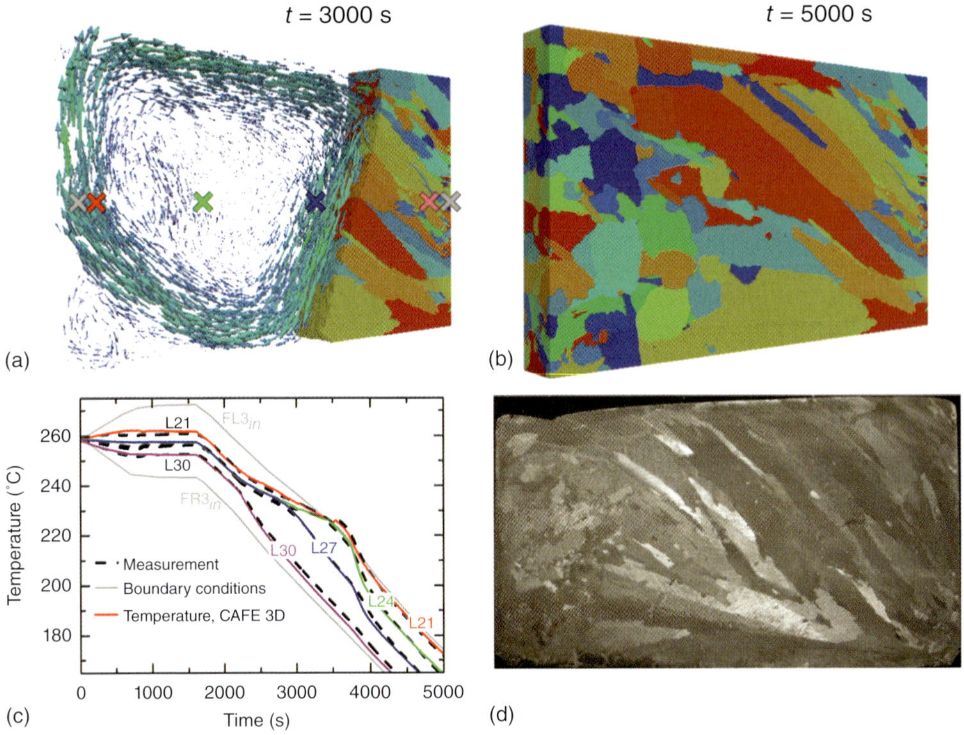

Figure 3.12 3D CAFE simulation of a benchmark experiment for a Sn–3wt%Pb ingot showing fluid flow and grain structure (a) during and (b) after solidification, in (c) temperature evolutions at locations identified by the crosses in (a) and (d) a metallographic observation of the solidified ingot. Domain size: $100 \times 60 \times 10\,\text{mm}^3$, number of CA cells: 7.5×10^6, computational time: 4 days at 64 processors.

progressively moves the convection loops toward the left-hand side of the cavity. When the undercooling is sufficiently large, nucleation of equiaxed grains takes place in the remaining liquid domain, leading to a columnar-to-equiaxed transition (CET) as observed in the final simulated structure as well as in the metallographic cross section.

Figure 3.13 shows a second illustration of the CAFE model dedicated to the GTAW (TIG) welding process. For such simulation, the crystallographic orientation of the grain structure of the base metal is randomly distributed. This is shown by the pole figures of the ⟨100⟩ directions for the grains located in the white box outside the weld bead. The grain structure is stored in the computer's ROM and progressively loaded and released in the RAM according to dynamic allocation strategies. This permits reaching a simulation domain representation of the process by using a very large total number of CA cells. Specific algorithms for parallel calculations are also developed for the CA method. The grain structure of the weld bead only develops from the preexisting and partially melted grains of the base metal. Epitaxial growth takes place so that the grains of the weld bead propagate

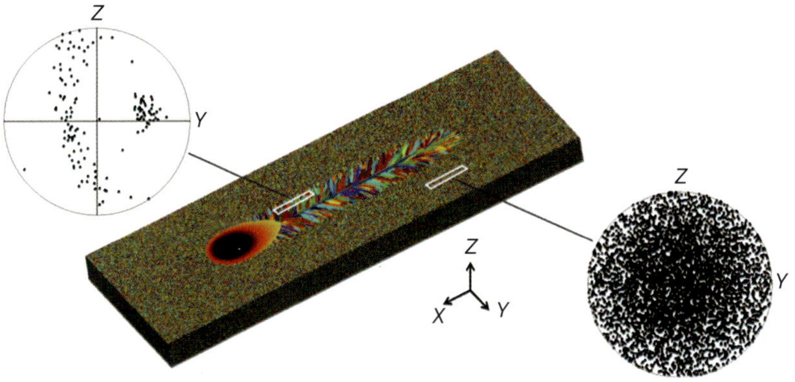

Figure 3.13 3D CAFE simulation of the grain structure during GTAW welding of a duplex stainless steel with heat source velocity, 1 mm/s, and power, 4500 W. Domain size: $180 \times 60 \times 12\,mm^3$, number of CA cells: 253×10^6, computational time: 10 h at 48 processors.

the crystallographic orientation of the partially melted grains. A high-temperature gradient exists in the melt, as illustrated by the color map of the temperature field in the teardrop-shaped weld pool deduced by the solution of the energy conservation equation with the FE method. As a consequence, growth competition in a strong temperature gradient takes place, and only few grains remain at the top surface of the weld bead. The colors used for the grains are randomly chosen and do not permit to estimate the texture of the grains in the weld bead. The second pole figure of the $\langle 1\,0\,0\rangle$ directions for the grains located in the box at the center of the weld bead is more illustrative with that respect. It reveals a strong fiber texture typical of a columnar grain structure generated by competitive growth during directional solidification in a strong temperature gradient.

3.5.7
Monte Carlo Potts Models

MC Potts models can be considered as a subset of CA models. Again, the microstructure is represented by discrete lattice points. Each lattice point can adopt different discrete states, for example, an identifier for the grain or the physical state like solid or liquid. The transition between the different states follows a probabilistic transition rule, and the likelihood for state change depends on the local neighborhood of a cell which eventually leads to self-organization of structures and cooperative movement of interfaces. Therefore, the formulation for such neighborhood interactions is a key ingredient of the model. As a simple example, a model for grain growth comprises the interaction energy γ for two neighboring sites i and j in its states s_i and s_j:

$$\gamma(s_i, s_j) = \begin{cases} 0 \text{ for } s_i = s_j \\ \frac{J}{2} \text{ for } s_i \neq s_j \end{cases} \tag{3.16}$$

Adding volume energy terms, the Hamiltonian of the system can be written as

$$E = \sum_{i=1}^{N} \sum_{j=1}^{Z} \gamma(s_i, s_j) - \mathrm{H} \sum_{i=1}^{N} s_i. \tag{3.17}$$

A two-state Hamiltonian $s_i = \{0, 1\}$ corresponds to the Ising model, and multiple states Q are referred to as Potts model: $s_i = \{0, Q\}$. For computing the microstructure evolution, the whole lattice is sampled in a random manner, and the probability for the state change $P(\Delta E)$ is computed for each cell. The cell either is keeping its old state or is set to the new one. A common choice for the probability transition function is the so-called Metropolis function:

$$\mathrm{P}(\Delta E) = \begin{cases} 1 \text{ if } \Delta E \leq 0 \\ \exp(-\Delta E/kT) \text{ if } \Delta E > 0 \end{cases} \tag{3.18}$$

The probabilistic algorithm leads to some issues in scaling the timescale. Although, compared to the processes on the atomistic scale, for example, an atom in the liquid is attaching to the solid crystal, such a Potts model is conceptually similar, but the frequency of atomic motion cannot be captured and therefore the timescale of the switching needs to be calibrated to phenomenological equations, for example, by the sharp interface model for grain coarsening. In the same way, artifacts caused by the symmetry of the numerical grid (lattice anisotropy) have to be handled in an appropriate manner.

Besides the simple generic model given earlier, many more refined energy functionals are available today, for example, appropriate to describe anisotropy, temperature-dependent kinetics, Zener pinning, recrystallization, Read–Shockley grain boundary energies, and many other phenomena.

The example in Figure 3.14 shows a MC Potts model simulation of grain coarsening. The different colors indicate different "grains," thus areas of identical states s_i. The initial structure has been generated by randomly setting the initial state s_i for each grid cell; thus the fine-grained structure at the left is already the result of self-organization, an inherent feature of the MC Potts model. The generic simulation shown in Figure 3.14 does not comprise any physical scale; instead the timescale is given by the number of MC iterations and length scale is given as number of cells.

3.5.8
Phase-Field and Multiphase-Field Models

In PF models [14], an order parameter ϕ is introduced to describe the microstructure evolution in time and space (please compare also the section about nucleation). ϕ can be considered not only as an identifier of the local phase state, for example, solid or liquid, but also as an identifier for different grains of the same phase, distinguished by its lattice orientation. By definition, ϕ is a real, continuous function of space and time and takes values in the range from 0 to 1. In this notation interfaces are represented by values $0 < \phi < 1$, while $\phi = 1$ and $\phi = 0$ mark

Figure 3.14 Example for grain growth simulation using a Monte Carlo Potts model. Time increases from left to right.

the bulk phases. By this construction, the scalar field $\phi(x,t)$ becomes a map of the microstructure. One order parameter field is sufficient to tackle, for example, single-phase solidification like dendritic solidification of a single grain. Whenever more phases or different grains of the same phase are present in the microstructure, additional order parameters are necessary for microstructure representation. One option is to add a further field variable, for example, an orientation field which describes the local lattice orientation and thus allows to distinguish between different grains. This requires only one more field variable.

A more general concept is followed by the so-called multiphase-field model (MPF). In the MPF concept, each object i of the microstructure is mapped by an individual PF parameter $\phi_i = \phi_i(x,t)$ with the additional constraint that at a given position in space x the sum over all NPF's equals 1, $\sum_{i=0}^{N} \phi_i(x) = 1$. In an MPF model, the grain orientation or the phase state is an attribute to the order parameter field ϕ_1 and has to be administered separately to assign proper interaction parameter for the grain pairs i and j, for example, for computing the driving force for the interaction between ϕ_i and ϕ_j.

The evolution equation for the microstructure is the temporal derivative $\partial\phi/\partial t$ which can be derived from a sound thermodynamic basis in combination with general principles of irreversible thermodynamics (see Chapter 4). The PF method therefore provides a physical-based model for free moving interfaces with predictive power. The evolution equation is derived from a free energy model of the microstructure. In contrast to a pure thermodynamic approach, the free energy F is modeled not only as a function of the thermodynamic state variables T, p, or c but also as a function of the PF variables ϕ_i and their gradients $\nabla\phi_i$ which account for the interfaces, that is, phase and grain boundaries. In this sense, the PF method extends equilibrium thermodynamics by incorporating a length and time scale. Different PF models usually differ by their underlying free energy model, for example, how the free energy contribution of the interfaces is modeled or which state variables are considered for the bulk energies. Equations 3.19 and 3.20 give an example for a free energy functional applicable to model isothermal solidification for a binary alloy kept at a constant external temperature T, the notation of the variables will be given later:

$$F = \int \mathrm{d}\vec{r} \left(\sum_{i=1}^{N} \phi_i f_i^{\mathrm{bulk}} \left(c_i, T\right) + f^{\mathrm{intf}} \left(\left\{ \phi_i, \nabla \phi_i \right\} \right) \right), \tag{3.19}$$

$$F = \int \mathrm{d}\vec{r} \left(1 - \phi_s\right) f_l \left(c_l\right) + \phi_s f_s \left(c_s\right) + \left(c - c_l - \phi_s \left(c_s - c_l\right)\right) \tilde{\mu} + \frac{4\sigma_{\mathrm{sl}}}{\eta} \left[\left(1 - \phi_s\right) \phi_s + \frac{\eta^2}{\pi^2} \nabla \phi_s{}^2 \right]. \tag{3.20}$$

From the free energy functional, the temporal evolution of ϕ is derived following a relaxation approach, which is equivalent to the assumption of maximum entropy production:

$$\frac{\partial \phi_i}{\partial t} = \sum_{i \neq j}^{N} M_{ij} \left(\frac{\delta F}{\delta \phi_j} \right), \tag{3.21}$$

$$\frac{\partial \phi_s}{\partial t} = \mu_{\mathrm{sl}} \left[\sigma_{\mathrm{sl}} \left\{ \nabla^2 \phi_s - \frac{\pi^2}{2\eta^2} \left(1 - 2\phi_s\right) \right\} - \frac{\pi^2}{8\eta} \Delta G_{sl} \right]. \tag{3.22}$$

Equations 3.21 and 3.22 are formulated in physical measurable quantities, which are interfacial energy σ, interface thickness η, and interface mobility μ as model parameters. The PF model for nucleation introduced in Section 3.5.3 is build up by model parameters ε, w, and Δf operating in the "energy space." A parameter matching by comparing the PF solution with a sharp interface solution leads than again to the physical observables. In Eq. 3.22, ΔG_{sl} is the local thermodynamic driving force which acts locally at the interface. ΔG_{sl} can be comprehensively computed based on thermodynamic databases or in terms of simplified approximations which are often used for fundamental investigations about microstructure formation. A characteristic feature of PF models is that the thickness of the diffuse interface η can be chosen much larger than the real physical interface thickness which greatly improves the numerical efficiency. Nevertheless, η needs to be small compared to the length scale which should be resolved by the microstructure simulation, like the dendrite tip radius or topological individual precipitates. Artifacts caused by thick interfaces can be corrected by different measures. The so-called thin interface limit and quantitative phase field are important keywords for this issue.

In order to compute the local driving force, the local composition and temperature have to be known, and the PF equation has to be solved concurrently with the evolution equation for the temperature field and/or the diffusion equations for the solutal fields. For solving the diffusion problem, the same physics and equations as described in Section 3.5.4 can be used. Depending on the phenomena, additional field equations may be added, for example, elastic stress and strain fields or electrical or magnetic fields. Coupling among the different equations can appear via the driving forces and/or by dependent material properties, for example, temperature-dependent kinetic coefficients or temperature- and composition-dependent diffusion coefficients.

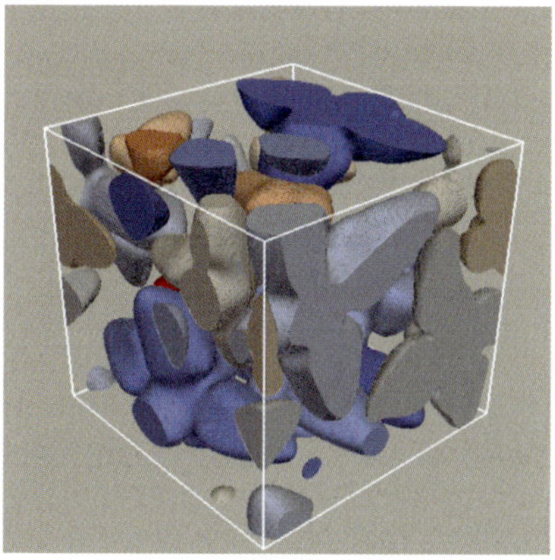

Figure 3.15 Phase-field simulation of the intermediate solidification morphology in a Al5wt%Cu alloy, RVE size $(200\,\mu m)^3$. Each dendritic grain is represented by an own phase-field parameter.

As a simulation example for the MPF model, a solidification microstructure for Ag–5wt%Cu is shown in Figure 3.15. Continuous cooling with $dT/dt = 0.8\,K/s$ was applied as thermal boundary condition. The different dendritic grains, each corresponding to a different PF, can be distinguished by color. The residual melt in between the solid dendrites is set transparent.

The PF method today has proven its potential for quantitative simulations in many areas of microstructure formation, for example, solidification (dendrites, eutectics, peritectics, etc.), solid-state phase transformations (austenite-to-ferrite phase transformation in steels, domain structure evolution in ferroelectric polycrystals, etc.), or heat treatment (grain growth, recrystallization or γ/γ' formation in Ni-based alloys). Application examples can be found, for example, in [15]. In Figure 3.16 the intermediate microstructure during isothermal static recrystallization of an Al alloy as simulated by an MPF model is shown. The dynamics of the simulated recrystallization follows JMAK kinetics as shown in Figure 3.3. It is interesting to note that in this case, the MPF model is not only restricted to recrystallization but also treats grain coarsening in the same run.

Figure 3.16 Phase-field simulation of recrystallization and grain growth. The color indicates the stored deformation energy (dislocation density). Recrystallized grains appear blue, and grain boundaries are rendered in gray.

3.5.9
Phase-Field Crystal Models

PF models as described in the previous section aim at a coarse-grained description of microstructures. Different phases and grains are discriminated by order parameters, but do not resolve the atomic structure. In contrast, PFC models capture the underlying lattice structures, including the ability to describe elastic and plastic deformations, dislocations, and grain boundaries, as well as different lattice structures. In this sense, PFC models are a link between PF models, which are suitable to describe processes on large length and timescales on the one hand, and MD simulations, which capture the motion of the individual atoms on short timescales on the other hand. PFC models have a similar spatial resolution as MD simulations, but on diffusive instead of vibrational timescales. This allows simulating processes on timescales, which cannot be accessed by atomistic methods.

The first and most basic PFC model is based on a very simple free energy expression

$$F = \int \mathrm{d}\vec{r} \left(\frac{\psi}{2} \left(-\varepsilon + \left(k_0^2 + \nabla^2\right)^2 \right) \psi + \frac{\psi^4}{4} \right), \tag{3.23}$$

which is supplemented by an EOM for the conserved density field:

$$\frac{\partial \psi}{\partial t} = D\nabla^2 \frac{\delta F}{\delta \psi}. \tag{3.24}$$

In this model ψ represents a time-averaged atom density, ε and k_0 are model parameters, and D is a diffusion coefficient. As will be shown in the following, ε corresponds to a temperature and k_0 to an inverse lattice constant. At a first glance, it is quite remarkable that such a simple model has attracted so much interest since 2002 when it was proposed by Ken Elder and coworkers for the first time. This is astonishing as the generating functional is not a new invention; it has been known in the pattern formation community already for a long time. Also, at this level it is not obviously linked to real physical parameters like temperature, crystal structures, and other material parameters. In fact, it is the minimalistic structure of the model which makes it so appealing in the search for generic material behavior.

The most basic property of the model is that it can lead to different phases. A liquid phase has a spatially constant order parameter Ψ. In this case, it is assumed that atoms can move around freely, such that the time-averaged density is spatially constant; this reflects the aforementioned diffusive timescales on which the processes are inspected, as compared to much faster scales of atomic vibrations. Apart from this rather simple phase, also solutions with periodic modulations of the atomic density appear, and they correspond to crystalline phases. The distance between the density peaks, that is, the lattice constant, is controlled by the parameter k_0. In equilibrium, the phases which minimize the free energy depend not only on "temperature" ε and the average density in the system, denoted as Ψ_0, but also on the system dimension. In one-dimensional (1D), only the constant (i.e., liquid) and a periodic (stripe) phase exist; in certain regions also phase coexistence is possible (see (a) of Figure 3.17). In two dimensions, apart from the stripe phase, also

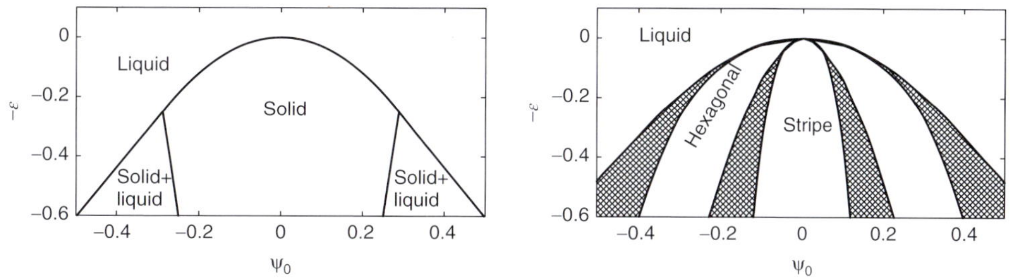

Figure 3.17 Phase diagrams of the one- and two-dimensional phase-field crystal model. The shaded regions in the 2D diagram (b) correspond to two-phase regions.

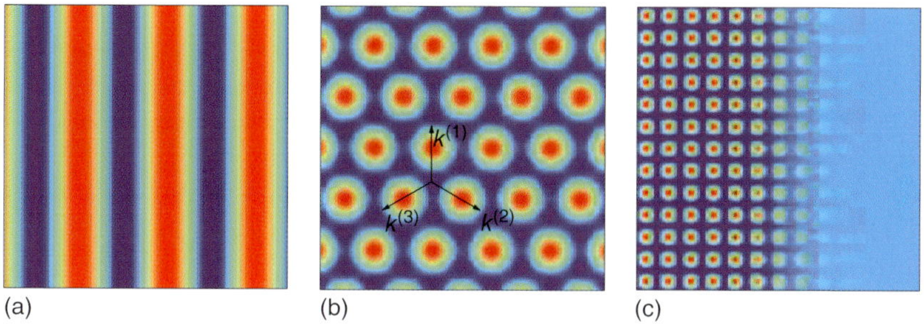

Figure 3.18 Snapshots of equilibrium density profiles in a phase-field crystal model. The density $\psi(x,y)$ is visualized by the color coding. Figure (a) shows a stripe phase in two dimensions. Figure (b) illustrates the hexagonal phase, including the reciprocal lattice vectors for an amplitude equation expansion. Figure (c) shows a cut through a three-dimensional situation with coexistence between a bcc and the melt phase. The ordering of the solid phase decays at the planar interface on a scale which is larger than the atomic spacing.

hexagonal ordering appears as equilibrium phase, and the corresponding phase diagram is shown in the right panel of Figure 3.17. Corresponding density profiles are sketched in Figure 3.18. In three dimensions, the phase diagram already gets more involved and contains also bcc, fcc, and hcp structures.

The simplicity of the model suggests that many improvements are possible to bring it closer to quantitative material modeling. Whereas the aforementioned basic PFC model contains only numerical parameters which do not easily allow a direct comparison to real physical parameters, this situation has changed since the connection between PFC and classical DFT (see Chapter 5) has been established. DFT itself is a microscopic theory, which describes first of all inhomogeneous complex fluids in equilibrium. Conceptually, it reduces the free energy of a many-body problem to a functional of the inhomogeneous one-body theory. In this sense it is similar to the electronic structure DFT, which reduces the need to know the high-dimensional many-body wave function of a quantum mechanical system to the knowledge of equivalent single-particle wave functions. The DFT is an equilibrium theory, but generalizations have been made toward dynamics in nonequilibrium systems, known as dynamical DFT (DDFT). Via a series of approximations, it is possible to derive the PFC model from DDFT. This allows relating the coefficients in the PFC model to the structure factor of the melt, the density, and elastic constants.

This derivation also shows the way how to improve the original PFC model in many respects. The visualization of the phases in Figure 3.18 suggests that the densities essentially follow a one-mode superposition of planar waves; this concept is further exploited in the amplitude equation (AE) formulation in the following. A two-mode model is an extension in the sense that it adds additional derivative terms, which allows describing, for example, fcc iron. A quantitatively better approximation of bcc materials is obtained by fully fitting the underlying direct correlation function of the melt phase near its maximum, which leads to a better agreement with MD simulations. More generally, this direct correlation function can be mimicked with several peaks in such a way that different crystal structures can be generated. The extension to alloys has also been pushed forward. For that, density fields for each atomic species are introduced, where apart from their own energetic contributions in the spirit of Eq. 3.23 also an interaction term is introduced, which influences the phase diagram. Also such models can be related to corresponding DDFT formulations.

The PFC has been applied to various materials science-related problems so far. This includes applications involving the thermodynamics and kinetics of solid–melt interfaces like dendritic and eutectic solidification and the related aspect of crystalline anisotropy, solute trapping, homogeneous and heterogeneous nucleation, growth of fractal structures, glass formation, epitaxial growth, surface ordering, and alloying. Also solid-state transformations, elasticity, plasticity, dislocation, and grain boundary dynamics can be studied, as well as fracture. Whereas the majority of applications are currently for metallic systems, also colloidal systems, liquid crystals, and foams have been studied.

There are still open issues in the PFC model. One of them is that the number of atoms, that is, of the density peaks, is not conserved. This means in particular that dislocation climb is easily possible without explicitly introducing vacancies in the descriptions. Another aspect is the appearance of bulk dissipation, for example, during grain rotation, whereas one would expect that translations and rotations should conserve the energy due to Galilean invariance. Here, one has to keep in mind that strictly speaking the PFC model is rather motivated from soft matter physics, where the density field describes colloidal particles being immersed in a surrounding liquid and therefore experience damping of the motion. Related to this is the fact that the aforementioned EOM (Eq. 3.24) leads also to a diffusive relaxation of the elastic degrees of freedom; however, the introduction of an additional acceleration term can help to overcome this restriction, as it introduces the sound speed as new parameter to the description.

A numerical implementation of the PFC model typically requires a high spatial resolution, as the density field has to be represented with subatomic accuracy. For this reason, the reachable system sizes are limited. The AE approach bypasses this limitation by approximating the density field as a superposition of plane waves:

$$\psi = \psi_0 + \sum_j A_j \exp\ (i\vec{k}_j \cdot \vec{r}) + \text{complex conjugate}, \tag{3.25}$$

where the amplitudes A_j vary on a scale, which is much larger than the atomic spacing proportional to $1/k_0$. The fast density oscillations are then captured by the exponential term, and the model can be reformulated in terms of the amplitudes, which vary on the scale of the interface thickness. It is possible to derive a free energy functional in terms of the amplitudes, which brings the concept of the PFC model much closer to a conventional PF model. For example, for the normalized 2D hexagonal lattice, the normalized reciprocal lattice vectors are $\vec{k}_1 = (0, 1)$, $\vec{k}_2 = (\sqrt{3}/2, -1/2)$, and $\vec{k}_3 = (-\sqrt{3}/2, -1/2)$. By inserting the one-mode approximation into the PFC free energy functional, one gets in terms of the amplitudes for slow amplitudes for the free energy density for k_0=1

$$\begin{aligned} f_{\text{AE}} = {} & \sum_{j=1}^{3} |L_j A_j|^2 + \sum_{j=1}^{3} (3\psi_0^2 - \varepsilon)|A_j|^2 + 6\psi_0(A_1A_2A_3 + A_1^*A_2^*A_3^*) \\ & + \frac{3}{2}(|A_1|^4 + |A_2|^4 + |A_3|^4) + 6(|A_1|^2|A_2|^2 + |A_1|^2|A_3|^2 + |A_2|^2|A_3|^2), \end{aligned} \tag{3.26}$$

with the differential operator $L_j = \nabla^2 + 2i\vec{k}_j \cdot \nabla$. Similar expressions can be obtained for other crystal structures. Indeed, the new free energy functional contains the same ingredients like a (generalized) gradient square and a double-well potential of an PF model. The advantage in comparison to an PF model is that the density profiles can be reconstructed according to Eq. 3.25, and therefore the atomic information is preserved. This includes the elastic effects as well as defect formation and grain boundaries. Also, the appearing parameters in the model are directly related to the ones in the underlying PFC model and the corresponding DFT description, which thus have a direct physical meaning. On the other hand, it should be mentioned that still open questions exist concerning the proper representation of high-angle grain boundaries. Also, the free energy functional typically does not describe all phases, as the crystal symmetries are selected by the reciprocal lattice vectors already in the representation of the one-mode approximation.

On the practical side, the emergence of the AE representation allows to model significantly larger systems, as the discretization scale is now determined by the interface thickness instead of the atomic spacing, and, as for the PFC model, general-purpose computing on graphics processing units turns out to be beneficial (Figure 3.19). Also, hybrid schemes, which use the full PFC model in the interface regions and AE representations in the bulk areas, are used successfully. Moreover, the AE expansion simplifies the model and allows obtaining structural insights for example, on phenomena like grain boundary premelting even on an analytical level.

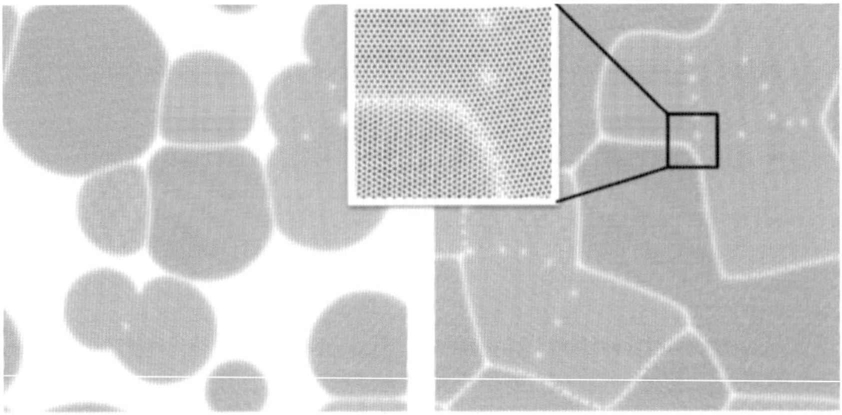

Figure 3.19 Polycrystalline solidification modeled with the two-dimensional amplitude equations. The density is reconstructed from the amplitudes according to Eq. 3.25. Initially, spherical seeds with different orientation and size are placed in the undercooled melt. During solidification, these nuclei grow and come together, but cannot merge due to the misorientation. After completion of the solidification, a polycrystalline structure has emerged, with grain boundaries and isolated dislocations between the grains, as magnified in the inset.

A standard model and implementation of the PFC and AE model is not yet available, as these concepts are still relatively new from point of view of materials science modeling and they are still under development at the moment. Presently, major advantages and limitations of the models have been worked out, and the matching to real physical parameters and situations has advanced significantly during the past years. First discussions in the direction toward a common code platform have been launched.

3.5.10 Crystal Plasticity

Even though functional materials become more and more important to date, the mechanical properties of structural materials are by far the most important ones. Predicting mechanical properties based on material microstructure is, therefore, a common task in computational materials science. Crystal plasticity (CP) is one approach to connect microstructure and mechanical properties. The basic idea of CP is to account for the crystalline structure of metals and many other structural materials.

3.5.10.1 Fundamentals

It is known since the 1930s [16, 17] that crystalline materials plastically deform by the motion of dislocations on well-defined slip systems; however, most

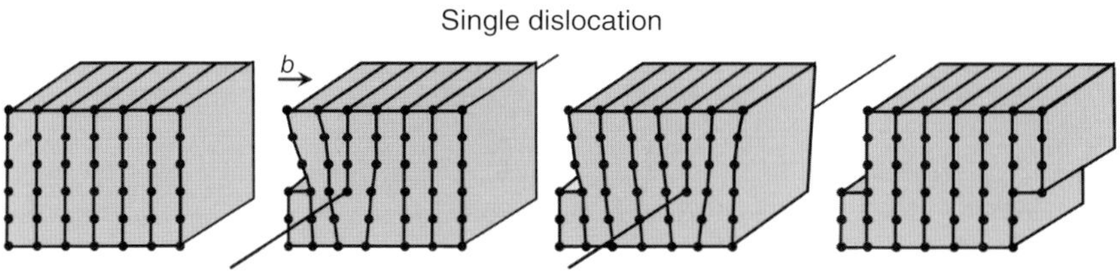

Figure 3.20 Shearing of a crystal by a single moving edge dislocation.

continuum-scale simulations still treat materials as (isotropic) homogeneous media. When a dislocation moves through a crystal, this movement results in a simple shear deformation while at the same time the crystal orientation is maintained (Figure 3.20).

The basic assumption of CP, therefore, is that any macroscopic plastic deformation can and must be described as a superposition of such elementary shear deformations on the slip systems.[1] In a small strain setting, this means that the plastic strain rate can be calculated as

$$\dot{\varepsilon} = \sum_{\alpha=1}^{n} M_{\text{sym}}^{\alpha} \dot{\gamma}^{\alpha} \tag{3.27}$$

Here M_{sym}^{α} is the symmetric part of the Schmid tensor of slip system α, which is calculated as dyadic product of a unit vector in slip direction and the slip plane normal. The $\dot{\gamma}^{\alpha}$ are the shear rates on the n slip systems, where the number n of slip systems depends on the crystal structure of the material (Figure 3.21).

To this end CP can be seen as a purely geometric construction, which however already fully accounts for the anisotropic nature of the plastic deformation of (single) crystals. Materials science enters the picture when calculating the shear rates based on an applied load. This is based on the constitutive equation of state, which connects the material state (still to be defined) and the boundary conditions with a resulting stress as material response:

$$\sigma = f(\dot{\varepsilon}, T, S). \tag{3.28}$$

Here, the deformation velocity $\dot{\varepsilon}$ and the temperature T represent the applied boundary conditions and S represents the material state. To complete the constitutive description, we finally need an evolution equation for the set of state variables

$$\dot{S} = g(\dot{\varepsilon}, T, S). \tag{3.29}$$

So far we still did not define what the state variables are. Based on the choice of state variables, we distinguish so-called phenomenological constitutive models and physics-based models. Most phenomenological models use critical stresses as state variables, while physics-based models use physical quantities like dislocation

1) We restrict ourselves to dislocation motion as the only plastic deformation mechanism at this point.

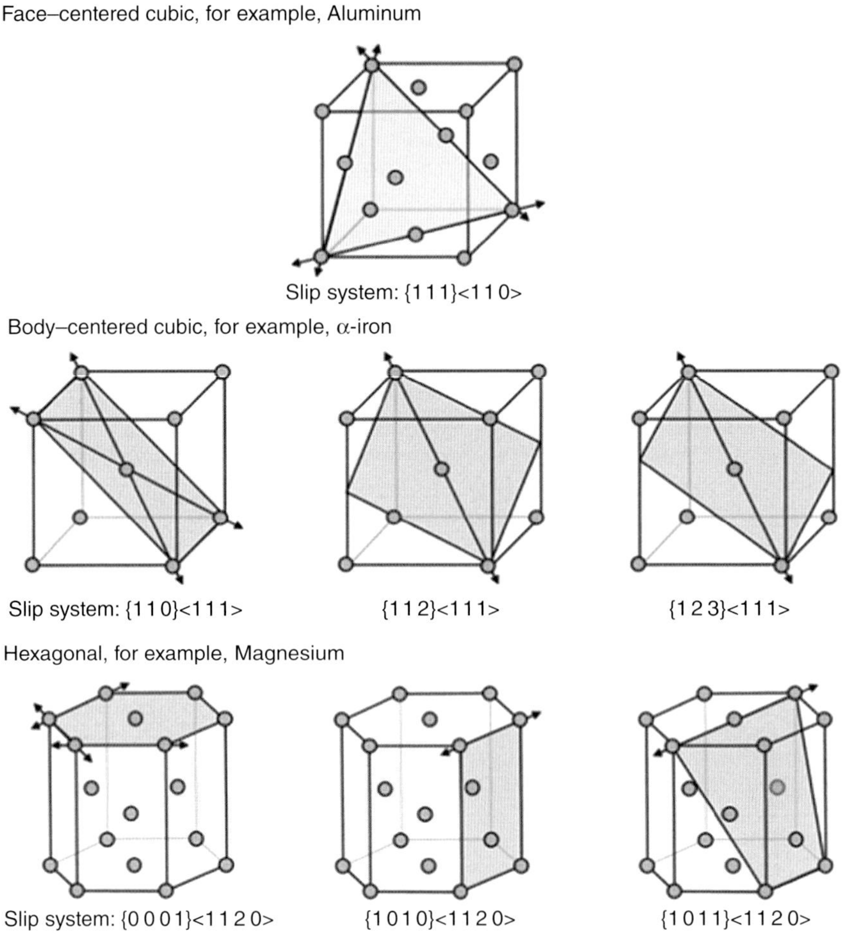

Figure 3.21 Slip system families for different crystal lattices.

densities as state variables. In view of ICME, physical quantities as state variables should be clearly preferred over phenomenological ones as they in general are much easier transferred from one submodel to another.

3.5.10.2 Texture Simulation

The first use of CP models in materials science was the prediction of macrotexture evolution. Texture is the frequency distribution of crystal orientations for the grains of a sample and usually changes during deformation. This is caused by the fact that the elementary shears caused by the dislocation motion also have a rotational part besides the strain part. Whenever this rotational part does not match the rotational part of the prescribed boundary conditions, a rotation of the grain will occur and thus change the crystal orientation.

The key problem in texture evolution prediction is the fact that the boundary conditions for the individual grains in a sample may be different from the boundary conditions imposed on the entire sample. Therefore, assumptions on these boundary conditions need to be made. The most common assumption is the so-called Taylor assumption, namely, that all grains deform just like the macroscopic sample. This leads to the so-called Taylor full constraints (FC) model [18]. However, this is a strong assumption, which results in a too stiff material response. One way to loosen this strict constraint is to relax certain strain components, that is, allow for deviations of some strain components (usually shears) of the individual grains from the strains of the full sample. This is the class of Taylor relaxed constraint (RC) models [19–21]. Another way of relaxing these strong constraints is to group several grains together and apply the sample boundary conditions only to the group of grains in an average sense. These models are termed grain cluster models [22–24]. Finally, there is the class of self-consistent models where each grain is embedded into a material with the average properties of all other grains [25]. A comparison of the different models can be found in [26].

3.5.10.3 **Constitutive Modeling (CP-FEM/CP-FFT)**

While texture simulation just covers the crystal orientations, the use of CP as full constitutive description in continuum-scale forming simulations has gained a lot of momentum in the last 20 years [27]. In most cases CP is implemented as material subroutine for commercial FEM solvers (CP-FEM [28, 29]). Recently spectral methods have been established as an alternative approach for solving the underlying mechanical boundary value problem (CP-FFT [30–32]).

As CP is essentially a single grain description, one distinguishes two kinds of CP simulations: (i) the so-called direct CP simulations and (ii) component-scale CP simulations. In (i) each point of the simulation model represents exactly one crystal orientation, that is, the individual grains of the material are spatially resolved (Figure 3.22). For component-scale simulations this is usually not possible as the typical size of a component is much larger than the single grains of the material (Figure 3.23). (ii) Therefore, this relies on homogenization assumptions similar to those used in texture simulation.

Direct CP simulations can use experimental or artificial microstructures as input. The first offers the opportunity for a direct comparison of simulations and experiment if the microstructure data is experimentally obtained in a nondestructive way. Such comparisons are especially useful for model validation and interpretation of complex experiments. Artificial microstructures can be obtained

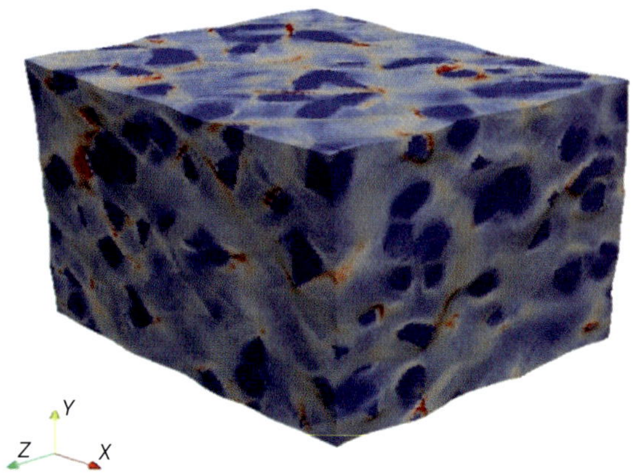

Figure 3.22 Strain distribution in a dual-phase steel obtained from a direct CP-FFT simulation.

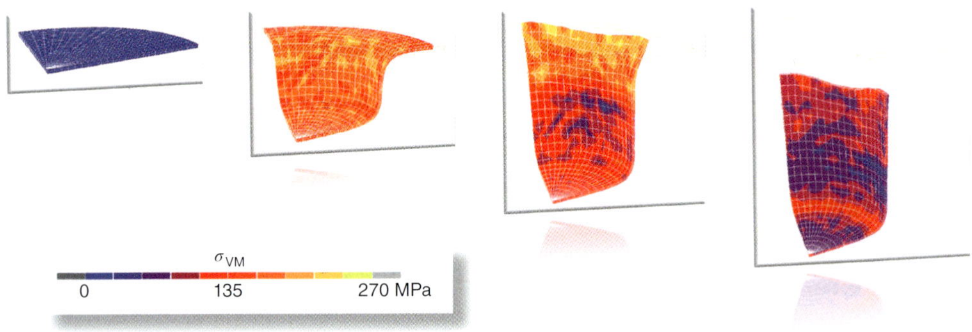

Figure 3.23 Component-scale CP-FEM simulation.

in many ways from simple Voronoi constructions to full-field microstructure evolution simulations. They are useful in determining the average typical properties of complex microstructures using RVEs. This kind of simulation is often referred to as virtual laboratory. One application is the replacement of experimental data necessary for the calibration of yield surface descriptions used in large-scale FEM forming simulations [33, 34].

3.6
Software Tools

Table of software codes related to microstructures

Name of code	Website	Type of code	Typical applications/ phenomena being treated	Remarks
Visualization/evaluation				
DP_MICRESS	www.micress.de	Freeware	Virtual EDX mappings, line scans, manual distance/angle measurements good for 2D and limited (orthogonal sections only) for 3D microstructures	Digital microstructures, native formats, and VTK export/conversion
ParaView	www.paraview.org	Freeware	Powerful general 3D viewer with numerous options/filters for import/export	VTK and many other formats available for both input and output
Experimental microstructures				
DREAM.3D	http://dream3d.bluequartz.net/	Open source/freeware	3D digital microstructures, reconstruction of serial sectioning EBSD images, reconstruction of light optical image stacks	Powerful tool for digitization of experimental data in 2D and 3D
GeoDict	www.geodict.de	Commercial	Imports CT and FIB-SEM data, filters and rotates, cuts out rectangular structures, and generates histogram and multiple thresholds. Converts .stl to voxel data. Export filters to Fluent, Abaqus, CAD, and more	

Table of software codes related to microstructures

Name of code	Website	Type of code	Typical applications/ phenomena being treated	Remarks
Simpleware	www.simpleware.com	Commercial	Processing of 3D image data (MRI, CT, micro-CT, FIB-SEM, etc.) and export models suitable for CAD, CAE, and 3D printing. Export of surface models/meshes; generation of CAE meshes, export of NURBS files and new physics modules +SOLID, +FLOW, and +LAPLACE to calculate effective material properties of scanned samples using homogenization techniques *(see Chapter 6)*	
Robo-Met	http://www.ues.com/content/robo-met3d-analytics	Commercial	3D digital microstructures from serial sectioning light optical microscopy	
Synthetic microstructures				
DREAM.3D	http://dream3d.bluequartz.net/	Open source/freeware	3D digital microstructures, generation of 3D microstructures based on statistical data. Includes also statistics generator module for multiphase structures, grain size distributions texture, and so on	Native format .dream3d is based on hierarchical data format HDF5, export to VTK possible

Table of software codes related to microstructures

Name of code	Website	Type of code	Typical applications/ phenomena being treated	Remarks
MBuilder	Discontinued, see remarks.	n/a	Microstructure Builder or MBuilder is a strategy to construct simulated 3D polycrystalline materials. The input is typically grain size and shape data as obtained from orthogonal images (optical or SEM) or 3D datasets. The output is a 3D voxel structure that matches the size and shape statistics provided at input. The voxel structures can be used directly as input to Monte Carlo simulations or can be converted to mesh structures for use in FE structural analysis	Now integrated in dream3d
Neper	http://neper.sourceforge.net/	Freeware	Synthetic microstructures/meshing	
MICRESS	www.micress.de	Commercial	Voronoi tesselation/deterministic positioning of elliptic or cubic grains	Part of the MICRESS software package; export to VTK is possible
Digimat	www.e-xstream.be	Commercial	Synthetic generation of 3D microstructures based on ellipsoids/prisms	Main purpose of Digimat is the determination of effective properties

Table of software codes related to microstructures

Name of code	Website	Type of code	Typical applications/ phenomena being treated	Remarks
Simpleware	www.simpleware.com	Commercial	Image processing and model generation; conversion of 3D image data into computational models for CAD, finite element analysis (FEA), computational fluid dynamics (CFD), and 3D printing	
GeoDict	www.geodict.com	Commercial	Structure generators for nonwovens and fiber-reinforced composites, cellulose paper structures, ceramics, dense sphere packings, standard woven structures, perforated and regular-array structures and periodic materials, pleated structures for filters, open-cell and closed-cell foams. Manual manipulation of voxelized 3D material models	
Diffusion modeling				
DICTRA	www.thermocalc.se	Commercial	Diffusion-controlled transformations; 1D multicomponent diffusion solver coupled to thermodynamic and mobility databases	
MICRESS	www.micress.de	Commercial	Multicomponent diffusion/phase equilibria 3D settings/coupling to thermodynamic and mobility databases	

Table of software codes related to microstructures

Name of code	Website	Type of code	Typical applications/ phenomena being treated	Remarks
Precipitation modeling				
TC-PRISMA	www.thermocalc.se	Commercial	Based on Kampmann – Wagner numerical model	
MatCalc	www.matcalc.tuwien.ac.at	Commercial	Based on maximum entropy production rate	
ClaNG	www.imm.rwth-aachen.de	Proprietary	Classical nucleation and growth. The statistic precipitation model "ClaNG" predicts the formation, growth, and dissolution of second-phase particles in aluminum alloys	
Potts model				
PGG3D	http://matforge.org/cmu/wiki/pgg3d	Open source	Classical Potts model PGG-3D is a synchronous parallel grain growth code to simulate material microstructure	
Phase-field modeling				
FiPy	http://www.ctcms.nist.gov/fipy/	Freeware	The FiPy framework includes terms for transient diffusion, convection, and standard sources, enabling the solution of arbitrary combinations of coupled elliptic, hyperbolic, and parabolic PDEs. Currently implemented models include phase-field treatments of polycrystalline, dendritic, and electrochemical phase transformations as well as a level set treatment of the electrodeposition process	

Table of software codes related to microstructures

Name of code	Website	Type of code	Typical applications/phenomena being treated	Remarks
MICRESS	www.micress.de	Commercial	The MICRostructure Evolution SImulation Software: multicomponent–multiphase-field code coupled to thermodynamic and kinetic databases; classical nucleation models; solidification, grain growth, recrystallization solid-state transformations, coupling the elastic stresses/strains and to fluid flow	Phase-field model, diffusion, thermodynamic minimization, thermomechanical equilibria
OpenPhase	http://www.openphase.de	Freeware	OpenPhase is a open-source software project targeted at the phase-field simulations of complex scientific problems involving microstructure formation in systems undergoing first-order phase transformations like solid–liquid and solid–solid transformations as well as the structural transformations like grain growth, or recrystallization	
PACE 3D	http://www.iaf.hs-karlsruhe.de/ice/ice/nestler/index.php?id=113	Proprietary	PACE 3D is a software package comprising tools and solvers like phase-field model for microstructure formation, Navier–Stokes equation, or a lattice Boltzmann solver for fluid flows	

Table of software codes related to microstructures

Name of code	Website	Type of code	Typical applications/ phenomena being treated	Remarks
RheoPlast	http://matforge.org/powell/wiki/RheoPlast	Freeware	Development is continued within FiPy since 2006	
Cellular automaton tools				
SphaeroSim	www.ikv.rwth-aachen.de	Proprietary	Simulation of spherulite growth in polymers	
CimLib®	http://www.cemef.mines-paristech.fr/	Proprietary	Simulation of grain structure Cellular automaton – finite element	
Phase-field crystal tools				
	n/a	Proprietary	A number of academic codes on phase-field crystal are currently around, but none of them seem publicly available	
Deformation modeling at the scale of the microstructure/crystal plasticity tools				
DAMASK	damask.mpie.de	Freeware and open source	Düsseldorf Advanced MAterial Simulation Kit (F. Roters, MPIE Düsseldorf) RGC, CP-FEM/FFT	
StrucSim	http://www.ibf.rwth-aachen.de/en/research-development/material-modelling/microstructure/	Proprietary	"StrucSim" implements microstructure simulation into simulation of forming processes. Additional to the plastomechanical forming process dynamic and static microstructure development (recrystallization processes) can be calculated	

Table of software codes related to microstructures

Name of code	Website	Type of code	Typical applications/ phenomena being treated	Remarks
T-Pack	http://www.ibf.rwth-aachen.de/en/research-development/material-modelling/microstructure/	Proprietary	"T-Pack": joint development by IBF and IMM of RWTH Aachen during collaborative research center 370 "integrative material modeling". Combines iteratively the hardening and softening model "3IVM+" and a Taylor texture model with the FE program Larstran/Shape	
(A)LAMEL			See, for example, [21, 22]	
GIA	http://www.muw.rwth-aachen.de/cms/Materialwissenschaft-und-Werkstofftechnik/Forschung/Institute/Institute/~bult/IMM/?lidx=1	Proprietary	Texture model "GIA"	Recently extended to hexagonal crystals and twin formation

Table of software codes related to microstructures

Name of code	Website	Type of code	Typical applications/ phenomena being treated	Remarks
MMSP	http://matforge.org/cmu/wiki/mmsp	Open source	Mesoscale Microstructure Simulation Project, Carnegie Mellon University, Pittsburgh. Typical MMSP applications include: grain growth and coarsening, precipitation reactions, crystal growth and solidification, lattice-based kinetic Monte Carlo, statistical mechanics: Ising model, classical Heisenberg model. Spinodal decomposition and other second-order transformations	
VPSC	http://public.lanl.gov/lebenso/	Freeware/limited circulation	The VPSC code is a multipurpose polycrystal plasticity research code based on the knowledge of the mechanisms of slip and twinning that are active in single crystals of arbitrary symmetry. VPSC can be used to predict the effective stress – strain response, texture evolution or anisotropy	
HMS	http://set.kuleuven.be/m2form/projects/hms_framework	Proprietary/academic cooperation	The HMS model exploits the finite element formulation to describe the macroscopic deformation of the material. At the macroscopic level, the properties of the material are approximated by domain specific models. The approximations are numerically derived from appropriate physics-based polycrystalline plasticity models, which operate at the microscale and provide homogenized response	

Table of software codes related to microstructures

Name of code	Website	Type of code	Typical applications/ phenomena being treated	Remarks
Abaqus CP-UMAT	http://www.columbia.edu/~jk2079/Kysar_Research_Laboratory/Single_Crystal_UMAT.html	Commercial with free subroutines	A User Material (UMAT) subroutine that models the deformation of single crystals which undergo plastic deformation can be downloaded for use with the commercial finite element code ABAQUS	
CP-FFT	http://public.lanl.gov/lebenso	Freeware/limited circulation	Treats 3D polycrystals deformation by dislocation glide. The FFT-based model provides an exact solution of the governing equations, has better performance than a finite element calculation for the same purpose and resolution, and can use voxelized microstructure data as direct input	
Microstructure databases				
MatDL	http://www.matdl.org/ http://nist.matdl.org/		NIST Computational File Repository, NIST Experimental Data Repository, NIST Structural Materials Data Demonstration Project, ASM RDA Demonstration Project: DTR/PID & MGI Infrastructure	

Table of software codes related to microstructures

Name of code	Website	Type of code	Typical applications/ phenomena being treated	Remarks
Further useful codes (unclassified)				
magpar	http://www.magpar.net/	Free, open source	Magpar is a finite element micromagnetics package which combines several features applicable to a variety of static and dynamic micromagnetic problems, includs uniaxial anisotropy, exchange, magnetostatic interactions, and external fields	Last changes to website in 2012
Materials Studio	http://accelrys.com/products/collaborative-science/biovia-materials-studio/	Commercial	Materials Studio is a modeling and simulation environment designed to predict and understand the relationships of a material's atomic and molecular structure with its properties and behavior. Applications comprise materials of all types, including pharmaceuticals, catalysts, polymers and composites, metals and alloys, batteries and fuel cells, and more	

Table of software codes related to microstructures

Name of code	Website	Type of code	Typical applications/ phenomena being treated	Remarks
WISETEX	http://www.mtm.kuleuven.be/Onderzoek/Composites/software/wisetex	Freeware	Textile composites; mesomodeling; internal geometry; permeability; micromechanics; finite elements; multilevel analysis	Description of code family: http://www.mtm.kuleuven.be/Onderzoek/Composites/Research/meso-macro/textile_composites_map/textile_modelling/downloads/textile-modelling-overview-paper-sl.pdf
MatForge	http://matforge.org	Open source	Provides a branded, trusted, noncommercial, and neutral site supporting open-source, collaborative materials code development. Comprises a collection of numerous codes related to microstructure formation (some of which are discontinued or merged into other codes). Several codes are explicitly listed in previous sections of the present table (e.g., FiPy, MMSP)	

References

1. Janssens, K.G.F., Raabe, D., Kozeschnik, E., Miodownik, M.A., and Nestler, B. (2007) *Computational Materials Engineering – An Introduction to Microstructure Evolution*, Elsevier Academic Press, ISBN-13: 978-0-12-369468-3.
2. Gránásy, L., Pusztai, T., Saylor, D., and Warren, J.A. (2007) Phase field theory of heterogeneous crystal nucleation. *Phys. Rev. Lett.*, **98**, 035703.
3. Härtel, A. *et al.* (2012) Tension and stiffness of the hard sphere crystal-fluid interface. *Phys. Rev. Lett.*, **98**, 226101.
4. Tóth, G.I. *et al.* (2011) Amorphous nucleation precursor in highly nonequilibrium fluids. *Phys. Rev. Lett.*, **107**, 175702.
5. Stearn, A.E. and Eyring, H. (1940) Absolute rates of solid reactions: diffusion. *J. Phys. Chem.*, **44** (8), 955–980.
6. Fick, A. (1855) On liquid diffusion. *Philos. Mag. J. Sci.*, **10**, 31–39.
7. Onsager, L. (1945) Theories and problems of liquid diffusion. *Ann. N.Y. Acad. Sci.*, **46**, 241–265.
8. Darken, L.S. (1948) Diffusion, mobility and their interrelation through free energy in binary metallic system. *Trans. AIME*, **175**, 184–201.
9. Stefan, J. (1889) Über einige Probleme der Theorie der Wärmeleitung. *Sitzungsber. Akad. Wiss. Berlin Math. Kl.*, **98**, 473–484.
10. Larsson, H. and Reed, R.C. (2008) On the numerical simulation of diffusion-controlled reactions under local equilibrium conditions. *Acta Mater.*, **56**, 3754–3760.
11. Liu, Z.-K., Höglund, L., Jönsson, B., and Ågren, J. (1991) An experimental and theoretical study of cementite dissolution in an Fe–Cr–C alloy. *Metall. Trans. A*, **22A**, 1745–1752.
12. Larsson, H. and Engström, A. (2006) A homogenization approach to diffusion simulations applied to alpha-gamma Fe–Cr–Ni diffusion couples. *Acta Mater.*, **54**, 2431–2439.
13. Bongartz, K., Lupton, D.F., and Schuster, H. (1980) A model to predict carburization profiles in high temperature alloys. *Metall. Trans. A*, **11A**, 1883–1893.
14. Provatas, N. and Elder, K. (2010) *Phase-Field Methods in Materials Science and Engineering*, Wiley-VCH Verlag GmbH, Weinheim. ISBN: 978-3-527-40747-7
15. Schmitz, G.J., Böttger, B., Eiken, J., Apel, M., Viardin, A., Carré, A., and Laschet, G. (2010) Phase-field based simulation of microstructure evolution in technical alloy grades. *Int. J. Adv. Eng. Sci. Appl. Math.*, **2** (4), 126–139.
16. Taylor, G.I. (1934) The mechanism of plastic deformation of crystals. Part I—theoretical. *Proc. R. Soc. London, Ser. A*, **145**, 362–387.
17. Taylor, G.I. (1934) The mechanism of plastic deformation of crystals. Part II—comparison with observations. *Proc. R. Soc. London, Ser. A*, **145**, 388–404.
18. Taylor, G.I. (1938) Plastic strain in metals. *J. Inst. Met.*, **62**, 307–324.
19. Honeff, H. and Mecking, H. (1981) Analysis of the deformation texture at different rolling conditions, in *Proceedings of ICOTOM 6*, vol. **1** (ed. S. Nagashima), The Iron and Steel Institute of Japan, pp. 347–355.
20. Kocks, U.F. and Chandra, H. (1982) Slip geometry in partially constrained deformation. *Acta Metall.*, **30**, 695–709.
21. Van Houtte, P. (1982) On the equivalence of the relaxed Taylor theory and the Bishop-Hill theory for partially constrained plastic deformation of crystals. *Mater. Sci. Eng.*, **55** (1), 69–77.
22. Van Houtte, P., Delannay, L., and Kalidindi, S.R. (2002) Comparison of two grain interaction models for polycrystal plasticity and deformation texture prediction. *Int. J. Plast.*, **18**, 359–377.
23. Crumbach, M., Goerdeler, M., and Gottstein, G. (2006) Modelling of recrystallisation textures in aluminium alloys: I. Model set-up and integration. *Acta Mater.*, **54**, 3275–3289.
24. Tjahjanto, D.D., Eisenlohr, P., and Roters, F. (2010) A novel grain cluster-based homogenization scheme. *Modell. Simul. Mater. Sci. Eng.*, **18**, 015006.

25. Lebensohn, R.A. and Tome, C.N. (1993) A self-consistent approach for the simulation of plastic deformation and texture development of polycrystals: application to Zirconium alloys. *Acta Metall. Mater.*, **41**, 2611–2624.
26. Li, S. and Van Houtte, P. (2002) Performance of statistical (Taylor, Lamel) and CP-FE models in texture predictions of aluminium alloys in cold rolling. *Aluminium*, **78**, 918–922.
27. Roters, F., Eisenlohr, P., Hantcherli, L., Tjahjanto, D.D., Bieler, T.R., and Raabe, D. (2010) Overview of constitutive laws, kinematics, homogenization and multiscale methods in crystal plasticity finite-element modeling: theory, experiments, applications. *Acta Mater.*, **58**, 1152–1211.
28. Roters, F., Eisenlohr, P., Bieler, T.R., and Raabe, D. (2010) *Crystal Plasticity Finite Element Methods in Materials Science and Engineering*, Wiley-VCH Verlag GmbH, Weinheim. ISBN: 978-3-527-32447-7
29. Kalidindi, S.R., Bronkhorst, C.A., and Anand, L. (1992) Crystallographic texture evolution in bulk deformation processing of fcc metals. *J. Mech. Phys. Solids*, **40**, 537–569.
30. Moulinec, H. and Suquet, P. (1998) A numerical method for computing the overall response of nonlinear composites with complex microstructure. *Comput. Methods Appl. Mech. Eng.*, **157**, 69–94.
31. Lebensohn, R.A. (2001) N-site modeling of a 3D viscoplastic polycrystal using fast Fourier transform. *Acta Mater.*, **49**, 2723–2737.
32. Eisenlohr, P., Diehl, M., Lebensohn, R.A., and Roters, F. (2013) A spectral method solution to crystal elastoviscoplasticity at finite strains. *Int. J. Plast.*, **46**, 37–53.
33. Kraska, M., Doig, M., Tikhomirov, D., Raabe, D., and Roters, F. (2009) Virtual material testing for stamping simulations based on polycrystal plasticity. *Comput. Mater. Sci.*, **46**, 383–392.
34. Zhang, K., Holmedal, B., Hopperstad, O.S., Dumoulin, S., Gawad, J., Van Bael, A., and Van Houtte, P. (2015) Multi-level modelling of mechanical anisotropy of commercial pure aluminium plate: crystal plasticity models, advanced yield functions and parameter identification. *Int. J. Plast.*, **66**, 3–30.

Further Reading

Nucleation

Kelton, K.F. and Greer, A.L. (2010) *Nucleation in Condensed Matter: Applications in Materials and Biology*, Elsevier, Amsterdam.

Greer, A.L., Brunn, A.M., Tronche, A., Evans, P.V., and Bristow, D.J. (2000) Modeling of inoculation of metallic melts: application to grain refinement of aluminum by Al-Ti-B. *Acta Mater.*, **48**, 2823–2835.

Oxtoby, D.W. (1991) Crystallization of liquids: a density functional approach, in *Les Houches 1989 Session LI: Liquids, Freezing and Glass Transition* (eds J.P. Hansen, D. Levesque, and J. Zinn-Justin), North-Holland, Amsterdam, pp. 145–192.

Gránásy, L., Podmaniczky, F., Tóth, G.I., Tegze, G., and Pusztai, T. (2014) Heterogeneous nucleation of/on nanoparticles: a density functional study using the phase-field crystal model. *Chem. Soc. Rev.*, **43**, 2159–2173.

Simmons, J.P., Shen, C., and Wang, Y. (2000) Phase field modeling of simultaneous nucleation and growth by explicitly incorporating nucleation events. *Scr. Mater.*, **43**, 935–942.

Cellular Automaton Models

Carozzani, T., Digonnet, H., and Gandin, C.-A. (2012) 3D CAFE modeling of grain structures: application to primary dendritic and secondary eutectic solidification. *Modell. Simul. Mater. Sci. Eng.*, **20**, 015010.

Carozzani, T., Gandin, C.-A., and Digonnet, H. (2014) Optimized parallel computing for cellular automaton–finite element modeling of solidification grain structures. *Modell. Simul. Mater. Sci. Eng.*, **22**, 015012.

Carozzani, T., Gandin, C.-A., Digonnet, H., Bellet, M., Zaidat, K., and Fautrelle, Y.

(2013) Direct simulation of a solidification benchmark experiment. *Metall. Mater. Trans. A*, **44**, 873–887.

Chen, S., Guillemot, G., and Gandin, C.-A. (2014) 3D coupled cellular automaton (CA)–finite element (FE) modeling for solidification grain structures in gas tungsten Arc welding (GTAW). *ISIJ Int.*, **54**, 401–407.

Phase-Field Crystal Model

Emmerich, H., Löwen, H., Wittkowski, R., Gruhn, T., Tóth, G., Tegze, G., and Gránásy, L. (2012) Phase-field-crystal models for condensed matter dynamics on atomic length and diffusive time scales: an overview. *Adv. Phys.*, **61**, 665–743.

4
Thermodynamics

Tore Haug-Warberg, Long-Qing Chen, Ursula Kattner, Bengt Hallstedt, André Costa e Silva, Joonho Lee, Jean-Marc Joubert, Jean-Claude Crivello, Fan Zhang, Bethany Huseby, and Olle Blomberg

4.1
Overview

Thermodynamic modeling software uses thermodynamic calculations to predict the properties and behaviors of materials under various conditions. At its most basic, for example, thermodynamic modeling software provides information such as the melting point of an alloy, but in practice, the software is used to answer much more complicated and time-consuming questions about the behavior of complex materials.

Thermodynamic modeling is currently one of the most mature areas of ICME (Integrated Computational Materials Engineering), and companies that produce software tools for thermodynamic modeling are deeply involved in initiatives in the European Union and the United States to advance the ICME vision, so thermodynamic modeling will likely continue to be an important part of many ICME projects.

This chapter gives an overview of the current theory and practice of thermodynamic modeling. Sections 4.2 and 4.3 provide an introduction to some basic thermodynamic concepts and theory. Section 4.4 then gives an overview of the so-called CALPHAD (CALculation of Phase Diagrams) method, which is a thermodynamic modeling approach frequently used when it comes to solving practical material design problems that involve multicomponent systems. In this approach, thermodynamic calculations provide the information on phases in stable or metastable equilibrium that are needed for predicting the properties of materials under a wide range of temperature, pressure, and composition conditions.

Section 4.5 focuses on thermodynamic data, data formats, and databases. Section 4.6 is concerned with how thermodynamic data and modeling can be integrated with modeling at larger scales. Section 4.7 then provides a set of examples of how thermodynamic modeling has been successfully used in various research and engineering projects. Section 4.8 contains a table of software

Handbook of Software Solutions for ICME, First Edition. Edited by Georg J. Schmitz and Ulrich Prahl.

tools and applications available for thermodynamic modeling, and the chapter concludes in Section 4.9 with a list of recommended further reading.

4.2 Basic Concepts and Principles

In thermodynamics, reality is simplified to the extent possible without sacrificing information about the static properties of the system. Thermodynamics builds on a *system* that is fully represented by a finite set of scalar, or sometimes tensorial, *state* variables. While these systems are adequate to describe the static properties of most solids, liquids, and gases, it is not a fully dynamic model and, therefore, has limitations; for example, natural fluctuations occurring within the system are not covered by the theory. A thermodynamic system has two extremes of complexity: *general* and *simple*. A *general* system description must aptly characterize micro- and nanostructures, anisotropic properties, and tensorial stress components, and maybe interaction with external fields, whereas a *simple* system, which is uniform throughout, is described by scalar state variables only and does not interact with external fields. In this chapter, the discussion is limited to simple systems.

A thermodynamic system is composed of the matter within a set space. Around the system is a *boundary*, or wall, and beyond that is everything else in the universe, known as the *surroundings*. An *open* system is one in which the system exchanges matter with its surroundings. By contrast, a *closed* system does not exchange matter with its surroundings. Furthermore, an *isolated* system exchanges neither matter nor energy. For practical reasons, the *closed* system is very important in the thermodynamics of material sciences. Note that the term control volume in fluid mechanics is often used synonymously with an *open* system in thermodynamics. The boundary is then called a control surface. An *adiabatic* boundary is in this respect equivalent to a perfect insulator, and a *diabatic* boundary is equivalent to a perfect heat conductor.

These are the main elements of the language used to underpin thermodynamic analyses of real-world systems. The dualism between the physical reality on the one side and abstract thermodynamic analysis on the other side is always at the center.

4.2.1 The Concept of the Thermodynamic State

The *state* of a thermodynamic system is determined by the system's properties – and vice versa. A *property* is here taken to be a state variable that is independent of the past, known as the history or the *path* of the system. It is the current state of the system that matters. The logic is seemingly circular, but from an experimental point of view the state is defined if, or when, all the independent thermodynamic properties have been measured.

When measuring, we often distinguish between *extensive* and *intensive* state properties. These terms relate to properties that, respectively, depend on the size of the system and those that are size independent. It is also common to distinguish between *external* state variables controlled by the experimentalist and *internal* state variables that are unattainable from the outside but still influenced by the external state variables and having well-defined values at equilibrium (explained later in this section).

One aid in revealing the properties of the system is the *process*, which is used to describe the changes that take place along a given path from one state to another. Hence, the path contains a complete description of the history of the process, or the sequence of state changes, if you like. A physical state change always implies the action of time, and the path is inherently time-dependent, but for a *stationary* process the state is time invariant and the path is reduced to a static step change from the input state to the output state. The same simplification applies to a process that has an indefinite time at its disposal. In this case, the state will approach its *equilibrium* value, which is the foundation of all the thermodynamic analysis laid out in this chapter. We shall see later that the energy of the system is being minimized in this particular state.

The number of state variables for a simple system is not universally fixed. In the simplest case, there are $C + 2$ such variables, where C is the number of independent chemical components (typically atoms) and the number 2 represents temperature and pressure. The practical application determines the number of chemical composition variables involved. Distilled water can, for example, be described using a single chemical component called H_2O under normal conditions, or using five components involving nine different chemical constituents in an isotope enrichment plant. Taking H_2 and O_2 into account, or perhaps H_3O^+ and OH^-, does not increase the degrees of freedom insofar as the overall stoichiometry of the system is equal to H_2O.

4.2.2
Fundamental Relations and Canonical State Variables

Thermodynamics is a phenomenological science, which means it is not supported by any strong theories of its own. Its foundation rests on numerous experimental facts forming a set of principles guiding our understanding on how to view a thermodynamic system. In the mid-18th century, there was an increasing body of observations indicating that a change in the internal energy of a closed system is the difference of two actions (nonexact differentials) called heat and work:

$$\mathrm{d}U_{\mathrm{c-s}} = \delta Q - \delta W.$$

This is known as the first law of thermodynamics, which states that the internal energy of a closed system does not change unless there is an interaction with the outside environment. At about the same time, it was also recognized that, for the so-called *reversible* state changes of simple systems, p and T serve as integrating factors for δW and δQ, turning the nonexact differentials into exact ones:

$$\delta W_{\text{rev}} = p\mathrm{d}V,$$
$$\delta Q_{\text{rev}} = T\mathrm{d}S.$$

The last of these relations is known as the second law of thermodynamics. Around 1873, Gibbs topped the knowledge with an exact understanding of open systems:

$$\mathrm{d}U = T\mathrm{d}S - p\mathrm{d}V + \sum_i \mu_i \mathrm{d}N_i.$$

He also recognized that U is a state function in S, V, and N_i. As a consequence, the differential is valid for all state changes, not only the reversible ones, given that the start and end points are equilibrium states. At the end of the century, it became clear that S is not only a state function but that it also has a definite zero, making it possible to integrate the differential into a *fundamental* form: $U = TS - pV + \sum_i \mu_i N_i$.

The integrated form of U is known as an Euler homogeneous function of order 1, and S, V, and N_i as the *canonical* variables of U. The factors T, p, and μ_i do not have clear-cut names, but we might refer to them as the thermal, mechanical, and chemical potentials of the system. Grouped together, TS, pV, and $\mu_i N_i$ are often called *conjugated* pairs of variables. In thermodynamics, we also refer to canonical potentials, meaning those that contain all thermodynamic information about the system. By applying Legendre transformations to internal energy, we can prove the existence of eight canonical energy potentials. The definition of the Legendre transformation ϕ of an arbitrary function f is rather simple:

$$\phi(\xi_i, x_j, x_k, \ldots, x_n) \equiv f(x_i, x_j, x_k, \ldots, x_n) - \left(\frac{\partial f}{\partial x_i}\right)_{x_j, x_k, \ldots, x_n} x_i. \tag{4.1}$$

But when applied to internal energy U, we get seven functions of general interest: Helmholtz energy A, enthalpy H, Gibbs energy G, grand canonical potential Ω, zero potential O, and two unnamed potentials given the symbols X and Y. An interesting property of the Legendre transform is that if we know one potential we know them all. This is not trivial as, for instance, $U(T, V, \mathbf{n})$ and $H(T, -p, \mathbf{n})$ do not have this property. For a thermodynamic function with $m = \dim(\mathbf{n}) + 2$ variables, there are $2^m - 1$ Legendre transforms. For a single-component system, this means that there are $2^3 - 1 = 7$ possibilities. By using Eq. (4.1) on each of the variables in turn, we get three of the transforms:

$$A(\tau, V, N) = U(S, V, N) - \left(\frac{\partial U}{\partial S}\right)_{V,N} S \equiv U - \tau S,$$
$$H(S, \pi, N) = U(S, V, N) - \left(\frac{\partial U}{\partial V}\right)_{S,N} V \equiv U - \pi V,$$
$$X(S, V, \mu) = U(S, V, N) - \left(\frac{\partial U}{\partial N}\right)_{S,V} N \equiv U - \mu N. \tag{4.2}$$

By using Eq. (4.1) on pairs of two variables, we can obtain three more transforms:

$$G(\tau, \pi, N) \equiv U - \pi V - \tau S,$$
$$Y(S, \pi, \mu) \equiv U - \pi V - \mu N,$$
$$\Omega(\tau, V, \mu) \equiv U - \tau S - \mu N. \tag{4.3}$$

Finally, by using Eq. (4.1) on all three variables in succession, we can obtain the null potential $O(\tau, \pi, \mu) \equiv U - \pi V - \tau S - \mu N$, but strictly speaking it is yet to be explained why this result requires that U is Euler homogeneous.

By examining the results carefully, we can prove that repeated Legendre transformation defines a closed algebraic loop, wherein the original information contained in U is retained, as shown in the following:

$$\begin{array}{ccc} U(S, V, N) & \xrightarrow{S} & A(\tau, V, N) \\ \uparrow -\tau & & \downarrow \tau \\ A(-\tau, V, N) & \xleftarrow{-S} & U(-S, V, N) \end{array}$$

.

This cycle represents, so to speak, the Pandora's box of thermodynamics: from it, an infinite palette of interconnected relations spring out – to the great dismay of many a student of physics, chemistry, and metallurgy. At this point, a second property of the thermodynamic potential comes to our rescue, namely, the Euler homogeneity of order 1.

In general, a stand-alone hypothesis in thermodynamics claims that the energy of a system is written as

$$U = \left(\frac{\partial U}{\partial X_1}\right) X_1 + \left(\frac{\partial U}{\partial X_2}\right) X_2 + \cdots + \left(\frac{\partial U}{\partial X_n}\right) X_n$$

provided all the *extensive* state variables X_i and all the *intensive* properties $(\partial U/\partial X_i)$ are known. It turns out that the difference between the extensive and the intensive functions fits into a more general classification scheme. They are, respectively, Euler homogeneous functions of order 1 and order 0. More precisely, the function $f(x_1, \ldots, x_n, \xi_{n+1}, \ldots, \xi_m)$ is homogeneous of order k in the variables $x_1, \ldots, x_n$ if the following criteria are satisfied:

$$F(X_1, \ldots, X_n, \xi_{n+1}, \ldots, \xi_m) = \lambda^k f(x_1, \ldots, x_n, \xi_{n+1}, \ldots, \xi_m),$$
$$X_i \equiv \lambda x_i. \tag{4.4}$$

This scaling law is known as Euler homogeneity of the order k. A mathematical analysis of Eq. (4.4) reveals several interesting results, of which we are going to mention only one:

$$\sum_{i=1}^{n} X_i \left(\frac{\partial F}{\partial X_i}\right)_{X_{j \neq i}, \xi_l} = kF. \tag{4.5}$$

Applying Eq. (4.5) to Eq. (4.2), and to Eq. (4.3), yields a table of Legendre–Euler equivalents of internal energy (restricted to a single-component system). The

Table 4.1 Euler integral forms of Legendre transformed internal energy.

Function	Variables	Derivatives	Euler form	Legendre form
U	S, V, N	τ, π, μ	$\tau S + \pi V + \mu N$	U
A	τ, V, N	$-S, \pi, \mu$	$\pi V + \mu N$	$U - \tau S$
H	S, π, N	$\tau, -V, \mu$	$\tau S + \mu N$	$U - \pi V$
X	S, V, μ	$\tau, \pi, -N$	$\tau S + piV$	$U - \mu N$
G	τ, π, N	$-S, -V, \mu$	μN	$U - \tau S - \pi V$
Ω	τ, V, μ	$-S, \pi, -N$	πV	$U - \tau S - \mu N$
Y	S, π, μ	$\tau, -V, -N$	τS	$U - \pi V - \mu N$
O	τ, π, μ	$-S, -V, -N$		$U - \tau S - \pi V - \mu N$

Euler form is in each case equivalent to the Legendre transform bearing the same function name, even though their functional forms look totally different (Table 4.1).

In the table, we have used $\pi \equiv -p$ in order to provide a consistent rule for the sign shifts of the derivatives: the derivative taken with respect to a transformed variable is always the negative of the untransformed variable.

4.2.3
Equations of State (EOS)

To describe the thermodynamic state, we need one or more functions that define the asymptotic behavior of the system (say a chemical mixture) in a consistent manner. Applying parameters to the functions renders flexibility to the engineer, while the asymptotes guarantee thermodynamic consistency. In this section, we have a brief look at a variety of useful concepts, such as ideal gas, virial expansion, the Debye phonon model, and van der Waals theory, but first we need to define the concept of an equation of state. Let us start with the differential of a known thermodynamic potential; for example, internal energy has

$$\mathrm{d}U = T\,\mathrm{d}S - p dV + \sum_{i=1}^{n} \mu_i dN_i, \tag{4.6}$$

in terms of the canonical variables S, V, and N_i. Now, if we know the fundamental state functions referred to as the *equations of state* of the system:

$$\begin{aligned} T &= T(S, V, \mathbf{n}), \\ p &= p(S, V, \mathbf{n}), \\ \mu_i &= \mu_i(S, V, \mathbf{n}), \end{aligned} \tag{4.7}$$

then Eq. (4.6) can be integrated using Euler's first theorem in Eq. (4.5) to give $U = TS - pV + \sum_{i=1}^{n} \mu_i N_i$. However, this will not work in practice because the equations of state in Eq. (4.7) rarely exist as explicit relations (for a photon gas it will). In practical work, it is, therefore, more convenient to start with

$$(\mathrm{d}G)_{T,p} = \sum_{i=1}^{n} \mu_i \mathrm{d}N_i,$$

or, alternatively,

$$(\mathrm{d}A)_T = -p\mathrm{d}V + \sum_{i=1}^{n} \mu_i \mathrm{d}N_i,$$

which can be integrated once $\mu_i(T, p, \mathbf{n})$ or, alternatively, $p(T, V, \mathbf{n})$ and $\mu_i(T, V, \mathbf{n})$ are known to us. By taking the ideal gas model as a reference, it is possible to obtain the residual functions $G^{\mathrm{r,p}}$ and $A^{\mathrm{r,v}}$ described in Section 4.3.1. From these, we can derive the Gibbs or Helmholtz energy of the mixture, and a consistent set of state functions can then be *derived* using partial differentiation.

4.2.3.1 Perfect Gas

The partition function for a particle-in-box without vibrational or rotational degrees of freedom yields the following set of canonical equations of state for Helmholtz energy:

$$\begin{aligned} \frac{p^{\mathrm{trans}} V}{NRT} &= 1, \\ \frac{\mu^{\mathrm{trans}}}{RT} &= -\frac{3}{2} \ln \left[\frac{2\pi M_w RT}{N_A{}^2 h^2} \left(\frac{V}{NN_A} \right)^{2/3} \right]. \end{aligned} \tag{4.8}$$

4.2.3.2 Harmonic Oscillator

The atoms in a molecule are held together by forces which, for small perturbations, allow the atoms to vibrate as if their centers were connected by elastic springs. The partition function for a *harmonic* oscillator of this kind having a characteristic frequency ν, quantum number j, and energy factor $\varepsilon = h\nu = h\omega/2\pi \equiv \hbar\omega$ yields

$$\frac{\mu^{\mathrm{vib}}}{RT} = \ln \left[2 \sinh \left(\frac{h\nu}{2kT} \right) \right].$$

4.2.3.3 Vibrations in Crystals

For vibrations in a crystal lattice, Einstein assumed that all atoms vibrate independently with the same frequency in all directions. Vibrations in three spatial directions simultaneously is easily accounted for by multiplying μ^{vib} and c_v^{vib} given earlier with a factor 3. More realistically, the distribution of the frequencies of vibration is a function of the frequency itself. Debye assumed that the distribution is proportional to ω^2 up to ω_{max}. This gives rise to the formula

$$\frac{\mu^{\mathrm{Debye}}}{RT} = 9 \left(\frac{T}{\theta_{\mathrm{D}}} \right)^3 \int_0^{\theta_{\mathrm{D}}/T} x^2 \ln(1 - e^{-x}) \mathrm{d}x,$$

where $\theta_{\mathrm{D}} \equiv \hbar\omega_{\mathrm{max}}/k$ is the so-called Debye vibrational temperature.

4.2.3.4 Virial Expansion of Gases

The virial expansion relates to thin gases. There is also an equivalent power series expansion for osmotic pressure in dilute solutions. Expanding p/RT in molar density $\rho = N/V$ yields

$$\frac{p^{\text{vir}}V}{NRT} = \sum_{k=1}^{\infty} B_k(T)\left(\frac{N}{V}\right)^{k-1} = 1 + B_2(T)\frac{N}{V} + \cdots. \tag{4.9}$$

This is the normal way to express the virial expansion of gases. The series converges rapidly at low density, but not at higher, that is, liquid-like, densities. Thus, for the study of vapor–liquid equilibria we need a more versatile equation of state described in the next section.

4.2.3.5 Van der Waals Fluid

For a given vector $\mathbf{n}$ consisting of the mole numbers $N_1, N_2, \ldots, N_n$, the van der Waals equation of state can be expressed as

$$p^{\text{VdW}} = \frac{NRT}{V - B(\mathbf{n})} - \frac{A(\mathbf{n})}{V^2}, \tag{4.10}$$

where $A(\mathbf{n})$ tells us something about the *attractive* forces between the molecules and $B(\mathbf{n})$ represents the *hard-sphere* volume of the mixture. By expanding Eq. (4.10) in the ideal gas limit $V \to \infty$ and comparing the coefficients with the ones in Eq. (4.9), we find that $B_2^{\text{VdW}} = b - a/RT$, $B_3^{\text{VdW}} = b^2$, $B_4^{\text{VdW}} = b^3$, and so on. Only B_2^{VdW} resembles the behavior of a real gas; the other coefficients are unrealistic. Even so, the van der Waals equation of state is our simplest starting point for equations that are going to model phase condensation phenomena.

4.2.4 Euler Integration of EOS into a Fundamental Relation

The first challenge is to generalize the ideal gas law in Eq. (4.8) into one that is similarly valid for a multicomponent system. In order to do that, we must write the ideal gas law in its extended form

$$p^{\text{ig}} = \sum_i \frac{N_i RT}{V}, \tag{4.11}$$

so that we can identify the components that constitute the mixture. It is an equation of state in the form $p(T, V, \mathbf{n})$, which eventually leads to Helmholtz energy $A(T, V, \mathbf{n})$, but in order to use the Euler method of integration in Eq. (4.5), we also need to know the equation of state $\mu_i^{\text{ig}}(T, V, \mathbf{n})$. This function cannot be obtained experimentally in the same way as p^{ig}, and must therefore be derived theoretically. The normal convention is to assume the pure component standard state $V_\circ/N_i^\circ = RT/p_\circ$, where $p_\circ = 1\,\text{bar}$. Referring to $\mu_i^\circ(T, p_\circ)$ as the standard state, the chemical potential of component i can be expressed as

$$\mu_i^{\text{ig}}(T, V, \mathbf{n}) = \mu_i^\circ(T, p_\circ) + RT \ln\left(\frac{N_i RT}{p_\circ V}\right). \tag{4.12}$$

Compared with the monoatomic potential in Eq. (4.8), this equation has a more general temperature dependency; it also relates to a multiple of components $i \in [1, n]$ and to the standard pressure $p_\circ$. It is now worth reiterating that using Euler's theorem to integrate Eq. (4.11) and Eq. (4.12) always works and that the Helmholtz energy of the gas can be expressed in the following general form:

$$A^{\text{ig}}(T, V, \mathbf{n}) = \pi^{\text{ig}} V + \sum_i N_i \mu_i^{\text{ig}} = -\sum_I N_i RT + \underbrace{\sum_i N_i \mu_i^\circ (T, p_\circ)}_{G^\circ(T, p_\circ, \mathbf{n})} + \sum_i N_i RT \ln\left(\frac{N_i RT}{p_\circ V}\right).$$

Note that the standard state is based on a conventional Gibbs energy function, because T and $p_\circ$ have been selected as the state variables of μ_i°, rather than T and $V_\circ$. The calculation of $\mu_i^\circ(T, p_\circ)$ is normally done in terms of a heat capacity integral:

$$\mu_i^\circ(T, p_\circ) = \Delta_f h_i^\circ + \int_{T_\circ}^{T} c_{\text{p},i}^\circ(\tau)\text{d}\tau - T\left[s_i^\circ(T_\circ, p_\circ) + \int_{T_\circ}^{T} \frac{c_{\text{p},i}^\circ(\tau)}{\tau}\text{d}\tau\right],$$

$$c_{\text{p},i}^\circ = R + c_{\text{v},i}^{\text{trans}} + c_{\text{v},i}^{\text{rot}} + \sum c_{\text{v},i}^{\text{vib}} + \cdots.$$

Condensed phases and complex gaseous phases are not readily applicable to a quantum mechanical description and, therefore, rely on empirical correlations for $c_{\text{p},i}{}^\circ$. Even polynomial formulae are used for this purpose, but that practice is questionable because polynomials will eventually diverge at high temperatures.

4.2.5
The Principle of Thermodynamic Equilibrium

In a physical system, the *external* state variables define a differentiable manifold on which a particular function of the state (yet to be defined) varies in terms of the *internal* state variables such that it eventually reaches a time invariant state called the equilibrium state. An alternative view of the physical picture is that the same function is allowed to vary with respect to all the state variables subject to a set of extensive *constraints*. The first view represents a generalization of the *extent of reaction* known from physical chemistry, while the latter leans toward the method of unknown Lagrange *multipliers* in optimization theory. The last one is the line of thought followed in this text.

Take it for granted, therefore, that the state is properly defined as a set of extensive variables making up one of the canonical potentials from the last section. Applied to a simple system, where each chemical component $i \in [1, n]$ is present in every phase $\pi \in [\alpha, \omega]$, the stationary conditions are equivalent to a spatial uniformity of T, p, and μ_i in the system. We shall now prove that these conditions are equivalent to minimizing the internal energy for a given total entropy, volume, and mole numbers:

$$U_{\mathrm{eq}} = \min_{\forall \mathbf{x}^{(\pi)}} U(\mathbf{x}^{(\alpha)}, \mathbf{x}^{(\beta)}, \ \dots \ , \mathbf{x}^{(\omega)}),$$
$$\text{subject to :}$$
$$\mathbf{x}^{(\alpha)} + \mathbf{x}^{(\beta)} + \cdots + \mathbf{x}^{(\omega)} = \mathbf{x}_{\circ}. \tag{4.13}$$

Here, $\mathbf{x}^{(\pi)} \equiv \begin{pmatrix} S^{\pi} & V^{\pi} & N_1^{\pi} & \cdots & N_n^{\pi} \end{pmatrix}$ is a list of extensive state variables for the arbitrary phase π. In all stationary states, the differential of U must vanish for all *feasible* variations in S^{π}, V^{π}, and N_i^{π}:

$$(\mathrm{d}U)_{S,V,\mathbf{n}} = \sum_{\pi=\alpha}^{\omega} \mathrm{d}U^{\pi} = \sum_{\pi=\alpha}^{\omega} \left(T^{\pi}\mathrm{d}S^{\pi} - p^{\pi}\mathrm{d}V^{\pi} + \sum_{i=1}^{n} \mu_i^{\pi}\mathrm{d}N_i^{\pi} \right) = 0.$$

From these $n+2$ balance equations, we can eliminate $\mathrm{d}S^{\omega}$, $\mathrm{d}V^{\omega}$, and $\mathrm{d}N_i^{\omega}$ for all the components $i \in [1, n]$. Substituted into the differential of U

$$(\mathrm{d}U)_{S,V,\mathbf{n}} = \sum_{\pi=\alpha}^{\omega-1} \left[(T^{\pi} - T^{\omega})\,\mathrm{d}S^{\pi} - (p^{\pi} - p^{\omega})\mathrm{d}V^{\pi} + \sum_{i=1}^{n} (\mu_i^{\pi} - \mu_i^{\omega})\mathrm{d}N_i^{\pi} \right] = 0.$$

In the neighborhood of an equilibrium point, the quantities $\mathrm{d}S^{\alpha}, \dots, \mathrm{d}S^{\omega-1}$ and $\mathrm{d}V^{\alpha}, \dots, \mathrm{d}V^{\omega-1}$ and $\mathrm{d}N_{\alpha}^{1}, \dots, \mathrm{d}N_{n}^{\omega-1}$ are truly independent, thus representing the internal state variables mentioned in the beginning of this section. If $(\mathrm{d}U)_{S,V,\mathbf{n}} = 0$, then it must be true that

$$\begin{aligned} T^{\alpha} &= T^{\beta} = \cdots = T^{\omega}, \\ p^{\alpha} &= p^{\beta} = \cdots = p^{\omega}, \\ \mu_1^{\alpha} &= \mu_1^{\beta} = \cdots = \mu_1^{\omega}, \\ &\vdots \\ \mu_n^{\alpha} &= \mu_n^{\beta} = \cdots = \mu_n^{\omega}. \end{aligned}$$

When relating these results with the differentials of Eqs (4.2) and (4.3), it dawns on us that the exact same equilibrium conditions are reached by minimizing any of the potentials with only slight changes to the constraints:

$$\min_{\mathbf{s},\mathbf{v},\mathbf{n}} U \Leftrightarrow \min_{\mathbf{s},\mathbf{n}} (H)_p \Leftrightarrow \min_{\mathbf{v},\mathbf{n}} (A)_T \Leftrightarrow \min_{\mathbf{n}} (G)_{T,p} \Leftrightarrow \cdots \Leftrightarrow \min_{\mathbf{v}} (\Omega)_{T,\mu}.$$

Minimizing internal energy with respect to entropy, volume, and mole numbers is, therefore, equivalent to minimizing Helmholtz energy at given T with respect to both volume *and* mole numbers, or Gibbs energy at given T and p with respect to the mole numbers *only*. All the equations we have seen so far are valid for the generic phase equilibria problem but we would also like to know how to solve heterogeneous reaction equilibria. In order to achieve this major change in the system description, we only need to change the constraint specification from the form

$$\mathbf{x}^{(\alpha)} + \mathbf{x}^{(\beta)} + \cdots + \mathbf{x}^{(\omega)} = \mathbf{x}_{\circ},$$

which is used in Eq. (4.13), to

$$\mathbf{B}^{(\alpha)}\mathbf{x}^{(\alpha)} + \mathbf{B}^{(\beta)}\mathbf{x}^{(\beta)} + \cdots + \mathbf{B}^{(\omega)}\mathbf{x}^{(\omega)} = \mathbf{x}_{\circ},$$

where $\mathbf{B}^{(\alpha)}$, $\mathbf{B}^{(\beta)}$, and so on are constant (albeit phase specific) constraint matrices. In ordinary phase equilibria, $\mathbf{B}^{(\pi)} = \mathbf{I}$ for all $\pi \in [\alpha, \omega]$. In many heterogeneous systems involving molten metals, ceramic phases at high temperatures, aqueous electrolytes, and other fully dissociated systems, the formula matrices takes the more general form

$$\mathbf{B} = \begin{pmatrix} \mathbf{I}_{2\times2} & 0 \\ 0 & \mathbf{A}_{m\times n} \end{pmatrix},$$

where $\mathbf{A}$ is the atoms × substance matrix and $\mathbf{I}$ corresponds to the constraints in S and V. By substituting the new constraint formulation into Eq. (4.13), there is a new set of more general equilibrium conditions. In other words, the thermal and mechanical equilibria are unchanged but the chemical equilibrium is changed into

$$\mu^{\alpha} = \mathbf{A}^{\alpha}\lambda, \mu^{\beta} = \mathbf{A}^{\beta}\lambda, \cdots, \mu^{\omega} = \mathbf{A}^{\omega}\lambda,$$

where λ is a vector of Lagrange multipliers associated with the problem. For kinetically hindered reactions, $\mathbf{A}$ is not the atoms × substance matrix anymore. We should then think of the matrix as spanning the null space of the reaction stoichiometry matrix $\mathbf{N}$. The constraint situation does not change, however, and the equilibrium conditions are also unchanged if $\mathbf{A} \equiv \text{null}(\mathbf{N}^{\mathrm{T}})$ such that $\mathbf{AN} = 0$. About the role of $\mathbf{A}$ and $\mathbf{N}$, we should make the comment that minimizing U with respect to $\mathbf{x}^{(\pi)}$ in the row space of $\mathbf{B}^{(\pi)}$ calls for the use of Lagrange multipliers, while minimizing U in the column space of $\mathbf{N}$ leads to a generalization of the concept of extents of reactions.

4.3 Thermodynamic Modeling

The ideal gas serves as the theoretical limit for all real fluids, but for solid solutions and complex liquids there are other kinds of ideality. In other words, an ideal molecular mixture is similar to an ideal gas mixture at a fixed pressure, while, for example, an ionic liquid behaves closer to the so-called Temkin model where the cations and the anions mix on two separate sublattices. A similar concept also applies to solid solutions, but the number of possible spatial arrangements is not so obvious in this case. It is, therefore, customary to reserve the Gibbs and Helmholtz residual functions

$$G(T, p, \mathbf{n}) = G^{\mathrm{ig}}(T, p, \mathbf{n}) + G^{\mathrm{r,p}}(T, p, \mathbf{n}),$$
$$A(T, V, \mathbf{n}) = A^{\mathrm{ig}}(T, V, \mathbf{n}) + A^{\mathrm{r,v}}(T, V, \mathbf{n})$$

to fluids, while solid solutions and complex liquids are better served by an excess Gibbs energy function:

$$G(T, p, \mathbf{n}) = G^{\mathrm{id}}(T, p, \mathbf{n}) + G^{\mathrm{ex}}(T, p, \mathbf{n}).$$

Notice that the excess function refers to a real physical state while the residual functions depend on a hypothetical ideal gas reference state. The residual function must, therefore, take phase condensation into account, whereas the excess function tacitly can rely on direct observations of the condensed phase.

4.3.1
Gibbs and Helmholtz Energy Residuals

The canonical variables of Gibbs energy include temperature and pressure. This makes it possible to compare the chemical potentials of the components in a real fluid with those of the same components in the ideal gas state when the phases are both at the same *temperature, pressure*, and *composition*. Using what we already know about G^{ig} allows us to define the residual Gibbs energy as

$$G^{\text{r,p}}(T, p, \mathbf{n}) \equiv G(T, p, \mathbf{n}) - G^{\text{ig}}(T, p, \mathbf{n}). \tag{4.14}$$

As $p \to 0$, all fluids approach virtually ideal properties in the sense of the pressure of the fluid approaching NRT/V when the volume increases to infinity. By taking the proper limits, it is possible to rewrite Eq. (4.14) as

$$G^{\text{r,p}}(T, p, \mathbf{n}) = \int_0^p (V - V^{\text{ig}})\text{d}\pi = \int_0^p \left(V(\pi) - \frac{NRT}{\pi} \right) \text{d}\pi.$$

Here, π stands for the integrated pressure, whereas the system pressure p is in the upper limit of the integral. The functional form of the chemical potential $\mu_i = (\partial G/\partial N_i)_{T,p,N_{j\neq i}}$ is valid for all proper Gibbs energy functions, including $G^{\text{r,p}}$. This allows us to define the residual chemical potential as

$$\mu_i^{\text{r,p}}(T, p, \mathbf{n}) \equiv \left(\frac{\partial G^{\text{r,p}}}{\partial N_i} \right)_{T,p,N_{j\neq i}}$$
$$= \int_0^p \left[\left(\frac{\partial V}{\partial N_i} \right)_{T,\pi,N_{j\neq i}} - \frac{RT}{\pi} \right] \text{d}\pi \equiv \int_0^p \left(\overline{v}_i - \frac{RT}{\pi} \right) \text{d}\pi,$$

where $\overline{v}_i \equiv (\partial V/\partial N_i)_{T,\pi,N_{j\neq i}}$ is defined as the *partial molar* volume of the component i. For historical reasons, it is customary to rewrite the residual potential as

$$RT \ln \varphi_i \equiv \mu_i^{\text{r,p}}(T, p, \mathbf{n}), \tag{4.15}$$

where $\varphi_i(T, p, \mathbf{n})$ is the *fugacity* coefficient of component i notwithstanding the rapid development of equations of state in the alternative form $p = p(T, V, \mathbf{n})$. Changing the free variable from p to V makes Helmholtz energy a better starting point than Gibbs energy. But φ_i measures the difference between a real fluid and its ideal gas approximation at a given pressure – not a given volume. This contradiction leads to a noncanonical description, which shall not be pursued any further here. If, however, we redefine the residual Helmholtz energy as

$$A^{\text{r,v}}(T, V, \mathbf{n}) \equiv A(T, V, \mathbf{n}) - A^{\text{ig}}(T, V, \mathbf{n}),$$

the variables are again canonical, and we can use the same approach as for residual Gibbs energy at the beginning of this section. The alternative departure function can be written as

$$A^{\mathrm{r,v}} = \int_{\infty}^{V} \left(\frac{NRT}{v} - p(v) \right) \mathrm{d}v,$$

which is perfectly well suited for building consistent thermodynamic frameworks.

4.3.2
Excess Gibbs Energy

Although the integration of a complete set of equations of state was successful in Section 4.2.4, it will not be helpful when dealing with solid phases. Solid phases cannot be formed from the gas phase in a continuous manner like a liquid can (not according to our present knowledge). For all solid phases and many complex liquids, we must, therefore, rely on a different set of formulae. The most straightforward procedure is to jump-start from an Euler integrated form of Gibbs energy called the excess Gibbs energy $G^{\mathrm{ex}} \equiv Ng^{\mathrm{ex}}$. This is an obvious choice because G^{ex} depends only on mole numbers, and not on volume and mole numbers in combination, as $A^{\mathrm{r,v}}$ does.

Before we enter the discussion of excess Gibbs energy models, we must first define what an ideal mixture looks like. There is no need for a physical model at this point – only a set of conditions that holds irrespective of the model we are going to use later. According to IUPAC, there is no mixing term for the enthalpy and volume of the ideal mixture. These two properties depend solely on the pure component values $h_i^\star$ and $v_i^\star$. Recognizing H and V as partial derivatives of G/T yields the expressions

$$G = \sum_i m_i N_i,$$

$$\left(\frac{\partial \left(G^{\mathrm{im}}/T \right)}{\partial (1/T)} \right)_{p,\mathbf{n}} = H^{\mathrm{im}} = \sum_i N_i\, h_i^*(T,p) \equiv H^*,$$

$$T\left(\frac{\partial \left(G^{\mathrm{im}}/T \right)}{\partial (p)} \right)_{1/T,\mathbf{n}} = V^{\mathrm{im}} = \sum_i N_i v_i^*(T,p) \equiv V^*.$$

It follows that the properties of G^{im}/T must be quite similar to H^{im} and V^{im}, possibly with the exception of an unspecified nonlinear term in $\mathbf{n}$ that is not coupled to T or p. The same argument applies to G^{im}. From physical chemistry, the nonlinear term is known as the ideal entropy of mixing S^{im}. This completes the physical picture of the mixture model, but in order to elucidate better how the many contributions depend on the composition of the mixture we shall look at the chemical potential μ_i rather than G itself. The Gibbs energy of the mixture is anyhow given as

$$\mu_i(T,p,\mathbf{n}) = \mu_i^\star(T,p_\circ) + \int_{p_\circ}^{p} v_i^\star(T,\pi)\mathrm{d}\pi$$
$$-T\bar{s}_i^{\,\mathrm{im}}(\mathbf{n})$$
$$+\mu_i^{\mathrm{ex}}(T,\mathbf{n}) + \int_0^p [\bar{v}_i(T,\pi,\mathbf{n}) - v_i^\star(T,\pi)]\mathrm{d}\pi.$$

In this equation, $\mu_i^\star(T,p_\circ)$ stands for the standard state chemical potential, $v_i^\star$ is the pure component molar volume, $\bar{v}_i$ is the partial molar volume in the mixture, $\bar{s}_i^{\,\mathrm{im}}$ is the partial molar entropy in the ideal mixture, and μ_i^{ex} is the excess chemical potential at zero pressure. The integral form, where each line in the previous equation is provided a separate symbol, is

$$G(T,p,\mathbf{n}) = G^\star(T,p,\mathbf{n}) + G^{\mathrm{im}}(T,\mathbf{n}) + G^{\mathrm{ex}}(T,p,\mathbf{n}).$$

The rest of this section is devoted to excess Gibbs energy models at fixed temperature and pressure. Thermodynamically, it is sufficient to make sure that G^{ex} is an Euler homogeneous function of the first order or, equivalently, that G^{ex} is homogeneous of order zero. However, G^{ex} must of course also be physically acceptable. In particular, and valid for all nonelectrolytes, the activity coefficient of the substances must be finite at all concentrations. But this statement is premature since we have not properly defined what exactly an activity coefficient is. We shall, therefore, rephrase the statement so that g^{ex}/x_ix_j must be nonzero and finite in each pure component corner in the composition space. Hence, for each binary ij the following must be true:

$$\lim_{\substack{x_i\to 0\\ x_j\to 1}} \frac{G^{\mathrm{ex}}}{x_ix_j} = K_{ij};\qquad K_{ij}\in\langle 0,\infty\rangle.$$

The simplest possible model is the regular mixture for which K_{ij} is constant over the entire composition range. For a binary system, we get

$$G^{\mathrm{ex,reg}} = x_1x_2K_{12}.$$

A generalization leads to K_{12} as a function of the composition. The expressions of Margule and Redlich–Kister are of polynomial functions of this type:

$$G^{\mathrm{ex,R-K}} = x_1x_2\sum_{n=0}^{0,1,2,\ldots,N} K_{12}^{(n)}(x_1-x_2)^n \simeq x_1x_2\mathbb{P}_N(x_1-x_2).$$

By normalizing the fractions x_1 and x_2 in a different way, by saying

$$\phi_1 \equiv \frac{q_1N_1}{\Phi}\ \text{and}\ \phi_2 \equiv \frac{q_2N_2}{\Phi}\ \text{where}\ \Phi \equiv q_1N_1 + q_2N_2,$$

and making allowance for $g^{\mathrm{ex}} \equiv G^{\mathrm{ex}}/\Phi$, we get the Wohl expansion, which is also used in the nonconfigurational part of the Flory–Huggins model for monomer–polymer mixtures. All of these models are in a sense polynomial models. Some mixtures behave strongly as a nonpolynomial, however, and a rational function of two polynomials of the orders N and $M > N$ might perform better. The Van Laar and Hildebrand–Scatchard models are the simplest models of this type. On general terms, we can write

$$g^{\text{ex,VLaar}} = x_1 x_2 \mathbb{R}_{N,M}(x_1 - x_2).$$

Another line of thought leading to the so-called local composition models was developed in the 1960s. The models are based on the idea that the composition of a mixture is not entirely uniform throughout the bulk phase – not on a microscopic level. There is no hard theory to support this idea either, but it has an appealing physical basis for sure. Progress in this field lead to three well-known models. These are the Wilson model

$$g^{\text{ex,Wilson}} = -x_1 \ln(x_1 + \Lambda_{12} x_2) - x_2 \ln(x_2 + \Lambda_{21} x_1),$$

which actually is an extension of the Flory–Huggins model. Then, there is the nonrandom two-liquid model of Renon and Prausnitz

$$g^{\text{ex,NRTL}} = x_1 x_2 \left(\frac{\tau_{21} G_{21}}{x_1 + x_2 G_{21}} + \frac{\tau_{12} G_{12}}{x_2 + x_1 G_{12}} \right),$$

and finally the UNIversal QUAsiChemical (UNIQUAC) model by Abrams and Prausnitz, which similar to the Flory–Huggins model consists of a combinatorial term and a residual term. At this point, the expressions are becoming quite complex and we leave it for the interested reader to look them up yourself.

All the excess functions we have seen so far are written for binary systems. This is done on purpose in order to illustrate the basic principles of modeling, but we can still define the activity coefficient based on the assumption that a multicomponent model is at our disposal:

$$RT \ln \gamma_i \equiv \mu_i^{\text{ex}}(T, p, \mathbf{n}) = \left(\frac{\partial G^{\text{ex}}}{\partial N_i} \right)_{T,p,N_{j \neq i}}. \tag{4.16}$$

There is a close link between the fugacity coefficient in Eq. (4.15) and the activity coefficient in Eq. (4.16). The former is a measure for the deviation of the real state from an ideal gas at the same $T, p, \mathbf{n}$, whereas the latter uses the pure substance i at the same T, p as a reference. Hence, by calculating the fugacity coefficient of the pure substance as $\varphi_i^\star(T, p)$, we get

$$RT \ln \gamma_i \equiv \frac{\varphi_i(T, p, \mathbf{n})}{\varphi_i^\star(T, p)},$$

but just like in the case of the fugacity coefficient there is no real need for the activity coefficient in the analysis of thermodynamics of mixtures. The prime function is, and will always be, the excess Gibbs energy. All mixture properties of interest can be derived from G^{ex} by differentiation.

4.4 The CALPHAD Approach

"CALPHAD," short for "CALculation of PHAse Diagrams," is a form of thermodynamic modeling where the future state of a system is predicted based on the laws of thermodynamics. However, the foundations and scope of the approach have expanded since its inception, so CALPHAD is now synonymous with the computer coupling of phase diagrams and thermochemistry.

CALPHAD-based calculations provide data about which phases will be present (and at what concentrations) in a system that has reached a state of equilibrium under certain temperature and composition conditions. Calculation results are typically presented as diagrams that show which phases are present in a multicomponent system given varying conditions that are presented along the X- and Y-axes. However, the calculations can also result merely in information about a single equilibrium or in a diagram that plots a specific property of the system (e.g., the molar volume of a specific phase) under various conditions.

The theoretical basis of CALPHAD is thermodynamics. According to this theory, the so-called Gibbs energy of a system evolves, other things being equal, steadily toward its minimum value. At the minimum Gibbs energy value, the system will remain in a stable equilibrium state and will only change if it is perturbed from the outside in some way.

The calculations are based on databases that contain a set of parameters that describe thermodynamic functions of all phases of a system. The parameters are set by database developers who optimize the parameter values in light of a wide range of weighted experimental measurements as well as theoretical calculation results (*ab initio* data).

If the Gibbs energy is accurately described for all phases of a system as functions of composition, temperature, and pressure, then these thermodynamic functions can be used to calculate all equilibria, phase diagrams, driving forces, and thermodynamic properties (e.g., entropy, heat capacity, or thermal expansivity). This means that they can be used to predict what will happen when the material is prepared or processed in various ways. The challenge for developers of CALPHAD models and databases is to find out what the correct Gibbs energy functions are.

Databases are primarily based on the information from binary and ternary material systems, but by using relatively simple models of phases and data about the "lattice stabilities" of elements and their binary interactions, it is possible to extrapolate from a calculation to make predictions about higher-order systems. The utility of the CALPHAD-based calculations depends on there being reliable and comprehensive databases for the systems that one is interested in.

Note that the "CALPHAD approach" can refer to three different aspects of thermodynamic modeling: first, scientific research and model development for understanding, for example, binary and ternary systems; second, database development and data optimization with respect to certain concrete systems; and finally, the applied use of CALPHAD to make predictions about real materials and in various applied settings.

4.4.1 History

The history of CALPHAD goes back to the 1930s when Carl Hugo Johansson, a physicist at the Royal Institute of Technology (KTH) in Stockholm, published an assessment of part of an Fe–C system. This may have been the very first modern CALPHAD assessment. In part inspired by Johansson's paper, Larry Kaufman, PhD student, and Professor Morris Cohen, his supervisor, evaluated a model of the Fe–Ni system in the 1950s at MIT. In 1970, Kaufman and Harold Bernstein published a book with descriptions of the Gibbs energy expressions for many more phases of binary and ternary alloy systems. The book demonstrated the power and practical use of the CALPHAD methodology.

In 1968, Hillert published a paper that demonstrated the predictive power that could be achieved if experimental data about thermodynamic properties and phase diagrams were used with computer programs. Before this, applying the CALPHAD methodology involved drawing phase diagrams with a ruler, based on Gibbs energy curves calculated by hand or with a desk calculator.

The use of computers to do calculations really began in the 1970s. During this decade, there was also significant development of solution models and calculations techniques. CALPHAD-based calculations could now, for example, take magnetic contributions into account. CALPHAD was also developed to make use of the results of theoretical *ab initio* calculations. This made it possible to generate data points for thermodynamic properties and phase diagrams for which experimental data were missing. In addition, computer software for fitting predictions of models to a variety of different types of data, by using the least squares method, were developed in the late 1970s and the 1980s.

In the 1970s, an international community of CALPHAD researchers and practitioners also emerged. The first and second CALPHAD meetings were held in 1973 and 1974 and the scientific journal CALPHAD was launched in 1977. In 1979, CALPHAD research groups in Europe founded the organization Scientific Group Thermodata Europe (SGTE). Crucial for the development and adoption of CALPHAD was SGTE's publication of a free and open unary database for a large number of pure elements in 1991. Before this common database, different research groups often used different data for the elements. SGTE has also published two commercially available databases: a substance database, which contains data for inorganic compounds and gases, and a solutions database, which contains data on binary and ternary systems (as well as some systems of an even higher order). SGTE has been updating all of these databases continuously. A subset of the solutions database that only contains data about binary systems is also freely available from the organization. In the 1980s and 1990s, the use of the CALPHAD methodology continued to spread in academia as well as industry. Several powerful software packages were released during this time (e.g., Thermo-Calc and FactSage). These were typically developed by research groups in academic settings, but in the 1990s and 2000s, the software packages became the basis for commercial companies that started to sell both software tools and thermodynamic databases.

Since its inception in the 1950s, the foundations and scope of the CALPHAD approach have expanded considerably. Coupled with information about the kinetic properties of different material systems, a temporal dimension can be added to simulations. Various kinds of diffusion phenomena can be simulated. Today, it is the most widely used methodology for solving practical material design problems in many industries.

In this current decade, CALPHAD has become part of efforts to revolutionize materials engineering associated with ICME and the Materials Genome Initiative (MGI), a major funding scheme that was announced in 2011 by the US President Barack Obama.

Today, CALPHAD is an established tool in material science and materials engineering in academic as well as industrial settings.

4.4.2
Crystallography and Models of Phases

The CALPHAD method is distinguished from mere curve fitting by physically based model descriptions of the phases. These model descriptions are the foundation for reliable extrapolation of CALPHAD descriptions of binary and ternary systems to higher-order systems and are crucial for coupling thermodynamic calculations with other simulations such as diffusion processes. For development of a good model description, the structural properties of the phase must be taken into account. In the case of crystalline phases, this means that the crystal structure of these phases should serve as a blueprint for modeling the composition dependence of the thermodynamic functions.

In a crystalline solid phase, the atoms are arranged in a regular, periodic pattern. The smallest unit to describe this structure is the unit cell. The relative arrangement of the atoms within this unit cell is the crystal structure. The symmetry of the unit cell is described by the space group, and the position of the atoms are described by the Wyckoff position. The space group symmetry determines the point symmetry of the Wyckoff position and its multiplicity, that is, how many atoms are generated by the symmetry in the unit cell. Each space group has one general Wyckoff position where each of the coordinates can have any value between zero and less than unity. Depending on the space group, these can range from none to many special positions where the values of the coordinates are constrained. These constraints are determined by the symmetry of the space group such as a fixed value or that two or all three must have the same value or are offset by a fixed value. If all coordinate values are fixed, this Wyckoff position can only be occupied once in the crystal structure, otherwise there is no theoretical limit to how often a Wyckoff position with different coordinate values can be occupied.

The Compound Energy Formalism (CEF, also called sublattice model) is very well suited for the description of ordered phases but is also suited for the description of the gas phase, regular solutions, and liquids with short-range order. The basis of this formalism is that a phase consists of different sublattices that are occupied by a number of species, which can be atoms, molecules, ions, or vacancies. The two simplest cases within the CEF are a description consisting of only one sublattice, which describes the gas phase and regular solutions, and the description of a phase with only one species on each sublattice, which describes a stoichiometric compound.

The first step in selecting a model description is based on the crystal structure where initially one sublattice is assigned to each Wyckoff position and the site occupation is noted. The multiplicity of the site becomes the stoichiometric factor of this sublattice. Some phases, such as oxides or salts, can be highly ordered, while some may show a lesser degree of ordering, such as the σ phase in the Cr–Fe system. The second step is to examine how these occupations change on deviation of stoichiometry, that is, whether substitution occurs, vacancies are found, or interstitials are formed. Ideally, this is done both for the system being modeled and for other systems in which this phase occurs to ensure that the selected model will be compatible with higher-order systems. This analysis provides an idea of the elements that occur on each sublattice. The crystal structures of many phases in metallic systems are fairly simple and may have only three or fewer Wyckoff sites. However, for phases with complex crystal structures that have four or more Wyckoff sites, it may be useful to simplify the description by combining two or more of the Wyckoff sites into one sublattice (Table 4.2).

In the past, one driver for simplifications was to reduce the numbers of model parameters that need to be determined. The majority of these model parameters are those describing the end-member phases. End-member phases are formed when each sublattice is occupied by only one kind of species, that is, atom, ion, or vacancy, and one usually represents the phase with ideal stoichiometry while the remaining end members are hypothetical and are assumed to be metastable. For an n-component phase where all elements occur on all k sublattices, the number of end members is n^k. Today, density functional theory (DFT) calculations allow one to obtain values for these hypothetical end members. However, another reason for considering the simplification of a description is that, with an increasing number of elements occurring in a specific phase, the calculation will become increasingly sensitive to the quality of the start values of the species composition on each of the sublattices.

In the early days of using the CEF, the crystal structure may have been used as a guide for the model description, but the model selection was finally governed by the convenience of simplicity that allowed it to cover the homogeneity range of the phase. This practice may work well for the description of binary and ternary systems but problems may be encountered when the descriptions of a large number of systems are combined into a database for multicomponent systems. These problems are usually model incompatibility between the descriptions of different systems or unrealistic extrapolation behavior in a higher component system. Therefore, simplifications must be selected with great care and often criteria such as preferred occupation, coordination number, coordination polyhedron, and point symmetry must be used. Instead of the combination of sublattices,

Table 4.2 Frequently used models for common solid phases.

Phase	Model	Comment
Fcc_A1	$(A,B)_1(Va,C)_1$	Substitutional (A,B) solution with C interstitials
Bcc_A2	$(A,B)_1(Va,C)_3$	Substitutional (A,B) solution with C interstitials
Hcp_A3	$(A,B)_1(Va,C)_{0.5}$	Substitutional (A,B) solution with C interstitials
Fcc_L10, Fcc_L12	$(A,B)_{0.25}(A,B)_{0.25}(A,B)_{0.25}(A,B)_{0.25}(Va,C)_1$	Four-sublattice, order/disorder formalism*
Bcc_B2, Bcc_D03	$(A,B)_{0.25}(A,B)_{0.25}(A,B)_{0.25}(A,B)_{0.25}(Va,C)_3$ $(A,B,Va)_{0.25}(A,B,Va)_{0.25}(A,B,Va)_{0.25}(A,B,Va)_{0.25}(Va,C)_3$	Four-sublattice, order/disorder formalism*, (A,B,Va) includes structural vacancies
Hcp_D019	$(A,B)_{0.75}(A,B)_{0.25}(Va,C)_{0.5}$	Two-sublattice, order/disorder formalism
C14_Laves	$(A,B)_1(A,B)_{1.5}(A,B)_{0.5}$	Three sublattices, constraint may be employed if there is no ordering between the second and third sublattices
C15_Laves	$(A,B)_1(A,B)_2$	
Sigma	$(A,B)_{10}(A,B)_{16}(A,B)_4$	Also simplified to $(A,B)_1(A,B)_2$, $(A,B)_{10}(A,B)_{16}(A)_4$
Mu	$(A,B)_1(A,B)_2(A)_4(A,B)_6$	
Chi	$(A,B)_5(A,B)_{12}(A,B)_{12}$	Also simplified to $(A)_5(A,B)_{12}(A,B)_{12}$

*If only $L1_2$ or B2 ordering is described, a two-sublattice, order/disorder formalism can be used.

the number of model parameters can also be reduced by introducing constraints between them. For example, experimental data may indicate ordering between two sublattices in a phase in specific systems while there is no such indication in other systems. This can be taken into account by the selection of appropriate constraints. Another way of introducing constraints is the definition of so-called exchange energies. These exchange energies can be defined for one species substituting for another on a specific sublattice or for an ionic species being replaced by the same element but with a different charge on the same sublattice.

Model selection should also reflect the relationship between different but related structures. The most common cases are order/disorder forms and the filling of vacant sites with interstitials. Important order/disorder phases that should be treated by special model descriptions are those that are based on the fcc, bcc, and hcp structures. These phases are modeled with multiple sublattices to describe the ordered variants. The cases that are usually described are the

$L1_2$ and $L1_0$ ordering of the fcc, the B2 and $D0_3$ ordering of the bcc and the $D0_{19}$ ordering of the hcp. The model description usually consist of two parts, one describing the disordered phase and the other describing the contribution from ordering. Constraints for the model parameters of the ordering part are derived from the number of first and second, for bcc ordering, nearest neighbor pairs. This treatment ensures that the phases can undergo a second-order phase transformation. The cluster site approximation (CSA) is another formalism using sublattices that can be used for describing order/disorder phases based on the fcc and hcp structures. The CSA is based on the tetrahedron approximation used in the cluster variation method in conjunction with a generalized quasichemical method.

The relationship between different crystal structures is also an important factor that needs to be considered for model selection. Different prototypes do not necessarily mean that the phase must be modeled as different phases, for example, if the prototypes are different but the space group and Wyckoff sequence are identical, as is the case for the $D0_3$ ($AlFe_3$) and the $L2_1$ ($AlCu_2Mn$, Heusler) structures. Both structures have the same space group $Fm\bar{3}m$ with the same Wyckoff sequence (a b c), which means that the two structures are not different at all. The only difference is that the $D0_3$ is a binary prototype while the $L2_1$ is a ternary prototype where the c and b positions in one case are occupied by the same elements and in the other case by different elements. Therefore, these two phases should be described as the same phase. Another case is where two crystal structures have the same space group but a continuous transition from one structure type to another structure type is possible. Many of these cases are the filling of interstitial sites. For example, many carbides and nitrides have the same close-packed structure as the pure metal with interstitial sites fully or partially filled. These phases should, therefore, be modeled as one phase. However, it may be useful to describe identical phases with independent descriptions if it is unlikely that they will form homogeneity ranges in a higher component system, such as a halide and a carbide, which both have the rock salt structure.

4.4.3 Models of Composition Dependence

4.4.3.1 Ionic Sublattice Model

The ionic two-sublattice liquid (I2SL) model was developed to be used when there is a tendency for ionization in the liquid, which happens in liquid oxides and sulfides, for example. The same model can be used both for metallic and oxide melts. At low levels of oxygen, the model becomes equivalent to a substitutional solution model between metallic atoms. Two sublattices are assumed, one containing charged cations and another containing charged anions, neutrals, and vacancies. The complication for an ionic liquid is that the number of sites on the cation sublattice and anion sublattice varies with the composition to preserve electroneutrality.

In I2SL, the number of sites on each sublattice is a function of composition in order to maintain electroneutrality. The sublattice description can be written as

$$(C_i^{\nu_i^+})_P(A_j^{\nu_j^-}, \mathrm{Va}, B_k^0)_Q,$$

where $C_i^{\nu_i^+}$ are cations, $A_j^{\nu_j^-}$ are anions with charges ν_i^+ and ν_j^-, respectively, and B_k^0 are neutral species. It is assumed that the vacancies Va have an induced negative charge with a valency equal to the average valency on the cation sublattice. The number of sites on each sublattice, P and Q, is given by the average charge on the other sublattice, that is,

$$Q = \sum_i \nu_i^+ y_{C_i}$$

$$P = \sum_j \nu_j^- y_{A_j} + \left(\sum_i \nu_i^+ y_{C_i} \right) y_{Va} = \sum_j \nu_j^- y_{A_j} + Q y_{\mathrm{Va}}.$$

In the syntax of a commonly used thermodynamic database format, the phase-type code Y is attached to a phase name after a colon, for example, LIQUID:Y, meaning that the phase is of this type.

4.4.3.2 Associate Model

The associate model can be treated within the default framework. The use of the model implies that additional constituents are added such that a phase obtains additional internal degrees of freedom. This may be needed for describing experimental data for the liquid phase, for example. The existence of molecules as constituents sometimes needs to be postulated, but the lifetime of such a molecule may be so short that it cannot be found as an independent entity. The notion of *associate* has, therefore, been introduced to capture tendencies for ordering around certain compositions. The existence of associates is supported if there is a tendency for covalent bonding between the elements.

4.4.3.3 Modified Quasichemical Model

This model is similar to the associate model, but different with respect to the configurational entropy contribution.

4.4.4 Model of Nanosize Effect

As the size of a nanoscale system decreases, the surface effect increases in significance. This means that the contribution of surface Gibbs energy must be considered when calculations are done on nanoscale systems. The surface Gibbs energy of a nanoscale system is described as a function of the particle size.

Accordingly, the total Gibbs energy of an alloy system is composed of the bulk and surface terms:

$$G^{\mathrm{Total}} = G^{\mathrm{Bulk}} + G^{\mathrm{Surf}}.$$

For a nanoparticle (NP) with radius r, the surface Gibbs energy is given by

$$G^{\mathrm{Surf(NP)}} = \frac{2\sigma V}{r}.$$

For a nanowire (NW) with radius r and length l ($r \ll l$), the surface Gibbs energy is expressed by

$$G^{\mathrm{Surf(NW)}} = \frac{\sigma V}{r}.$$

It is likely that the surface Gibbs energy of a nanofilm (NF) with thickness t is

$$G^{\mathrm{Surf(NF)}} = \frac{(\sigma + \sigma^{I})V}{t},$$

where σ is the surface tension, σ^{I} is the interface tension (including interface stress), and V is the molar volume. The surface tension of liquid and solid alloy can be computed using Butler's equation with known surface tension and density of pure metals as well as the bulk excess Gibbs energy from the scientific literature.

The expression of the Gibbs energy of a nanoscale system is not so different from that of bulk system. For example, the Gibbs energy of an alloy nanoparticle can be described by

$$G^{\mathrm{Total(NP)}} = \sum X_i G_i^{o\mathrm{(NP)}} + RT \sum X_i \ln X_i + G^{\mathrm{EX(NP)}}.$$

For simplicity, we may assume that the excess Gibbs energy has a Redlich–Kister-type composition dependence:

$$G^{\mathrm{EX(NP)}} = \sum_{i} \sum_{<j} X_i X_j \sum_{v=0}^{m} L_{ij}^{(v)} (X_i - X_j)^{v} + \sum_{i} \sum_{<j} \sum_{<k} X_i X_j X_k L_{ijk} + \cdots.$$

Here, the parameter $L_{ij}^{(v)}$ is expressed by

$$L_{ij}^{(v)} = f_1 \left(\frac{1}{r}\right) + f_2 \left(\frac{1}{r}\right) T + f_3 \left(\frac{1}{r}\right) T \ln T + \cdots.$$

The function f_m ($m = 1, 2, 3, \ldots$) is simply given as a linear function of inverse of the particle's radius:

$$f_m = a_m + \frac{b_m}{r}.$$

Therefore, when the particle radius becomes infinite, the excess Gibbs energy becomes the same as the Gibbs energy of the bulk system.

The thermodynamic parameters of an alloy nanoparticle system can be reoptimized by following this procedure:

1) Collect measurements of thermophysical properties such as surface tension and molar volume of pure liquids and solids as well as a thermodynamic data set about the bulk system.
2) Determine the surface tension of pure solid phase from the melting behavior of nanoparticles. (Note that a correction factor for the surface tension of pure solid due to surface orientation and edge/corner effect is needed.)

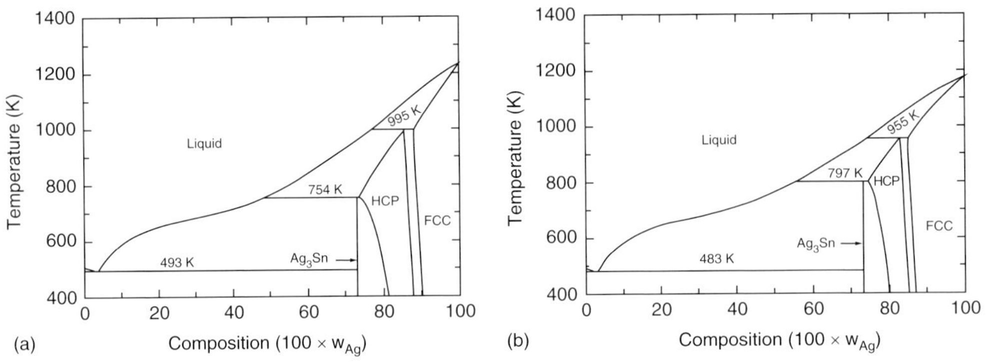

Figure 4.1 The calculated phase diagram of Ag–Sn alloy: (a) bulk and (b) nanoparticle with $r = 20$ nm. (Lee and Sim [1]. Reproduced with the permission of Elsevier.)

3) Determine the molar volume of the various pure solid phases and intermetallic compounds including hypothetical ones.
4) Calculate the surface tension of liquid and solid alloys using Butler's equation.
5) Determine the standard Gibbs energy of pure elements as a function of the particle size.
6) Optimize the excess Gibbs energy as a function of the particle size.

Figure 4.1 compares the calculated bulk and nanophase diagrams of Ag–Sn alloy. The validity of the calculated phase diagrams may be confirmed by experimental data. *In situ* TEM or DSC analyses are frequently used to determine the phase transformation of a small-scale system. For more details on modeling and experiments, the reader is referred to the references section.

4.4.5
CALPHAD Databases

The utility of CALPHAD-based simulation tools depends on reliable databases being available for the kinds of material systems being investigated. A database is typically focused on a certain class of alloys or other materials. It is important to realize that a CALPHAD database is not a collection of tagged and interlinked data. Rather, it contains the parameter values of the Gibbs energy functions defined by the thermodynamic models of the phases that are included in the database. It can also contain additional information, such as the sources of different parameters, instructions for its use, limitations, and so on. The database is stored in a plain text file format, although commercial databases are almost always encrypted before they are sent to the customer. The file size is relatively small; even a large multicomponent database rarely consists of more than 1 or 2 MB of data.

CALPHAD databases are available for a broad range of applications, including steels and Fe alloys, Al, Mg, Ni, and Ti alloys, cemented carbides, solders, noble alloys, nuclear applications, salts, oxide ceramics, slags, and aqueous solutions. However, there are still large uncharted areas for new materials and less extensively investigated materials. This even includes particular combinations of alloying elements in otherwise well-known alloy systems. If one considers that 80 elements could be of some practical use, then there are about 3000 possible binary systems and 80 000 possible ternary systems. Of the binary systems, a large fraction have at least been tentatively investigated experimentally and a reasonable fraction have been extensively investigated and/or thermodynamically modeled. Of the ternary systems, only a small fraction have been modeled thermodynamically, and for a large fraction, there is no knowledge at all.

Most databases have a particular application or group of applications in mind, such as steels. The database must then include all binary systems and several ternary systems involving Fe. The database should also contain all other binary systems, with the exception of impurity elements (such as S, P, or O in steel) where the most important binary systems would be sufficient. The database should also contain further important ternary and higher-order systems. The fraction of ternary systems included in the database can be used as a measure of the (potential) quality of the database. The database might not include all phases from the included data sets, but only the phases that are relevant for the application in mind. This can be helpful because the user is not confronted with a large number of irrelevant phases, but has the consequence that many phase diagrams from the included data sets will not be correctly reproduced. Also, which phases are relevant may not always be obvious.

A CALPHAD database is "constructed" from data sets for the unary (elemental), most binary, important ternary, and a few higher-order systems. Most of the binary and ternary data sets are taken from the open literature where they are published as thermodynamic assessments (or evaluations). In a large database, data sets are compiled from a larger number of different sources. Creating a high-quality database is not just a matter of collecting these individual data sets, but requires adjustments and reoptimization of some of the data sets, unification of incompatible models, identification of missing parameters, and extensive checking. These issues are discussed in more detail in Section 4.4.6. The level of quality and scope of individual data sets can be very different. The quality of a database is strongly dependent on the quality of the data sets included, the fraction of ternary and higher-order systems, and on the care with which it has been constructed.

Since the Gibbs energy functions for the elements are used in several different data sets coming from different sources, it is of utmost importance that everyone uses the same Gibbs energy functions. Otherwise, it would not be possible to combine different data sets in databases. Today, the SGTE unary database, originally published by Dinsdale [2] and regularly updated, is used almost universally as a source for the element data sets.

Although the pressure is a thermodynamic state variable, the Gibbs energy models used in CALPHAD databases usually do not include a pressure dependence except for the gas phase, which is modeled as an ideal gas. The ideal gas model is reasonable for ambient pressure (or lower) but ceases to be so at very moderate pressure. A few CALPHAD databases include molar volumes for the condensed phases. These molar volumes are modeled as temperature dependent, but not pressure dependent, and can be used to calculate the pressure dependence of phase equilibria between condensed phases at moderate pressure. Full pressure dependence requires the use of pressure-dependent molar volumes, that is, the use of an equation of state. This poses a number of modeling issues and, not least, a problem of a lack of experimental data. Full pressure modeling is, thus, very expensive and there is not much interest in the area of materials development and processing. So far, thermodynamic databases with full pressure modeling have only been developed in the geological community.

4.4.6
Database Development and Parameter Optimization

This section describes how data sets are derived and some of the issues encountered when they are used to develop a database. The description (i.e., the Gibbs energy functions with their determined parameter values) of a lower-order system (binary, ternary, or, rarely, quaternary) that has been individually modeled is called a data set. The process of deriving the Gibbs energy models and determining the parameter values is usually called a thermodynamic assessment or evaluation. This includes a critical review of all available experimental data and *ab initio* data relevant for the system, coding the data in a form that can be used by an optimization program, performing the optimization (or otherwise determine the parameter values), and documenting the results, typically in the form of a publication.

There are a few very specific circumstances that render the CALPHAD method its usefulness and it is important to be aware of those when modeling a system or building a database:

- The phase diagram is very sensitive to the relative Gibbs energies of the phases present, but not sensitive to the absolute values of the Gibbs energy (or rather the enthalpy and entropy). Therefore, a key point in the CALPHAD method is to model both phase diagram data and thermodynamic data at the same time. If one type of data is missing or inaccurate, the derived parameter values will not be reliable.
- Alloy thermodynamics is dominated by lattice stabilities and pair interactions. Therefore, (thermodynamic) extrapolation from binary to ternary and higher-order systems for the included solution phases will be reasonable, provided the binary parameters are reasonable. Consequently, there are rather few phases that are truly quaternary or higher order. A carefully constructed CALPHAD database will give reasonable results, for example, a 10-component system that has never been investigated experimentally.
- The Gibbs energy for each phase is available for the complete concentration range for which its Gibbs energy model is defined, irrespective of whether the

phase is stable for a specific composition. This means that the Gibbs energy can be calculated for any conceivable metastable (or even unstable) state. This is needed for the simulation of phase transformations.

The process of performing an assessment is described in detail by Lukas *et al.* [3] and Schmid-Fetzer *et al.* [4]. While Lukas *et al.* [3] give a broad introduction to the basic machinery, Schmid-Fetzer *et al.* [4] concentrate on the many pitfalls that can be encountered. The latter also present a procedure to check data sets with respect to correctness, reasonability, accuracy, and safety.

Starting with the modeling of a binary system, the first step is to collect as much relevant information about the system as possible. This will mostly be publications from scientific journals but will usually also include conference papers, reports, theses, and further so-called "gray" sources. Once sufficient information has been collected one has to decide which phases to include and which models to use. Terminal solution phases will usually be one (or more) of fcc, bcc, and hcp (in addition to the liquid). These phases are always modeled as solid solutions and should, therefore, include both elements. For many common intermetallic phases, such as Laves, σ, and μ, models have been defined, but there is not always agreement on which of the suggested models should be used. There is a tendency to use more complex models with several sublattices to better reflect the crystal structure of the phase. Such models generate a large number of model compounds whose Gibbs energies cannot be determined from experiment, but have to be estimated, or, better, calculated by *ab initio* methods. For ordered phases based on fcc, bcc, and hcp, order–disorder models have been defined. These models require particular care in their handling. Intermetallic phases with a narrow and uncertain composition range should be modeled as stoichiometric compounds. Then, the experimental data points must be coded in such a way that they can be used for the parameter optimization. This means that it should be possible to calculate the corresponding equilibrium using the current set of parameter values. The difference between the calculated and measured value divided by an uncertainty is the error that will be used in a least square minimization. The parameter values are varied using a minimization algorithm until a minimum is found. This is what is called parameter optimization.

Very often it will be found that data are in conflict. This means that one or the other sets of data can be correct, but not both and also not the mean. It is, therefore, important to discard one of them from the optimization, but it can be difficult to determine which one should be discarded. In the end, a data set has been derived, which reproduces the phase diagram and thermodynamic properties with sufficient accuracy and provides reliable extrapolations. The data set and the modeling process then need to be carefully documented, ideally in such a way that someone else could improve the data set when new data or better models become available.

In ternary systems, in contrast to many binary systems, there are rarely sufficient data available to optimize all relevant model parameters. Usually, only a partial optimization is possible and some parameters have to be estimated and fixed beforehand. The general procedure for modeling a ternary system starts with the three binary data sets. These are combined and an extrapolation of the solution phases, typically liquid, fcc, bcc, and/or hcp, is made. This will often give an indication of whether the binary interactions are reasonable. Missing binary interactions need to be identified and given reasonable values or included in the ternary optimization. For example, the Cr–Mo system does not contain an fcc phase, but the Cr–Mo–Ni system does. If a binary fcc interaction in the Cr–Mo system is not determined, the Cr–Mo–Ni system can still be modeled (by including a ternary fcc interaction), but extrapolations to higher-order systems will not be good. An early version of the Thermo-Calc steel database could not accurately reproduce fcc–bcc equilibria for duplex stainless steels, which contain Cr, Mo, and Ni, until the missing fcc interaction in the Cr–Mo system had been identified. When binary intermetallic phases dissolve the third element, this information has to be added to the model of the intermetallic phase. This generates further model compounds whose Gibbs energies must be determined. As a last step, any ternary phases should be added. If a ternary phase is modeled with a solution model, it will most probably generate binary model compounds. It is then important to make sure that they do not become stable in a binary system. A calculated ternary phase diagram is also a very good guide for further experimental work.

The general procedure to construct a CALPHAD database is not so different from modeling a ternary system. Several of the steps are in common, such as adding missing binary interactions, unifying compounds that belong to a single phase into a single model, and determining Gibbs energies for model compounds not previously included. In contrast to binary and ternary data sets that are determined once and often not changed further, a database is usually extended and improved over a longer period of time.

All databases need to be checked extensively. There are two types of checking to be done. The first type is to recalculate all included lower-order systems (data sets) in order to ensure the integrity and formal correctness of the database. This is needed to catch any misprints and phases that might appear in systems where they do not belong. The second kind of testing is to verify the database against a set of selected alloys. This can also be used for the improvement of the database. Any discrepancies found need to be traced back to the relevant low-order (binary or ternary) systems. These systems should then be modified or remodeled accordingly. However, it can be very difficult to identify the low-order systems actually causing a mismatch for a complex alloy; see the previous example of the Cr–Mo system and duplex stainless steels. When changing a low-order system, it is necessary to at least check all higher-order systems where this system is included. If, for example, Fe–Cr is changed, then all ternary systems with Fe and Cr will have to be checked and possibly modified. When a change is made to an element, this will influence a large number of systems in the database. Thus, developing a multicomponent CALPHAD database requires considerable care and extensive checking.

4.4.7 Phase Names

Finding an appropriate schema for naming is important for associating data with the phases they are describing and for retrieving experimental and computational data from conventional databases. Finding an appropriate naming schema is challenging because a phase name can serve as an identifier in both types of databases: conventional and CALPHAD type. Inside a thermodynamic database, the entries are tied together by the names of the phases. Although an abstract identifier would serve for this purpose, the phase name is also important information for the user. One could follow the example of the International Chemical Identifier (InChI) by constructing a string describing the various aspects that are needed to unambiguously describe a phase. However, the InChI can be lengthy and includes more information than needed, for example, in CALPHAD databases. For the user, it is important that the phase name is brief and unique and that the phase it describes can easily be identified.

In the early days of alloy phase diagrams, phases were named using Greek letters. In unary systems, letters were assigned from lower temperatures to higher temperatures, and in binary systems from the left- to the right-hand side of the system. This cannot work for a database that has even 10 elements since it will contain the descriptions of 45 binary systems and all their phases. This is easily illustrated by looking at the use of the letter γ. It describes the austenitic phase in Fe alloys, the Ni phase in superalloys, as well as the TiAl phase in Ti–Al-based alloys, the $Mg_{17}Al_{12}$ phase in light-weight alloys, and the Cu_5Zn_8 phase in brass. On the other hand, the use of specific Greek letters has made those letters commonly recognized names for some of the topologically close-packed phases, such as σ, μ, or χ, that have large homogeneity ranges and rarely occur at ideal stoichiometries. The use of chemical formulae is also not practical because the same phase may occur in a number of the constituent binary subsystems of a multicomponent system; for example, Laves phases with the ideal stoichiometry AB_2 occur in many of the binary systems that are relevant for superalloys. It is obvious that the use of a generalized chemical formula such as AB_2 is not feasible as there can be many different phases with the same ideal stoichiometry.

It would be convenient if a phase could be identified by a short name that is independent of the system. The identification of the crystal structure of phase could serve as a phase name, but the use of space group and Wyckoff sequence are not very intuitive. An early crystallographic data collection, Strukturbericht, devised a symbol system for crystal structures. This system uses letters for classes of compounds and numbers, with and without subscripts, to distinguish the different structures within the classes, for example, A for the elements and B for AB compounds. It turned out that this system was not sustainable given the wide variety of different crystal structures; thus, it was abandoned. However, the symbols for the simpler structures, such as A1 for fcc or A2 for bcc, are still popular and are used in many CALPHAD databases as phase names.

The IUPAC Redbook on Nomenclature of Inorganic Chemistry does not consider the use of Greek letters or the Strukturbericht designation as acceptable since they do not convey the necessary information and are not self-explanatory and recommends the use of the Pearson symbol for the identification of the structure of a phase. However, the Pearson symbol is not unique, for example, the symbol *cF8* belongs to a number of different crystal structures. Although a crystal structure can be uniquely identified by its space group and its Wyckoff sequence, it gives no further information on the character of a phase, for example, whether the phase belongs to the salt or carbide branch of a particular structure. In geosciences, it is customary to use the mineral name for phase identification. In a certain way, one can view the CALPHAD practice of developing phase names using a mixture of Greek letters, Strukturbericht symbols, mineral names, and stoichiometric formulae as a hybrid approach to provide phase names that are useful for the database developer and the user by being sufficiently intuitive to make it clear which phase is described by the name.

It has become practice to use the term "fcc_A1" to describe disordered fcc solid solutions and the term "fcc_L12" for an ordered variant of this structure. A problem arises if these two phases are described as one phase that can be ordered or disordered, as this may result in the easy distinction of the two phases being lost to the user. In this case, it is useful if the software determines the variant of the phase for the user. The same applies to compounds that are based on interstitial solutions in the phases of the pure elements, such as cubic and hexagonal nitrides. It should be noted that both structures must have the same space group to allow this continuous transition. For example, in the case of the fcc_A1 structure, the filling of the interstitial octahedral site results in a continuous transition to a B1_carbide structure. Here again, the software should determine which variant of the phase is encountered. It may be cumbersome to introduce all these provisions for the variants of a phase, but it enables the user to quickly recognize a phase without having to analyze the composition details of a phase. Table 4.3 presents the suggestions for phase names and name variations for distinguishing different phase variants.

A short, easily recognizable phase name may be user friendly, but it has little value for linking the phase to contents of non-CALPHAD databases. The challenge is that one phase in a CALPHAD database will likely be represented as a number of different phases in other databases. One solution to this problem would be to add identifier records for the phases that are described by this single description. This record could be constructed using the concept of the InChI and could contain records for formula, a general tag for the aggregate state, SOL, LIQ, or GAS and tags with prototype (PT), Pearson symbol (PS), Strukturbericht designation (SD), space group (SG, using the number), Wyckoff sequence (WS), and site occupation (SO) of the Wyckoff positions and could look similar to one of the following examples: Fe/SOL/PT:Cu/PS:cF4/SD:A1/SG:225/WS:a/SO:a:Fe describes pure fcc Fe; FeN0.0897/SOL/PT:FeNx/PS:cF4/SD:A1/SG:225/WS:ab/SO:a:Fe:b:N describes the interstitial solution of N in fcc Fe;

Table 4.3 Suggestions for names of common crystalline phases in CALPHAD-type databases.

Phase name	Strukturbericht designation	Prototype	Chemical formulae	Comment
Fcc_A1	A1	Cu	Al, Ni, Cu, ...	
Bcc_A2	A2	W	V, Cr, Mo, ...	
Hcp_A3	A3	Mg	Be, Mg, Ti, ...	
Fcc_B1	B1	NaCl	TiC, NbC, TaC, ...	
B1_Oxide	B1	NaCl	CaO, FeO, TiO ...	
Bcc_B2	B2	CsCl	AlNi, AlFe, FeTi, ...	Ordered bcc
B2_Halide	B2	CsCl	CsCl, TlBr, KI	"Independent" B2
FCC_L10	$L1_0$	AuCu	AlTi, MnNi, FePt, ...	Ordered fcc
HCP_D019	$D0_{19}$	Ni_3Sn	$AlTi_3$, Co_3W, $CoZr_3$, ...	Ordered hcp
C14_Laves	C14	$MgZn_2$	Cr_2Ti, $HfMn_2$, Ni_2U, ...	
C15_Laves	C15	$MgCu_2$	Cr_2Ti, Co_2Ti, Ni_2Y, ...	
Mu	$D8_5$	Fe_7W_6	Co_7Mo_6, Fe_7Nb_6, Ni_7Ta_6, ...	
Sigma	$D8_b$	CrFe	$AlNb_2$, Cr_3Mn, Cr_2Re_3, ...	
Me3Si	...	PTi3	AsNb3, GeZr3, SiTi3, ...	Limited to the combination of transition metal with metalloid
Al3Fe	...	Al3Fe	Al3Fe, Al13Ru4	Only two compounds known with this structure
Al2Fe	...	Al2Fe	Al2Fe	No other chemistry known
Al3Co	...	...	Al3Co	Structure unknown

TiC0.957/SOL/PT:NaCl/PS:cF8/SD:B1/SG:225/WS:ab/SO:a:Ti:b:C describes the off-stoichiometric compound TiC (note that site b, in contrast to the previous examples, is almost filled); NiAs/SOL/PT:NiAs/PS:hP4/SD:B81/SG:194/WS:ac/SO:a:Ni:c:As describes the prototype NiAs; Ni2In/SOL/PT:Ni2In/PS:hP6/SD:B82/SG:194/WS:acd/SO:a:Ni:c:In:d:Ni describes the prototype Ni_2In (note that the difference to the NiAs structure is that and additional site (d) is filled); and Au0.62Pd0.74Sn/SOL/PT:NiAs/PS:hP4/SD:B81/SG:194/WS:acd/SO:a:Au,Pd:c:Sn:d:Pd describes a phase with NiAs/Ni_2In structure where the d site is partially filled. For identifying the liquid, instead of describing the crystal structure, one would identify the type of liquid (LT) and whether special species (SP) are identified: AgAu/LIQ/LT:regular/SP:Ag,Au; MnO/LIQ/LT:ionic/SP:Mn+2,Mn+3:O-2,Va-q. The description of the gas phase would be a simple list of the species occurring in it: As:/GAS/SP:As,As2,As4.

It should be noted that these suggestions currently do not cover amorphous or quasicrystalline phases.

4.4.8
Reference States

When interpreting the values of some thermodynamic variables, it is important to know which reference states the variable values are expressed relative to. These properties include Gibbs and Helmholtz energy values as well as chemical potentials, and activities are always expressed relative to some reference state. If reference states have not been properly defined for the values of these conditions, they will be meaningless.

In substance or alloy solution CALPHAD databases, the data for each phase are described by Gibbs energies relative to the "Standard Element Reference" (SER), that is, enthalpies that the pure elements have in their most stable phase at a temperature of 298.15 K and a pressure of 100 kPa (1 bar). The Gibbs energy of the SER state is referred to as G-HSER and is equal to H298. Note that while the Gibbs energy of a phase has both enthalpy and entropy contributions, the entropy of a phase in the SER state is set to the entropy at 0 K and, according to the third law of thermodynamics, is zero.

Using G-HSER as a reference state is convenient since all data in a database can be expressed and relative to G-HSER so that data can be combined in thermodynamic calculations. All thermodynamic functions can be calculated directly from one or more derivatives of the Gibbs energy expression.

4.4.9
Database Formats

A thermodynamic database, which stores the Gibbs energy function of every phase of interest in the system, is an essential part of CALPHAD-based thermodynamic calculations. The most widely used thermodynamic database format is the TDB format, which is described here in detail.

There are also other well-defined database formats. For instance, the Lukas program uses the DB format for phase diagram calculations. Details on this database format can be found in the book *Computational Thermodynamics: The Calphad Method* [3]. FactSage has developed Compound databases and Solution databases on oxides, salts, and other systems. The FactSage database format is described in [2, 5].

A TDB database file is a plain text file that typically consists of four sections: Elements, Functions, Type_Definitions, and Phases. Figure 4.2 illustrates the structure of a typical TDB file, this one describing the Al–Ni binary system with the key words highlighted in blue. Figure 4.2 is included for the purpose of illustration and is not a full database.

At the beginning of the file, the Elements section defines every component included in the database. Next to each element, the following information is listed: the stable structure at 298 K (for Al in the Figure 4.1, this is FCC_A1), the atomic weight (26.982 g/mol), the enthalpy (4577.3 J/mol), and the entropy (28.322 J/mol K) of the element in its most stable state at 298 K.

```
Element AL        FCC_A1      26.982   4577.3   28.322 !
Element NI        FCC_A1      58.69    4787     29.796 !
Element VA        VACUUM      0        0        0 !

Function GHSERAL  298.15   -7976.15+137.093*T-24.3672*T*ln(T)-0.00188466*T**2-8.77664e-007*T**3+74092*T**(-1);  700  Y
                           -11276.2+223.048*T-38.5844*T*ln(T)+0.018532*T**2-5.76423e-006*T**3+74092*T**(-1);  933.6  Y
                           -11278.4+188.684*T-31.7482*T*ln(T)-1.231e+028*T**(-9);  6000   N !
Function GLIQAL   298.15   11005-11.8419*T+7.934e-020*T**7+GHSERAL;     933.6   Y
                           10482.4-11.254*T+1.231e+028*T**(-9)+GHSERAL;   6000   N !
Function GBCCAL   298.15   10083-4.813*T+GHSERAL; 6000 N !
Function GHSERNI  298.15   -5179.16+117.854*T-22.096*T*ln(T)-0.0048407*T**2;     1728   Y
                           -27840.7+279.135*T-43.1*T*ln(T)+1.12754e+031*T**(-9);  6000   N !
Function GLIQNI   298.15   16414.7-9.397*T-3.82318e-021*T**7+GHSERNI;      1728   Y
                           18290.9-10.537*T-1.12754e+031*T**(-9)+GHSERNI;   6000   N !
Function GBCCNI   298.15   8715.08-3.556*T+GHSERNI;   6000   N !
Function U1ALNI   298.15   -14556.4+2.945*T;   6000   N !

Type_Definition   &   GES A_P_D   BCC_A2   Magnetic  -1   0.4!
Type_Definition   N   GES A_P_D   L12_FCC  dis_part DISORDER ,,,!
Type_Definition   (   GES A_P_D   FCC_A1   Magnetic  -3   0.28!

Phase LIQUID % 1 1 !
Constituent LIQUID :AL,NI:!
Parameter G(LIQUID,AL;0)        298.15   GLIQAL;   6000   N !
Parameter G(LIQUID,NI;0)        298.15   GLIQNI;   6000   N !
Parameter G(LIQUID,AL,NI;0)     298.15   -197088+30.353*T;   6000   N !
Parameter G(LIQUID,AL,NI;1)     298.15   5450;     6000   N !
Parameter G(LIQUID,AL,NI;2)     298.15   54624-11.383*T;     6000   N !

Phase AL3NI1 % 2 0.75 0.25 !
Constituent AL3NI1 :AL:NI:!
Parameter G(AL3NI1,AL:NI;0)   298.15   -40463+5.312*T+0.75*GHSERAL+0.25*GHSERNI;    6000   N !

Phase AL3NI2 % 3 3 2 1 !
Constituent AL3NI2 :AL:AL,NI:NI,VA:!
Parameter G(AL3NI2,AL:AL:NI;0)      298.15   -41219+2.792*T+5*GBCCAL+GBCCNI;     6000   N !
Parameter G(AL3NI2,AL:NI:NI;0)      298.15   -428055+71.808*T+3*GBCCAL+3*GBCCNI;   6000   N !
Parameter G(AL3NI2,AL:AL:VA;0)      298.15   30000-3*T+5*GBCCAL;    6000   N !
Parameter G(AL3NI2,AL:NI:VA;0)      298.15   -356836+66.079*T+3*GBCCAL+2*GBCCNI;   6000   N !
Parameter G(AL3NI2,AL:AL,NI:NI;0)   298.15   -38157;    6000   N !
Parameter G(AL3NI2,AL:AL:NI,VA;0)   298.15   -23319;    6000   N !

Phase DISORDER % 1 1 !
Constituent DISORDER :AL,NI:!
Parameter G(DISORDER,AL;0)      298.15   +GHSERAL;    6000   N !
Parameter G(DISORDER,NI;0)      298.15   +GHSERNI;    6000   N !
Parameter G(DISORDER,AL,NI;0)   298.15   -16367+12*U1ALNI;      6000   N !

Phase L12_FCC %N 2 0.75 0.25 !
Constituent L12_FCC :AL,NI:AL,NI:!
Parameter G(L12_FCC,AL:AL;0)      298.15   0;    6000   N !
Parameter G(L12_FCC,NI:AL;0)      298.15   3*U1ALNI;   6000   N !
Parameter G(L12_FCC,AL:NI;0)      298.15   3*U1ALNI;   6000   N !
Parameter G(L12_FCC,NI:NI;0)      298.15   0; 6000 N !
Parameter G(L12_FCC,AL,NI:AL;0)   298.15   6*U1ALNI;   6000   N !
Parameter G(L12 FCC,AL,NI:NI;0)   298.15   6*U1ALNI;   6000   N !
```

Figure 4.2 Partial database of the Al–Ni binary system showing the typical TDB format.

In the Function section, the Gibbs energies of each pure element in each of its crystal structures are listed. The Gibbs energy of an element with a certain crystal structure is typically represented as a segmental function of temperature. The lower and upper temperature limits of each segment are specified at the start and end of each function segment. The "Y" after the upper segment limit indicates that there is another temperature segment above this upper limit. An "N !" indicates the end of the segmental function specification. These Gibbs energies are specified relative to the element's SER state (see Section 4.4.8). In the Function section, database developers can also define functions that can be used directly in the Phase section.

The Type_Definition section defines other features related to a particular phase. Simple characters are used to identify which Type_Definitions are used by a particular phase description. In Figure 4.2, "&" is used to define the magnetic parameters of the BCC_A2 phase and "N" specifies that L12_FCC has a disordered Gibbs energy contribution.

The major body of a TDB file consists of the Phase section. In this section, thermodynamic models are defined and model parameters are listed for each phase in the system. Details are given in the following for the phases listed in Figure 4.2.

The description of each phase starts with the key word Phase followed by the phase name. The "%" separates the phase name from the phase description, that is, from information about sublattices. If the phase has other features than the sublattices that need to be described, such as magnetic or disordered contributions, then the symbol defined in the Type_Definition should directly follow % without any space in between (see the section that defines the L12_FCC phase in Figure 4.2). If this phase has no such additional feature, then there should be a space after %.

The LIQUID phase in Figure 4.2 is described as a substitutional solution phase. The first "1" after % indicates that the phase has one sublattice and the second "1" means the fraction of this sublattice is 1. The next line, which starts with the key word "Constituent," defines the constituents of each sublattice. After the phase name, the constituents in each sublattice are specified with a colon (":") separating the sublattices. A comma is used to separate species within the specification of a sublattice. A percent sign can be appended to the constituent symbol to indicate that it is a major constituent of this sublattice.

After the Constituent line, the model parameters follow.

The first two Parameter lines for the LIQUID phase represent the reference state. The next three lines define the excess Gibbs energy terms. For more information, see Section 4.3.2. In Figure 4.2, the Gibbs energy of LIQUID (J/mol) is

$$\begin{aligned} G_{\text{LIQUID}} = & x_{\text{Al}} \times \text{GLIQAL} + x_{\text{Ni}} \times \text{GLIQNI} + RT(x_{\text{Al}} \ln x_{\text{Al}} + x_{\text{Ni}} \ln x_{\text{Ni}}) \\ & + x_{\text{Al}} x_{\text{Ni}} [(-197\,088 + 30.353 \times T) + 5450(x_{\text{Al}} - x_{\text{Ni}}). \\ & + (54\,624 - 11.383 \times T)(x_{\text{Al}} - x_{\text{Ni}})^2]. \end{aligned}$$

Note that the functions GLIQAL and GLIQNI are defined in the Function section and can thus be directly used in the Gibbs energy expressions.

The phase following LIQUID in Figure 4.2 is AL3NI1. This phase has two sublattices. The fraction of the first sublattice is 0.75 and the fraction of the second is 0.25. The second line specifies that AL occupies the first sublattice and that NI occupies the second sublattice. The Parameter line gives the Gibbs energy of AL3NI1 (J/mole atoms) as

$$G_{\text{AL3Ni1}} = 0.75 \times \text{GHSERAL} + 0.25 \times \text{GHSERNI} - 40463 + 5.312 \times T.$$

The third phase in Figure 4.2 is AL3NI2. For this phase, the stoichiometries (rather than the fractions) of its three sublattices are specified as 3, 2, and 1, respectively.

These stoichiometries are given in moles, which means that the AL3NI2 definition concerns a phase with 6 moles of atoms. Hence, the Parameters listed later specify the Gibbs energy of 6 moles of atoms of this phase. The Constituent line specifies that AL occupies the first sublattice, AL and NI occupy the second sublattice and NI and vacancy (VA) occupy the third sublattice. AL3NI2 is an ordered intermetallic phase. Its Gibbs energy is described by the CEF. In the phase definition, a comma is used to separate species in the same sublattice, while a colon is used to separate species belonging to different sublattices. The following Parameter lines give the Gibbs energy of 6 mole atoms of AL3NI2 as

$$\begin{aligned}
G_{\mathrm{AL3NI2}} &= y^{\mathrm{I}}_{\mathrm{Al}} y^{\mathrm{II}}_{\mathrm{Al}} y^{\mathrm{III}}_{\mathrm{Ni}}(-41\,219 + 2.792 \times T + 5 \times \mathrm{GBCCAL} + \mathrm{GBCCNI}) \\
&+ y^{\mathrm{I}}_{\mathrm{Al}} y^{\mathrm{II}}_{\mathrm{Ni}} y^{\mathrm{III}}_{\mathrm{Ni}}(-428\,055 + 71.808 \times T + 3 \times \mathrm{GBCCAL} + 3 \times \mathrm{GBCCNI}) \\
&+ y^{\mathrm{I}}_{\mathrm{Al}} y^{\mathrm{II}}_{\mathrm{Al}} y^{\mathrm{III}}_{\mathrm{Va}}(30\,000 - 3 \times T + 5 \times \mathrm{GBCCAL}) \\
&+ y^{\mathrm{I}}_{\mathrm{Al}} y^{\mathrm{II}}_{\mathrm{Ni}} y^{\mathrm{III}}_{\mathrm{Va}}(-356\,836 + 66.079 \times T + 3 \times \mathrm{GBCCAL} + 2 \times \mathrm{GBCCNI}) \\
&+ y^{\mathrm{I}}_{\mathrm{Al}} y^{\mathrm{II}}_{\mathrm{Al}} y^{\mathrm{II}}_{\mathrm{Ni}} y^{\mathrm{III}}_{\mathrm{Ni}}(-38\,157) + y^{\mathrm{I}}_{\mathrm{Al}} y^{\mathrm{II}}_{\mathrm{Al}} y^{\mathrm{III}}_{\mathrm{Ni}} y^{\mathrm{III}}_{\mathrm{Va}}(-38\,157) \\
&+ RT[3 \times y^{\mathrm{I}}_{\mathrm{Al}} \ln y^{\mathrm{I}}_{\mathrm{Al}} + 2 \times (y^{\mathrm{II}}_{\mathrm{Al}} \ln y^{\mathrm{II}}_{\mathrm{Al}} + y^{\mathrm{II}}_{\mathrm{Ni}} \ln y^{\mathrm{II}}_{\mathrm{Ni}}) + (y^{\mathrm{III}}_{\mathrm{Ni}} \ln y^{\mathrm{III}}_{\mathrm{Ni}} + y^{\mathrm{III}}_{\mathrm{Va}} \ln y^{\mathrm{III}}_{\mathrm{Va}})].
\end{aligned}$$

Note that since AL is the only species that occupies the first sublattice, $y^{\mathrm{I}}_{\mathrm{Al}} = 1$ in the previous equation.

The DISORDER phase definition is actually part of the definition of Phase L12_FCC. This is evident from the fact that the phase name L12_FCC is followed by %N. In the Type_Definition, "N" is defined so that it can be used to specify that L12_FCC has a disordered contribution from DISORDER. The Gibbs energy of L12_FCC (J/mole atoms) is written as

$$G_{\mathrm{L12_{FCC}}} = [G_{\mathrm{L12_{Ord}}}(y^s_i) - G_{\mathrm{L12_{Ord}}}(y^s_i = x_i)] + G_{\mathrm{DISORDER}}.$$

For the L12_FCC phase in Figure 4.2, we have

$$\begin{aligned}
G_{\mathrm{L12_Ord}} &= y^{\mathrm{I}}_{\mathrm{Ni}} y^{\mathrm{II}}_{\mathrm{Al}}(3 \times \mathrm{U1ALNI}) + y^{\mathrm{I}}_{\mathrm{Al}} y^{\mathrm{II}}_{\mathrm{Ni}}(3 \times \mathrm{U1ALNI}) \\
&+ y^{\mathrm{I}}_{\mathrm{Al}} y^{\mathrm{I}}_{\mathrm{Ni}} y^{\mathrm{II}}_{\mathrm{Al}}(6 \times \mathrm{U1ALNI}) + y^{\mathrm{I}}_{\mathrm{Al}} y^{\mathrm{I}}_{\mathrm{Ni}} y^{\mathrm{II}}_{\mathrm{Ni}}(6 * \mathrm{U1ALNI}) \\
&+ RT[0.75 * (y^{I}_{Al} \ln y^{\mathrm{I}}_{\mathrm{Al}} + y^{\mathrm{I}}_{\mathrm{Ni}} \ln y^{\mathrm{I}}_{\mathrm{Ni}}) + 0.25 \times (y^{\mathrm{II}}_{\mathrm{Al}} \ln y^{\mathrm{II}}_{\mathrm{Al}} + y^{\mathrm{II}}_{\mathrm{Ni}} \ln y^{\mathrm{II}}_{\mathrm{Ni}}).
\end{aligned}$$

$$\begin{aligned}
G_{\mathrm{DISORDER}} &= x_{\mathrm{Al}}\mathrm{GHSERAL} + x_{\mathrm{Ni}}\mathrm{GHSERNI} + x_{\mathrm{Al}} x_{\mathrm{Ni}}(-16367 + 12U1\mathrm{ALNI}) \\
&+ RT(x_{\mathrm{Al}} \ln x_{\mathrm{Al}} + x_{\mathrm{Ni}} \ln x_{\mathrm{Ni}}).
\end{aligned}$$

Other quantities, such as magnetic transition temperature, Bohr magnetons, molar volume, or compressibility, can be described in a TDB file. However, while the TDB format can be used with several software packages, such as Thermo-Calc and PANDAT, there are minor differences in how the TDB files are interpreted by the different software. For example, PANDAT uses some additional key words and Property names in TDB files. Figure 4.3 shows an example of a TDB file that contains the PANDAT-specific key word Optimization to define a variable that is to be optimized. Following this key word is the variable's name, its lower limit,

```
Element          AL                FCC_A1     26.982       4540        28.3 !
Element          VA                VACUUM          0          0           0 !
Element          ZN                HCP_A3      65.39       5657       41.63 !

Optimization L0 0 10465.55; 20000 N !
Optimization L0T -10 -3.39259; 0 N !

Phase LIQUID % 1 1 !
Constituent LIQUID :AL,ZN:!
    Parameter G(LIQUID,AL;0) 298.15   G_AL_LIQUID; 6000 N !
    Parameter G(LIQUID,ZN;0) 298.15   G_ZN_LIQUID; 6000 N !
    Parameter G(LIQUID,AL,ZN;0) 298.15   L0+L0T*T; 6000 N !

    Parameter VM(LIQUID,AL;0) 298.14  1.02E-005+1.12E-009*T; 2000 N !
    Parameter VM(LIQUID,ZN;0) 298.14  1.92E-005+2.43E-009*T; 2000 N !

    Parameter ACTIVATIONENERGY(LIQUID,AL;0) 298.14  20340+22.5*T; 2000 N !
    Parameter ACTIVATIONENERGY(LIQUID,ZN;0) 298.14  6437+25.8*T; 2000 N !

    Parameter SURFACETENSION(LIQUID,AL;0) 298.14  1.1-1.6E-4*T; 2000 N !
    Parameter SURFACETENSION(LIQUID,ZN;0) 298.14  0.4161-7E-5*T; 2000 N !
```

Figure 4.3 A TDB file with PANDAT-specific key words.

initial value, and upper limit. Additional property parameters, such as molar volume (VM), activation energy (ACTIVATIONENERGY), and surface tension (SURFACETENSION) can be defined in a PANDAT-specific TDB file.

4.4.10
Extensions

Initially, the CALPHAD method was established as a tool for treating thermodynamics and phase equilibria of multicomponent systems. Since then, the method has been successfully applied to diffusion mobilities in multicomponent systems, creating the foundation for simulation of diffusion processes in these systems. In principle, it should be possible to describe temperature, pressure, and composition dependence of any phase-specific property using the CALPHAD approach. Recently, the CALPHAD method has been expanded to other phase-based properties, including molar volumes and elastic constants, and has the potential to treat electrical and thermal conductivities and even two-phase properties, such as interfacial energies. The calculation of some of these quantities has already been implemented in a number of software. The general challenge for describing these properties is that, unlike the thermochemical quantities, most of them can be anisotropic depending on the crystal system of the phase. Although it will be sufficient to use an average value for the specific property for treating the bulk, this is not adequate for the treatment of textured or thin film materials.

The suitability of the CALPHAD method for modeling diffusion processes was recognized early on. The underlying hypothesis is that diffusion in a crystalline phase with an equilibrium vacancy concentration occurs by a vacancy exchange mechanism. The phenomenological diffusion coefficients are then expressed in terms of a diffusion mobility function and a thermodynamic factor. The thermodynamic factor is determined using an existing multicomponent thermodynamic CALPHAD database. The conjecture is that the thermodynamic factor has the strongest composition dependence and provides the correct behavior of the multicomponent off-diagonal diffusion coefficients. Because of this linkage, it is imperative that the thermodynamic model reflects the phase as closely as possible. For example, ignoring the occurrence of structural vacancies of a phase will be detrimental to an accurate description of the diffusivities in this phase. This is illustrated in Figure 4.4 for the B2 phase of the Ni–Al system. In Figure 4.4a, the phase is modeled as a simple substitutional solution [7] and in Figure 4.4b with the sublattice model taking into account the structural vacancies that occur on the Al-rich side of this phase [8]. Although both models reproduce the phase diagram, the effect on the thermochemical factor (Figure 4.4c and d) is pronounced. For the same reason, the composition dependence of the diffusion mobilities [6] is modeled with the same model as the thermodynamics. Results from calculations with both models are shown for the tracer and chemical diffusivities in Figure 4.4e and f, respectively.

Use of the CALPHAD method to develop diffusion mobility databases has become essential in numerical simulations of diffusion processes for multicomponent alloys where composition- and temperature-dependent diffusion coefficient matrices are needed for each point after each time step. These simulations are carried out using a finite difference method (FDM), finite element method (FEM) or the phase field method. The diffusion simulation codes are linked via software libraries or are directly linked to the thermodynamics code. For example, the FDM code DICTRA is built on top of the thermodynamics code Thermo-Calc while the FDM phase field code MICRESS is linked via a software library to the same thermodynamics code. Most of the thermodynamic CALPHAD software has a module for the assessment of the thermodynamic model parameters; however, only a few are available for the assessment of diffusion mobilities, namely, DICTRA and Pandat.

Although molar volume and bulk modulus or the elastic constants are parameters needed for the description of the pressure dependence of the Gibbs energy, descriptions of these quantities have been added only fairly recently to CALPHAD-type databases and are only available for a few base elements. Consistent description of these quantities will result in equation of state descriptions of a phase as function of composition.

In addition to thermodynamics and diffusion mobilities, the CALPHAD method can also be applied to the description of thermal conductivity, electrical conductivity, optical properties such as refraction and reflectivity, viscosity, and plasticity including dislocation migration energies. In metallic phases, where the heat is conducted by electrons, thermal and electrical conductivities are connected via the Wiedemann–Franz law. In insulator phases, where the heat is conducted by phonons, the thermal conductivity is linked to heat capacity

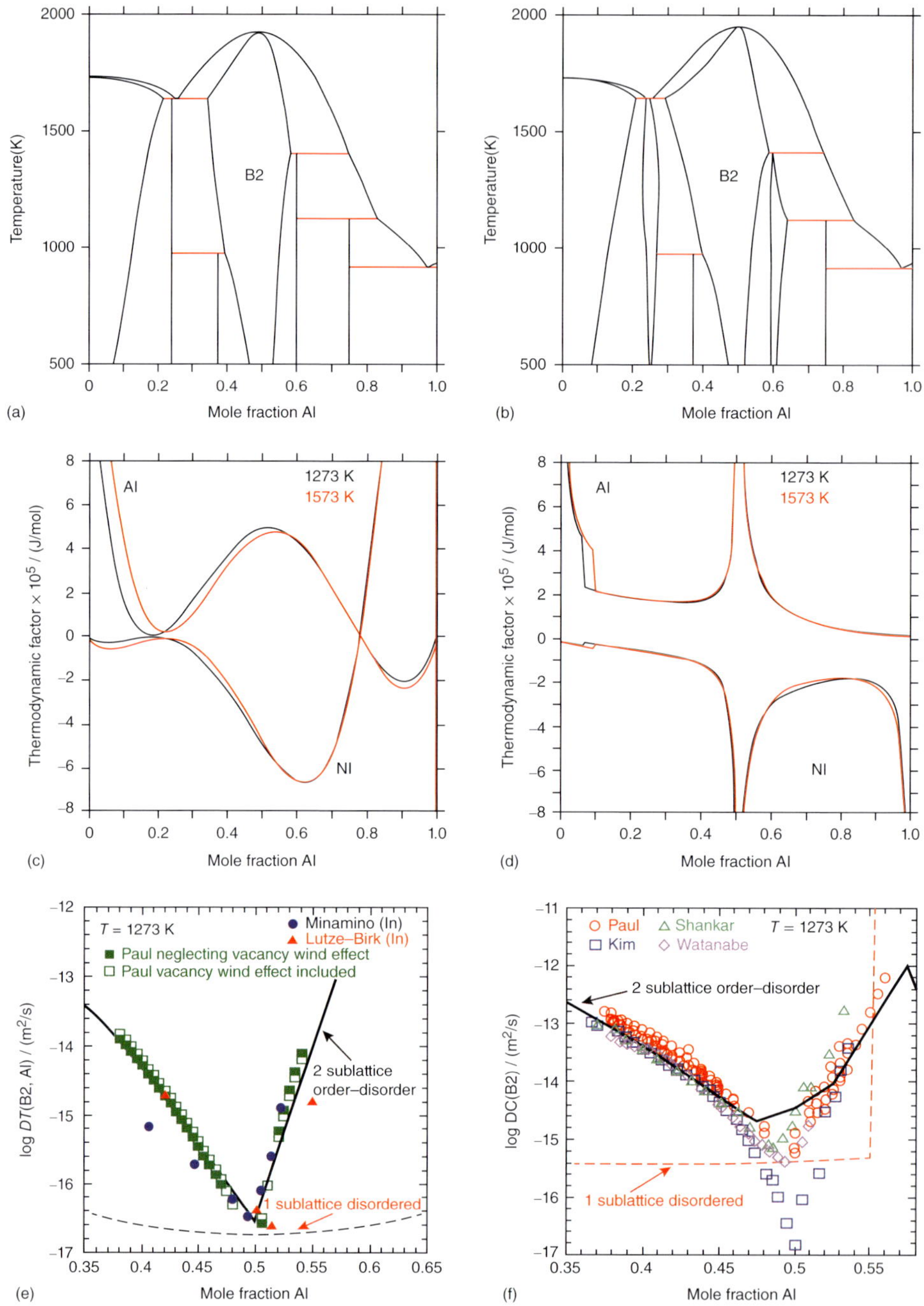

Figure 4.4 Model selection and diffusion modeling using a one-sublattice substitutional model (a,c) and a two-sublattice model considering structural vacancies (b,d) for the B2 phase in the Ni–Al system. (a,b) Phase diagram, (c,d) thermodynamic factor, (e) tracer, and (f) chemical diffusivities using both models. Diffusion data are referenced in Campbell [6].

and density. However, models for the description of these properties within the framework of the CALPHAD method still need to be developed.

For realistic modeling of microstructure evolution and properties, interphase properties are needed in addition to the properties of the individual phases. Examples of the needed interphase quantities are surface tension, interfacial energies, and grain boundary diffusion. The surface tension of liquid alloys can be derived from the thermodynamic quantities of the liquid using Butler's equation. Interfacial energies are important for the simulation of precipitation and coarsening processes. Currently, most simulation tools for these processes use composition-independent interfacial energies due to the lack of sufficient data collection. Although reasonable estimates can be obtained using thermodynamic approaches for composition-dependent interfacial energies, it may be desirable to obtain these quantities from first-principles approaches. Several models are available to evaluate nucleation rates from the thermodynamics and additional input parameters and have been implemented in precipitation calculation software. Grain boundary diffusion has been modeled using an approach similar to the one used for bulk diffusion.

Although the CALPHAD method has great potential for the description of properties beyond thermodynamics and diffusion mobilities, the development of models for use within the CALPHAD framework is still in its infancy.

4.4.11 Limitations and Challenges

The CALPHAD method is well established for the modeling of the phase relationships and diffusion processes in multicomponent, multiphase alloys and has been extensively used for alloy and process development. Although the method is powerful in predicting the properties of systems with more than three components, it is not a truly predictive method. The prediction of the properties of a multicomponent system is based on the extrapolation of the descriptions of the constituent binary and ternary systems. Information on the phases, phase boundaries, and thermochemical quantities is needed for the development of the CALPHAD descriptions of these constituent systems. These data are mostly experimental data, which are supplemented by data from first-principles calculations (*ab initio*), especially where they are lacking or cannot be determined by experiments, such as the quantities of the hypothetical end-member phases needed for the CEF. If the occurrence of a phase is not known from experiments or predictions from first-principles calculations, it cannot be predicted with the CALPHAD method. However, advances in first-principles calculations and classic atomistic simulations have made it possible to obtain most of the needed information.

The two major challenges the CALPHAD method is facing are the development of models and databases for the description of other phase-based and interphase properties and the development of verification and validation methods.

Development of models and databases for other properties is essential for the application of the CALPHAD method beyond its initial use for evaluating whether a candidate alloy composition would produce the desired phases and avoid unwanted phases at the temperatures of interest and for selecting a temperature regime for processing. Combining results from CALPHAD calculations with models for other important alloy characteristics, such as creep resistance, castability, and cost, allows for computational testing of thousands of candidate alloy compositions and the selection of a few promising candidate alloys for experimental evaluation. The CALPHAD calculations can be implemented in systematic searches using, for example, a genetic algorithm or mesh-adaptive direct search algorithm to find new candidate alloy compositions.

CALPHAD calculations using the same models and parameters and the same external conditions with different software should give the same results if the criteria for convergence are similar. It should be noted here that uncertainty of results from the CALPHAD calculations depends on the databases used and how reliably the original experimental data used in the development of these databases are reproduced. Currently, no method for uncertainty quantification of the results from a CALPHAD database is available. The reliability of the results is most commonly expressed by plotting experimental data versus calculated results under the same conditions. If experimental information is lacking and data are supplemented with results for DFT calculations, differences of a few kilojoules per mole in these DFT energies can produce topologically different phase diagrams with phases that appear/disappear as stable ones depending on the input data set used. Similar effects can be obtained if different sets of optimized CALPHAD parameters are obtained from experimental data after weighting individual data sets differently. Mathematical methodologies to better evaluate such uncertainties are desirable but challenging (if at all possible). The first logical step toward verification of CALPHAD descriptions is to perform a sensitivity analysis of the model parameters that are being adjusted during the optimization of the thermodynamic description of a binary or ternary system. In addition to uncertainty quantification, it is desirable to develop a set of test cases to ensure that the results are consistent with accepted codes. Although the results from calculations should agree once the convergence criteria are fulfilled, it is not always guaranteed that the true set of equilibrium phases has been found. Development of protocols for the validation of CALPHAD software would provide additional assurance of the reliability of the results from CALPHAD calculations.

4.5
Deriving Thermodynamics from Ab Initio Calculations

Ab initio calculation of various thermodynamic data has developed impressively in recent years. More and more scientists are able to perform calculations, and high performance computing facilities have been considerably increased. This means that an incredible number of calculated data have become widely available recently.

For thermodynamic modeling, in particular done in the frame of the CALPHAD method, calculated data can be used in the same way as experimental information. Among first-principles approaches, the DFT is able to find the ground-state energy of a given compound with known structure and composition, which allows one to calculate its heat of formation at 0 K. Moreover, with additional statistically derived approaches, the mixing enthalpy of disordered solid solution can be calculated. Furthermore, the phonon calculation yields vibrational contribution (entropy, heat capacity, etc.) and allows one to estimate energetic stability at a finite temperature.

After a short introduction of the DFT methodology, the different thermodynamic properties that can be obtained by *ab initio* calculations will be detailed. More about DFT and other models at the electronic, atomistic, and mesoscopic scales is found in Chapter 5.

4.5.1 DFT Methodology

DFT is among the most widely used methods in condensed-matter physics and computational chemistry and has played a significant role in the development of new materials. This is not only due to the increasing performance of supercomputers but also, and perhaps mainly, due to the availability of relatively inexpensive high-performance workstations.

Consider an ordered and periodic compound with a known crystallographic structure and element distribution in the lattice. Its electronic structure results from a many-electron problem (n ions, N electrons) which cannot be solved analytically. A first theorem states that the ground-state properties of this system are uniquely determined by an electron density ρ that in turn depends on only three spatial coordinates. Basically, the theorem reduces the many-body problem of N electrons with $3N$ spatial coordinates to only three variables. This is achieved through the use of functionals of the electron density. The ground-state energy is then obtained by minimizing $E[\rho]$:

$$E[\rho] = \underbrace{K[\rho] + U[\rho]}_{F[\rho]} + \int V_{\text{ext}}(\boldsymbol{r})\rho(\boldsymbol{r})\mathrm{d}\boldsymbol{r},$$

where $K[\rho]$ is the kinetic energy, $U[\rho]$ is the electron–electron interaction, and V_{ext} is the electron–ion interaction.

Because of the electron–electron interaction, $K[\rho]$ cannot be directly expressed. It has, therefore, been suggested that one should first, as an estimation, consider an auxiliary noninteracting system in order to solve the so-called mono-orbital Kohn–Sham equation:

$$\left[-\frac{h^2}{2m_{\text{e}}}\nabla + V_{\text{ext}} + V_{\text{H}} + V_{\text{XC}}\right]\psi_i = \varepsilon_i\psi_i,$$

where V_{H} is the Hartree contribution describing the electron–electron Coulomb repulsion and V_{XC} is the so-called exchange-correlation potential. All the unknown many-body electron–electron interactions are now gathered in this latter term. This means that the accuracy of the DFT results heavily depends on the choice of the XC functional. V_{XC} can be expressed by the LDA (local density approximation), where it is assimilated to the energy of a homogeneous electron gas of a density $\rho(r)$. The GGA (generalized gradient approximation), which includes the density and its first derivative, is considered to be better for systems including strong variation of electronic density.

The Kohn–Sham equations can be solved by a self-consistent resolution done automatically in many available codes. Some of these codes, such as WIEN2K, take an all-electron description into account. These are called "full-potential" codes. Other codes, such as VASP, Quantum-Espresso, or Abinit, use "pseudopotentials." The Kohn–Sham equations are solved using the only valence electrons in a pseudopotential generated by the frozen core electrons and the nucleus contributions. By using plane waves as the basis set, a pseudopotential approach can be very efficient and fast.

4.5.2 Heat of Formation

Enthalpies of formation of a given compound are easily obtained from the difference between its calculated total energy and the sum of the weighted energies of the constituting pure elements in their stable reference states. Where experimental data are lacking, these differences are of great interest for CALPHAD modeling and other computational techniques that depend on the availability of thermodynamic data. Besides carrying out one's own calculations, it is also possible to use available databases with stored calculated results (see, e.g., http://oqmd.org/, https://materialsproject.org/, or http://www.aflowlib.org/).

As an example, the experimental enthalpies of formation, the values assessed after thermodynamic modeling, and the calculated heats of formation of ordered compounds in Mo–Pt system are compared in Figure 4.5. Note the tight fit between calculated values and experimental measurements, especially in the case of compounds close to their stoichiometric composition ($MoPt_2$, $B19$). But the figure also shows that the calculated values and the experimental measurements differ for compounds that have a significant nonstoichiometry accommodated by substitution or vacancy mechanism ($A15$, $D0_{19}$). For these compounds, the high-throughput approach used with large databases should be considered with care. As an alternative, one can simulate defects in the structure to obtain both the right composition and the corresponding enthalpy for these compounds. This can be achieved by using different approaches such as supercells or SQS. One advantage of the high-throughput approach can also be illustrated with the figure, namely, that one can obtain predictions of stable ordered structure at low temperature, here on the Pt-rich side ($MoNi_4$, Pt_8Ti).

Note that if the Kopp–Neumann rule applies (the heat capacity of a compound is an average of the heat capacity of the constituting elements), the enthalpies of

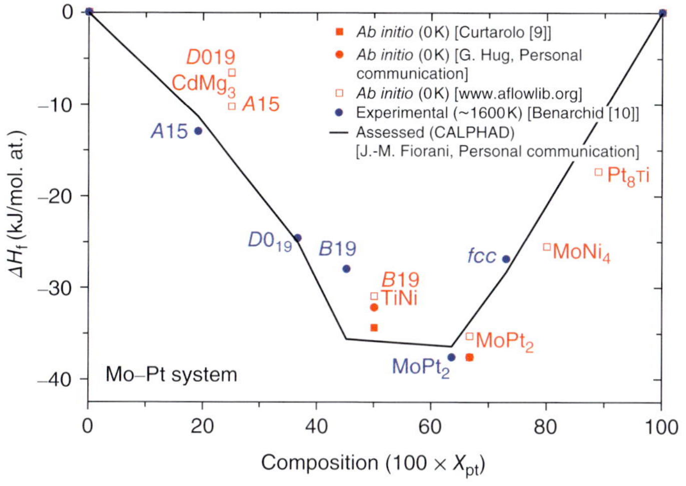

Figure 4.5 Experimental, thermodynamically assessed, and DFT-calculated enthalpies of formation in Mo–Pt system (mole refers to 1 mole of atoms).

formation of compounds are temperature independent and the enthalpies of formation calculated at 0 K are still valid at high temperatures.

Finally, one of the most interesting contributions of DFT to CALPHAD modeling is the possibility to estimate the enthalpy of formation of metastable compounds (end members in the sublattice model) for which empirical values are missing.

4.5.3 Mixing Enthalpy

Contrary to what has been described earlier, not only ordered compounds but also disordered solutions can be calculated using DFT. Useful data such as enthalpies of mixing can be estimated by the cluster expansion method (CEM) or coherent potential approximation (CPA). A simpler and more efficient approach has been developed in recent years: the special quasirandom structure (SQS) method. This consists of generating supercells with a limited number of atoms presenting neighbor correlation functions as close as possible to the ones in a fully random structure. For given compositions in the simplest structures (bcc, fcc, hcp), binary (16 atoms) and ternary (24 or 32 atoms), structures have already been generated and are available in the literature. For other compositions or other structures, dedicated tools are available to generate the SQS supercells (MCSQS in ATAT package).

As an example, Figure 4.6 shows the calculated mixing enthalpies in the bcc phase for two different systems (Mo–Nb and Cr–Fe) by CPA and SQS compared with experimental data. Close agreement between calculated values and experimental measurements can be observed in both cases whatever the sign of the mixing enthalpy.

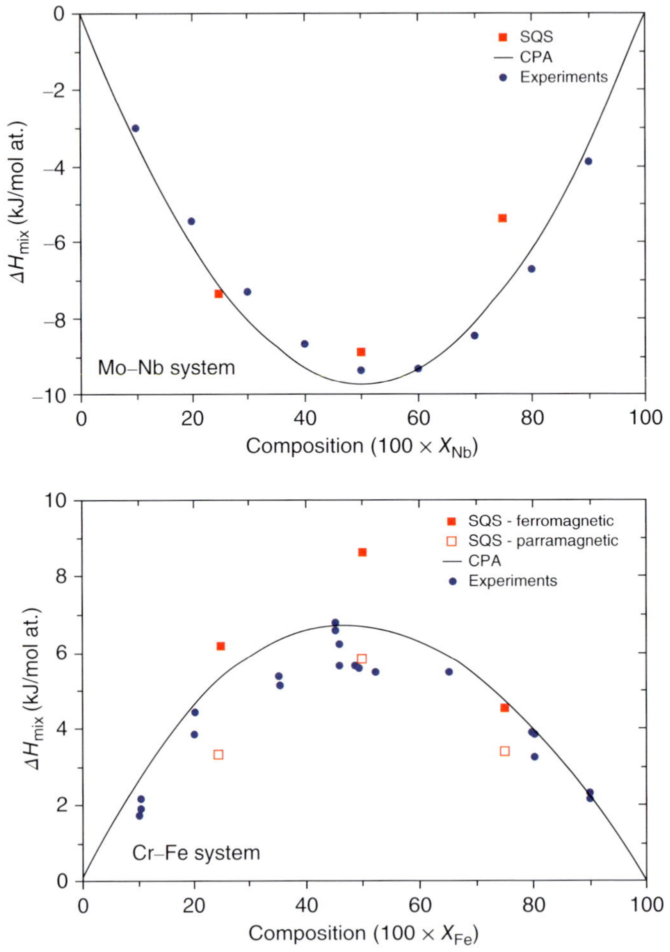

Figure 4.6 Experimental and calculated enthalpy of mixing in Mo–Nb and Cr–Fe bcc structure (redrawn from [11]).

4.5.4
Lattice Vibrations

DFT also allows for the calculation of force matrix in distorted structures, after slight atom movements have been imposed (Hellmann–Feynman theorem). This may be applied to the calculation of elastic or vibrational properties.

The phonon calculation can be numerically performed using different approaches: by perturbation theory (DFPT, density functional perturbation theory)/linear response methods or the supercell approach/direct method.

Two kinds of approximation may be used. In the harmonic approximation, the free energy is estimated at a constant volume. Its first derivative gives the vibrational entropy and its second derivative the heat capacity at constant volume. Using the quasiharmonic approximation, the free energy is expressed as a function of volume and allows one to obtain the C_p, thermal expansion, and all the elastic moduli as a function of temperature.

Different codes are available to perform phonon calculations: phonon, phonopy, and phon.

Figure 4.7 presents a comparison of calculated and experimental elastic constants and bulk modulus as a function of temperature. Both data are important for a number of applications in ICME, in particular the extrapolations to temperatures where no experimental data can be measured. On the other hand, both C_p and vibrational entropy may be used in the frame of CALPHAD modeling.

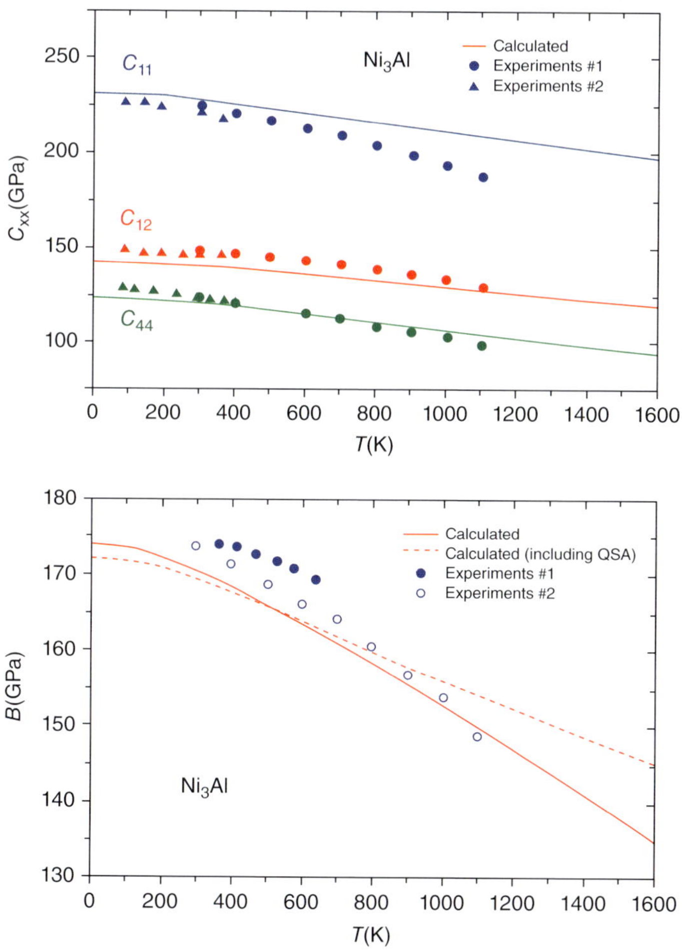

Figure 4.7 Experimental and calculated elastic constants and bulk modulus of Ni_3Al as a function of temperature (redrawn from [12]).

4.6
Use of Thermodynamics at Larger Scales

Thermodynamics is the fundamental basis for essentially all phenomenological theories of materials, as well as mesoscale and continuum computational models of materials processes including phase transformations, chemical reactions, electronic and atomic/ionic transport, and microstructure evolution. A complete thermodynamic description of a system, that is, a fundamental equation of thermodynamics, contains all the thermodynamic information of the system including not only the energies or free energies of all possible thermodynamic states but also all the equations of states as well as all the linear properties of the material such as heat capacity, thermal expansion coefficients, compressibility, and so on. An example is the internal energy as a function of all extensive parameters such as entropy, volume, or strain, number of moles of chemical species, surface/interface area, polarization, charge, magnetization, etc., $U(S, V$ or $e, N1, N2, \ldots, Nn, A, Q, P, M, \ldots)$. A fundamental equation more often used in practice is the Gibbs free energy as a function of temperature, pressure, and/or stress and extensive variables such as the number of moles of each species in the system, amount of surface area, and others, that is, $G(T, p, s, Ni, A, \ldots)$. However, essentially all existing thermodynamic databases are mainly comprised of the Gibbs free energies of individual homogeneous phases as a function of temperature and composition under ambient or zero pressure while ignoring all other thermodynamic parameters. Therefore, the discussion on the use of thermodynamics in this section will tacitly assume that thermodynamic information means $G(T, x1, x2, \ldots xn)$ for a multicomponent single-phase system. Although the main application of such thermodynamic functions is to obtain the phase equilibria of a given materials system, that is, the equilibrium states of the system at different temperatures and compositions, this section focuses on how such thermodynamic information can be used in theory and modeling of microstructure development. In addition to the Gibbs free energy function, this section also outlines the additional thermodynamic properties that will be needed for a more complete description of microstructure stability and evolution, for example, in computational microstructure models.

A microstructure refers to an inhomogeneous distribution of structural features that include phases of different compositions and/or crystal structures, grains of different orientations, domains of different structural variants or magnetizations, as well as structural defects such as interphase boundaries, grain boundaries, domain walls, cracks, surfaces, and dislocations. For a more detailed discussion of the term microstructure, refer to Chapter 3. The thermodynamic stability of a microstructure is determined by the total Gibbs free energy of a system, G, which may include the bulk chemical free energy density of each individual phase, the energies of defects, such as grain boundaries, surfaces, interfaces, and dislocations, as well as the electrostatic, elastic, and magnetic interaction energies. In a sharp-interface description, that is, assuming the interfaces separating different phases or different domains within the same phase are mathematically sharp, a microstructure may be considered as a composite of individual single crystallites of different crystal structures/orientations/compositions and the interfaces between them. The total Gibbs free energy of a microstructure in this sharp-interface description can be written as

$$G = \sum_{i=1}^{p} V_i g_i \sum_{i=1}^{q} S_i \gamma_i + \sum_{i=1}^{m} L_i E_i^d + E_{\text{elec}} + E_{\text{elast}} + E_{\text{mag}}, \tag{4.17}$$

where g_i is the bulk chemical free energy density of ith phase with volume V_i, γ_i is the specific interfacial energy of ith interface (grain boundary, surface, or interphase boundary) with interfacial area S_i, E_i^d is the dislocation core energy per unit length of ith dislocation with length L_i, and E_{elec}, E_{elast}, and E_{mag} are the total electrostatic, elastic, and magnetic energies, respectively. The molar Gibbs free energy (or chemical potential) as a function of temperature and composition for a phase from CALPHAD types of databases can be directly converted to the free energy per unit volume.

In a diffuse interface or phase field description, the total free energy of an inhomogeneous microstructure in Eq. (4.17) is replaced by

$$F = \int_V f_l^o \, \mathrm{d}V + \int_V [f_l - f_l^o + f_g] \mathrm{d}V + \int_V f_e \, \mathrm{d}V, \tag{4.18}$$

where f_l^o and f_l are the equilibrium and nonequilibrium local bulk chemical free energy densities that are functions of composition or phase field variables, f_g is the gradient energy density, and f_e represents the total energy due to elastic, electrostatic, and magnetic long-range interactions.

Microstructure evolution takes place to reduce its total Gibbs free energy. The driving force could be the reduction in the bulk chemical free energy (the first term in Eq. 4.17), leading to the phase changes or transformations or chemical reactions. A phase transformation may lead to a change in the number of phases, for example, a phase separation or decomposition reaction, or to a simple change of crystal structure for one of the phases to another with lower bulk chemical free energy density. A chemical reaction results in a change in chemical components and thus compositions. Therefore, the Gibbs free energies from thermodynamic databases can be used to determine the relative stability of possible thermodynamic states of a material system, and thus can be used quantify the thermodynamic driving forces for material processes such as phase transformations or chemical reactions under a given set of processing conditions.

The driving force for microstructure evolution can also be the reduction in the total interfacial energy (the second term in Eq. 4.17), leading to microstructure coarsening, that is, the average spatial scale of domains, particles, or grains in a microstructure increases as a function of time. In this case, the thermodynamic data necessary to model microstructure coarsening is the interfacial energy – grain boundary energy or interphase boundary energy or surface energy. Interfacial energy and its anisotropy are not only important in determining the morphologies of a particle, but also the driving force for microstructure coarsening. At finite temperatures, the change in grain boundary energy with

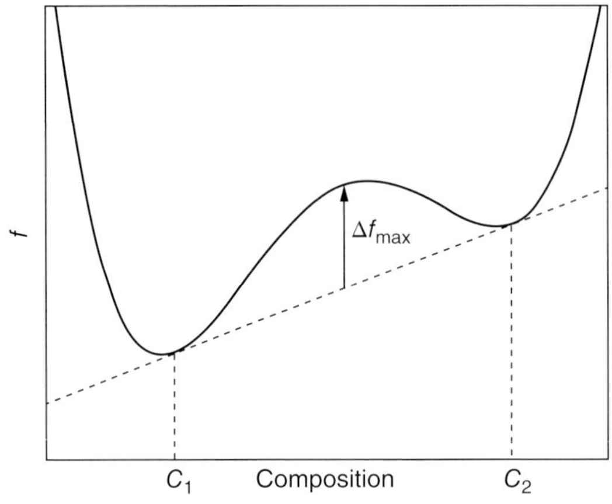

Figure 4.8 The schematic of free energy as a function of composition. Δf_{max} is the maximum difference between the nonequilibrium free energy and the equilibrium free energy (the dashed common tangent).

temperature and the entropy contribution, mainly the vibrational entropy contribution, has to be included to obtain the interfacial free energy as a function of temperature. One useful approach to estimate interfacial energy is to use Gibbs free energy of a homogeneous system through the following expression:

$$\sigma = \frac{1}{2} l \Delta f_{\max}, \tag{4.19}$$

where l is the interfacial width and $\Delta f_{\max}$ is the maximum difference between the nonequilibrium free energy as a function of composition and the equilibrium free energy of a two-phase mixture (Figure 4.8).

In a solid microstructure, there can be local stresses generated, which arise from strains produced due to thermal expansion coefficient mismatch between different phases in a microstructure or the lattice mismatch between new phases of a phase transformation and the parent phase or the lattice match among the different domains of the new phase. Different regions in a solid microstructure impose mechanical constraint on each other, and thus when the lattice parameters change in any region due to temperature or composition variations or due to a structural phase transformation, local elastic strains and stresses will develop. Microstructure will evolve to relax the total elastic strain energy and thus the total Gibbs free energy. The microstructure will also evolve under an applied stress field. For example, a ferroelastic domain wall, also called a twin wall, can be moved by an applied stress. Under a mechanical load, dislocations may also be generated and moved, that is, a system may undergo plastic deformation, to release the strain or strain energy in a crystal. To model the strain- or stress-field microstructure evolution, the lattice parameters of individual phases as a function of temperature and composition is an important piece of information. Similar to the temperature dependence of lattice parameter of a material characterized by its thermal expansion coefficient, the composition dependence of lattice parameter is typically described by a composition expansion coefficient, $e_o = 1/a(\mathrm{d}a/\mathrm{d}c)$.

A constant or composition-independent e_o is typically referred to as the Vegard's law.

Microstructure changes can be driven by the reduction in electrostatic or magnetostatic interaction energies and the corresponding electric and magnetic fields. To incorporate these electric and magnetic contributions, the electric and magnetic properties will be necessary to compute the corresponding thermodynamic contributions, that is, the electrostatic energy and the magnetic energy.

In summary, in order to predict the microstructure evolution of a materials system, it requires the information on the thermodynamic driving forces (Eq. 4.17) determined from the bulk free energy density for an individual phase, the interfacial (grain boundary, surface, and interphase boundary) energies, and lattice parameters and elastic constants of individual phases for obtaining the elastic energy density, as well as electrical and magnetic properties if electrostatics and magnetic materials are involved.

4.7
Applications and Success Stories

Until about the end of the first half of the 20th century, process and materials development had been based mostly on the knowledge of trends, basic concepts, and extensive experimental programs. This route, as previously discussed elsewhere in this book, involves some intuition on the outcome of the process. However, several processes and materials defy, to some extent, the "extrapolations" based on intuition. The development of the argon–oxygen decarburization (AOD) process in the 1970s [13, 14] and the development of the Sandvik-owned SAF 2507 duplex stainless steel in 1983 [15] can be considered milestones in the use of modeling to support material and process development. Since then, computational thermodynamics has been an essential part of the toolbox in the design of stainless and specialty steel [16–20]. Considerable progress has been made in process development by coupling thermodynamic models for liquid oxide slags and metal.

So-called inclusion engineering – engineered control of nonmetallic inclusions in alloys – has emerged as a success story of the application of computational thermodynamics in steel processing by making it possible to design a proper slag for processing a given steel. For example, computational thermodynamics has been used to eliminate aluminates and spinel inclusions, and to minimize the content of alumina inclusions in bearing steels (such as 52100 or 100Cr6) [21]. This involves simulating processing with slags of varying basicity (CaO/SiO_2), alumina and MgO contents, and evaluating the effects on the steel content of Ca, Mg, Al, and O, for a given basic composition of Cr, C, Si, and Mn. Figure 4.9 shows the calculated effect of varying the MgO content of a slag of constant basicity on the content of Al, O, and Mg present in the steel. After performing equilibrium calculations for a range of compositions, a slag was selected and the actual results were compared with the calculated values. The agreement with respect to Al and O content is shown in Figure 4.10. Similar calculations are now routine for spring steel

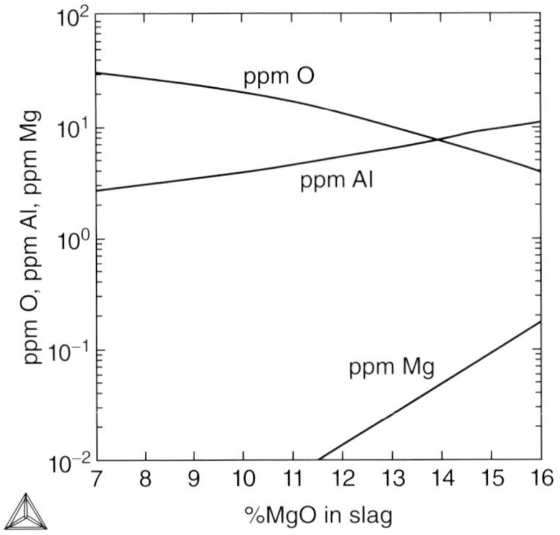

Figure 4.9 Effect of %MgO in slag with constant $\%Al_2O_3$ and B = $\%CaO/SiO_2$ on Al, O and Mg content in bearing steel in equilibrium with slag. Calculated with Thermo-Calc® and SLAG2 database at 1540°C.

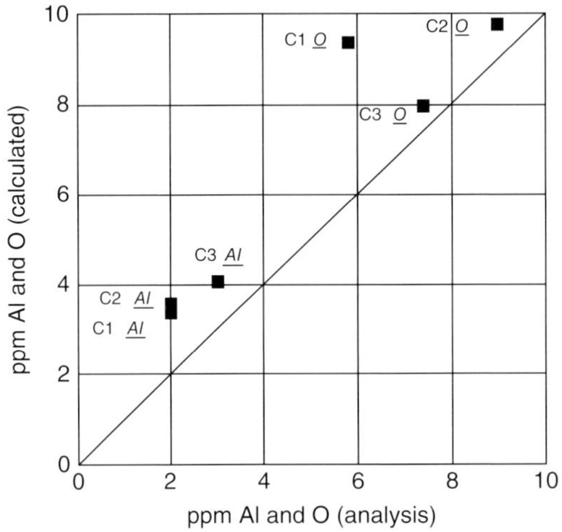

Figure 4.10 Comparison of calculated and measured Al and O content in bearing steel after ladle furnace processing. Calculated with Thermo-Calc® and SLAG2 database.

and tire cord steels [22, 23]. The counterintuitive result that the basicity should be kept at lower values than usual for ladle metallurgy is an important result.

Similarly, the calculations of the amount of calcium to be added to liquid steel to prevent valve clogging and guarantee inclusion modification are now widely used [24–26], eliminating the trial-and-error approach that was used until the end of the last century. Many reviews of various aspects of the application of the technique in steelmaking are available [27–29]. Since many variables influence the

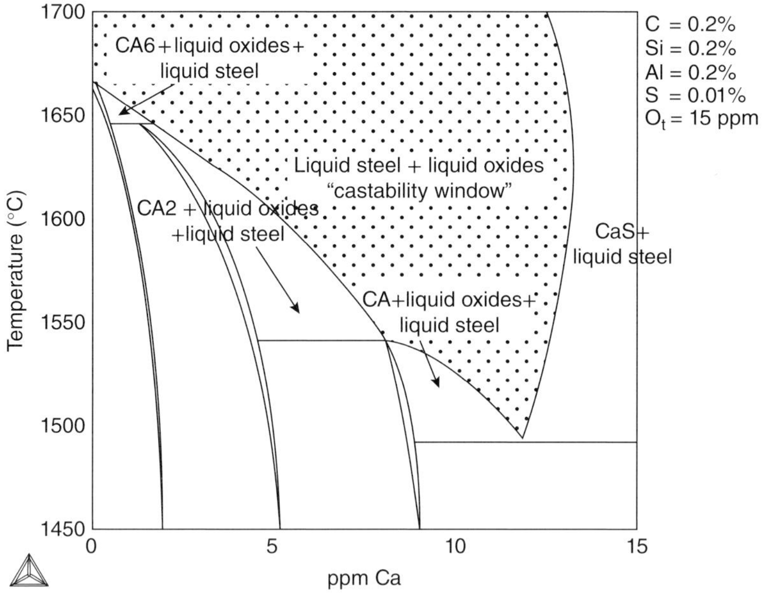

Figure 4.11 Temperature versus calcium content diagram for a steel with the composition listed on the right top. The region where liquid inclusions are formed (the so-called castability window) is clearly visible. The resemblance to the $CaO-Al_2O_3$ diagram makes it easier for the user to understand the results. Calculated with Thermo-Calc® and SLAG3 database.

clogging phenomena, various two-dimensional views are used to visualize the so-called castability window, that is, the range of compositions in which low-melting-point inclusions will be formed. Figure 4.11 is a particularly interesting way of presenting the data, since it strongly resembles the topology of the $CaO-Al_2O_3$ pseudobinary diagram, classically used in the qualitative interpretation of clogging and inclusion modification. Currently, one of the challenges in inclusion modification and clogging prevention is the control of the process effectiveness during actual steelmaking so that corrections and process changes can be immediately performed. It was recently shown that by measuring soluble oxygen (which is an easy and rapid measurement on the shop floor with the use of disposable electrolytic cells), one can check the effectiveness of the process, even for Al-killed steels [30]. Figure 4.12 shows the results of several heats controlled in this way compared with the castability window calculated for the steel in question. The good agreement is evident.

Alloy design has also greatly benefited from the use of computational thermodynamics. This is true also in the case of the development and manufacture of "standard" products. The benefits of using microalloying in various steels have been optimized by computational thermodynamic techniques [31–33], including simulations of precipitation in nonequilibrium conditions [34]. For example, while designing a multiphase flat steel for automotive applications, Murari *et al.* [35] first simulated the possible effects of the main substitutional elements and of

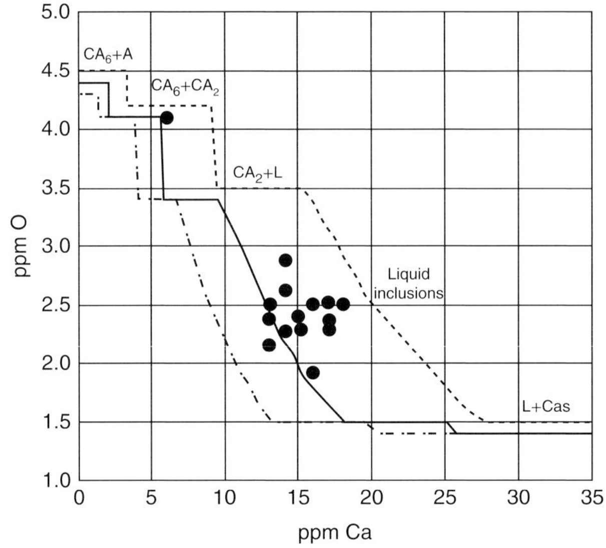

Figure 4.12 Calculated soluble oxygen in steel containing 0.025%Al, 0.01%S, and various values of Ot (15, 20 and 30ppm, left to right) as a function of Ca content at 1540°C, compared with experimental measurements. Non-metallic phases present indicated for each region of composition. Calculated with Thermo-Calc® and SLAG3 database.

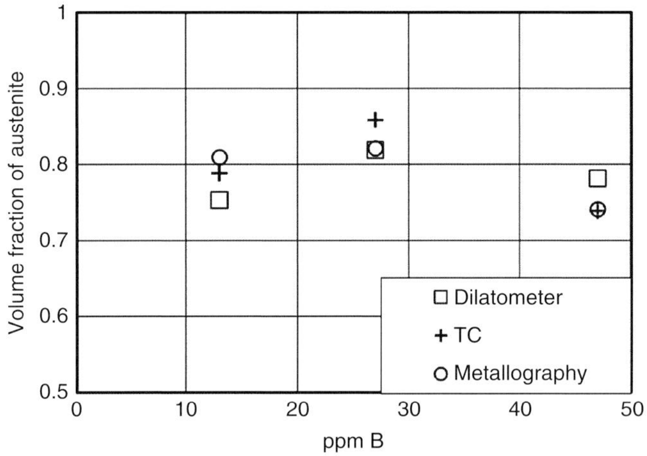

Figure 4.13 Comparison of the effect of boron content the volume fraction of austenite at an intercritical heat treatment temperature of 820°C obtained by three different methods for a commercial multiphase C, Mn, Si, Al, Nb, Ti, N, and B steel. Calculations (TC) performed with Thermo-Calc® and TCFE6 database.

the interstitials carbon and nitrogen on the volume fraction and carbon content in the austenite present during intercritical treatment. To optimize the mechanical properties, the effects of boron additions were then studied. Figure 4.13 shows the effect of the boron content on the austenite volume fraction at the intercritical temperature treatment of 820°C. The effect is shown as it was obtained by the three methods of computational thermodynamics, dilatometry, and metallography of quenched specimens. The agreement of the results obtained by these methods is promising from the point of view of alloy design.

These and other examples of applications of computational thermodynamics can be found in the reports of the Ringberg meetings [22, 36, 37]. They provide clear proof of the maturity of computational thermodynamics as a part of the ICME toolbox (Figure 4.14).

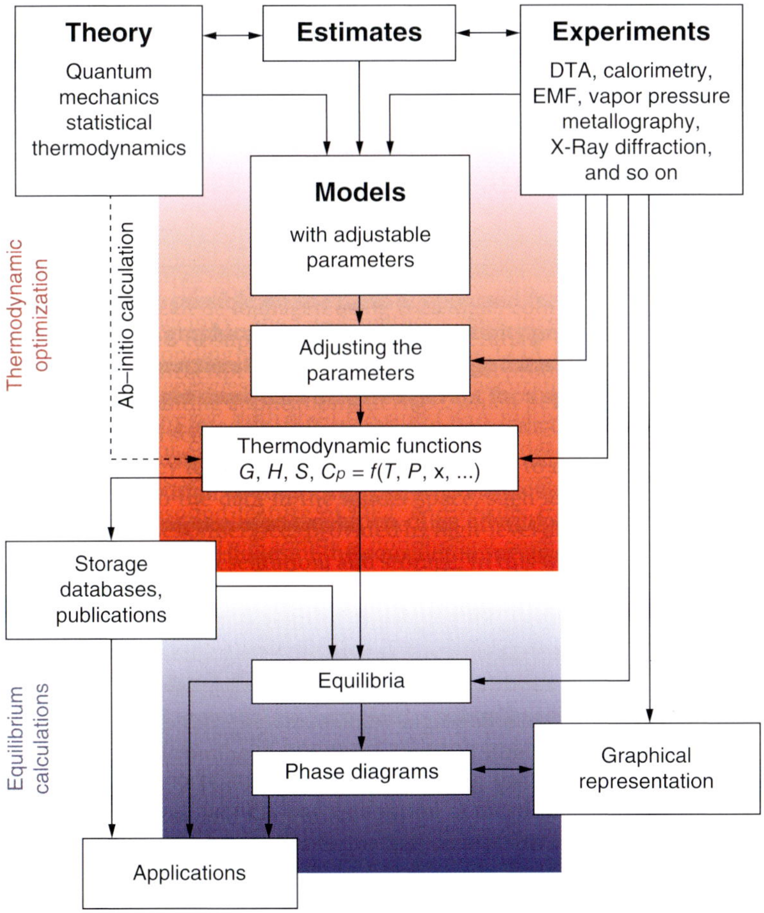

Figure 4.14 Diagram illustrating the CALPHAD methodology. By integrating the results of theoretical calculations and experiments, the adjustable model parameters are optimized and stored in databases ("Thermodynamic Optimization"). Using these databases, reliable phase diagrams can then be calculated ("Equilibrium Calculations").

4.8 Software Tools

Table 4.4 lists the software tools and applications available for thermodynamic modeling.

Table 4.4 Software tools and applications available for thermodynamic modeling.

Name of code	Website	Type of code	Short description
Aspen Plus	www.aspentech.com/products/aspen-plus.aspx	Commercial	Chemical process optimization software
CaTCalc	https://staff.aist.go.jp/k.shobu/CaTCalc/	Commercial	Software that determines chemical thermodynamic equilibrium by the Gibbs energy minimization
Chemical Equilibrium with Applications (CEA)	www.grc.nasa.gov/WWW/CEAWeb	Public domain	Calculates chemical equilibrium product concentrations from any set of reactants and determines thermodynamic and transport properties for the product mixture
CHETAH	www.astm.org/BOOKSTORE/PUBS/DS51F.htm	Commercial	A resource for predicting both thermochemical properties and certain reactive chemical hazards associated with a pure chemical, a mixture of chemicals, or a chemical reaction
CP2K	http://www.cp2k.org/	Open source	A program to perform atomistic and molecular simulations of solid state, liquid, molecular, and biological systems
CULGI	www.culgi.com	Commercial	A general-purpose modeling platform including all relevant chemical modeling tools, from molecular representation to continuum level
FactSage	www.factsage.com	Commercial	A fully integrated database computing system in chemical thermodynamics
FluidProp	www.asimptote.nl	Commercial	A package for the calculation of fluid properties implementing advanced thermodynamic and transport models giving accurate and fast predictions

(continued overleaf)

Table 4.4 (*Continued*).

Name of code	Website	Type of code	Short description
Gemini	http://thermodata.online.fr/	Commercial	Software for thermochemical equilibria calculation by minimization of the total Gibbs energy of the system under either constant pressure or volume conditions
GEMS	http://gems.web.psi.ch/	Free for academia and research	Gibbs energy minimization software for geochemical modeling
gPROMS	http://www.psenterprise.com/gproms.html	Commercial	Platform technology for process modeling with a link to thermodynamic libraries
HSC CHEMISTRY	www.outotec.com/en/Products--services/HSC-Chemistry	Commercial	Software designed for various kinds of chemical reactions and equilibria calculations as well as process simulation
JMatPro	www.sentesoftware.co.uk	Commercial	Simulation software that calculates a wide range of materials properties for alloys and is particularly aimed at multicomponent alloys used in industrial practice
LAMMPS Molecular Dynamics Simulator	http://lammps.sandia.gov/	Open source	A classical molecular dynamics code
MatCalc	http://matcalc.tuwien.ac.at/	Commercial	A software project for computer simulation of phase transformations in metallic systems
MegaWatSoft	www.megawatsoft.com	Commercial	Calculation of thermodynamic and transport properties of water and steam, carbon dioxide, ammonia, and humid air
MELTS	http://melts.ofm-research.org/		A software package designed to model phase (mineral, rock, and liquid) relations during melting and crystallization
MODELICA	https://www.modelica.org/		A nonproprietary, object-oriented, equation-based language to conveniently model complex physical systems that may link to thermodynamic libraries

(*continued overleaf*)

Table 4.4 (*Continued*).

Name of code	Website	Type of code	Short description
MTDATA	www.npl.co.uk/science-technology/mathematics-modelling-and-simulation/mtdata	Commercial	A software tool for the calculation of phase equilibria and thermodynamic properties
Multiflash	http://www.kbcat.com/infochem-software/flow-assurance-software-multiflash	Commercial	Multiflash is a comprehensive PVT and physical properties package that allows for modeling and solving the phase behavior of complex mixtures and pure substances
OpenCalphad	www.opencalphad.com	Open source	An informal international collaboration of scientists and researchers interested in the development of high-quality software and databases for thermodynamic calculations for all kinds of applications
Pandat	www.computherm.com	Commercial	An integrated computational tool developed on the basis of the CALPHAD (CALculation of PHAse Diagram) approach for multicomponent phase diagram calculation and materials property simulation
Perple_X	www.perplex.ethz.ch		A thermodynamic calculation package suitable for rapidly creating phase diagrams of all types
Simulis Thermodynamics	www.prosim.net	Commercial	A calculation server for thermophysical properties and phase equilibria calculations on pure components and mixtures
THERMOCALC	www.metamorph.geo.uni-mainz.de/thermocalc		A thermodynamic calculation program that uses an internally consistent thermodynamic data set to undertake thermobarometry and phase diagram calculations for metamorphic rocks

Table 4.4 (Continued).

Name of code	Website	Type of code	Short description
Thermo-Calc	www.thermocalc.com	Commercial	A powerful software package for the calculation of thermodynamic and phase equilibria. In conjunction with suitable thermodynamic databases, assessed using the CALPHAD approach, Thermo-Calc can be used for a wide variety of applications
ThermoData-Engine (TDE)	http://trc.nist.gov/	Commercial	Provides critically evaluated thermodynamic and transport property data on demand using an expert system to extract data from the TRC Source data archive
T-MATS	www.grc.nasa.gov/WWW/cdtb/software/t-mats.html	Open source	A graphical thermodynamic simulation package built in MATLAB/Simulink (The MathWorks, Inc)
VMGSim	http://www.virtualmaterials.com/	Commercial	Process simulator for the oil, gas, and chemical industries

References

1. Lee, J. and Sim, K.J. (2014) General equations of CALPHAD-type thermodynamic description for metallic nanoparticle systems. *Calphad*, **44**, 129–132.
2. Dinsdale, A.T. (1991) SGTE data for pure elements. *Calphad*, **15**, 317–425.
3. Lukas, H.L., Fries, S.G., and Sundman, B. (2007) *Computational Thermodynamics: The CALPHAD Method*, Cambridge University Press, Cambridge.
4. Schmid-Fetzer, R., Andersson, D., Chevalier, P.Y., Eleno, L., Fabrichnaya, O., Kattner, U.R., Sundman, B., Wang, C., Watson, A., Zabdyr, L., and Zinkevich, M. (2007) Assessment techniques, database design and software facilities for thermodynamics and diffusion. *Calphad*, **31**, 38–52.
5. Bale, C.W. *et al.* (2009) FactSage thermochemical software and databases — recent developments. *Calphad*, **33**, 295–311.
6. Campbell, C.E. (2008) Assessment of the diffusion mobilities in the γ' and B2 phases in the Ni-Al-Cr system. *Acta Mater.*, **56**, 4277–4290.
7. Miettinen, J. (2005) Thermodynamic description of the Cu-Ni-Al system at the Cu-Ni side. *Calphad*, **29**, 40–48.
8. Ansara, I., Dupin, N., Lukas, H.L., and Sundman, B. (1997) Thermodynamic assessment of the Al-Ni system. *J. Alloys Compd.*, **247**, 20–30.
9. Curtarolo, S., Morgan, D., and Cederb, G. (2005) Accuracy of ab initio methods in predicting the crystal structures of metals: a review of 80 binary alloys. *Calphad*, **29**, 163–211.
10. Benarchid, M., David, N., Fiorani, J.-M., Vilasi, M., and Benlaharche, T. (2009) Enthalpies of formation of Mo-Pt alloys

by high temperature direct reaction synthesis calorimetry. *J. Chem. Thermodyn.*, **41**, 383–385.

11. Jiang, C., Wolverton, C., Sofo, J., Chen, L.-Q., and Liu, Z.-K. (2004) First-principles study of binary bcc alloys using special quasirandom structures. *Phys. Rev. B*, **69**, doi: 10.1103/physrevb.69.214202.
12. Wang *et al.* (2010) *J. Phys. Condens. Matter*.
13. Krivsky, W.A. (1973) The linde argon-oxygen process for stainless steels: a case study of major innovation in a basic industry. *Metall. Trans.*, **4**, 1439–1447.
14. Andersson, N.Å.I., Tilliander, A., Jonsson, L.T.I., and Jönsson, P.G. (2013) Fundamental decarburisation model of AOD process. *Ironmaking Steelmaking*, **40** (5), 390–397.
15. Widmark, H. (1991) Importance of Mats Hillert's research work on industrial applications. *Scand. J. Metall.*, **20**, 72–78.
16. Calliari, I., Zanesco, M., and Ramous, E. (2006) Influence of isothermal aging on secondary phases precipitation and toughness of a duplex stainless steel SAF 2205. *J. Mater. Sci.*, **41** (22), 7643–7649.
17. Wessman, S., Karlsson, L., Pettersson, R., and Östberg, A. (2012) Computational thermodynamics study of the influence of tungsten in superduplex stainless weld metal. *Weld. World*, **56** (11–12), 79–87.
18. Olson, G.B. (2013) Genomic materials design: the ferrous frontier. *Acta Mater.*, **61** (3), 771–781.
19. Huth, S., Krasokha, N., and Theisen, W. (2009) Development of wear and corrosion resistant cold-work tool steels produced by diffusion alloying. *Wear*, **267** (1–4), 449–457.
20. Farina, A.B. *et al.* (2014) Use of thermodynamic modeling to the process optimization and development of new alloys. *Tecnol. Metal. Mater. Miner.*, **11** (4), 1–8 (in Portuguese).
21. Monnot, J., Heritier, B., and Cogne, J.Y. (1986) Inclusion cleanness and fatigue resistance of rolling bearing steels. *Rev. Metall. - CIT*, **83** (10), 761–770.
22. Costa e Silva, A., Ågren, J., Clavaguera-Mora, M.T., Djurovic, D., Gomez-Acebo, T., Lee, B.-J. *et al.* (2007) Applications of computational thermodynamics — the extension from phase equilibrium to phase transformations and other properties. *Calphad*, **31** (1), 53–74.
23. Holappa, L., Nurmi, S., Louhenkilpi, S., and Antola, T. (2007) Thermodynamic evaluation for inclusion formation in high carbon and spring steels. Clean Steel Conference 2007, Balatonfüred, Hungary.
24. Costa e Silva, A. (2008) Calcium and magnesium thermodynamics in steel and its impacts on secondary steelmaking: a computational thermodynamics approach. *Rev. Metall. - CIT*, **105** (4), 181–193.
25. Holappa, L., Hamallinen, M., Liukkonen, M., and Lind, M. (2003) Thermodynamic examination of inclusion modification and precipitation from calcium treatment to solidified steel. *Ironmaking Steelmaking*, **30** (2), 111–115.
26. Oertel, L. and Costa e Silva, A. (2000) Application of thermodynamic modeling to slag-metal equilibria in steelmaking. *Calphad*, **23** (3–4), 379–391.
27. Jung, I.-H. (2010) Overview of the applications of thermodynamic databases to steelmaking processes. *Calphad*, **34** (3), 332–362.
28. Costa e Silva, A. (2013) Application of computational thermodynamics to steelmaking and processing - keynote Lecture. Proceeding of the 19th IAS Steel Conference, Rosario, Santa Fe, Argentina, IAS, pp. 1–21.
29. Costa e Silva, A. (2006) Thermodynamic aspects of inclusion engineering in steels. *Rare Met.*, **25** (5), 412–419.
30. Mercier, M. (2014) Dynamic CaSi addition. M.Sc. thesis. UFMG, Belo Horizonte, Brazil (in Portuguese).
31. Akamatsu, S., Hasebe, M., Senuma, T., Matsumura, Y., and Akisue, O. (1994) Thermodynamic calculation of solute carbon and nitrogen in Nb and Ti added extra-low Carbon Steel. *ISIJ Int.*, **34** (1), 9–16.
32. Frisk, K. (2008) Thermodynamic modelling of multicomponent cubic Nb, Ti and V carbides/carbonitrides. *Calphad*, **32** (2), 326–337.

33. Inoue, K., Ishikawa, N., Ohnuma, I., Ohtani, H., and Ishida, K. (2001) Calculation of phase equilibria between austenite and (Nb, Ti, V)(C, N) in microalloyed steels. *ISIJ Int.*, **41** (2), 175–182.

34. Carvalho, R.N., Costa e Silva, A., and Barbosa, R. (2008) Precipitation of carbonitrides in steels microalloyed with Nb and Ti during hot rolling of seamless tubes. 3rd International Conference on Thermomechanical Processing of Steels, Padua, Italia, Associazione Italiana di Metallurgia, Rome.

35. Murari, F.D., Costa e Silva, A., and Avillez, R.R. (2011) Effect of Boron on the microstructure and mechanical properties of cold rolled multiphase steels. Program and Abstracts Book CALPHAD 2011, Rio de Janeiro, Brasil, ABM, p. 121.

36. Kattner, U., Erickson, G., Spencer, P., Schalin, M., Schmid-Fetzer, R., Sundman, B. *et al.* (2000) Report of Group 4 & 5: process modeling and applications- 1997 Ringberg Workshop. *Calphad*, **24** (1), 55–94.

37. Agren, J., Hayes, F.H., Höglund, L., Kattner, U., Legendre, B., and Schmid-Fetzer, R. (2002) Applications of computational thermodynamics. *Z. Metallkd.*, **93** (2), 128–142.

Further Reading

Baroni, S., Giannozzi, P., and Testa, A. (1987) Greens function approach to linear response in solids. *Phys. Rev. Lett.*, **58**, 1861. Article (Bar87).

Campbell, C.E., Kattner, U.R., and Liu, Z.-K. (2014) The development of phase-based property data using the CALPHAD method and infrastructure needs. *Integr. Mater. Manuf. Innovation*, **3**, 12.

Dinsdale, A.T. (1991) SGTE data for pure elements. *Calphad*, **15**, 317–425.

Dove, M.T. (1993) *Introduction to Lattice Dynamics*, Cambridge University Press, New York.

Gąsior, W. (2014) Viscosity modelling of binary alloys: comparative studies. *Calphad*, **44**, 119–128.

Gonze, X. and Lee, C. (1997) Dynamical matrices, born effective charges, dielectric permittivity tensors, and interatomic force constants from density-functional perturbation theory. *Phys. Rev. B*, **55**, 10355–10368, Article (Gon97).

Grimvall, G. (1999) *Thermophysical Properties of Materials*, Enlarged and revised edition, Elsevier, Amsterdam.

Kohn, W. (2003) *Nobel Lectures, Chemistry (1996–2000)*, World Scientific Publishing, Singapore, Conference (Kohn).

Lee, J., Lee, J., Tanaka, T., and Mori, H. (2009) In situ atomic-scale observation of melting point suppression in nanometer-sized gold particles. *Nanotechnology*, **20**, 475–706.

Lee, J.G., Mori, H., Lee, J., Tanaka, T., and Penttila, K. (2005) Phase diagrams of nanometer-sized particles in binary systems. *JOM*, **57**, 56–59.

Lee, J., Park, J., and Tanaka, T. (2009) Effects of interaction parameters and melting points of pure metals on the phase diagrams of the binary alloy nanoparticle systems: a classical approach based on the regular solution model. *Calphad*, **33**, 377–381.

Lee, J., Tanaka, T., Lee, J.G., and Mori, H. (2007) Effect of substrates on the melting temperature of gold nanoparticles. *Calphad*, **31**, 105–111.

Lukas, H.L., Fries, S.G., and Sundman, B. (2007) *Computational Thermodynamics: The CALPHAD Method*, Cambridge University Press, Cambridge.

Park, J. and Lee, J. (2008) Phase diagram reassessment of Ag–Au system including size effect. *Calphad*, **32**, 135–141.

Parlinski, K., Li, Z.Q., and Kawazoe, Y. (1997) First-principles determination of the soft mode in cubic ZrO$_2$. *Phys. Rev. Lett.*, **78**, 4063–4066. Article (Par97).

Sanchez, J. (2010) Cluster expansion and the configurational theory of alloys. *Phys. Rev. B*, APS, **81**, 224202, Article (San10).

Sanchez, J.M., Ducastelle, F., and Gratias, D. (1984) Generalized cluster description of multicomponent systems. *Physica A*, **128**, 334–350. Article (San84).

Saunders, N. and Miodownik, A.P. (1998) *CALPHAD, (Calculation of Phase Diagrams): A Comprehensive Guide*, Pergamon, Oxford.

Sim, K. and Lee, J. (2014) Phase stability of Ag–Sn alloy nanoparticles. *J. Alloys Compd.*, **590**, 140–146.

Singh, D.J. and Nordstrom, L. (2006) *Planewaves, Pseudopotentials and the LAPW Method*, Springer.

Sundman, B., Lu, X.-G., and Ohtani, H. (2015) The implementation of an algorithm to calculate thermodynamic equilibria for multi-component systems with non-ideal phases in a free software. *Comput. Mater. Sci.*, **101**, 127–137.

Tanaka, T. and Hara, S. (2001a) Thermodynamic evaluation of binary phase diagrams of small particle systems. *Z. Metallkd.*, **92**, 467–472.

Tanaka, T. and Hara, S. (2001b) Thermodynamic evaluation of nano-particle binary alloy phase diagrams. *Z. Metallkd.*, **92**, 1236–1241.

Tang, C., Sung, Y.-M., and Lee, J. (2012) Nonlinear size-dependent melting of the silica-encapsulated silver nanoparticles. *Appl. Phys. Lett.*, **100**, 201903.

Tester, J.W. and Modell, M.M. (1996) *Thermodynamics and Its Applications*, 3rd edn, Prentice Hall.

Togo, A., Oba, F., and Tanaka, I. (2008) First-principles calculations of the ferroelastic transition between rutile-type and CaCl2-type SiO2 at high pressures. *Phys. Rev. B*, **78**, 134106 (Tog08).

van de Walle, A., Asta, M., and Ceder, G. (2002) The alloy theoretic automated toolkit: a user guide. *Calphad*, **26**, 539–553.

van de Walle, A., Tiwary, P., de Jong, M., Olmsted, D., Asta, M., Dick, A., Shin, D., Wang, Y., Chen, L.-Q., and Liu, Z.-K. (2013) Efficient stochastic generation of special quasirandom structures. *Calphad*, **42**, 13–18. Article (Wal13).

Series of papers from Ringberg Workshops on CALPHAD modelling: (a) 1995) Unary data for elements and other end-members of solutions. *Calphad*, **19**, 433–571; (b) 1997) Solution modelling. *Calphad*, **21**, 139–285; (c) 2000) Applications of computational thermodynamics. *Calphad*, **24**, 15–94; (d) 2001) Thermodynamic modelling and applications. *Z. Metallkd.*, **92**, 513–562, **93** (2002) 128–142; (e) 2007) Thermodynamic modeling and first-principles calculations. *Calphad*, **31**, 2–74; (f) 2014) Ringberg unary workshop 2013. *Phys. Status Solidi B*, **251**, 9–129.

Zunger, A., Wei, S.H., Ferreira, L.G., and Bernard, J.E. (1990) Special quasirandom structures. *Phys. Rev. Lett.*, **65**, 353. Article (Zun90).

5
Discrete Models: Down to Atoms and Electrons

Seyed Masood Hafez Haghighat, Ignacio Martin-Bragado, Cláudio M. Lousada, and Pavel A. Korzhavyi

5.1
Overview and Definitions

To adequately describe the macroscopic behavior of material, information about the microscopic mechanisms underlying the macroscopic phenomena is often required. This information is available from experimental and theoretical studies, but nowadays computational modeling has become an indispensable tool for materials research.

This chapter reviews some computational methods and software tools available for modeling materials at small scales where the matter exhibits its discrete nature. Discrete models operate on different levels, ranging from the electronic structure and atomistic level to the mesoscopic level at which the discrete objects are molecules or crystal defects rather than the individual atoms.

The cohesive properties of matter form at the *level of electronic structure*, where the interaction of electrons and nuclei and the fundamental quantum mechanical (QM) laws of their motion are exactly known. However, the exact equations are mathematically tractable only for very small systems, and therefore approximations are introduced in order to reduce the number of variables. In the following, the procedure to reduce the number of variables in the description of a system is referred to as *coarse-graining*.

At the *atomistic level*, the electronic degrees of freedom are effectively integrated out (and converted into the form of effective potentials, interatomic interactions, or "force fields" (FFs)) so that the model considers the motion of atoms or molecules interacting via the FFs. For massive atoms and high temperatures, the QM equations of motion may be replaced with classical Newton's equations to further simplify the description to the level of molecular dynamics (MD). Furthermore, if one is interested in long-term dynamics of the system rather than in describing the exact trajectories on the short timescale of atomic vibrations, one can switch to a statistical mechanical description of the system as it is done in the Monte Carlo (MC) method.

Handbook of Software Solutions for ICME, First Edition. Edited by Georg J. Schmitz and Ulrich Prahl.

At the *mesoscopic level*, one no longer keeps track of every atom in the system, but considers the homogeneous part of the system as some effective medium, while its inhomogeneous parts are considered as discrete objects (e.g., lattice defects in crystals, solute atoms in solvents). These objects form, interact with each other, migrate, transform, and annihilate according to complex rules whose exact form is unknown and must be established from lower-level modeling. The laws of motion in coarse-grained models typically include friction forces (damping related to fluctuations) which are not present in QM or MD models.

At each step of coarse-graining, one gets fewer variables to deal with, but the complexity is transferred into the model formulation. Thus, the FFs and other model parameters are generally undefined; they have to be determined and represented in a form that can carry the essential physical information from lower-level modeling.

In this chapter, we adopt a top-down approach and start our consideration at the mesoscopic level, where the main discrete objects of our interest will be dislocations, because of their role as mediators of plastic deformation. Approaches to dislocation modeling are considered in Section 5.2. The main methods of atomistic modeling, MD and MC, are considered in Section 5.3, and QM electronic structure methods in Section 5.4. The effective potentials, interatomic interactions, and FFs, which serve the purpose of passing the physical information between the different levels of modeling, are briefly discussed in Section 5.5. Section 5.6 lists software packages and integrated modeling platforms that are available for small-scale discrete modeling.

5.2
Discrete and Semidiscrete Mesoscopic Models in Materials Science

Discrete and semidiscrete models can predict the evolution of material microstructure at mesoscale, where it is nontrivial using only nanoscale modeling methods. These models integrate experimental and computational solutions to provide information, parameters, and connecting key mechanisms, through space and time, to macroscale modeling techniques. In this section the method of use for modeling dislocation plasticity in crystalline materials, the so-called discrete dislocation dynamics (DDD), is introduced. MC method is then described by introducing the Metropolis MC, in particular the Potts model, and object kinetic Monte Carlo (OKMC). While the Potts model describes the evolution of microstructure that is not necessarily at atomistic scale (see also Chapter 3), the OKMC considers the evolution of more complex objects as defect clusters in material microstructure.

5.2.1
Discrete Dislocation Dynamics

Dislocation is a crystal line defect that conveys the plasticity of crystalline materials. Atomic displacement due to a dislocation is described by Burgers vector, which is obtained by performing the so-called Burgers circuit around the dislocation line. It may have edge, screw, or mixed character depending on the angle between the Burgers vector and its line direction. Dislocation moves by applying the Peierls–Nabarro threshold stress, which is much smaller than the theoretical shear stress of perfect crystal.

DDD is a simulation technique for modeling the dynamics, reactions, and far-field interactions of dislocations embedded in a linear continuum on the basis of the elasticity theory of dislocations. This technique has now reached a mature level such that complex dislocation mobility and interaction phenomena can be described at high details. Among the different mean-field continuum-based crystal plasticity simulation techniques, DDD model can provide insights specifically into the single dislocation level. DDD simulations have also been used for modeling dislocation interactions with obstacles; however, only limited efforts were made to tackle confined dislocation plasticity due to microstructure interfaces. This technique offers important insights that help to answer the fundamental questions in crystal plasticity, such as the origin of the complex dislocation patterns that emerge during plastic deformation and the relationship between microstructure and mechanical strength.

5.2.1.1 Types of the DDD Techniques

The existing simulation methods in this area classify into two types: (i) the network model and (ii) the nodal model. The difference comes mainly from how to discretize dislocation lines. In the network model, there is a discrete lattice upon which dislocation segments are placed and their mobility law is thus implicit, whereas in the nodal model dislocation segments can take any location and their mobility law is explicit. Nodal model approximates the dislocation curves by series of mixed straight segments of an arbitrary length and orientation, whereas in the network model space is discretized and dislocation segments can have only screw or edge character. Although the nodal model has advantages in the numerical precision, it is much complex in dealing with topological aspects of segments, since it involves more degrees of freedom in segment types as compared to the network model.

In the case of the network model, there has been an attempt to increase numerical accuracy by introducing one more segment type to the edge–screw model, which is called pure-mixed model. This model incorporates additional line directions, that is, ± 60 characters, as mixed property segments. The model aims at an accurate description of a curved dislocation line with a minimum number of segments. Figure 5.1 depicts the discretization methods for a curved dislocation line used in the network model, edge–screw or pure-mixed, and the nodal model.

5.2.1.2 DDD Methodology

The methodology of the DDD simulation consists of two main stages: creating a simulation cell with starting dislocation microstructure, steps 1–2, and modeling dislocation dynamics using classical mechanics and the elasticity theory of dislocations, steps 3–4:

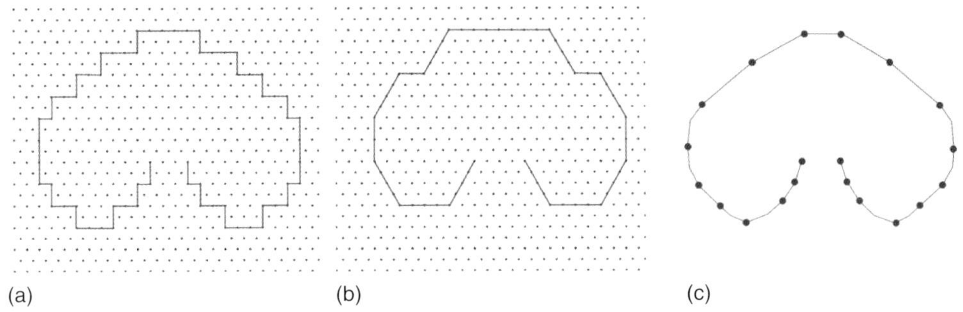

Figure 5.1 Methods of dislocation line discretization in (a) network edge–screw, (b) network pure-mixed, and (c) nodal models. (Reproduced with the permission from the Grenoble Institute of Technology.)

1) A simulation cell containing subcells is created as an embedding crystal for dislocations, where they are treated as purely elastic defects.
2) Dislocation lines are randomly located in the simulation cell and decomposed into straight segments having character from pure edge to pure screw. Each segment is characterized by its Burgers vector, glide plain normal, and line direction.
3) The effective force on each segment is computed from (i) the applied stress, (ii) the stress field of dislocation segments from the same subcell, and (iii) the mean stress field of other dislocations from neighboring subcells. The force vector on each segment is calculated through the Peach–Koehler relationship

$$\boldsymbol{f}_{\mathbf{pk}} = (\sigma_t \cdot \boldsymbol{b}) \times \boldsymbol{u},$$

where f_{pk} is the force per unit length of dislocation, $\boldsymbol{b}$ is the Burgers vector of the segment, $\boldsymbol{u}$ is the segment unity vector, and σ_t is the local total stress tensor due to the three sources (i–iii).
4) The position of a dislocation segment is defined by calculating the speed of the segment, according to simple kinematical relationship, and local events that may occur during its displacement such as direct annihilation, junction formation, cross slip, climb, and interaction with defects. The speed of the segment, $\boldsymbol{v}$, is given by

$$\boldsymbol{f}_{\mathbf{pk}} = \boldsymbol{B} \cdot \boldsymbol{v},$$

where $\boldsymbol{B}$ is the dislocation drag tensor, which depends on the temperature and dislocation character of a particular material. Starting from an initial set of segments at a time t, the segment position at the time $t + \delta t$ is predicted with the kinematical relationship

$$\boldsymbol{r}(t + \delta t) = \mathbf{r}(t) + \delta t \cdot \mathbf{v}(t),$$

where $\boldsymbol{r}$ is the segment position at a specific time. The influence of inertia on the dislocation motion can be neglected since the effective mass of a dislocation is very small. The simulation time step, δt, is approximated from the mean free path of dislocations, which is the mean gliding distance between two dislocation reactions. Definition of the simulation time step is varied depending on the modeling technique. It can be a fixed value that is given at the beginning of the simulation or can be defined at each simulation time step based on the speed of the segments and possible local events.

5.2.1.3 Boundary Conditions

The most common boundary conditions used in DDD simulations is periodic boundary condition (PBC), which accounts for the bulk microstructure and eliminates the unwanted surface effects that are otherwise inevitable due to the finite size of the simulation cell. In the PBC, when a dislocation segment crosses the simulation cell boundary, a replica of it emerges in the simulation cell at the equivalent position in the opposite boundary. There is, however, a known problem associated with PBC regarding the self-reaction of portions of dislocation loops with their replicas after a certain glide path over the extended simulation volume. Two solutions have been employed to control this artifact and prescribe a minimum value for the self-reaction distance of the order of the relevant mean free path. The first is based on the adequate choice of the dimensions of the simulation cell, which is no longer cubic but orthorhombic, and the second is through introducing a nonzero climb mobility for all dislocations to reduce their self-reaction by climbing to other crystallographic glide planes.

Free surface boundary conditions are also used in DDD simulations for studying the scaling effects on the mechanical properties of single crystals. It is used to investigate the sample size effects in single crystalline micropillars, whiskers, thin films, and nanowires that are nontrivial to investigate using only experimental techniques. In these boundary conditions, image forces are applied on the dislocation segments due to free surface, where the segments are able to be annihilated through it.

5.2.1.4 Simulation Inputs

The inputs of DDD simulations are divided into three data groups. The first group provides the initial configuration of the discretized dislocations including all specific characters of the segments. All relevant properties of the dislocation segments such as geometry, Burgers vector, glide plane, connection to the neighboring segments, and any constraints on the segment are given through a file that is created prior to the simulation. The second data group gives the material properties and the parameters of the mobility law of dislocations. The third data group contains the simulation control inputs, which gives the deformation conditions, loading conditions, segments remeshing criteria, and writing periodicity of output files.

5.2.1.5 Applications of the DDD Technique

With the DDD simulations, one can acquire the collective and detailed behavior of dislocations during a deformation process. This technique allows to trace single dislocation movement and its interaction with other dislocations and defects

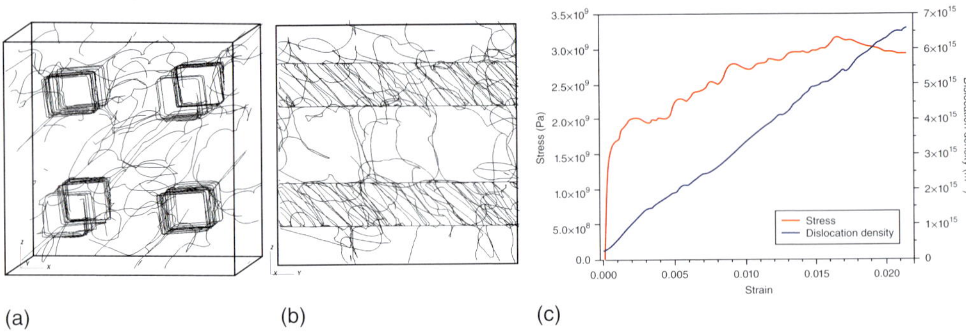

Figure 5.2 (a and b) Dislocation propagations in carbide-strengthened microstructure of a typical high-Mn high-Al steel containing irregular arrangement of carbides in austenitic matrix and (c) stress and dislocation density as a function of plastic strain obtained using the DDD simulations.

and to observe the details of dislocation mechanisms. Using DDD simulations, the evolution of dislocation density, dislocation flux at different regions of the simulation cell, and activation of specific glide system can be examined. Depending on the deformation conditions, the mechanical quantities such as stress, strain, and strain rate are calculated at each time step.

To demonstrate the applicability of the DDD simulations in the study of dislocation mechanisms and the resulting stress–strain behavior, an example simulation was carried out for a typical microstructure of lightweight high-Mn high-Al steels using the ParaDiS code. The microstructure of this alloy contains an irregular arrangement of cuboidal carbides that are oriented along the $\langle 100 \rangle$ directions in austenitic matrix. These alloys do not show a regular particle size and spacing along the three different $\langle 100 \rangle$ directions. This irregular arrangement of the particles can strongly affect the mechanical properties of these alloys. Figure 5.2(a) and (b) shows the deformed microstructure with a few nanometers particle spacing along $y = [0\,1\,0]$ direction and 35 nm spacing along the $[1\,0\,0]$ and $[0\,0\,1]$ directions. It appears that dislocations are propagated within the wide channels, where they can wrap around the carbides without being able to penetrate into the narrow channels between the closely packed carbides. The stress response and dislocation density as a function of strain obtained from the DDD simulation are given in Figure 5.2c.This simulation shows a yield stress of around 1.5 GPa. Strain hardening is also seen in this graph, which relates to the interaction of dislocations with the particles as well as to the formation of junctions.

5.2.1.6 **Drawbacks of the DDD Technique**

Despite rapid development of the DDD technique in the past decade, it has yet some drawbacks that limit the applicability of this technique for rigorous studying of dislocation plasticity. Here are some of those limitations:

1) Boundary conditions are either PBC to account for the bulk of material or simple free surface in cubic or cylindrical simulation cell. It is yet not possible

to define complex boundary conditions similar to what is used in the finite element simulations.

2) Achieving plastic strains more than a few percent is nontrivial, since the calculation time is not linearly related to the plastic strain. With increasing dislocation density, reaching to high plastic strains, the simulation demands significant amount of calculation power.
3) Simulations may only be carried out in single crystals. Despite some efforts on implementing internal microstructure interfaces in the DDD technique, a global solution has not yet been developed to handle microstructure defects such as grain boundaries, twin boundaries, and interphases.
4) Implementing dislocation mobility laws and obtaining relevant drag coefficient, B, for the dislocation segments of different characters are not straightforward. Drag coefficient should be obtained for specific alloy or crystal structure and depends on the temperature and the dislocation character. Drag coefficient is obtained through either experiments, for example, *in situ* transmission electron microscopy (TEM) measurements and mechanical test approximations, or atomistic simulation of dislocation dynamics. In the latter case, developing many-body interatomic potentials for complex alloy compositions is nontrivial, which hinders obtaining the drag coefficient.

5.2.2 Monte Carlo Method

MC method refers to broad range of modeling techniques that solve a problem using randomization and statistical physics. Two types of this method, namely, Metropolis MC and OKMC, are described as mesoscale modeling techniques.

5.2.2.1 Types of the MC Method

Metropolis MC samples space and generates configurations according to the physical-statistical distribution of the states of discrete space. In the particular case of material microstructure, this technique explicitly considers the interactions among the individual grains by discretizing the microstructure in order to construct an image of the grain ensemble in micrometer length scale. Energy rules are then defined to govern the migration of grain boundary segments and to compute the evolution of individual grains as a function of time.

Metropolis MC is based on the thermodynamics of a system and takes into account only the initial and final energy states and thus cannot predict kinetics. An alternative for that is the kinetic Monte Carlo (KMC) method that is used for studying kinetics of slow processes. As compared to the Metropolis MC method, the KMC method has a different approach in the evolution of system. In the Metropolis MC method, we decide whether to accept a move by considering the energy difference between two states, whereas in the KMC the transition rate is achieved based on the energy barrier between the states. It is indeed an alternative to MD simulation at the nanoscale, where modeling of slow thermally activated processes, such as diffusion-controlled mechanisms, may not be done using reasonable simulation power.

The OKMC is developed for studying defect evolution in solid-state at scale of hundreds of nanometers. This technique is based on the KMC approach, which is described in Section 5.3. It models the evolution of an object (defect) in a system using its associated transition rates and actions. This model requires a rate manager to calculate the time evolution and to pick up the performing event and a space manager to control space translations and to carry out neighbors searching and interactions. It has successfully been used for the modeling of defect evolutions, for example, interstitial and vacancy loops, He clusters, and vacancy concentrations, due to irradiation of high-energy particles in metals or for ion implantation in both metals and semiconductors. Extended defects of any shape and capture volume can also be studied using this method. An explicit description of all atoms would, however, limit the system size that can be handled in the OKMC modeling.

5.2.2.2 **Methodology of Potts Model**

We describe here the particular Metropolis MC algorithm, Q-state Potts model, which is used for the simulation of metal microstructure. The Potts model is used for the modeling of recrystallization and grain growth in single-phase metals and also in the presence of second-phase particles (see also Chapter 3). A computer image of polycrystalline microstructures is created by mapping the continuum grain microstructure on a discrete lattice. This is done by dividing the microstructure into small elements and placing the center of these elements in the lattice sites. The algorithm of the simulation for the grain growth is as follows:

1) A two- or three-dimensional $N \times N$ ($\times N$) lattice is generated.
2) A random value, S_i, between 1 and Q is assigned to each lattice site, where S_i namely defines the crystallographic orientation of a specific grain. Grain boundary thus exists between the neighboring sites having different orientations. A site surrounded by other sites of the same orientation is then regarded as grain interior.
3) In order to incorporate second-phase particles into the model, an orientation ($S_i = -1$) was assigned to individual lattice sites equal to the volume fraction of the second-phase particles. This orientation differs from the orientation of the grains.
4) The motion of grain boundary migration is simulated by random selection of a lattice site and changing its crystallographic orientation to another random orientation.
5) The total system energy is associated with grain boundary that is calculated using the following equation:

$$E = \frac{J}{2} \sum_{i=1}^{N} \sum_{j=1}^{nn} (1 - \delta_{S_i S_j}),$$

where J is a positive constant setting the scale of the grain boundary energy per unit length. The first and second summations are calculated over N sites of the system and over nearest neighbors of the particular site i, respectively, and δ_{SiSj} is the Kronecker delta.

6) If the energy change due to the reorientation attempt is less than or equal to zero, it is accepted; otherwise, it is rejected:

$$p(\text{conversion}) = \begin{cases} 1 & \Delta E \leq 0 \\ 0 & \Delta E > 0 \end{cases}$$

This reorientation attempt is not made for a site occupied by a particle.

7) Steps 4–6 are repeated up to the number of lattice sites N that is defined as one Monte Carlo step (MCS).
8) To analyze the evolution of simulated microstructure, an image processing and analyzing software may be used.

5.2.2.3 **Time Conversion**

The results of MC simulations are generated in MCS time intervals. Although some of the considered parameters such as normalized grain size can be assessed by MCS because of its time independency, it is, however, essential to convert the MCS to real time step in order to compare the simulation results with the results of experimental and analytical models. This is done using the rate-controlling mechanisms during the microstructure evolution. In this conversion, the real time in seconds for 1 MCS is obtained by

$$1(\text{MCS}) = \frac{d^2}{6DQ}(\text{s}),$$

where D is the grain boundary diffusion coefficient, d is the simulation lattice constant, and Q is the state of the simulation lattice. To convert MCS to real time, the length conversion is necessary to calculate d value. Comparing the length in the simulated microstructure (in pixel) with that of the experimental microstructure (in microns), 1 pixel was substituted with the appropriate length scale to obtain d as a parameter of the simulation lattice.

5.2.2.4 **Modeling Inputs**

Input parameters of the Potts model are:

1) Material parameters: The grain boundary energy, molar volume, and lattice constant of the material
2) Thermodynamic quantities: The temperature and the gas constant
3) Modeling variables: The starting grain microstructure, the lattice Q-states, and the lattice constant of the model, which is not necessarily equal to that of the material

5.2.2.5 **Applications of the MC**

In the last three decades, the MC technique has particularly been employed in the microstructure design of metallic materials for predicting their microstructure, since it plays an important role in the physical and mechanical properties of materials.

Industrial production of metals consists of different manufacturing process such as solidification, heat treatment, and thermomechanical processes, where the grain deformation, recrystallization, and grain growth induce the final microstructure and consequently the mechanical properties. The MC method is able to model the static and dynamic recrystallization, grain growth, and pinning effects of second-phase particles. This method has also been employed to model the microstructure of thin films, texture formation in electrodeposition, magnetic switching in nanowires, fracture, and failure of materials.

5.3 Atomistic Simulations: Models and Methods

Two atomistic methods are briefly reviewed in this chapter. KMC is first introduced as a technique able to link the nanoscale and mesoscale domains but relying heavily on atomistic parameterizations. Then, the powerful MD technique is explained, and some links between both techniques are also hinted.

5.3.1 Kinetic Monte Carlo

Given a list of mechanisms in a system and the frequency at which these mechanisms happen, KMC will provide the time evolution of the system.

5.3.1.1 Introduction

The KMC algorithm is a nondeterministic method to evolve a system in time. It is able to follow the dynamics of a system, which might be out of equilibrium. It assumes that there are different states in the system and that the transition between those states (mechanisms) depends only on the initial and final states. These transitions are the input parameters of any KMC algorithm and are usually expressed as Arrhenius laws using transition state theory:

$$\nu_{i\to j} = \nu_0 \exp\left(-\frac{E}{k_B T}\right),$$

with ν_0 being the preexponential factor (prefactor) and E the activation energy for the process.

The parameters required to describe a system using KMC are obtained from different sources. Many times, values are obtained from *ab initio* and/or MD simulations. Controlled experiments can also be designed to obtain particular KMC parameters. Finally, it is also possible to do parameterization studies of general experiments and obtain parameters by calibration.

5.3.1.2 The KMC Algorithm

Once all the transition rates $\nu_{i\to j}$, which for simplicity are called ν_k, are known, the KMC direct method algorithms is performed as follows:

1) Compute the total cumulative rates: $R_i = \sum_{k=1}^{i} \nu_k$ for $i = 1, \ldots, N$, N being the total number of transitions in the system.
2) Obtain two random numbers, r and s, in the interval (0,1).
3) Perform the transition j, chosen as $R_{j-1} < rR_N < R_j$.
4) The total simulated time is increased by $\Delta t = \frac{\ln(1/s)}{R_N}$.
5) Update the rate list and recalculate the cumulative rates R_i.
6) Repeat step 1 until the desired total simulated time is achieved.

As can be seen from the previous description, the KMC algorithm is very generic, being concerned about time dependencies only. The needed state and transitions require further definition. These extra definitions make the KMC algorithm highly customizable because, depending on the level of accuracy that is expected from the model, more or less complex states and transitions can be defined allowing different degrees of model description. In many cases, space dependencies can be introduced during such definition.

5.3.1.3 **KMC Drawbacks**

The major KMC drawbacks are as follows:

1) It requires a long parameterization set.
2) It can discover very complex interplays between defined mechanisms, but it cannot discover mechanisms that were not a priori introduced or accounted for.
3) It is a sequential algorithm, complex to scale up using parallelism.

Some of the solutions to the cited problems are as follows:

1) Atomistic simulations (*ab initio*, MD) can be run to provide parameter sets. They can be complemented with values from dedicated experiments or parametric studies.
2) Self-learning KMC codes that run MD simulations to find the different transition paths and energetics have been developed but at the cost of CPU efficiency. Extra mechanisms can always be added ad hoc.
3) Different strategies for parallelization and synchronization of KMC codes are being actively researched. Simple parallel simulations without domain communication are always possible as a way to improve statistics.

5.3.1.4 **KMC Applications**

The applications of KMC to materials sciences are many, but in this work we want to focus on just two: crystal growth and diffusion/reaction systems. Examples of implementations of these mechanisms and their results can be found, for example, in the MMonCa Open Source distribution.

5.3.1.4.1 **Crystal Growth**

The simulation of crystal growth is a typical KMC application (Figure 5.3). Growth is simulated by creating different mechanisms that expand an initially existing

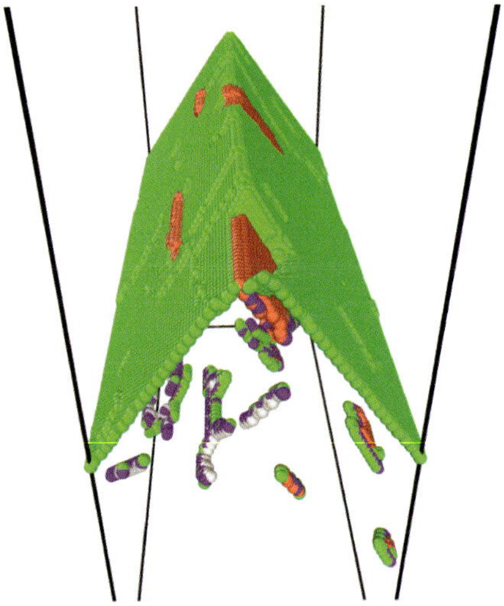

Figure 5.3 Kinetic Monte Carlo simulation of the 3D interface between growing crystalline silicon (bottom) at the expense of amorphous one (top) simulated with MMonCa. The material at the sides (left and right) does not provide a template for recrystallization, forcing the arrow-shaped interface. Green atoms are substrate ones, while red islands simulate nanotwins. Some defects produced by the twins can be seen in the crystalline phase represented with purple–white atoms. Atoms not at the interface or in defects are removed for clarity.

lattice. Such lattice can be idealized, representing a continuum volume of matter, or be a very detailed atomistic description of the crystal. In the first case, cubes or prisms can be implemented representing small volumes. In the second, atoms in their lattice positions are included.

In any case, an initial description of how the crystal grows has to be provided in terms of transitions and transition rates. The most common mechanisms implemented are as follows:

- **Adhesion**. Extra matter or atoms from a different phase (solid, liquid, or gas) attach to the existing substrate. Dependencies depending on the local configuration can be added. Such dependencies are usually used to implement different rates for nucleation and island formation and for the growing of such islands.
- **Migration**. It is common to simulate the migration of atoms on the surface and even the interactions of such atoms. Such interactions can provide alternative mechanisms for nucleation. Formation energies at different localities are usually accounted for, allowing the diffusion of atoms to move matter from low formation energy areas to high formation energy ones, and evolving the crystal to a more stable one, usually by faceting.
- **Desorption** or etching. This mechanism simulates the removal of atoms that are weakly bonded to the substrate.

5.3.1.4.2 Diffusion/Reaction

KMC is especially apt for the simulation of diffusion/reaction systems. The diffusion is implemented by assigning migration events to the mobile species of the system, where they move randomly (random walk). In this case, the transition rate is just a frequency to perform random hop between initial and final positions of the diffusing entities (interstitials, vacancies, etc.). If the hopping distance λ is kept constant, the relation between Fick diffusivity D and hopping frequency ν is $\nu = {}^{D}\!/_{2N\lambda^2}$, where N is the dimension of the system being simulated. Once the diffusion is implemented in the system, the reaction part is also easy to implement. The forward reactions are implemented by extending every diffusion hop to look for neighbors after each hop. If a suitable reacting particle or compound is found, an interaction takes place, and final products are created replacing the initial ones. Interaction barriers can be simulated by adding a probability of interaction P, usually implemented as $P = \exp(-E_{\text{barrier}}/k_B T)$. Backward reactions are implemented as an additional transition mechanism for each compound, where they can revert to their original state. This simple scheme allows a very realistic simulation of the competition between diffusion, formation, and breaking of compounds that, depending on the relative frequencies of each transition, will lead to realistic simulations of both diffusion-limited and reaction-limited systems.

The following table summarizes the most important mechanisms and parameters to simulate diffusion/reaction systems.

Mechanism	Simulated by	Transition rate	Parameters
Diffusion	Random hop	From diffusivity	Diffusivity. Hopping distance
Formation ($A + B \rightarrow C$)	Replacement of $A + B$ by C if A(B) sees B (A) after diffusion	None. Mechanism implemented after diffusion and energetic barriers are considered	Formation energy of initial and final compounds. Energetic barriers. Capture radius
Breakup (C -> A+B)	Replacement of C by $A + B$	From a prefactor and energetic differences and barriers	Formation energy of initial and final compounds. Energetic barrier. Decomposition prefactor

Typical examples of KMC simulations of diffusion/reaction systems are simulation of damage irradiation of metals (for nuclear energy applications), simulation of ion implantation and annealing of semiconductors (for microelectronic applications), simulation of segregation, nucleation, and decomposition in alloys (for material sciences), and so on.

5.3.2 Molecular Dynamics

The starting point for MD is a well-defined microscopic description of a physical system. Then, the method will calculate the time evolution of the system by numerically solving the classical equations of motion.

5.3.2.1 Introduction

MD is a deterministic simulation method that calculates the equilibrium and transport properties of n-body physical systems. It is a classical algorithm in the sense that it uses classical mechanics to formulate the equations of motion of the system and to solve them: it uses a Newtonian or equivalent formalism. In MD, the equations of the system are discretized and solved, generating a path in the phase space. Later, the phase space can be processed obtaining different properties: formation energies, diffusivities, capacities, strains, and so on. The simulation is generated in a "step-by-step" way, following an iterative scheme, where each step simulates the incremental time Δt, with a typical value of 10^{-15} s. Such small time step is the cause of one of the main MD limitations: it can only simulate very small time ranges. Even 1 billion (10^9) MD iterations will advance the system only by 1 μs.

5.3.2.2 Equations of Motion

In general, an MD system will be an N-particle problem described by the Hamiltonian

$$H = \frac{1}{2}\sum_i \frac{p_i^2}{m_i} + U(r_1, r_2, \ldots, r_n),$$

where r_i is the position of particle i and U represents the total potential energy of the system. With the aforementioned Hamiltonian, the Newton equations are easily expressed as

$$p_i = m_i \frac{\mathrm{d}r_i}{\mathrm{d}t}, \quad \frac{\mathrm{d}p_i}{\mathrm{d}t} = F_i,$$

where the forces F_i are, as usual, obtained from the potential by differentiation.

5.3.2.3 Integration Schemes

An integrator is an algorithm aiming to numerically solve the set of mathematical equations of motion introduced in Section 5.3.2.2. Integrators introduce errors due to the numerical process involved, but such errors can be made negligible by the proper selection of the integrator and the Δt. Usually, more accuracy involves the use of more information (i.e., information from previous states, rather than just from the current states), implying a further use of computer resources, both memory and computing time.

One of the most used integrators, providing a good trade-off between resources and accuracy, is the Verlet integrator, which can be written as

$$\begin{aligned} r_i^{n+1} &= r_i^n + v_i^n \Delta t + \frac{F_i^n}{2m} \Delta t^2 \\ v_i^{n+1} &= v_i^n + \frac{F_i^{n+1} + F_i^n}{2m} \Delta t \end{aligned},$$

where the superindices n and $n+1$ are the magnitudes (position r, velocity v, force F) at the time step n and $n+1$, respectively.

5.3.2.4 Potential

With MD being a numerical algorithm to solve a classical physical system, all the information of the system is embedded in the current state and the potential U. Two assumptions are implied in the description of systems by a classical potential: First, the nucleus is assumed to be a point particle ruled by Newtonian equations. Second, electrons are assumed to follow nucleus almost instantaneously, allowing them to be treated separately (the Born–Oppenheimer approximation). Potentials can be theoretical ones (such as a Coulomb interaction), empirical ones, or a combination. In practice, the potential $U = U(r_1, r_2, \ldots, r_n)$ of the system as a whole is not known, and two more simplifying assumptions are done: The first approximation is that the total potential can be described as an addition of potentials between smaller units of the system – pair potentials, ternary potentials, and so on. The second one assumes that the potential between particles that are farther than some cutoff radius is negligible. Using the first approximation, it is then possible to write the potential U, considering, for instance, only pair interatomic forces as $U \approx \sum_{i<j} U(r_{ij})$, considerably simplifying the problem of finding a suitable potential for the material in consideration. The second approach is related mostly with computational efficiency. If potentials were to see particles at any distance, then for a system of N particles there would be N^2 interactions, making the system hardly scalable in sizes and introducing global dependencies for each particle. By limiting the maximum distance at which a particle interacts with others, a practical limit to the number of neighbors, n, is set, so that the total number of interactions is nN, linear rather than squared. This also introduces other programming conveniences: it facilitates the implementation of PBC and makes system evolution local, a very convenient property to easily implement efficient parallel algorithms.

There are many empirical potentials available in the literature and offered by common MD applications. A complete description of them would be out of the scope of this book. The usual MD simulations for ICME use empirical potentials of which two main types can be defined: pair potentials and many-body potentials. Pair potentials assume that the total internal energy of the system can be reduced to an addition of the internal pair energy:

$$U(r_1, \ldots, r_n) \approx \sum_{i=1,\, j<i}^{n} U(r_{ij}),$$

being one the most remarkable examples of the Lennard–Jones potential. N-body potentials extend the contributions beyond pairs:

$$U(r_1, \ldots, r_n) \approx \frac{1}{2}\sum_{i,j=1}^{n} U(r_{ij}) + \frac{1}{3}\sum_{i,j,k}^{n} U(r_{ij}, r_{ik}, \theta_{ijk}) + \cdots$$

having as an example the Stillinger–Weber and Tersoff potentials for Si, C, and Ge systems.

Another well-known type of n-body potentials are the embedded atom method (EAM) potentials. These are of particular interest for metals. EAM potentials are described as

$$U(r_1, \ldots, r_n) = \sum_i F_i(\rho_{h,i}) + \frac{1}{2}\sum_{\substack{i,j \\ i \neq j}} \phi_{ij}(r_{ij}),$$

where ϕ_{ij} is the short-range pair potential, r_{ij} is the distance between atoms i and j, $\rho_{h,i}$ is the total host electron density at atom i, and F_i is the embedding energy, the energy of an atom in a uniform electron gas relative to the atom separated from the electron gas. The importance of these potentials relies on the relative accuracy and large availability of EAM and modified EAM (MEAM) potentials for many metals and several metallic alloys.

5.3.2.5 Boundary Conditions

Atomistic simulation techniques are limited to very small domain sizes (nanometers) and cannot reach the sizes of real macroscopic systems. Consequently, an atomistic simulation is always embedded in a small simulation domain. The border of such domain is always a singularity requiring a special treatment. One simple solution would be to assume that there is no material at the other side, and consequently, the border is actually a free surface. Although this might be useful for specific applications, the small simulation domains would imply that the surface-to-bulk relation would be too high for the average case, making the simulation results useless to understand the properties of common bulk materials. Mirror condition and PBC are the two solutions to this problem, of which PBC is by far the most used.

Let us assume that the simulation cell is defined, in one particular axis, to range from 0 to s. If we are at x, being x close to the border 0, mirror conditions are imposed by assuming that the material at $-x$ is exactly the same as at x. On the contrary, PBC imply that the properties of the material at $-x$ are the same as at $s-x$. While mirror "reflects" the material, periodic "repeats" it.

The use of boundary conditions is very common, but it imposes restrictions on the minimum size of simulation cells. In particular, if the total size in any axis is smaller than twice the cutoff radius, particles will see themselves through the PBC. Also, in PBC, one has to be aware that when simulating systems with n defects rather than perfect bulk materials, the repetition of the system infinitely in the space produces a system with a defect concentration of $n/\Delta V$, ΔV being the simulation cell size, imposing increasingly bigger cell sizes for decreasing defect concentrations.

5.4
Electronic Structure Methods

The modern advancements in the understanding of the structure of matter have their foundations in the development of quantum mechanics, also known as the quantum theory. The achievements of quantum mechanics and its impact in our lives since the early days are countless. Quantum mechanics has a fundamental difference from the classical mechanics based on Newton laws, in the sense that the former is probabilistic in nature in contrast with the deterministic nature of classical physics.

The Schrödinger Equation: Quantum mechanics has its conceptual foundation on the Schrödinger equation, formulated in 1925 and published in 1926 by the Austrian physicist Erwin Schrödinger. This equation took a whole new approach to the underlying nature of matter in the sense that it is probabilistic in nature. The time-independent Schrödinger equation is defined as

$$E\Psi = \widehat{H}\Psi, \tag{5.1}$$

where Ψ is a unique wave function that exists for any collection of N particles, E is the energy of the system, and $\widehat{H}$ is the Hamiltonian operator that acts upon the wave function and returns E. Whenever Eq. (5.1) is valid, Ψ is an eigenfunction and E an eigenvalue. In the case of Eq. (5.1), E is the energy of the system because the operator applied in the wave function is the Hamiltonian operator. However, E can be any other property of the system, provided that the corresponding operator is used on the unique wave function. The fact that the Schrödinger equation is an eigenvalue equation means that there is an underlying analogy between this equation and matrix algebra. Taking such analogy, Ψ becomes an N-element column vector, $\widehat{H}$ is an $N \times N$ square matrix, and E is a scalar constant. One important feature of the wave function Ψ is that its product with its complex conjugate, $|\Psi^* \Psi|$, gives a probability density. For real space, the probability density of finding a particle with wave function Ψ is the integral of $|\Psi|^2$. Within this conceptual definition, Ψ has to obey some constraints: the normalized integral of $|\Psi|^2$ over the whole space has to be unity – there is a probability of 1 of finding the particle in the whole space – this implies that $|\Psi|^2$ has to be integrable and Ψ has to be continuous and single valued. For a collection of atoms such as a molecule or material, the general Hamiltonian $\widehat{H}$ for the system of particles described by (5.1) is expressed as

$$\widehat{H} = -\sum_{I=1}^{P} \frac{\hbar^2}{2M_I}\nabla_I^2 - \sum_{i=1}^{N} \frac{\hbar^2}{2m}\nabla_I^2 + \frac{e^2}{2}\sum_{I=1}^{P}\sum_{J\neq 1}^{P} \frac{Z_I Z_J}{R_I - R_J} + \frac{e^2}{2}\sum_{i=1}^{N}\sum_{j\neq i}^{N} \frac{1}{r_i - r_j} - e^2 \sum_{I=1}^{P}\sum_{i=1}^{N} \frac{Z_I}{R_I - r_i}, \tag{5.2}$$

where $R = \{R_I, I = 1, \ldots P\}$ is a set of coordinates for P nuclei with charges Z_I and masses M_I, $r = \{r_i, i = 1, \ldots N\}$ is a set of coordinates for N electrons, and m and e represent the mass and charges of the electron, respectively. For a system of nuclei and electrons, applying (5.2) to the wave function Ψ according to (5.1) outputs the energy of the system. This task is very complex and no analytical solution exists for Eq. (5.1) – except, for example, for hydrogenoid atoms He^+, Li^{2+}, Be^{3+}, and B^{4+} – because of the QM nature of the particles and because of the many-body problem that the Hamiltonian in (5.2) implies.

The movements of nuclei and electrons are correlated due to the Coulomb interactions that the nuclei exert on the electrons and vice versa. Because the mass of the electron is much smaller than that of the nucleus and as such the response time of the nucleus to a perturbation caused by an electromagnetic field is much longer than that of the electrons, the sums involving nuclei and electrons in the Hamiltonian of (5.2) can be decoupled following the Born–Oppenheimer approximation which is adiabatic in nature. The resulting expression for the Hamiltonian (5.2) is then

$$H_{\text{el}} = \sum_{i=1}^{N} \frac{1}{2} \nabla_i^2 - \sum_{i=1}^{N} \sum_{I=1}^{P} \frac{Z_I}{r_{iI}} + \sum_{i=1}^{N} \sum_{j \neq i}^{N} \frac{1}{r_{ij}}. \tag{5.3}$$

This expression is now an electronic Hamiltonian (H_{el}) in the sense that it depends solely on the motion of N electrons in the field of P point charges (the fixed nuclei). This means that in order to solve the Schrödinger equation (5.1), one can fix the spatial coordinates of the nuclei and determine the electronic wave function. The obtained wave function for the electrons is then used with the Hamiltonian of Eq. (5.3) to compute the total energy of the system. Then, for new sets of nuclear coordinates, this procedure is repeated over and over again until an optimal positioning of the P nuclei in space is found – when these are subject to the Coulomb forces created by the N electrons or any other external potential. This removes the problem of the correlation between nuclei and electrons. But it is still necessary to deal with the correlation between electrons.

The Electronic Structure Problem: The determination of the electronic wave function to be used with Eq. (5.3) which in this way gives the total energy of the system via the Schrödinger equation is a complex task because of the intricate physical nature of the electrons. Besides the mathematical constraints that the wave function has to obey in order to provide us with a finite probability density for the positions of electrons in real space, there are other details involving the physical nature of electrons that have to be incorporated in the approximation to the real wave function of the system. Electrons are fermions and as such obey Fermi–Dirac statistics. This means that the total electronic wave function has to be antisymmetric, that is, it has to change sign whenever the spatial coordinates of two electrons are exchanged. This is known as the exchange interaction and it is the underlying cause of the Pauli exclusion principle, and it is also the exchange component (Fermi hole) of the so-called exchange–correlation (XC) hole. Additionally from a Coulomb repulsion perspective, the movements of electrons are correlated in the sense that if one electron at a given instant occupies a set of spatial coordinates, no other electron will occupy the same position at that instant – this is the correlation component of the XC hole. This simple sounding principle when applied to systems of many electrons – more than two – leads to a complex many-body problem with no analytic solution.

5.4.1
Approximate Solutions to the Electronic Wave Function

The determination of an approximate solution for Ψ is a formidable task, and a series of approximations have to be employed. Remarkably, there are no rules that state from what Ψ should be constructed: only that Ψ should obey the constraints mentioned earlier and that whatever ingredients we use to construct Ψ from, these ingredients, or their combination, should grasp the correct quantum statistics that govern the particles involved, that is, correlation, exchange, spin, invariance, and so on. For practical matters, the determination of the wave function Ψ is done using a collection of mathematical functions that collectively can approximate the real Ψ. Since the real Ψ is not known, the mathematical procedure to compute Ψ from the collection of functions denominated the "basis set" is called the variational principle. In this principle, Ψ is computed from the basis set which is subject to an optimization of parameters until a minimum of energy is found. According to the variational principle, this energy minimum will be the best solution for Ψ given by the specific basis set.

We still lack two important details here: (i) What is and how to construct a basis set? (ii) Which mathematical formalism should be used to optimize the interactions between the component functions of the basis set in order to find the best approximation for Ψ? Since we are interested in applying quantum mechanics to atoms or other collections of nuclei and electrons whose interactions lead to the diversity of materials and molecules we have around us, a good way to determine an initial guess to the basis set is to use collections of basis functions for the individual atoms that constitute a system of interest. Following this, these basis functions will be able to interact with each other, and they will be subject to a search for the optimal interaction that will lead to a minimum of energy that corresponds to the best solution for Ψ given by that basis set according to the variational principle. One of the first solutions to the construction of a basis set was proposed following the principle known as linear combination of atomic orbitals (LCAO). This is expressed as

$$\phi = \sum_{i=1}^{N} a_i \varphi_i, \tag{5.4}$$

where ϕ is a guess wave function that approximates Ψ; φ_i is a set of N basis functions, that is, the basis set; and a_i is a coefficient associated of the ith basis function. The spatial positioning of the basis set does not have to follow a specific requirement. An intuitive place to center each basis function is on each of the atoms that constitute the system. However, this does not have always to be the case, and some electronic structure calculation methods benefit from using nonatom-centered basis functions. The basis functions for each atom to be used on (5.4) are constructed in terms of basis functions for the electrons. This also means that there will be some freedom in terms of ways to approach the construction of basis functions, even if one has to obey certain constraints. For example, in one of its simplest approximations, ϕ can be built following a combination of one-electron basis functions – following the Hückel method. This method assumes that the total energy of a many-electron system can be determined just by summing the individual energies of the occupied one-electron orbitals. Many details are missing in such approach, and more sophisticated methods were developed post-Hückel theory.

5.4.1.1 Hartree–Fock Theory

Expressing the electronic Hamiltonian in terms of one-electron basis functions means that the only terms to be considered in H_{el} are the electronic kinetic energy and the electron–nuclear attraction. This transforms the Hamiltonian (5.3) into

$$H_{\text{el}} = \sum_{i=1}^{N} h_i,\ h_i = -\frac{1}{2}\nabla_i^2 - \sum_{I=1}^{P} \frac{Z_I}{r_{iI}}, \tag{5.5}$$

where h_i is a set of N one-electron Hamiltonians for the system of N electrons and P nuclei. The one-electron Hamiltonians have to be eigenfunctions of the one-electron wave functions ψ_i in order to satisfy the one-electron Schrödinger equation

$$\epsilon_i \psi_i = h_i \psi_i, \tag{5.6}$$

where ψ_i is a one-electron wave function of the ith orbital and ϵ_i is the energy of the ith one-electron orbital. Because (5.6) is an eigenfunction equation, this implies orthonormality of the basis functions involved. As such, the wave function of the system Ψ to be used with Eq. (5.1) can, according to the Hartree method, be approximated as a product of the one-electron wave functions as

$$\Psi_{\text{HP}} = \psi_1 \psi_2 \psi_3 \ \cdots \ \psi_N. \tag{5.7}$$

This yields what is called the Hartree product wave function Ψ_{HP}. The Hartree Hamiltonian of Eq. (5.5) does not account for electron–electron repulsion, but this can be added as an interaction potential that is expressed as the integral of the probability density for the electrons over the whole space.

In order for the wave function to be coherent with quantum statistics, it must be able to account for spin and antisymmetry of the particles. A result of the Hartree product approximation is that the particles are treated as distinguishable, though, electrons are indistinguishable. The inclusion of these details – spin and exchange – can be achieved by constructing the Hartree product as an antisymmetrized wave function, which is done by using the permutation operator when expressing the Hartree product in terms of spin orbitals. Thus, for a system of two electrons designated electron 1 and 2, which can have two spins states each – α

and β – we express the wave function in terms of a Slater determinant. This is called the Hartree–Fock (HF) approximation and has the form

$$\Psi_{\mathrm{HF}}(x_1,x_2,\ldots,x_N)=\frac{1}{\sqrt{N!}}\begin{vmatrix}\varphi_1(x_1) & \varphi_2(x_1) & \ldots & \varphi_N(x_1)\\ \varphi_1(x_2) & \varphi_2(x_2) & \ldots & \varphi_N(x_2)\\ \vdots & \vdots & \ddots & \vdots\\ \varphi_1(x_N) & \varphi_2(x_N) & \ldots & \varphi_N(x_N)\end{vmatrix}$$
$$=\mathrm{SD}\{\varphi_1(x_1)\varphi_2(x_2)\ldots\varphi_N(x_N)\},\tag{5.8}$$

where x includes both the spin and spatial orbitals of the electron; $\varphi_i(x_1)$ represents the ith one-electron spin orbital, composed of spatial and spin components; and $\frac{1}{\sqrt{N!}}$ is a normalization factor. This Slater determinant has N electrons occupying N spin orbitals without distinguishing between the particles, includes exact exchange, and accounts for the Pauli exclusion principle. In this way, it incorporates the correct quantum statistics that govern electrons. The corresponding HF Hamiltonian becomes

$$H_{\mathrm{HF}}=\sum_{i=1}^{N}h_i+\frac{1}{2}\sum_{i=1}^{N}\sum_{j=1}^{N}(J_{ij}-K_{ij})+V_P,\tag{5.9}$$

where J and K are the Coulomb and exchange integrals, respectively, and V is the interaction energy between P nuclei. The Coulomb operator is diagonal, while the exchange operator it is not. Since the solution to the HF one-electron equations is not unique, it is possible to find a new set of orbitals φ_i' that are eigenfunctions of the Fock operator. This is known as the canonical representation of the orbitals and permits the interpretation of the Lagrange multipliers in terms of the energies of the set of one-electron orbitals ϵ_i. Because of this, within HF theory it is possible to relate the ionization energy and the electron affinity with the energies of the orbitals where the electron is removed from or added to, respectively. This is known as the Koopmans theorem.

5.4.1.2 Post-Hartree–Fock Methods

5.4.1.2.1 Single-Reference Methods

Even though HF theory can account for the correct quantum statistics of the electrons, it still lacks an important feature of the whole electron system, the correlation energy. In the HF theory, correlation is present only as an average repulsion between electrons, but not explicitly in terms of electron–electron interaction. It is possible to add the correlation energy on the HF energy as $E_{\mathrm{c}}=E_{\mathrm{tot}}-E_{\mathrm{HF}}$, where E_{c} is the correlation energy, E_{tot} is the total energy, and E_{HF} is the HF energy. The correlation energy is one of the components of the energy that presents the biggest challenge for its determination. The three most widely used ways of accounting for correlation are (i) using combinations of several Slater determinants, (ii) adding correlation as a perturbation (using perturbation theory) over the HF solutions, and (iii) employing correlated wave functions.

Multideterminantal wave functions Ψ_{md} are constructed according to

$$\Psi_{md}(x_1, x_2, \ldots, x_N) = C_0\Psi_{HF}(x_1, x_2, \ldots, x_N) + \sum_{i=1}^{N_{SD}} C_i\Psi_i(x_1, x_2, \ldots, x_N), \quad (5.10)$$

where N_{SD} is a number of Slater determinants and Ψ_{HF} is the HF solution to the ground-state configuration. This considers the energy of the occupied orbitals. The determinants Ψ_i consider unoccupied one-electron states. In this method, electrons are promoted (excited) from the occupied Ψ_{HF} states to the unoccupied Ψ_i states. If only one-electron excitations are considered, the extra determinants get the name *single* (S); if two electrons are considered, the extra determinants are called *double* (D); and so on up to *triple* (T) and *quadruple* (Q). This is known as the configuration interaction (CI). It is very accurate, but the cost of the computations escalates as N^{10} for CISDTQ which is close to full CI.

Another way to add correlation is as a perturbation which is applied on the HF wave function. The starting point is to express the Hamiltonian of the system, $\hat{H}$, as an unperturbed reference Hamiltonian $\hat{H}_0$ and to this apply a time-independent perturbation, $\Delta\hat{H}$ as ($\hat{H} = \hat{H}_0 + \lambda\Delta\hat{H}$), where λ is a parameter that measures the strength of the perturbation. When this is done, the HF wave function becomes a first-order perturbation. The most common way of adding perturbation terms is using the Møller–Plesset theory. Here the reference Hamiltonian used as the reference for the perturbative expansion is expressed as a sum of the Fock operators $\hat{F}_i$ as

$$\hat{H}_0 = \sum_{i=1}^{N} \hat{F}_i = \sum_{i=1}^{N} \left[\hat{h}_i + \sum_{j=1}^{N} \left(\hat{J}_{ij} - \hat{K}_{ij} \right) \right], \quad \text{and } \Delta\hat{H} = \frac{1}{2}\sum_{i \neq j}^{N} \hat{v}_2(i, j) - \sum_{i \neq j}^{N} \hat{v}_2(\hat{J}_{ij} - \hat{K}_{ij}). \quad (5.11)$$

Different amounts of correlation can be added using second-order (MP2), third-order (MP3), or fourth-order (MP4) methods. As a rule of thumb, an MP4 computation has an accuracy close to that of a CISD.

Another way of adding the perturbation terms is via resummation perturbative methods also denominated coupled clusters (CCs). The resulting CC wave function Ψ_{CC} is expressed as

$$\Psi_{CC} = e^{\hat{T}}\Psi_{HF}, \; \hat{T} = \hat{T}_1 + \hat{T}_2 + \cdots + \hat{T}_N, \quad (5.12)$$

where $\hat{T}$ is an operator that acts upon the HF wave function and generates all the excited Slater determinants up to a certain order. Depending on the order of terms for $\hat{T}$ reached in (5.12), the method is called *singles* (CCS), *singles* and *doubles* (CCSD) *singles doubles*, and *triples* (CCSDT). CCSD has a computational effort close to that of MP3 (N^6), while for CCSDT it becomes N^8. If the excitations are included perturbatively rather than exactly, these appear as the character in parenthesis CCSD(T). Here the single and double excitations are included exactly, but the third-order excitations are included perturbatively. As a rule of thumb, the CC calculations are slightly more accurate than the equivalent CI results.

5.4.1.2.2 **Multireference Methods**

The methods described up to now are based on the fact that correlation can be added as a correction term to the HF energy. In HF theory the ground-state wave function is computed using a single Slater determinant; for some cases, this is not a proper description of the system. In such case it is necessary to resort to multiconfigurational self-consistent field (MCSCF) methods. These methods use multiple determinants as in the case of the methods described earlier, but with the difference that, here, the optimization of the orbitals is done considering the multideterminantal character of the wave function. In MCSCF, the determinants together with the coefficients are variationally optimized. There are no big advantages of MCSCF over CI unless the HF solution (single determinantal) to the ground-state wave function is not correct. Cases where the single determinantal solutions are not able to describe the system are, for example, those where the spin multiplicity changes during the course of a chemical process.

5.4.1.2.3 **Explicitly Correlated Methods**

Correlation can be included directly in the trial wave function by what are called the explicitly correlated methods (ECM). These methods account for the fact that there is a discontinuity in the first derivative of the real wave function at the origin. The ECM wave function includes a function of the electron–electron distance. These conditions are constraints to build the ECM in terms of Slater orbitals that are components of an expansion of CI type. Quantum Monte Carlo (QMC) is such an ECM method that has different developments based on different approaches such as variational, diffusion, and the Green's function MC. The QMC wave function Ψ_{QM} has the form

$$\Psi_{\mathrm{QM}} = D_{\alpha} D_{\beta} \prod_{i<j} f(r_{ij}), \tag{5.13}$$

where D_{α} and D_{β} are determinants of the α and β spins. The correlation function $f(r_{ij})$ can include three body terms, and as long as there is no spin coupling in the Hamiltonian, there are no errors which are included due to this approximation. The integrals of the wave function are evaluated using an MC scheme. Because these methods are very accurate and scale as N^3, they are being considered by some authors as a possible standard for the future of electronic calculation methods.

5.4.2 **Density Functional Theory (DFT)**

It is possible to make an analogy between the real space probability density obtained by multiplying the wave function with its complex conjugate and the real space electronic density. While in order to tackle the electronic structure problem via the first approach it is necessary to determine the wave function of the system, following the second approach it is only necessary to compute the electronic density when the electrons are subjected to an external potential – that of the static nuclei. The second approach is known as density functional theory (DFT), and it has computational advantages when compared with the wave function approach. DFT relies on the premise that there is a true functional of the electron density $F[\rho]$ – which is directly dependent on the wave function – and any property of interest can be determined by employing appropriate operators on this functional. DFT is based on the two theorems by Hohenberg–Kohn: (i) the external potential of an electron system is determined by the electronic density, and (ii) for a nonnegative density normalized to the number of electrons N, a variational energy (E_v) can be defined as a functional of the density, and E_v is the ground state of the system. The practical usage of DFT is based on the introduction of orbitals as suggested by Kohn and Sham (KS). The KS equations consider the separability of the classical electrostatic energy (Hartree term) from the exchange and correlation contributions. The KS model is closely related to the HF method in what concerns the kinetic, electron–nuclear, and Coulomb electron–electron energies. In KS theory, the kinetic energy is calculated by assuming noninteracting electrons – as in HF theory described in Section 5.4.1.1. In the same way as in HF theory, this approach provides ~99% of the correct answer; the difference between the exact kinetic energy and that calculated by assuming noninteracting orbitals is small. The remaining kinetic energy is accounted for with an exchange–correlation term, and a general DFT energy expression can be written as

$$E_{\mathrm{DFT}}[\rho] = T_s[\rho] + V_{\mathrm{ext}}[\rho] + E_{\mathrm{NN}}[\rho] + E_{\mathrm{X}}[\rho] + \widetilde{E}_{\mathrm{C}}[\rho], \tag{5.14}$$

where $T_s[\rho]$ is the noninteracting kinetic energy functional; $V_{\mathrm{ext}}[\rho]$ is the interaction energy with external potentials, such as nuclei; $E_{\mathrm{NN}}[\rho]$ is the electron–electron interaction energy; $E_{\mathrm{X}}[\rho]$ is the exchange energy; and $\widetilde{E}_{\mathrm{C}}[\rho]$ is an averaged correlation term. In DFT all the energy components could – at least conceptually – be expressed as functionals of the density, though, in reality this is not the case. The challenge with DFT is to find functions which can serve as good approximations of the true functional of the density. Once again there is no ideal prescription on how the true functional of the density can be best approximated, and different approaches exist to tackle this problem. The $E_{\mathrm{X}}[\rho]$ and $\widetilde{E}_{\mathrm{C}}[\rho]$ components constitute the biggest challenge for the computation of $E_{\mathrm{DFT}}[\rho]$.

5.4.2.1 Local Density Approximation (LDA)

The simplest practical approximation to $F[\rho]$ was done by taking the XC energy as that for a uniform electron gas. This approach considers the inhomogeneous electron density as being locally homogeneous and uses the XC hole known for the homogeneous system. One might question the validity of such approach because chemical bonding is very far from being a uniform electron density. But for solid metals, for example, the electron density can be approached to a good extent considering a local homogeneous density. Because the only descriptor used for the inhomogeneous electron density is the homogeneous local density at each spatial coordinate, this approach is known as the local density approximation (LDA). The local spin density approximation (LSDA) is an extension of the LDA with the difference that the α and β spin densities are computed separately. For the case of closed-shell systems, the LDA is equal to the LSDA. The LDA methods, in spite of being capable of describing properties important for solid-state physics (SSP), always tend to produce electronic densities which are more homogeneous than the true densities and to overestimate the binding energy in molecules and the cohesive energies of solids. Some examples of such functionals are PW92, VWN, and PZ81.

5.4.2.2 Generalized Gradient Approximation (GGA)

In order to account for the inhomogeneities of the density, an approach was developed where the gradient of the density was accounted for by including higher-order derivatives of the density. The generalized gradient approximation (GGA) is a method where the density gradient is also accounted for by an expression of the following form for the XC energy:

$$E_{XC}[\rho] = \int \rho(\boldsymbol{r})\epsilon_{XC}[\rho(\boldsymbol{r})]F_{XC}[\rho(\boldsymbol{r}), \nabla\rho(\boldsymbol{r}), \nabla^2\rho(r), \ldots]d\boldsymbol{r}, \tag{5.15}$$

where F_{XC} is the factor that modifies the LDA expression taking into account the variation of the density in the vicinity of the considered spatial coordinate. By considering the gradient of the density, the GGA methods are semilocal in the sense that they grasp the physics of the density not only at a given spatial coordinate but also in its vicinity. The GGA methods have increased accuracy for describing chemical bonding when compared to the LDA methods. They also provide better binding energies for solids and more accurate electronic structure-based properties. However, both the LDA and GGA methods are affected by the self-interaction error and show reduced accuracy for describing semiconductors and transition states where chemical bonds are broken. The self-interaction error is a physically unsound feature of DFT and results from the fact that each electron interacts with itself adding a nonphysical component to the XC hole. This erroneous component causes severe limitations of DFT for describing some systems. Band gaps in semiconductors are usually underestimated when computed with GGA methods. To this class of functionals belong the BLYP, PBE, and RPBE, among others.

5.4.2.3 Meta-GGA Methods

The next step for adding more physical descriptors of the density was taken by introducing the fourth-order terms on the gradient expansion of the XC energy. This is the foundation of the meta-GGA approach. Here, semilocal information is incorporated via the Laplacian of the density which results in a kinetic energy density component to the energy. In general the meta-GGA approach improves the atomization energies when compared with GGA, but frequencies and geometries are in general worsened. Examples of these functionals are TPSS and M06-L.

5.4.2.4 Hybrid DFT Hartree–Fock

We have seen in Section 5.4.1.1 that a great advancement in wave function methods was achieved with the introduction of HF theory. This is also because HF theory employs a Slater determinant in order to solve the equations of the one-electron orbital expansion. An advantage of this approach is that the Slater determinant is self-interaction free. In order to solve the issue of the self-interaction error in DFT, the logical way was to use the HF method to solve the KS equations. Because these methods mix the HF methodology from wave function theory with DFT, they are known as hybrid methods. The hybrid methods employ the Slater determinant expansion of one-electron KS orbitals in order to account for the exchange component of the energy. This means that besides canceling exactly for the self-interaction error even in its nonlocal components, the HF exchange will account for nonlocal terms of the energy, leading to a class of functionals which are not anymore limited by the semilocality of the gradient expansion. These methods use DFT correlation and HF exchange in the following way:

$$E_{\mathrm{XC}}^{\mathrm{Hyb}} = \alpha E_{\mathrm{XC}}^{\mathrm{HF}} + (1-\alpha)E_{\mathrm{X}}^{\mathrm{DFT}} + E_{\mathrm{C}}^{\mathrm{DFT}}, \tag{5.16}$$

where the HF component of the energy can be added as a percentage of the exchange energy of the functional via the parameter α. Some functionals such as the PBE0 use 25% HF exchange, while others such as the M06-HF use 100% of the HF exchange component to the energy. The latter are fully self-interaction error-free, but the effect of the incorporation of too much nonlocal HF exchange does not always lead to improvements. In fact, too much HF exchange in the functional is detrimental, for example, for the description of metallic systems. Because DFT uses density as the descriptor of the system, the physical significance of a given component of the expansion of the density is not always evident. The hybrid functionals show good performance for materials with localized electrons and for free electron metals. The description of bond breaking in transition states is greatly improved with hybrid DFT. In general these methods underestimate the atomization energies of transition metals. Examples of hybrid functionals are PBE0, HSE03, and M06.

5.4.2.5 Van der Waals Corrected DFT

One of the shortcomings of pure DFT is its incapability of accounting for dynamic correlation. While an average static correlation can be accounted for to some extent as $\widetilde{E}_{\mathrm{C}}[\rho]$ in Eq. (5.14), the dynamical correlation which is at the origin of van der Waals (vdW) dispersion interactions is not accounted for in LDA and GGA functionals. The dispersion interactions are not related with exchange either. While in wave function methods the dynamical correlation can be added as a correction to the HF solution as described before, for DFT this problem does not have an obvious solution. However, the empirical hybrid meta-GGA functionals such as the Minnesota (M06, M011, etc.) are able to grasp vdW interactions. Other approaches to add the vdW corrections are achieved, for example, with the nonlocal functionals that depended explicitly on r, such as the vdW-DF-04 and vdW-DF-10. A popular way to overcome this problem with low computational burden employs a force field (FF) that accounts for the interaction potential due to vdW forces as a Lennard–Jones potential. The vdW dispersion energy should decay as $E_{\mathrm{vdW}} = -C_6/R^6$ in the long-range limit. The DFT plus FF approaches are known as the DFT-D, and several developments for the D term have been proposed such as the D, D2, and D3. Some of the issues with the FF correction to the energy are, for example, the lack of shielding for the dispersion interactions between two atoms which are located far apart but have other atoms in between and also when redox chemistry leading to noninteger formal oxidation states takes place.

5.5
Potentials, Force Fields, and Effective Cluster Interactions

Connection between the many-particle quantum mechanics and the effective one-particle DFT description is established via the XC functional. The existence of this functional is proven in DFT, but its exact form is unknown and must be approximated. Many approximate XC functionals (or potentials) have been proposed, such as the local and semilocal functionals based on the LDA or the GGA, as well as hybrid functionals including some part of nonlocal HF exchange or vDW corrections. Various XC functionals are collected in a library called *Libxc (*http://www.tddft.org/programs/octopus/wiki/index.php/Libxc), which is maintained by the European Theoretical Spectroscopy Facility (ETSF) (http://www.etsf.eu/).

In the quantum chemistry (QC) research community, a library of standardized basis sets for all elements of the periodic table is kept at the *Basis Set Exchange* database (https://bse.pnl.gov/bse/portal). The data are provided in the formats of the main QC codes (see software list in Section 5.6).

The step between the electronic structure calculations and atomistic simulations is often made by means of artificially constructed, so-called effective interatomic interaction potentials (or FFs). There are different types of such potentials or FFs, and there are different techniques for deriving them. Commercial and public databases and libraries of interatomic potentials and FFs are available. Some libraries are connected with software packages such as AMBER (http://ambermd.org/), CHARMM (http://en.wikipedia.org/wiki/CHARMM), GROMOS (http://www.gromos.net/), and GULP (http://nanochemistry.curtin.edu.au/gulp/). One should also mention the Interatomic Potentials Repository at the National Institute of Standards and Technology (NIST), United States (http://www.ctcms.nist.gov/potentials/).

For simulations of atomic and magnetic order–disorder transitions in alloys and also for the construction of the corresponding phase diagrams, one uses effective interatomic potentials defined on a lattice. Such potentials can be derived from electronic structure calculations using two complementary approaches, one based on cluster expansion or so-called structure inversion method (SIM) and the other based on perturbation theory or generalized perturbation method (GPM). Software tools for SIM include the following codes: ATAT (http://www.brown.edu/Departments/Engineering/Labs/avdw/atat/),

CLUPAN (http://sourceforge.net/projects/clupan/), and UNCLE (DOI: 10.1088/0965-0393/17/5/055003). Screened GPM for interatomic and magnetic exchange interactions is implemented in the EMTO package; see Section 5.6.

5.6
Software Tools in the Area of Discrete Modeling

Scientific communities of researchers and code developers have emerged long ago, connecting people who used similar methods or approaches: electronic structure (or band structure) communities, communities of atomistic (or molecular) modelers who used methods such as MD or MC, and so on. The structure of these scientific communities was rather inhomogeneous and evolved with time. For example, the electronic structure community originated from two fields:

- QC, where the researchers developed so-called *ab initio* methods such as the HF and post-HF methods and applied them mainly to study finite-size systems such as molecules and clusters
- SSP, where the researchers developed methods employing the effective one-electron approach based on DFT and applied these methods mainly to study periodic crystalline systems

The two communities actively exchanged opinions, methods, and research and programming tools to become more and more united and mutually dependent.

There are communities connecting the developers and users of some particular methods and software packages (practically, for every popular code, see the following lists). There are also networks of scientists working on the applications of small-scale models in certain important areas of science such as spectroscopy, irradiation damage, mechanical properties of solids, thermodynamics, and so on. While listing all networks and communities is impossible, we give some important European examples (electronic structure):

- Psi-k network for *ab initio* studies of solids and surfaces (http://www.psi-k.org/)
- Centre Européen de Calcul Atomique et Moléculaire (CECAM) (http://www.cecam.org/)
- ETSF (http://www.etsf.eu/)
- CCP9 – Collaborative Computational Project for the Study of the Electronic Structure of Condensed Matter (http://www.ccp9.ac.uk/)

Multiscale material modeling has been one important activity uniting scientists from different small-scale and continuum modeling communities to develop methods and software tools enabling the data transfer between modeling approaches operating at different length scales and timescales. Since the announcement of the Materials Genome Initiative (MGI) http://en.wikipedia.org/wiki/Materials_Genome, this activity has intensified in the United States to produce new research tools and results databases. One example is the *Materials Project* database https://www.materialsproject.org/ empowered by toolkit Python

Table 5.1 List of major DDD simulation software.

No.	Name of code	Information or reference	Software developer or provider	License	Description
5201	ParaDiS	http://paradis.stanford.edu/	Lawrence Livermore National Laboratory, CA, USA	Free	A C++ code based on the nodal model that uses the MPI library for parallel calculations. Modeling of bcc and fcc structures is possible
5202	micro3d	http://www.cmm.wsu.edu/Downloads/Downloads.html	Washington State University, WA, USA	Free	A Fortran code based on the nodal model for large-scale parallel simulations. Modeling of bcc and fcc structures is possible
5203	TRIDIS	http://www.numodis.fr/tridis/	Laboratoire SIMAP, Grenoble INP, CNRS, France	Free	A serial code based on the network model. Modeling of fcc, bcc, and hcp structures is possible
5204	microMegas	http://zig.onera.fr/mm_home_page/	Laboratoire d' Etude des Microstructures, CNRS-ONERA, France	Free	A Fortran code based on the network model, which can run on clusters. Modeling of different crystal structures – namely, fcc, bcc, hcp, and dc – is possible

Table 5.2 Molecular modeling software including molecular dynamics (MD), Monte Carlo (MC), and other small-scale methods, in some cases combining them with electronic structure and continuum models such as implicit water (IW) treatment.

No	Name of code	Information or reference	Software developer or provider	License	Description
5301	Abalone	http://en.wikipedia.org/wiki/Abalone_(molecular_mechanics)	Agile Molecule (http://www.biomolecular-modeling.com/Abalone/index.html)	Free	Biomolecular modeling and protein folding, MD, MC, replica exchange method (REM), IW
5302	ACEMD	http://pubs.acs.org/doi/abs/10.1021/ct9000685	Acellera Ltd. (http://www.acellera.com/acemd)	Commercial/free	MD with CHARMM or Amber force fields (FF)
5303	ADUN	http://dx.doi.org/10.1002%2Fjcc.20312	Computational Biochemistry and Biophysics Laboratory, UPF, Barcelona, Spain (http://adun.imim.es/)	GPL	MD and quantum mechanics (QM), molecular mechanics (MM) using CHARMM, Amber FFs
5304	AMBER	http://en.wikipedia.org/wiki/AMBER	AMBER community (http://ambermd.org/)	Commercial	Modeling toolkit and FF database
5305	Ascalaph Designer	http://en.wikipedia.org/wiki/Ascalaph_Designer	Ascalaph project (http://www.biomolecular-modeling.com/Products.html)	GPL	MD simulations of DNA, proteins, hydrocarbons, and so on
5306	ASE	https://wiki.fysik.dtu.dk/ase/	CAMD, DTU, Lyngby, Denmark (http://www.fysik.dtu.dk/english/Research/CAMD/About-CAMD)	GPL	Atomic Simulation Environment: GUI to many electronic structure codes, MD, QM, simulations for solids and molecules
5307	CHARMM	http://en.wikipedia.org/wiki/CHARMM	Martin Karplus (http://en.wikipedia.org/wiki/Martin_Karplus), CHARMM developer network, Accelrys (http://accelrys.com/)	Commercial	Chemistry at HARvard Macromolecular Mechanics: MD and MM tools and FF database
5308	COSMOS		COSMOS Software (http://www.cosmos-software.de/)	Commercial/Free	QM, MM, FF set, molecule builder and viewer

5309	CP2K	http://en.wikipedia.org/wiki/CP2K	CP2K (http://cp2k.org/ development team; http://sourceforge.net/p/cp2k/_members/)	GPL	Open-source QM, MD, MM simulations of solid state, liquid, molecular, and biological systems
5310	Culgi		Culgi BV (http://www.culgi.com/)	Commercial	Atomistic simulations and mesoscale methods
5311	Desmond	http://en.wikipedia.org/wiki/Desmond_(software)	DE Shaw Research (http://deshawresearch.com/resources.html), Schrödinger (http://www.schrodinger.com/)	Commercial/free	High-performance MD. Comes with a comprehensive GUI for building, visualizing, and reviewing results as well as calculation setup and launching
5312	Discovery Studio, Materials Studio	http://accelrys.com/products/discovery-studio/, http://accelrys.com/products/materials-studio/	Accelrys/Biovia (http://accelrys.com/)	Commercial	Software environments for QM, MD, and other chemical and materials science simulations
5313	GROMACS	http://en.wikipedia.org/wiki/GROMACS	GROMACS developers' network	GPL	High-performance MD
5314	GROMOS	http://en.wikipedia.org/wiki/GROMOS	GROMOS network (http://www.gromos.net/), BIOMOS b.v (biomos@igc.phys.chem.ethz.ch)		Modeling software for biomolecules
5315	GULP		Julian Gale, Curtin University, Australia (http://nanochemistry.curtin.edu.au/gulp/)	Academic	General utility lattice program for MM, MD, and FF set

(*continued overleaf*)

Table 5.2 (*continued*)

No	Name of code	Information or reference	Software developer or provider	License	Description
5316	LAMMPS	http://en.wikipedia.org/wiki/LAMMPS	Sandia National Labs http://lammps.sandia.gov/	GPL	High-performance MD tools and utilities
5317	MacroModel	http://en.wikipedia.org/wiki/MacroModel	Schrödinger (http://www.schrodinger.com/)	Commercial	Multiscale modeling tools, Maestro GUI, interfaces, databases, and analysis
5318	MAPS		Scienomics (http://www.scienomics.com/products/molecular-modeling-platform/)	Commercial	Building visualization and analysis tools; interfaces to simulation engines
5319	Materials Simulator	https://staff.aist.go.jp/h.ogawa/Hsim/indexE.html	Hiroshi Ogawa, AIST, Tsukuba, Japan (https://staff.aist.go.jp/h.ogawa/Hsim/indexE.html)	Free	Online tools for building atomistic models of grain boundaries (GBstudio), dislocations (DLstudio), and nanoparticles (NPstudio)
5320	MedeA		Materials Design (http://www.materialsdesign.com/)	Commercial	Combines experimental databases with computational engines (VASP, LAMMPS, GIBBS, MOPAC, Gaussian, Phonon) with prediction, analysis, and visualization tools in an integrated flowchart environment
5321	MMonCa	http://www.materials.imdea.org/MMonCa	Dr. I. Martin-Bragado. IMDEA Materials Institute (http://www.materials.imdea.org/)	Apache2 (open source)	KMC tool that provides: * Process simulation * Crystal growth * Damage irradiation evolution * Microstructure simulation
5322	MOE		Chemical Computing Group (http://www.chemcomp.com/)	Commercial	Molecular Operating Environment

5323	ORAC	http://www.chim.unifi.it/orac/	ORAC team at CEA, France, and Florence University, Italy (http://www.chim.unifi.it/orac/#toc2)	Free	MD/MM for biomolecular systems
5324	NAMD + VMD	http://en.wikipedia.org/wiki/NAMD, http://en.wikipedia.org/wiki/Visual_Molecular_Dynamics	Beckman Institute (http://www.ks.uiuc.edu/Research/vmd/)	Academic	Massive parallel MD code and viewer/analyzer
5325	NWChem	http://en.wikipedia.org/wiki/NWChem	NWChem Development Consortium (http://www.nwchem-sw.org/)	GPL	High-performance QM/MM simulation software
5326	PHON	http://chianti.geol.ucl.ac.uk/~dario/pubblicazioni/COMPHY3792.pdf	Dario Alfè, University College London, UK (http://www.homepages.ucl.ac.uk/~ucfbdxa/phon/)	Free	Module for phonon spectrum and thermal property calculations. Interface to an electronic structure code to get the force constants
5327	Phonopy	http://phonopy.sourceforge.net/	Atsushi Togo, Kyoto University, Japan (http://atztogo.users.sourceforge.net)/(http://atztogo.github.io/)	Free	An open-source package of phonon calculations based on the supercell approach. Interface to VASP, Wien2K, Abinit, QE, and FHI-aims
5328	RedMD	http://dx.doi.org/10.1002%2Fjcc.21223	ICM, University of Warsaw, Poland (http://bionano.icm.edu.pl/software/redmd)	GPL	Reduced MD. Package for coarse-grained simulations
5329	StruMM3D (STR3DI32)		Exorga, Inc. (http://www.exorga.com/)	Commercial/free	MM software with molecule builder and viewer and analyzer

(*continued overleaf*)

Table 5.2 (*continued*)

No	Name of code	Information or reference	Software developer or provider	License	Description
5330	SCIGRESS	http://en.wikipedia.org/wiki/Scigress	SCIGRESS (http://www.scigress.com/)	Commercial	QM/MM/MD, builder, and analyzer
5331	TINKER	http://en.wikipedia.org/wiki/TINKER	Washington University (http://dasher.wustl.edu/tinker/)	Free	Software tools for molecular design
5332	VEGA ZZ		Vega ZZ network (http://www.vegazz.net/)	Academic	3D viewers and analyzers, MOPAC and NAMD interfaces, database engine, parallel design
5333	VLifeMDS		vLife Sciences Technologies (http://www.vlifesciences.com/)	Commercial	Complete MM package and toolkit for combinatorial studies
5334	WHAT IF		Radboud University, The Netherlands (http://swift.cmbi.ru.nl/whatif/)	Commercial	Visualizer and GUI to GROMACS
5335	YASARA		YASARA community (http://yasara.org/)	Commercial	Yet Another Scientific Artificial Reality Application: portable MM code

Table 5.3 Electronic structure codes with origins in quantum chemistry.

No	Name of code	Information or reference	Software developer or provider	License	Description
5401	ACES II and ACES III	http://www.qtp.ufl.edu/ACES/	University of Florida, Gainesville, FL, USA (aces@qtp.ufl.edu)	GPL	Advanced concepts in electronic structure theory: DFT, HF, and post-HF
5402	ADF	http://www.scm.com/	Dr. S. J. A. van Gisbergen, info@scm.com CEO, Scientific Computing & Modelling NV, Vrije Universiteit, Amsterdam, The Netherlands	Commercial	Amsterdam Density Functional
5403	CASINO (QMC)	http://www.tcm.phy.cam.ac.uk/~mdt26/casino2.html	Dr. Richard Needs (rn11@cam.ac.uk), TCM Group, Cavendish Lab., Cambridge, UK	Academic	Quantum Monte Carlo program
5404	CFOUR	http://www.cfour.de/	Prof. Jürgen Gauß (gauss@uni-mainz.de), Mainz, Germany	Academic	Coupled-Cluster techniques for Computational Chemistry
5405	COLUMBUS	http://www.univie.ac.at/columbus/	Hans Lischka (Hans.Lischka@univie.ac.at), University of Vienna, Austria	Academic	Computational chemistry software suite, related to DALTON
5406	CRYSTAL	http://www.crystal.unito.it/	Prof. Roberto Dovesi (roberto.dovesi@unito.it), University of Torino, Torino, Italy	Commercial/academic	Quantum chemistry (QC) program for crystals
5407	DALTON	http://daltonprogram.org/	Prof. Trygve Ulf Helgaker, Universitet et i Oslo, Norway	Academic	Molecular electronic structure program suite

(*continued overleaf*)

Table 5.3 (*continued*)

No	Name of code	Information or reference	Software developer or provider	License	Description
5408	DIRAC	http://wiki.chem.vu.nl/dirac/index.php/Dirac_Program	Prof. Trond Saue (trond.saue@irsamc.ups-tlse.fr), Université Paul Sabatier (Toulouse III), France	Academic	Relativistic QC program for molecules
5409	ErgoSCF	http://www.ergoscf.org/	Elias Rudberg, Emanuel H. Rubensson, and PawelSalek, info@ergoscf.org	GPL	QC program for large-scale self-consistent field calculations
5410	FHI-aims	https://aimsclub.fhi-berlin.mpg.de/	Prof. Matthias Scheffler (scheffler@fhi-berlin.mpg.de), Fritz Haber Institute of the Max Planck Society, Berlin, Germany	Commercial	Full-potentialall-electron code, numeric basis functions
5411	FreeON	http://en.wikipedia.org/wiki/FreeON	Matt Challacombe (mchalla@lanl.gov), Los Alamos National Laboratory, USA	GPL	Experimental, open-source (GPL) suite of programs for linear scaling quantum chemistry, formerly known as MondoSCF
5412	Firefly/PC GAMESS	http://classic.chem.msu.su/gran/gamess/index.html	Dr. Alex A. Granovsky (gran@classic.chem.msu.su), Moscow State University, Russia	Academic	QC program for x86 and x86-64 processors
5413	GAMESS (UK)	http://www.cfs.dl.ac.uk/	Prof. Martyn F. Guest (guestmf@cardiff.ac.uk), Cardiff University, Wales, UK	Commercial/ Academic (UK)	General-purpose *ab initio* molecular electronic structure program for DFT, HF, and post-HF

5414	GAMESS (USA)	http://www.msg.ameslab.gov/gamess/	Prof. Mark S. Gordon, Iowa State University, Ames, IA, USA	Academic	—
5415	GAUSSIAN	http://www.gaussian.com/	Gaussian Inc., Wallingford, CT, USA., (custserv@gaussian.com)	Commercial	QC program released in 1970 by John Pople *et al.*
5416	Jaguar	http://www.schrodinger.com/	Schrödinger Inc., (info@schrodinger.com)	Commercial	QC package for both gas and solution phases
5417	MOLCAS	http://www.molcas.org/	Dr. Valera Veryazov (molcas@molcas.org), Lund University, Sweden	Commercial	*Ab initio* quantum chemistry software package for molecules
5418	MOLPRO	http://www.molpro.net/	Prof. Hans-Joachim Werner (werner@theochem.Uni-Stuttgart.de), Universität Stuttgart, Germany	Commercial	Quantum Chemistry Package
5420	MPQC	http://www.mpqc.org/	Prof. Edward Valeev (mpqc@valeyev.net), Virginia Tech, Blacksburg, USA	LPGL	Massively Parallel Quantum Chemistry
5421	NWChem	http://www.nwchem-sw.org/	Bert de Jong (Wibe.DeJong@pnl.gov), Pacific Northwest National Laboratory, Richland, WA, USA	Educational	Northwest Computational Chemistry Package
5422	ORCA	http://www.cec.mpg.de/forum/	Prof. Dr. Frank Neese, c/o Secretary Petra Steiger (petra.steiger@cec.mpg.de), MPI für Chemische Energiekonversion, Postfach, Germany	Academic	An *ab initio*, DFT, and semiempirical SCF-MO package

(*continued overleaf*)

Table 5.3 (continued)

No	Name of code	Information or reference	Software developer or provider	License	Description
5423	PSI	http://www.psicode.org/	Prof. Henry F. Schaefer (ccq@uga.edu), University of Georgia, Athens, GA, USA	GPL	Open-source suite of *ab initio* quantum chemistry programs designed for serial or parallel simulations of molecular properties with more than 2500 basis functions
5424	PyQuante	http://pyquante.sourceforge.net/	Prof. Rick Muller (rmuller@sandia.gov), Sandia National Labs, Albuquerque, NM, USA	BSD	Python Quantum Chemistry: an open-source suite of programs for developing quantum chemistry methods
5425	Q-Chem	http://www.q-chem.com/	Q-Chem, Inc., Pleasanton, CA, USA	Commercial	*Ab initio* quantum chemistry package for accurate predictions of molecular structures, reactivities, and vibrational, electronic, and NMR spectra (DFT/HF calculations and post-HF correlation methods)
5426	Quantemol-N	http://quantemol.com/	Quantemol Ltd., University College London, London, UK., (info@quantemol.com)	Academic/commercial	Code to compute observables of electron – molecule collisions such as: • Elastic cross sections • Electronic excitation cross sections •Electron impact dissociation • Ionization cross sections • Scattering reaction rate

5427	Spartan	http://en.wikipedia.org/wiki/Spartan_(software)	Wavefunction, Inc., Irvine, CA, USA (http://www.wavefun.com/)	Commercial	Application containing codes for molecular mechanics, semiempirical methods, HF, DFT, and post-HF models
5428	Siam Quantum	https://sites.google.com/site/siamquantum/	Dr. Teepanis Chachiyo (teepanis@kku.ac.th), KhonKaen Univ., KhonKaen, Thailand	GPL	An *ab initio* quantum chemistry program for molecules within HF and MP2 method
5429	TeraChem	http://en.wikipedia.org/wiki/TeraChem	PetaChem, LLC, Los Altos Hills, CA, USA (http://petachem.com/)	Commercial	QC, *ab initio* MD code to run on graphics processing units (GPUs)
5430	TURBOMOLE	http://www.turbomole.com/	TURBOMOLE GmbH, Karlsruhe, Germany, contact@turbomole-gmbh.com	Commercial	Program package for large-scale quantum chemical simulations of molecules, clusters, and periodic solids using Gaussian basis sets and predictive electronic structure methods such as time-dependent density functional theory (TDDFT), second-order Møller–Plesset theory, and explicitly correlated coupled cluster (CC) methods

Table 5.4 Electronic structure codes with origins in solid-state physics.

No	Name of code	Information or reference	Software developer or provider	License	Description
5431	Abinit	http://www.abinit.org/	Prof. Xavier Gonze (xavier.gonze@uclouvain.be), Université catholique de Louvain, Belgium	GPL	DFT pseudopotential, plane-wave code
5432	CASTEP	http://www.castep.org/	Prof. Mike Payne (mcp1@cam.ac.uk), TCM Group, Cavendish Lab., Cambridge, UK	Commercial/academic	CAmbridge Serial Total Energy Package
5433	CPMD	http://www.cpmd.org/	Dr. Alessandro Curioni (cur@zurich.ibm.com), Zurich Research Laboratory, Switzerland	Academic	Car–Parrinello molecular dynamics, linked to QE
5434	DACAPO	https://wiki.fysik.dtu.dk/dacapo/dacapo	CAMD, Technical University of Denmark (http://www.fysik.dtu.dk/english/research/camd/)	GPL	DFT total energy program, linked to ASE
5435	ELK	http://elk.sourceforge.net/	Dr. Kay Dewhurst (dewhurst@mpi-halle.de), MPI of Microstructure Physics, Germany	GPL	DFT full-potential code, related to EXCITING
5436	EXCITING	http://exciting-code.org/	Dr. Pasquale Pavone (pavone@physik.hu-berlin.de), Humboldt-Universität Berlin, Germany	GPL	Full-potential all-electron DFT code

5437	FLEUR	http://www.flapw.de/	Prof. Dr. Stefan Blügel (S.Bluegel@fz-juelich.de), Peter Grünberg Institute (PGI-1) and Institute for Advanced Simulation (IAS-1), Jülich, Germany	Academic	The Jülich FLAPW code family, full-potential all-electron DFT
5438	FPLO	http://www.fplo.de/	Dr. Klaus Koepernik (k.koepernik@ifw-dresden.de), IFW Dresden, Germany	Academic	Full-potential local-orbital minimum-basis code. Full relativism, alloy disorder
5439	HiLAPW	http://home.hiroshima-u.ac.jp/fpc/manuals/HiLAPW/HiLAPW.html	Prof. Tamio Oguchi (oguchi@hiroshima-u.ac.jp), Hiroshima University, Japan	Academic	Hiroshima Linear Augmented Plane Wave
5440	JDFTx	http://jdftx.sourceforge.net/	Prof. Tomás Arias (muchomas@ccmr.cornell.edu), Cornell University, Ithaca, NY, USA	GPL	Joint density functional theory, formerly DFT++
5441	LM Suite	http://www.lmsuite.org/	Mark van Schilfgaarde (http://www.kcl.ac.uk/nms/depts/physics/people/academicstaff/van-Schilfgaarde-.aspx) King's College London, UK	Academic	Offers a range of first principles codes with a linear muffin-tin orbital (LMTO) basis set to describe the electronic structure of different materials. Also includes an empirical tight-binding code TBE
5442	ONETEP	http://www2.tcm.phy.cam.ac.uk/onetep/	Prof. Peter Haynes (p.haynes@imperial.ac.uk), Imperial College London, UK	Commercial/academic	Order-N Electronic Total Energy Package

(*continued overleaf*)

Table 5.4 *(continued)*

No	Name of code	Information or reference	Software developer or provider	License	Description
5443	OpenAtom	http://charm.cs.uiuc.edu/OpenAtom/	Prof. Laxmikant V. Kale (kale@illinois.edu), Siebel Center for Comp Sci., Urbana, IL, USA	Academic	Car–Parrinello Ab Initio Molecular Dynamics software
5444	QE: Quantum ESPRESSO	http://www.quantum-espresso.org/	Prof. Stefano Baroni (baroni@sissa.it), International School of Advanced Studies (SISSA), Trieste, Italy	GPL	Modular Package involving PWscf, PWCOND, PHONON, CPMD, YAMBO, etc.
5445	VASP	http://www.vasp.at/	VASP team, Computational Materials Physics, Vienna, Austria (vasp.materialphysik@univie.ac.at)	Commercial	Vienna Ab Initio Simulation Package is a computer program for atomic-scale material modeling, for example, electronic structure calculations and quantum mechanical molecular dynamics, from first principles
5446	WIEN2k	http://www.wien2k.at/	Prof. Karlheinz Schwarz (kschwarz@theochem.tuwien.ac.at), Dr. Peter Blaha (pblaha@theochem.tuwien.ac.at), TU Wien, Austria	Commercial	Code for electronic structure calculations of solids using density functional theory (DFT). It is based on the full-potential (linearized) augmented plane-wave ((L)APW) + local orbitals (lo) method. WIEN2k is an all-electron scheme including relativistic effects

5447	RSPt	http://fplmto-rspt.org/	Prof. John M. Wills (jxw@lanl.gov), Los Alamos National Laboratory, USA; Prof. Olle Eriksson (olle.eriksson@physics.uu.se), Uppsala University, Sweden	Academic, open source	Relativistic spin-polarized (test) – a full-potential LMTO code for spin-polarized and/or spin-orbit calculations with several LDA and GGA functionals implemented
5448	TB-LMTO	http://www2.fkf.mpg.de/andersen/	Ole Krogh Andersen (oka@fkf.mpg.de), MPI for Solid State Research, Germany	Academic	The Stuttgart TB-LMTO program
5449	LR-LMTO; LmtART	http://www.physics.ucdavis.edu/~mindlab/MaterialResearch/Scientific/index_lmtart.htm	Prof. Sergei Savrasov (savrasov@physics.ucdavis.edu), University of California, Davis, CA, USA	Academic	Linear response LMTO; LmtART – Full-Potential LMTO Program
5450	SPR-KKR	http://ebert.cup.uni-muenchen.de/	Prof. Dr. Hubert Ebert (Hubert.Ebert@cup.uni-muenchen.de), Ludwig-Maximilians-Universität München, Germany	Academic	Full-potential all-electron code for ordered and disordered alloys
5451	EMTO	http://docs.snic.se/wiki/EMTO	Prof. Levente Vitos (leveute@kth.se) and Prof. Andrei Ruban (ruban@kth.se), KTH Royal Institute of Technology, Stockholm, Sweden	Academic	Exact Muffin-tin Orbital method. All-electron code for ordered and disordered alloys

(*continued overleaf*)

Table 5.4 (continued)

No	Name of code	Information or reference	Software developer or provider	License	Description
5452	BigDFT	http://bigdft.org/	Dr. Luigi Genovese (luigi.genovese@cea.fr), Institut Nanosciences et Cryogénie, SP2M/L_Sim, Grenoble, France	GPL	DFT massively parallel electronic structure code
5453	CP2K	http://www.cp2k.org/	Dr. Matthias Krack (matthias.krack@psi.ch), OHSA/D07, Villigen PSI, Switzerland	GPL, open source	Molecular dynamics, including QUICKSTEP for DFT calculations
5454	DFT++	http://jdftx.sourceforge.net/	Prof. Tomás Arias (muchomas@ccmr.cornell.edu), Cornell University, Ithaca, NY, USA	GPL	New name: JDFTx
5455	MADNESS	http://code.google.com/p/m-a-d-n-e-s-s/	Prof. Robert J. Harrison (Robert.Harrison@stonybrook .edu), Inst. for Advanced Comput.Science, Stony Brook University, NY, USA	GPL	Multiresolution Adaptive Numerical Environment for Scientific Simulation
5456	CONQUEST	http://www.order-n.org/	Dr. David Bowler (david.bowler@ucl.ac.uk) and Emer. Prof. Mike Gillan (m.gillan@ucl.ac.uk), University College London, UK	Academic	Linear scaling, or O(N), DFT electronic structure code
5457	DFTB+	http://www.dftb-plus.info/	Dr. Bálint Aradi (balint.aradi@bccms.uni-bremen.de), Universität Bremen, Germany	Academic/commercial	Density functional-based tight binding

5458	DMol3	http://people.web.psi.ch/delley/dmol3.html	Dr. Bernard Delley (bernard.delley@psi.ch), WHGA/123, Villigen PSI, Switzerland	Commercial	DFT code for molecules, surfaces, and crystals with numerical basis
5459	GPAW	https://wiki.fysik.dtu.dk/gpaw/	Prof. Karsten Wedel Jacobsen (kwj@fysik.dtu.dk), CAMD, DTU, Lyngby, Denmark (http://www.fysik.dtu.dk/english/research/camd/)	GPL	Grid-based projector-augmented wave method
5460	Octopus	http://www.tddft.org/programs/octopus/	Carlo Andrea Rozzi (carloandrea.rozzi@unimore.it), CNR-INFM Università di Modena e Reggio Emilia, Modena, Italy	GPL	Pseudopotential *ab initio* electronic structure calculations within density functional theory (DFT), in its time-dependent form (TDDFT)
5461	OpenMX	http://www.openmx-square.org/	Prof. Taisuke Ozaki (t-ozaki@jaist.ac.jp), Japan Advanced Institute of Science and Technology (JAIST), Japan	GPL	Open-source package for Material eXplorer
5462	PLATO	http://homepages.lboro.ac.uk/~masdk/plato.html	Dr. Steven Kenny (S.D.Kenny@lboro.ac.uk), Loughborough University, Loughborough, Leicestershire, LE11 3TU, UK	Academic	Package for Linear Combination of Atomic Orbitals
5463	SIESTA	http://departments.icmab.es/leem/siesta/	Prof. Pablo Ordejón (pablo.ordejon@cin2.es), UAB, Bellaterra (Barcelona), Spain	Academic	Spanish Initiative for Electronic Simulations

Materials Genomics (*pymatgen*) (http://pymatgen.org/), an open-source Python library for materials analysis.

In Europe the European Materials Modelling Council (EMMC) has been formed (http://www.emmc.info/). The availability of high-performance computing facilities also naturally connects the material modeling networks with the networks in computer science, such as the e-Science (http://en.wikipedia.org/wiki/E-Science#Europe). The structure of these networks is constantly developing.

Software packages and individual programs for DDD simulations, electronic structure calculations, and atomistic simulations (including some multiscale and mesoscale modeling tools) are listed in Tables 5.1–5.4. The software may be available on a commercial basis (commercial) or freely distributed (free) or in some cases under a GNU Public License (GPL) or is available only to academic or educational institutions (academic). The lists are by no means complete, but are just to provide information on the diversity of the field and to list the main software developers (companies, communities, and some prominent individuals).

Further Reading

Argon, A. (2008) *Strengthening Mechanisms in Crystal Plasticity*, Oxford University Press, Oxford.

Arsenlis, A., Cai, W., Tang, M., Rhee, M., Oppelstrup, T., Hommes, G., Pierce, T.G., and Bulatov, V.V. (2007) Enabling strain hardening simulations with dislocation dynamics. *Modell. Simul. Mater. Sci. Eng.*, **15**, 553–595.

Becker, C.A. *et al.* (2013) Considerations for choosing and using force fields and interatomic potentials in materials science and engineering. *Curr. Opin. Solid State Mater. Sci.*, **17**, 277–283, http://www.ctcms.nist.gov/potentials (accessed 11 May 2016).

Bulatov, V. and Cai, W. (2006) *Computer Simulations of Dislocations*, Oxford University Press, Oxford.

Cramer, C. (2002) *Essentials of Computational Chemistry: Theories and Models*, John Wiley & Sons, Inc.. ISBN: 0471485527

Finnis, M. (2003) *Interatomic Forces in Condensed Matter*, Oxford University Press.

Frenkel, D. and Smit, B. (1996) *Understanding Molecular Simulation: From Algorithms to Applications*, Academic Press, San Diego, CA.

Gale, J.D. (2005) Interatomic potential models for ionic materials, in *Handbook on Materials Modelling, Part A. Methods* (ed. S. Yip), Springer, Dordrecht, pp. 479–497.

Jensen, F. (1998) *An Introduction to Computational Chemistry*, John Wiley & Son, Ltd. ISBN: 0471984256

Justo, J.F. (2005) Modeling covalent bond with interatomic potentials, in *Handbook on Materials Modelling, Part A. Methods* (ed. S. Yip), Springer, Dordrecht, pp. 499–507.

Kohanoff, J. (2006) *Electronic Structure Calculations for Solids and Molecules Theory and Computational Methods*, Cambridge University Press. ISBN: 9780521815918

Kubin, L. (2013) *Dislocations*, Mesoscale Simulations and Plastic Flow, Oxford University Press, Oxford.

LAMMPS Open Source MD Simulator, http://lammps.sandia.gov (accessed 11 May 2016).

Messerschmidt, U. (2013) *Dislocation Dynamics During Plastic Deformation*, Springer, Heidelberg.

Mishin, Y.M. (2005) Interatomic potentials for metals, in *Handbook on Materials Modelling, Part A. Methods* (ed. S. Yip), Springer, Dordrecht, pp. 459–478.

MMonCa Open Source KMC Simulator, http://www.materials.imdea.org/MMonCa (accessed 11 May 2016).

Parr, R.G. and Weitao, Y. (1994) *Density-Functional Theory of Atoms and Molecules*,

International Series of Monographs on Chemistry, vol. **16**, Oxford University Press. ISBN: 0195092767

Pippan, R. and Gumbsch, P. (2010) *Multiscale Modelling of Plasticity and Fracture by means of Dislocation Mechanics*, Springer, Wien, New York.

Raabe, D. (1998) *Computational Materials Science: The Simulation of Materials Microstructures and Properties*, Wiley-VCH Verlag GmbH, Weinheim.

Raabe, D., Roters, F., Barlat, F., and Chen, L.-Q. (eds) (2004) *Continuum Scale Simulation of Engineering Materials*, Wiley-VCH Verlag GmbH, Weinheim.

Rosso, L. and de Baas, A.F. (eds) (2012) *What Makes a Material Function? Let Me Compute the Ways ... Modelling in FP7 NMP Programme, Materials projects*, Third version, Publications Office of the European Union, Luxembourg, http://ec.europa.eu/research/industrial_technologies/pdf/modelling-brochure_en.pdf (accessed 11 May 2016).

Ruban, A.V. and Abrikosov, I.A. (2008) Configurational thermodynamics of alloys from first principles: effective cluster interactions. *Rep. Prog. Phys.*, **71**, 046501.

Sanchez, J.M. (2010) Cluster expansion and the configurational theory of alloys. *Phys. Rev. B*, **81**, 224202.

Shin, C.S. (2004) 3D discrete dislocation dynamics applied to dislocation-precipitate interactions. PhD thesis. INP Grenoble & Séoul National University.

Szabo, A. and Ostlund, N.S. (1996) *Modern Quantum Chemistry: Introduction to Advanced Electronic Structure Theory*, Dover Publications. ISBN: 0486691861

Tadmor, E.B. and Miller, R.E. (2014) *Modeling Materials: Continuum, Atomistic, and Multiscale Techniques*, Cambridge University Press, Cambridge.

Tadmor, E.B. and Miller, R.E. (2014) Modeling materials: continuum, atomistic, and multiscale techniques, Chapter 5, in *Empirical Atomistic Models of Materials*, Cambridge University Press, Cambridge, pp. 237–303.

Wikipedia Force field (chemistry), http://en.wikipedia.org/wiki/Force_field_(chemistry) (accessed 11 May 2016).

Wikipedia List of Molecular Modelling Software, http://en.wikipedia.org/wiki/List_of_software_for_molecular_mechanics_modeling (accessed 11 May 2016).

Wikipedia List of Quantum Chemistry and Solid-state Physics Software, http://en.wikipedia.org/wiki/List_of_quantum_chemistry_and_solid-state_physics_software (accessed 11 May 2016).

Yip, S. (ed.) (2005) *Handbook on Materials Modelling, Part A. Methods*, Springer, Dordrecht.

6
Effective Properties

Ludovic Noels, Ling Wu, Laurent Adam, Jan Seyfarth, Ganesh Soni, Javier Segurado, Gottfried Laschet, Geng Chen, Maxime Lesueur, Mauricio Lobos Thomas Böhlke, Thomas Reiter, Stefan Oberpeilsteiner, Dietmar Salaberger, Dieter Weichert, and Christoph Broeckmann

6.1
Computational Homogenization Methods and Codes: An Overview

6.1.1
Review of Homogenization Methods for Heterogeneous Materials

With the emergence of complex materials, the numerical simulations of structures require insight into the material behavior at the lower scale and into the material structure in order for the numerical analyses to remain accurate. As direct numerical simulations accounting explicitly for this material structure are unreachable because of the overwhelming computational cost, multiscale methods were developed in order to enrich structural simulations with results from analyses conducted at the material structural scale.

Homogenization is a key ingredient of multiscale methods for heterogeneous materials. In homogenization-based approaches, the macroscale problem is defined as the structural scale of interest and is characterized by a characteristic length l_{macro} linked to the gradient of the structural loading (see Figure 6.16.1). This macroscale problem Ω defines a boundary value problem (BVP), which is solved using a numerical method that considers homogenized material properties. In contrast to one-scale approaches, multiscale approaches extract those homogenized properties at each (macro) material point $\mathbf{X}$ of interest from the resolution of a mesoscale BVP. This mesoscale BVP is defined on a mesoscale volume ω, which aims at representing the heterogeneities of the material characterized by a size l_{micro} (see Figure 6.1).

In the multiscale homogenization method depicted in Figure 6.1, the homogenized macro stress $\langle\sigma\rangle$ and macro strain $\langle\varepsilon\rangle$ tensors correspond to the average value of, respectively, the local stress σ and strain ε on the mesoscale volume ω. The consistency of the method is ensured by defining "energetically consistent" boundary conditions (BCs) in such a way that the deformation

Handbook of Software Solutions for ICME, First Edition. Edited by Georg J. Schmitz and Ulrich Prahl.

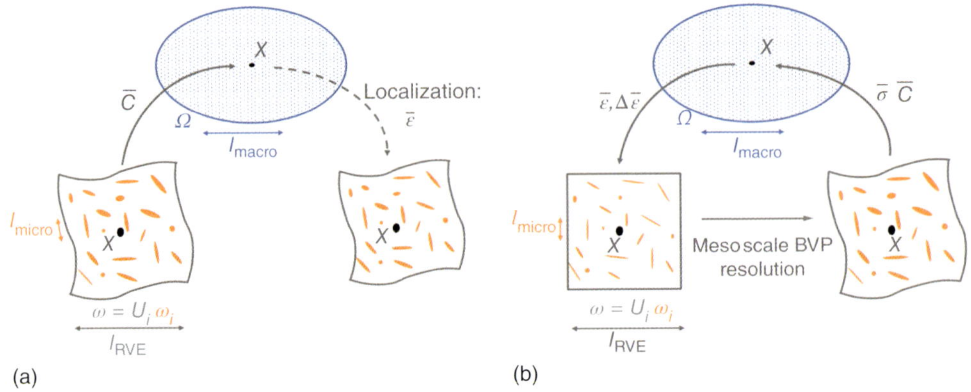

Figure 6.1 The definition of a homogenization-based multiscale analysis. (a) In the "sequential" approach the mesoscale homogenized properties are obtained from prior resolution of the mesoscale boundary problem. The local stress field can be evaluated afterward through a "localization" step. (b) In the "concurrent" approach the resolutions of both the mesoscale and macroscale boundary value problems (BVP) are coupled and involve "downscaling" and "upscaling." (Reproduced with the permission of e-Xstream Engineering.)

energy at the macroscale correspond to the deformation energy at the microscale (Hill–Mandell condition), see [1], for example, one important condition is the statistical representativity of this mesoscale volume ω, so that it is entirely typical of the whole mixture on average [2]. Practically, if the size of the mesoscale volume element ω is large enough, the homogenized properties depend neither on the volume element definition nor on the energetically consistent BCs type applied on it [3]. In this case, the mesoscale volume element is said to be representative and called "representative volume element" or RVE of size l_{RVE}. From these definitions, one can deduce the length-scale separation conditions:

$$I_{\text{macro}} \gg I_{\text{RVE}} \text{ to ensure the relevance of homogenized properties;}$$
$$I_{\text{RVE}} \gg I_{\text{micro}} \text{ to ensure the RVE existence.} \tag{6.1}$$

Besides bounds of the material properties of the homogenized material, for example, the Hill–Voigt upper bound [2, 4] (constant strain field over the RVE) and the Hill–Reuss lower bound [2, 5] (constant stress field over the RVE) for linear materials, or again the narrower Hashin–Shtrikman bounds obtained from a variational principle for isotropic materials [6], among others, the homogenized response can be obtained from the resolution of the mesoscale BVP defined on ω. The mesoscale homogenized properties can be deduced from prior resolutions of this mesoscale boundary problem, in which case the method is called "nonconcurrent" or "sequential" and only upscaling (from the microscale to the macroscale) is required, see Figure 6.1a. The local (microscale) stress field evolution can in some cases be evaluated through a "localization" step. This approach is particularly efficient in the case of linear or reversible microscale material behaviors, proportional monotonic loadings such as tensile tests. In the more general case of history-dependent microscale material behaviors, it might be required to conduct the resolutions of both the mesoscale and macroscale BVPs in a coupled way, which involves the "downscaling" of the macroscale strain to define the BCs to be applied on the RVE, and the "upscaling" of the homogenized stress and material tensors, for example, see Figure 6.1b. Such a coupled framework is called "concurrent" multiscale resolution.

For both sequential and concurrent homogenization-based multiscale approaches, while the macroscale BVP is usually solved using the finite element approach, the mesoscale BVP can be solved using different methods. Extensive reviews of the different homogenization methods have been conducted by Kanouté *et al.* [7], Geers *et al.* [8], and Llorca *et al.* [9], for example. In the following only the main ideas of those methods will be discussed, and we refer to those references for details. Afterward we will discuss the new developments arising in homogenization methods.

The unit-cell methods, in which a RVE is finely meshed and from which the mesoscale material properties can be extracted, is probably the simplest "sequential" analysis. This method was extensively developed in the context of reinforced elastoplastic composite materials [10–16], among many others.

In the method of (sub)cells or generalized method of cells (GMCs), periodic unit cells of unidirectional fiber-reinforced matrix composite materials are discretized by parallelepiped subcells. Local equilibrium can then be solved in an average way in each subcell, while compatibility of force and displacement expressed in an average way at the interfaces, and completed by periodic cell BC, lead to closed-form equations, allowing an efficient semianalytical resolution of the problem, including in the nonlinear range. The method was first developed using linear approximation for the displacement field within each subcell, see the review by Aboudi [17], before being extended to the use of a second-order displacement field [17] in order to improve the prediction of the local stress and strain fields.

Another semianalytical method for elastoplastic composite materials is the transformation-field analysis (TFA), which divides the RVE into subdomains having at most one phase (but a phase can be divided in several subdomains) [18]. The key points of this incremental approach are (i) to consider in each subdomain a local eigen-strain tensor representative of the plastic strain, thermal effects allowing the local constitutive law to be rewritten in a linear form, and (ii) to establish relations between the local and global fields through transformation influence tensors, which depend solely on the linear elastic properties of the phases. In the original TFA formulation the local fields are considered piece-wise uniform, and in the case of two-phase composite materials represented by two domains, the transformation influence tensors can be obtained in a closed-form solution [19] from the Eshelby [20] tensor (solution of a single inclusion in an infinite matrix) through a Mori–Tanaka (MT) method (one inclusion is interacting with a homogeneous medium undergoing the matrix deformation [21, 22]) or a self-consistent method (one inclusion is embedded in a fictitious matrix possessing the homogenized material stiffness which is sought [23, 24]). However, for nonlinear heterogeneous materials, an accurate solution requires a subdivision of each phase in several subdomains due to the nonuniformity of the plastic strain within the phases, yielding extra computation cost and reducing the method efficiency. Moreover, the transformation influence tensors have now to be evaluated by the finite element resolution of periodic unit cells in which each subdomain has a constant deformation state. The nonuniform transformation-field analysis (NTFA) was developed in order to avoid the subdivision of one phase into subdomains [25]. In this formalism, the eigenstrain tensor is formulated as a finite combination of nonuniform modes. In order to reduce the number of internal variables, it is possible to couple the different modes supported by the same phase. The main difficulty of this method is to define a priori the nonuniform modes.

The mean-field homogenization (MFH) approach is an efficient semianalytical framework for the modeling of multiphase composites in which the homogenized strain (stress) field follows from the average value of the strain (stress) in the phases. The equations are completed by a strain concentration tensor relating the average strain in the inclusions to the average strain in the matrix. This strain concentration tensor is obtained by the extension of the single inclusion solution of Eshelby [20] to multiple inclusions interacting in an average way in the composite material, usually either through the MT [21, 22] or on the self-consistent scheme [23, 24]. In the nonlinear range, the linear equations are applied on a so-called linear comparison composite (LCC) [26–31], which is a virtual composite material whose constituents linear behaviors match the linearized behavior of the real constituents at a given strain state.

Several definitions of this LCC exist:(i) the incremental tangent formulation [32–35] considers linearized relations between the stress and strain increments of the different constituents around their current strain states, (ii) the affine method considers the total strain field instead of strain increments during the homogenization process in combination with a stress polarization tensor [31, 36–41], and (iii) the secant method joins the origin to the current strain/stress state [42]. On the one hand, in the secant method, the linearized law is pseudoelastic in terms of total stress and strain, which limits the method to monotonic and proportional loading paths. On the other hand, both the affine and incremental tangent formulations have shown to predict acceptable to accurate results in the case of nonlinear behavior of multiphase materials providing the tangent operator to be made isotropic (as the isotropic nature is lost during the plastic flow) during the homogenization process [34, 43, 44]. To avoid this isotropization step, an incremental-secant MFH approach for elastoplastic materials was developed [45] in which, at a given strain/stress state of the composite material, a virtual unloading step is applied on the composite level before applying a secant approach from the residual strains obtained upon unloading. Besides avoiding the isotropization step, this method naturally handles nonmonotonic and nonproportional loadings as the incremental and affine formulation. Note that MFH was also developed in order to be applicable for mechanical analysis involving large strains.

Most MFH methods only consider first-moment statistical values, that is, the average stress tensor, to predict the plastic flow in the different phases. This can lead to poor predictions in the elastoplastic case, mainly for short fibers composite materials, as the plastic yield is not accurately captured [46]. To improve the prediction, the second-moment statistical values can also be considered [47–51]. It was shown in [48] that the variational homogenization forms pioneered by Ponte Castañeda [29] correspond to a second-order secant formulation, which was called modified secant. Recently, incremental variational formulations, which also correspond to a second-moment estimation, were proposed [52–57]. Finally the iterated variational homogenization [58, 59] was developed for porous material. In this formulation the microstructure is iteratively constructed by the addition of the matrix phase in an incremental way, starting with a high volume fraction of inclusions and ending with the targeted final volume fraction. This way of proceeding significantly improves the estimates for nonlinear composite materials as the local fields are represented more accurately.

In order to avoid the assumptions lying behind the semianalytical methods, and to provide a more integrated multiscale method than the unit-cell method, the RVE can be solved numerically in a concurrent way with the macroscale analysis. The basic idea originates from the Voronoï cell finite element method (VCFEM) [60, 61]. In this method, the macrostructure is divided into Voronoï cells which are defined by considering as germination points the heterogeneities. The more general computational homogenization, also called FE^2, method associates to each macroscale Gauss point a finite element (FE) model of the RVE which has to be solved (iteratively in the nonlinear range) for each macroscale iteration. The mesoscale BVP is defined from the internal variables history of the RVE and from the macroscale deformation tensor. The resolution of this mesoscale BVP provides the homogenized macroscale stress tensor $\overline{\sigma}$ as well as the consistent RVE material operator $\overline{C}$. The basic ideas of the computational homogenization approach have been presented in [1, 62–67] as a nonexhaustive list. The main advantage of the FE^2 method is its ability to consider general microstructures, general material behaviors, as well as large deformations. In order to include a size effect, the FE^2 method has been extended to macroscale generalized continua, leading to second-order multiscale computational homogenization (SMCH) [68–71], which provides a systematic way to couple the generalized continuum at the macroscale with the classical continuum at the microscale. The FE^2 method is quite general and can easily be used in parallel (as the microproblem resolutions are independent from each other) but suffers from a high computational cost particularly in the nonlinear case as each macroscale iteration requires several iterations of the RVE problem.

Another difficulty of the FE^2 method is to provide a RVE mesh embedding complex microstructures. To avoid this burden, the fast Fourier transforms (FFT) method was developed so that the RVE can be discretized as data sampled from a regular grid, possibly obtained from digital images of the real microstructure [72]. In this method, material properties are first assigned to each sampling point pixel in the 2D case or voxel in the 3D case of the grid. The phases are assumed to be perfectly bound and in order to close the mesoscale BVP, the periodic BC is applied. A preliminary problem is then defined as a homogeneous linear elastic body subject to a polarization tensor within the RVE. The polarization tensor depends on the difference of the real material tensor at the sampling point and of the material tensor of the preliminary problem. The iterative resolution of this last relation in the real space and of the preliminary problem expressed in the Fourier space converges to the solution of the mesoscale BVP, without requiring the computation of a stiffness matrix. The method was also extended to the nonlinear case in [73]. The convergence of the scheme strongly depends on the choice of the material tensor of the preliminary problem, and the initial FFT method [72] cannot handle problems with the presence of rigid inclusions or voids. Several alternative schemes have been proposed to deal with these issues as the use of augmented Lagrangian [74] for example.

Instead of defining the multiscale problem from the average stress and strain fields on the RVE combined to energetically consistent BC (see Figure 1.1) for periodic microstructure (or when it can be assumed as such), the multiscale equations can be derived using the asymptotic homogenization method (see [75] for a review). This method relates the stretch coordinate vector y representative of the repeated structure to the local coordinates x thought $y = x/\in$, which represent the length scale separation. At this point, it is assumed that the perturbation of a field is also periodical on Y, leading to a decomposition such as $\sigma\varepsilon(x) = \sigma 0(x,y) + \varepsilon\sigma 1(x,y) + \cdots + \varepsilon k \sigma k(x,y)$, where $\sigma j(x,y)$ is periodic on the RVE sides for $j > 0$, and similarly for the other fields. The governing equations at the different scales can now be obtained by, on the one hand, applying the chain rule on the balance equation

$$\frac{\mathrm{d}}{\mathrm{d}x}\cdot\sigma^{\varepsilon} = \frac{\partial}{\partial x}.\sigma^{\varepsilon} + \frac{1}{\varepsilon}\frac{\partial}{\partial y}.\sigma^{\varepsilon}$$

and, on the other hand, by splitting the resulting equations into the terms of the same order. As a result, when applying the constitutive relations, the stress tensor at order n depends on the displacement at orders n and $n + 1$. Using the periodic BC and averaging the equations of different orders yield for order zero the definition of the so-called homogenization functions and for order 1 the macroscale governing equation in terms of the macrostress. The macrostress is defined as the volume average of 0 $(x; y)$ on Y and is expressed in terms of the homogenization functions, which are obtained using the finite element method. Asymptotic homogenization can consider successively different scales, although most of the time only two-scale methods are used. Note that the theoretical background of the VCFEM is also the asymptotic homogenization derivation. Finally, the asymptotic homogenization method can be reformulated with an eigen-strain tensor requiring eigen-strain influence functions to account for plastic deformation [76], in which case an incremental method is required. This last case can be seen as a combination of asymptotic homogenization with the TFA approach.

The first multiscale homogenization methods were mainly developed not only for nonlinear elastoplastic or viscoplastic materials but also for thermomechanical problems [18, 33, 54, 61, 77–79], for example, and have been shown to lead to accurate and predictions. Despite all these achievements, the challenges related to multiscale homogenization methods remain numerous. Being by no mean nonexhaustive, we intend to discuss some of them in what follows.

In the last years, the homogenization methods were extended to account for material failure. In the context of crack propagation, the main difficulties not only lie in the methodology required to propagate the discontinuities through the different scales but also on the loss of representativity of the mesoscale volume element because of the presence of the microcracks [80–82]. This loss requires adequate formulations to recover the method consistency such as using appropriate BCs [80, 83, 84], using the concept of failure zone averaging [81], considering periodic structures [82, 85], or by propagating the discontinuities to the macro level by characterizing the loss of ellipticity at the subscale [86]. In [87], the reaction forces of the mesoscale BVP, which embeds crack propagation, are used to characterize the fracture properties of a crack at the macroscale. Clearly the concurrent resolution of the multiple nonlinear microscale problems with the nonlinear macroscale problem induces prohibitive costs. The unit-cell method was recently used to extract a mesoscale traction-separation law of carbon-fiber-reinforced epoxy matrix [88] and of polycrystalline materials [89] following the method proposed by Verhoosel *et al.* [87]. This mesoscale cohesive law can thus be readily used in macroscale simulation, however the crack loading modes are *a priori* assumed.

In the context of continuum damage mechanics, the GMC was extended to account for the damage evolution in each subcell [90]; however, only a unit cell was studied without accounting for strain softening. The asymptotic homogenization method using eigenstrain tensors proposed in [76] was extended to phase and interface nonlocal (by volume averaging) damage in [91]. Recently, a nonlocal MFH accounting for the damage evolution of the matrix phase of the composite material was developed using either an incremental tangent [92] approach or an incremental-secant MFH approach [93]. In the latter, it was shown to be possible to account for the elastic unloading of one of the composite material phases during the strain softening of the other phase. In order to avoid the strain/damage localization caused by the matrix material softening, an implicit nonlocal formulation [94–97] was adopted during the homogenization process, which allows the simulation of laminate structures to be in good agreement with the experiments, in particular concerning the damage evolution directions [98]. The iterated variational homogenization was considered in [99] to study the macroscopic response of porous plastic metals at finite strains. This method is able to capture the evolution of the porosity and the changes in the average shape and orientation of the pores. As a result the homogenized response can be seen as a plasticity model which can predict shear localization.

The different homogenization methods have recently been extended to multiphysics problems such as electromagnetic using asymptotic homogenization [100, 101], magneto-electro-thermoelasticity using mean-field homogenization [102], finite electromechanics using computational homogenization [103], thermo-electro-mechanics using the subcells method [104], magneto-electro-elasticity using the unit-cell method combined to an X-FEM representation

[105], and so on. However, most of them bear some limitations and the field is still under investigation.

The previously described homogenization methods were developed for problems in which there is no dynamic effect confined to the microscale. In the case of timescale separation, some works have investigated multiscale homogenization analyses which takes into account microstructural inertia effects, by using computational homogenization [106], peridynamic formulation [107], and asymptotic homogenization [108]. Note also the derivation of the problem equations proposed in [109].

Although allowing accurate predictions for a large range of material behaviors and microstructural organizations, the concurrent computational multiscale approach implies a high computational costs in the nonlinear range as Newton–Raphson iterations have to be conducted at the two scales in a coupled way. This has motivated the development of more computationally efficient methods. In the linear range a decoupled resolution of the mesoscale BVP allows extracting the mesoscale homogenized material tensor that can be used during the macroscale simulation. The material map method [110, 111] extends this concept to the elastic nonlinear range by precomputing discrete homogenized stress–strain states. The parameters domain of the mesoscale constitutive law is then meshed with multidimensional elements to interpolate the values of the precomputed stresses, allowing the material model to be used by the macroscale elements. This concept was enhanced, still in the elastic nonlinear range, by deriving the mesoscopic stress–strain relationship from a database containing the values of the effective strain density function [112]. In the phase/interface damage two-scale problem stated using the asymptotic homogenization method proposed in [91], the microscale Eigenstrain tensors and damage-related variables are decomposed as a linear combination of phase/interface shape functions. Similarly, their weighted averages are obtained using weight functions orthogonal to the shape functions. The phase/interface shape functions are defined on subdomains wider than the underlying RVE FE mesh, thus reducing the computation cost of the mesoscale BVP resolution. In this so-called eigendeformation-based reduced order homogenization method, the main problem is the definition of the adequate basis of the reduced size problem in order to ensure accuracy. This issue can be addressed by an optimization process [113]. Note that both TFA and NTFA approaches also belong to the reduced order homogenization category, and to some extend MFH too as the number of mesoscale variables is reduced.

Another problem related to the homogenization is the definition of the RVE. In order for the multiscale process to be accurate, the RVE geometry should be representative of the reality. One method is to construct the RVE from image acquisition [114], or to build the RVE using microstructure generation tools [115]. In both cases, the recourse to level set functions is a key ingredient. Because of the resulting complex microstructure, a resolution of the mesoscale BVP using the X-FEM method, avoiding the need for conforming mesh, can be envisioned [114].

Besides the definition of the microstructure geometry, the existence of the RVE is not always ensured as the mesoscale volume element can lack statistical representativity. Strictly speaking, a mesoscale volume element is representative if the homogenized results are not dependent on its size and on the (energetically consistent) BCs [3]. In case of nonperiodic microstructures, this requires Eq. (1.2) to be satisfied, which is not always possible; indeed, this could lead to RVE sizes either too large to be computationally affordable, or to large as compared to the size of the macroscopic loading variation, in which case Eq. (1.1) is not satisfied anymore. In this case, the mesoscale volume element is not statistically representative and different homogenized properties are obtained for different mesoscale volume element realizations (i.e., different random microstructures), sizes, or BCs. The mesoscale volume element is then called statistical volume element (SVE) [116].

To address this lack of statistical representativity, it is possible to study SVE of different sizes with different BCs and for each size to perform several realizations in order to extract mean homogenized properties and their dispersion [117]. One can thus use the average values for deterministic macroscopic simulations. This method requires a large number of realizations of the different SVEs to reduce the error on the average effective mesoscale properties.

Another approach is to propagate the uncertainties at the microscale to the macroscale, leading to stochastic multiscale methods. To this end, the realizations of the SVEs can be used to define a mesoscale random field. Computational homogenization methods have been applied in [118] to define the probability convergence criterion of masonry RVE, and in [119] to study the scale dependency of homogenization for random composite materials. In the work of [120], the simulations of SVEs are used to capture the stochastic properties of the parameters in the constitutive model with which the uncertainties are propagated to the structural scale using the stochastic FE method [121]. In [122], the resolution of composite material elementary cells is used to explicitly define a mesoscale potential with the aim of studying the uncertainties in the fibers geometry/distribution in the case of finite elasticity. Recently, a stochastic three-scale methodology has been developed [123] to study the mechanical response of small dimension structures made of polycrystalline materials. In this approach, (i) at the microscale several SVEs with random grain orientations are considered; (ii) at the mesoscale finite element simulations on different SVEs, defined from a larger material sample by using the window technique [124], lead to the distribution of the homogenized polycrystalline material properties and of their spatial correlation; and (iii) a random field of the (correlated) mesoscale elastic tensor is generated -the generation of mesoscale random field is discussed in [125–131] based on the information obtained from the SVE simulations and is used with the stochastic FE method to predict the probabilistic behavior at the structural scale. Besides the stochastic approaches based on random fields, stochastic multiscale analyzes have also been developed to account for random variables by using order reduction of asymptotic homogenization.

6.1.2 Homogenization in Industrial Application: Current State of the Art

The main driver for Integrated Computational Materials Engineering (ICME) in an industrial context is the difficulty of describing the influence of the material behavior on an ever increasing number of complex composite designs. Today the formerly independent topics of materials and manufacturing processes intercombine. The target is to minimize the amount of used material whereas the part performance is to be maximized.

The overall trend goes toward a computer-aided design approach rather than relying on costly experiments in each cycle of development. Given the above goals, this poses a real challenge and calls for advanced simulation approaches such as ICME. The overall constraint in industry is, however, that whatever new technology will arise – if it is to be applied productively, it must comply with the constraint that it shall be easy to use and fast in performance. This call for efficiency is the global mantra.

The accurate description of complexity and efficiency are targets though that point into opposite directions. The balance between the two basically determines today's state-of-the-art of using ICME in industrial applications. Depending on the topic, different solutions and strategies have emerged.

Micromechanical material modeling is the general link between the composite's microstructure and its effective properties. Thus, in-depth knowledge in micromechanics cannot be avoided if the target is to mirror realistically the performance of the part and correlating to its experimentaliy observed behavior. Therefore, composite design is a team effort in which material experts, part designers and manufacturers work in a synergetic fashion to achieve the goal.

6.1.2.1 Homogenization Technology

The choice of method for the extraction of effective properties depends on the material and manufacturing process of focus. A basic and meaningful distinction between composites can be made between metals and polymer-based materials.

Metals bear multiphase microstructures which are prone to complex changes upon manufacturing, including phase transitions. However, once the composition can be described and the properties of phases are known, it is relatively straightforward to compute a representative local stiffness and to map it onto the FE model to run analyses coupled to results from process prediction.

The main approach to derive effective properties for metals is to use FEA for the homogenization of RVEs. The challenge here is to come up with a proper description of the microstructure via a FE mesh and to add the knowledge about the properties of the single phases to the simulation model [132]. For the sake of efficiency, voxel-based meshing is quite commonly used.

Polymer-based composites usually follow the strategy to enhance stiffness and strength via fiber reinforcement. In comparison to polycrystalline metals, this leads to much simpler microstructures. In contrary, complexity for polymers most often arises from the fact that the matrix material exhibits nonlinear performance which transfers onto the global composite behavior. In addition, fiber–matrix interactions cause yet another difficulty, especially when it comes to the description of failure and fatigue.

Materials like short and long fiber-reinforced plastics, classical UD and as well as woven and braided composites are commonly modeled based on Eshelby-type of approaches. Depending on the need, those can be extended to cover nonlinearity as in MT homogenization [133].

A broad range of relevant material performances for industrial application can be covered this way, for example:

- (Thermo-)elasticity
- (Thermo-)elastoplasticity
- (Thermo-)viscoelasticity
- (Thermo-)elastoviscoplasticity

All this can be described in combination with strain rate and temperature-dependent failure models and even difficult performances like creep and fatigue can be covered. In summary, ready-to-use solutions for homogenization are implemented in commercial software and cover an acceptable range of material types and performances to describe [134].

6.1.2.2 Parameterization of Material Models

A major difference between an academic and an industrial approach lies in the way how micromechanical material models are parameterized. Homogenization can be summarized as the art to cleverly combine the performances of multiple phases to describe the end performance of a composite material, whereas the description of each of those phases is based on continuum mechanical models. Each of those bears a certain set of parameters. The question is just how can one obtain a suitable set of such parameters?

The *direct engineering* approach stands for a straightforward application of homogenization technology. To follow this strategy means to directly enter the measured performances of the *isolated phases* into the homogenization scheme. Composite properties are thus derived from an observed per-phase behavior. In real application this strategy results in a poor description of the composite performance on the macroscopic level. The reason is that this approach neglects the complexity of interaction in real composite materials. For example, perfect bonding between phases is most often assumed in the homogenization scheme whereas real interactions arise from the area of phase boundaries.

A practical workaround for this situation is to switch to the strategy of *reverse engineering*. It follows a different thinking in the sense that here homogenization technology is seen as a tool more to describe the anisotropic performance of a composite material rather than serving the purpose to deliver an exact physical description of the underlying interactions in the material. In *reverse engineering* a suitable micromechanical model is parameterized to match a maximum amount of anisotropic measurements of the composite under different loads. The parameters of the phase materials are varied in iterative steps until the global response of the micromechanical model matches the experimentally observed behavior of a composite best.

Henceforth, for each additionally computed microstructure (and/or load case) the obtained model will predict the best possible response of a composite. Of course a high-quality *reverse engineered* model will contain material parameters that reflect some physical sense. Yet the material response they represent does not necessarily match any longer the experimentally observable behavior of the isolated material.

The quality of material models in *reverse engineering* largely depends on the amount of input curves taken for the procedures. In addition, of course the physics as described by the homogenization scheme used still will determine the overall limits of what can be achieved. A quantitative match between the real composite response and some material model is a high level goal. Certainly it can be reached, but it requires large effort in terms of measurements in addition to in-depth knowledge in micromechanics.

The beauty of the approach is that material models with an already acceptable quality can be set up in a quick and straightforward fashion. The major advantage for industrial applications arises from this fact. Expert knowledge is not necessarily required. If combined with common sense for the physical description of the composite the material model can be set up based on a minimum amount of tests within a few minutes. Whole databases of generic material models can be set up following this idea [134].

Such efficient implementations are of great interest to the industries. Strategically, qualitative and semiquantitative material models can be used in an early design stage and major benefit can arise through inscale coupling which means that the impact of manufacturing influences can be investigated in a screening-like fashion.

The quantitative performance of materials is of higher interest in the late stages of design. Here the final composite design needs to be validated, and security aspects play the major role. For this, the material model must of course deliver the most realistic description of performance possible.

6.1.2.3 ICME Strategy

The above thoughts clearly show that the real challenge for ICME in an industrial context is to choose the right strategy to achieve a given goal. Efforts in terms of model setup and computation are to be kept low with the highest possible impact on the quality of results and additional information.

Thus in industry, it is important to deal with simulation tasks as direct as possible. It is thus more important to discuss the ICME strategy than the homogenization technology per se. The trick is to embed the right method in an applicable strategic background (Figure 6.2). The most important step toward this goal is a clear definition of the simulation target. Below, some possible distinction will be highlighted.

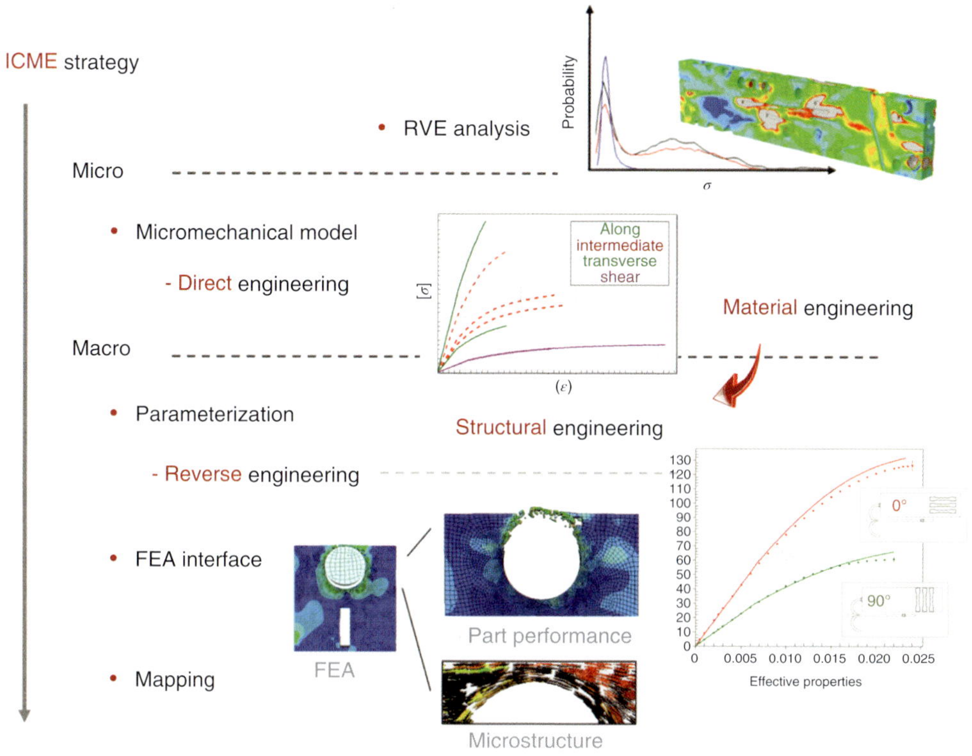

Figure 6.2 ICME strategy for using effective properties in an industrial context. Material engineering deals with the investigation of composites across the scales, with the final goal of delivering high-quality material models to describe the performance of promising material candidates. Structural engineering uses micromechanical modeling to compute effective properties. The final goal is in-scale coupling from manufacturing toward part performance which includes the mapping of local properties onto the part. (Reproduced with the permission of e-Xstream Engineering.)

6.1.2.4 Material Engineering

Material Engineering is the art of understanding composites in-depth, to innovate materials based on this knowledge and to follow a micromechanical approach in describing their real performance. Material engineers use simulations to manipulate composites on the microscopic scale and investigate the influence of those variations on the microscopic and macroscopic scales.

In industrial research this approach allows to gain deep insight into the materials themselves and to systematically understand mechanisms that dominate their macroscopic properties arising from the microscopic composition. Based on this, promising candidates for new composite materials can be identified. The approach follows the *direct engineering* strategy.

The next level of *material engineering* would then deal with the task to provide high-quality micromechanical material models. The setup of quantitative micromechanical models typically follows the *reverse engineering* strategy. The constituents' parameters are varied in a characteristic physical range to match the global performance of the material as observed in anisotropic measurements. Such models find further use in integrative simulations as described further in the following text.

6.1.2.5 **Structural Engineering**

Structural Engineering is the art of designing producable composite parts with low weight and best possible performances. The design of composite parts is a challenge due to the many influencing factors that impact the targeted performance. Composite material properties are highly sensitive to changes in microstructure. Such changes are due to the manufacturing of the part. Integrative simulations take into account the variation of local properties as arising from the manufacturing process.

Microstructure information is thus fed into a micromechanical material model to describe realistically the distribution of material properties over the part. Results from coupled analyses can give quantitative correlation with the performance of the part as observed in measurements. In the simulation strategy the micromechanical model serves basically as a tool to map the influence of manufacturing onto the material properties as found in the part.

For linear elastic performance mapping is straightforward. The stiffness property can directly be transferred in a *weak coupling* scheme. However, as soon as nonlinear performance must be covered things become more complex. Micromechanical models are usually implemented in a *strong coupling* scheme. In every step of analysis, for each integration point homogenization technology is run to compute the local material property at a given time. Whereas currently this strategy results in the highest possible quality of description, from a computational perspective the procedure is relatively inefficient, especially in explicit analyses.

An enhancement is offered by moving toward *hybrid coupling*. The main advantage of the micromechanical model is that it is capable of transforming any input information given on the microstructural level into a material property on the macroscopic scale. In *strong coupling*, this is tackled at once. In the *hybrid coupling* approach, the goal is achieved stepwise. The micromechanical model is used in a preprocessing step to compute the material performance under all possible microstructure arrangements covering different load scenarios. In that sense, it serves as a virtual laboratory that produces a wide range of test results in a very short time.

Based on those virtual tests, it is then possible to parameterize an effective, macroscopic model that is suited to describe the target performance of the polymer. Whereas the micromechanical model is based on homogenization and thus delivers results on each scale for each composite phase, the effective model reduces the information to the performance on the macroscopic scale and describes it dependent on the microstructure.

The implementation of such macroscopic models can be far more efficient than the direct application of micromechanics and results in a large gain in CPU performance. In the sense of ICME, the *hybrid coupling* basically separates the coupling across the scales and in-scale into subsequent steps. The in-scale step, the coupling of the manufacturing with the part performance, is the major interest in the industries. The cross-scales homogenization procedures are only a necessary tool that facilitates the mapping of the effective properties.

6.2 Finite Element-Based Homogenization

6.2.1 Effective Properties of Polycrystalline Materials

Polycrystalline materials are any kind of heterogeneous media in which the microstructure is formed by the aggregation of several volumes (grains) formed by one or more different crystalline materials (phases) and having each grain a particular shape, size, and orientation of the corresponding material lattice, (Figure 6.3a).

Polycrystal homogenization provides a bridge between micro- and macroscale by means of integration of the microscopic strain and stress fields within the different grains to obtain the macroscopic stresses and strains in the polycrystal. This kind of approach is applicable to problems with a clear separation of scales, that is, those in which the typical length scale associated with the gradients of the mechanical fields at the macroscale is large compared with the typical length scale of the polycrystalline microstructure (e.g., the grain or subgrain size). Within this framework, the influence of the microscopic features of the polycrystal (grain size, shape, and orientations as well as elastic constants and the CRSS of the different

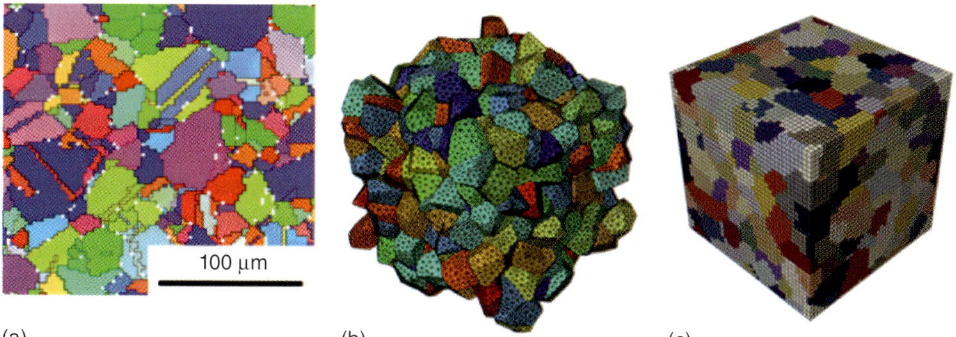

(a) (b) (c)

Figure 6.3 (a) Polycrystalline microstructure of an Mg alloy where colors represent grain orientations obtained by EBDS. (b) Voronoi-based RVE discretized by FE, and (c) Voxel-based RVE generated from Dream3D.

slip and twinning modes) on the macroscopic response can be taken into account. Polycrystal homogenization is a very complex, nonlinear problem that has been solved with two different approximations, namely, mean-field methods [135, 136] and computational homogenization [137–139].

In this chapter, attention will be focused on computational polycrystalline homogenization for being the most suited option to be used in an ICMEg environment. The computational homogenization techniques aim to predict the macroscopic behavior solving a boundary value problem on a RVE of the microstructure and integrating the microfields resulting of that problem. Several ingredients are fundamental in a computational polycrystalline homogenization framework: the phases' constitutive equations, the crystal plasticity model which is the common choice, the RVE, and the solving method.

6.2.1.1 Computational Polycrystalline Homogenization

The crystal plasticity (CP) model: CP model is a continuum constitutive equation describing the anisotropic deformation of a crystal produced by dislocation slip, twinning, or other micromechanical deformation mechanism. The basis of the model was established in the 1960s and 1970s by Hill and Rice [140–142]. A more detailed formulation is referred to their early work and to other reviews on crystal plasticity [143]. CP is not a standard model and therefore is not included in any commercial FE codes. Several subroutines have been developed for these codes, among them the Damask code [144] and the one developed by Segurado and coworkers [139].

The RVE: The RVE is a sample of the polycrystalline material that is entirely representative of the microstructure on average, and contains a sufficient number of grains for the apparent properties to be independent of the surface values of traction and displacement, see Figures 6.3b and c. RVEs should contain the information about grain size, shape, and orientation statistical distributions. Several noncommercial codes are available to generate synthetic RVEs, among them Neper [145] and Dream3D [146] are the most common approaches.

The boundary value solving methodology: Computational homogenization is based on solving a BVP problem on the RVE, normally subjected to periodic BCs. Two main approaches appear here, the classic one is the use of FE method to solve the problem [136, 139]. In this case both noncommercial homemade codes and commercial packages as Abaqus are used. The other approach that has been widely used in the last 5 years is based on the use of the FFT to solve by means of Green's function the periodic BVP of a heterogeneous media. This approach is based on the pioneer work of Suquet and Moulineq for composites and adapted to elastoviscoplastic materials by Lebensohn and others [138, 144].

6.2.2 Variation of the Effective Elastic Properties During $\gamma-\alpha$ Phase Transformation of a Low-Carbon Steel, Simulated by the Phase-Field Method

The numerical evaluation of varying thermomechanical properties of multiphase steels undergoing microstructural changes in manufacturing processes, for example, during heat treatment, is a complex task. On the one hand, the kinetics of the phase transformation has to be evaluated not only in terms of integral phases but one has to handle the full morphology and its changes in a complex 3D microstructure. On the other hand, from these evolving microstructures the material properties have to be derived accurately. Phase-field methods [147] are nowadays widely used to simulate microstructure formation and its evolution in time and space for multicomponent and multiphase materials like low-carbon steels. A second essential ingredient is the accurate evaluation of effective mechanical properties of different kinds of polycrystal steel microstructures and thus to establish a direct link between its microstructure and mechanical properties. To achieve this goal the asymptotic homogenization method has been extended to take into account eigen-strains induced by solid-phase transformations [148]. This approach offers the opportunity to calculate in one simulation all the effective thermoelastic properties of simulated 3D microstructures.

6.2.2.1 Phase-Field Simulation of the Austenite–Ferrite Phase Transformation in a Fe–C–Mn Steel

The austenite-to-ferrite transformation is an important solid-state reaction in steels. Dependent on their chemical composition, both phases may have a significant different molar volume, eigen-strains and eigenstresses play a role and influence the transformation. We consider here a graded steel disk, made of Fe–0.023C–0.17Mn–0.009Si and produced by sand casting. The cooling process is simulated thermally and local cooling curves have been extracted and transferred to the multiphase-field simulation. These curves define the heat extraction rate of the simulated RVE. The thermoelastic properties of ferrite and austenite together with their molar volumes are given in [148]. Temperature-dependent surface energies and mobility are adopted here. Moreover, the surface energy and the mobility of low angle grain boundaries (GBs) are reduced as function of the misorientation angle θ, according to the Read–Shockley model [149] for the surface energy and to the Humphreys model [150] for the GB mobility. In addition to these materials data, a model for ferrite nucleation is needed. In this simulation ferrite grains nucleate at the intersections of the triple lines at 823°C corresponding to a critical undercooling of 5°C. If this temperature is reached at a quadruple point, a ferrite nucleus is set.

A statistical homogenization procedure has been recently developed [151] in order to define the size of the initial austenitic 3D microstructure. This study

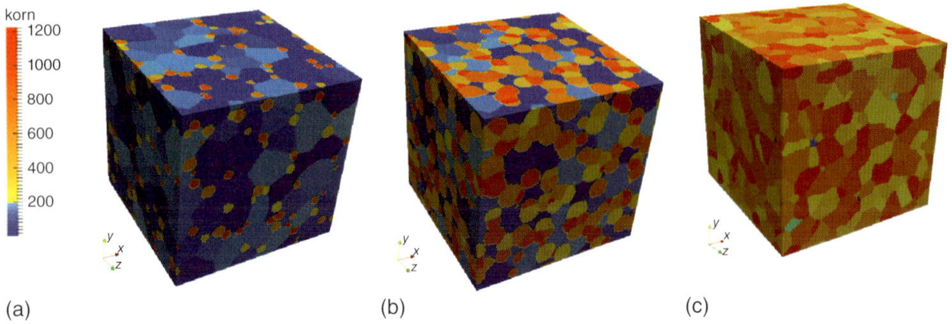

Figure 6.4 Sequence of simulated low-C steel microstructures during phase transformation. Grain numbers between 1 and 185 are austenitic grains (blue) and above this value are ferrite ones (orange, red). (a) $t = 38$ s, $T = 862$°C, $f_\alpha = 9.2\%$, (b) $t = 42$ s, $T = 844$°C, $f_\alpha = 49.7\%$, (c) $t = 120$ s, $T = 699$°C, $f_\alpha = 99.6\%$.

shows that a RVE of 125 grains, generated by a Voronoi tessellation and dote with orientations, distributed according to the Halton sequence [152], allows the prediction of effective isotropic elastic modules within a defined tolerance. This physical RVE has a size of $73.5 \times 73.5 \times 73.5\,\mu m^3$ and a grid spacing of 0.5 µm. Thus, the mean radius of an austenite grain is nearly 8 µm. In Figure 6.4, the evolution of the microstructure during the austenite-to-ferrite phase transformation, starting from 1000°C down to 697°C, is shown. Due to the low carbon content, the austenitic phase transforms into ferrite (837 grains) without cementite formation but 32 residual austenite grains with high carbon concentration remain. At the end of the transformation, the mean grain size is reduced to 4.47 µm.

6.2.2.2 Effective Thermoelastic Properties of the Fe–C–Mn–Si Steel Microstructures

At selected time steps of the $\gamma-\alpha$ phase transformation, the actual grain microstructure with its orientation, defined as a quaternion, are transferred to the homogenization tool HOMAT.

After cooling start at 1000°C, the Young and shear moduli of the pure austenite phase increase classically with the temperature decrease, as outlined in Figure 6.5a and b, respectively. But, during $\gamma-\alpha$ phase transformation, both modules decrease strongly in the temperature range [860 to 826°C]. After transformation again both moduli of the quasi pure ferrite microstructure increase with further cooling. The effective Young $E_{ii}(\mathrm{T})$ and shear moduli $G_{ij}(\mathrm{T})$ per structural direction present their maximum deviation from the respective mean value for the initial pure austenite state: $\Delta E_{\mathrm{m},2} = -0.95\%$; $\Delta G_{\mathrm{m},13} = -1.80\%$. The deviations are always less than the allowed tolerance (2%) and confirm that the effective macroscopic behavior of the polycrystal is quasi-isotropic during the phase transformation. The curves of the mean Young and shear moduli, $E_\mathrm{m}(\mathrm{T})$ and $G_\mathrm{m}(\mathrm{T})$, are of practical interest and transferred to the subsequent thermomechanical simulation of the cooling of the graded steel disc.

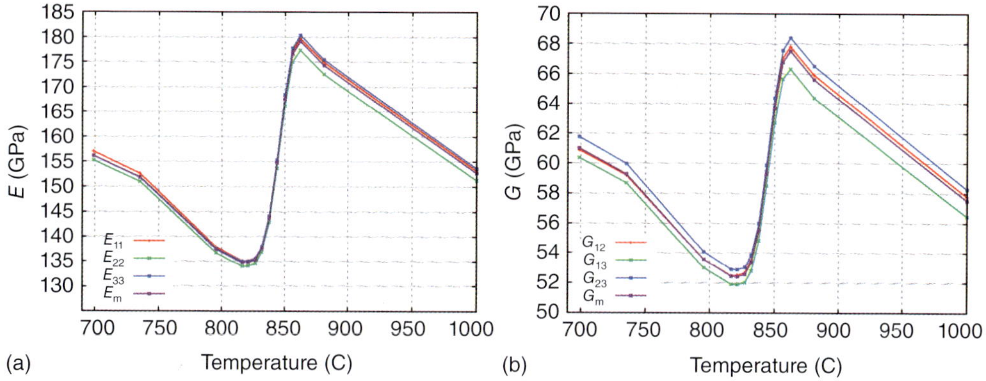

Figure 6.5 Variation of the effective Young moduli E_m and E_i (a) and of the effective shear moduli G_m and G_{ij} (b) with the cooling temperature.

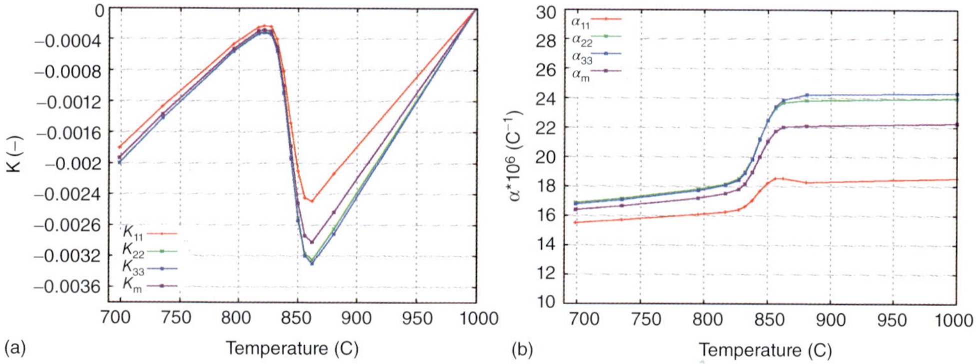

Figure 6.6 Variation of the effective volumetric eigen-strain components k_{ii} (a) and of the effective thermal expansion coefficients (b) with the cooling temperature.

Above 860°C, the effective volumetric eigen-strains k_{ii} of the austenite phase decrease with a constant slope due to thermal contraction of the austenite phase (see Figure 6.6a). Note that component k_{11} varies at least as its slope, given by α_{11}, corresponds to the lowest effective thermal expansion coefficient (see Figure 6.6b). During the $\gamma-\alpha$ phase transformation, all eigen-strain components increase as the molar volume of ferrite is larger than the austenite one. Below 825°C, the eigen-strains k_{ii} diminish again due to effect of the thermal expansion of the ferrite phase. Note that the residual orthotropy of the effective thermal expansion coefficient of the pure austenite is noticeable ($\Delta\alpha_{m,11} = -16.9\%$). But it reduces during transformation and becomes $\Delta\alpha_{m,11} = -5.4\%$ for the ferrite phase. This result underlines that the final ferrite microstructure is more isotropic than the initial austenite one and implies that the RVE size has to be defined for the initial austenite polycrystal by a stochastic homogenization scheme.

6.2.3
A Direct Method-Based Statistical Prediction of the Effective Strengths of Particulate-Reinforced Metal Matrix Composite

Particulate-reinforced metal matrix composites (PRMMCs) constitute an important class of existing structural composites. These materials combine favorable metallic properties such as ductility and toughness with ceramic properties such as high strength and high modulus. PRMMC are characterized by their microstructure where the granular-shaped brittle grains are surrounded by the ductile metal matrix. The distribution of the reinforcement phase is random, therefore the composite microstructure owns a remarkable geometrical complexity and causes serious difficulty for modeling the RVE. Meanwhile, due to the absence of the microstructural periodicity, the infinite domain of the material cannot be embodied by an individual RVE possessing only a finite size. For this reason, to enhance the objectivity of the prediction results, it requires the effective behavior to be obtained from a series of equivalent RVEs and interpreted in a sense of statistics.

In this study, the goal is to evaluate the effective behavior of PRMMCs with taking into account the statistical feature of the results. Beside the conventional effective behavior such as Young's modulus $\overline{E}$ and yield strength $\overline{\Sigma}^{Y}$, the behaviors to be considered also include strengths of the material to resist the monotonic and fatigue typed loading, that is, the ultimate strength $\overline{\Sigma}^{U}$ and endurance limit $\overline{\Sigma}^{\infty}$, respectively.

To overcome the difficulty of building RVE models for reflecting complex composite microstructure, the authors have introduced an image-based model algorithm in their previous study [153]. As demonstrated in Figure 6.7, using this

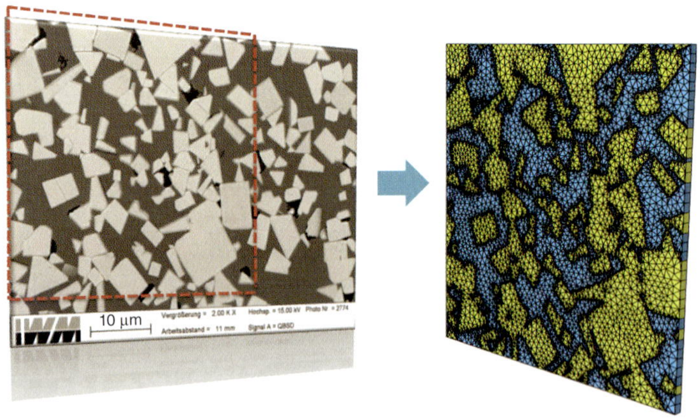

Figure 6.7 Conversion of SEM images into 2.5D RVE model.

algorithm, a large set of scanning electron microscope (SEM) images can be automatically converted into corresponding FE models obtaining an adaptive mesh. Since the models built from this algorithm have a small thickness, these models are called 2.5D models.

To calculate $\overline{\Sigma}^{U}$ and $\overline{\Sigma}^{\infty}$, without explicitly following the entire loading profile, the study employs a direct method-based numerical scheme constructed according to the renowned Melan's theorem [154]. By applying this numerical scheme to 50 statistically equivalent RVE models generated from WC-30 wt% Co microstructure, the associated structural problems are converted into nonlinear programming problems and then solved by means of interior point method. Results obtained from solving these optimization problems correspond to strengths of the material under different types of load combinations. By collecting result from each model and interpreting them statistically, the feasible load domain along with their confidence intervals can be presented as illustrated in Figure 6.8.

In summary, it has been shown how image analysis, nonlinear programming and homogenization techniques can be incorporated to give a fast evaluation of the strength of PRMMC materials. In our future study, based on the established techniques, advanced statistical learning techniques will be further applied to expose the relationship between the strength and composite microstructure.

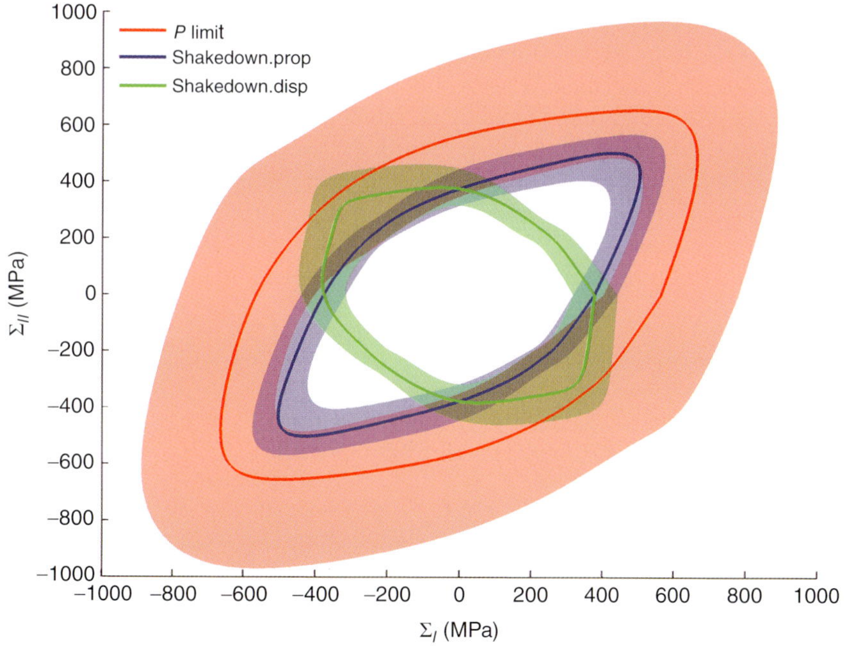

Figure 6.8 Load domains obtained from RVE samples of WC 30 wt% Co. (Figure reprinted with kind permission of e-Xstream engineering.)

6.2.4
Effective Elastic Properties of Semicrystalline Thermoplastic Microstructures of Injection-Molded Parts

Semicrystalline thermoplastics belong to an important group of materials in plastics processing. In contrast to pure amorphous materials, the polymer chains of semicrystalline plastics, like isotactic polypropylene (α-PP), are arranged in periodic patterns when they solidify. Plate-shaped arrangements of the molecules, called lamellae, develop through a folding process and are arranged as crystalline superstructures in the solidified melt. In most injection-molded parts, radial superstructures, called spherulites, are formed at the microscale. The spherulites are made of crystalline and amorphous phases having the same chemical but different physical properties. Because of the radial arrangement of the lamellae the distribution of the amorphous and crystalline domains is influenced by the size and shape of the spherulites. The difference of their material properties leads to effective elastic properties of the semicrystalline thermoplastics that depend locally on the formed microstructure during solidification.

To take into account the complex solidification process of semicrystalline materials, a new multiscale simulation approach has been developed, based on the platform concept introduced in [155]. This approach couples different software tools, as shown in Figure 6.9, to perform the simulation of the injection molding process at the macroscale, the solidification process at the microscale, the prediction of effective elastic properties and their interaction. The data exchange between the different programs was done by using the open source file format VTK.

From the coupled heat and mold filling simulation within COMSOL, the local temperature and velocity fields are extracted and transferred to the in-house 3D microstructure evolution program SphäroSim. Based on the cellular automata formalism, it specifies the nucleation of spherulite germs and describes their radial expansion rate [156]. Semicrystalline polymers, like α-PP, are built at the microscale by an agglomeration of spherulites. Each spherulite grows from its

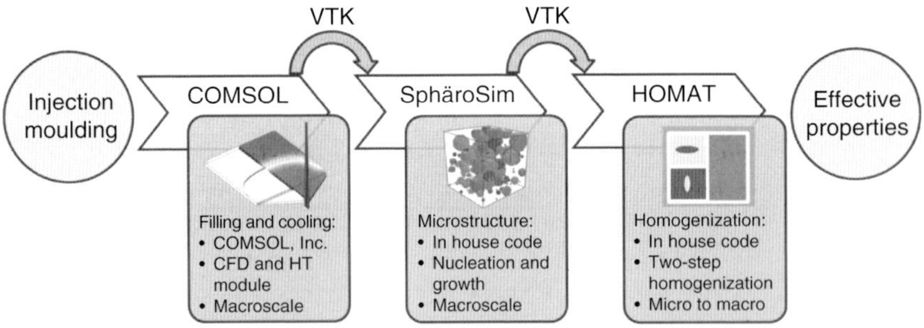

Figure 6.9 Integrative simulation chain for the prediction of effective in injection-molded parts.

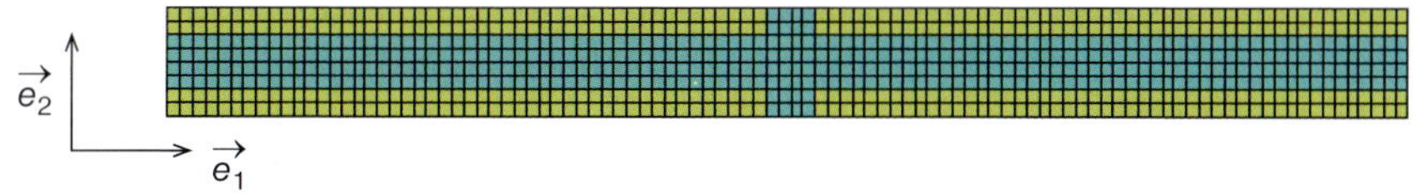

Figure 6.10 Crux-like design of the bilamellae of isotactic PP with crystalline (yellow) and amorphous phases.

monocrystalline center quasi radially and is built of lamellas of crystalline phase surrounded by an amorphous phase (see Figure 1 of Ref. [157]).

The homogenization of semicrystalline thermoplastics is delicate as the amorphous and crystalline phases are not separable and it is impossible to produce bulk test specimens of each phase. Therefore, to derive their effective elastic properties a novel two-level homogenization scheme has been developed, based on the asymptotic homogenization method [158]. At the nanoscale, the homogenization of the lamellae formed of the crystalline and amorphous phases is realized at first. Then at the microscale, the homogenization of the spherulites is performed by assuming a radial distribution of equivalent lamellae around the spherulite center.

Following Bedoui *et al.* [157], we assume the amorphous phase of the α-PP to be isotropic and in a rubbery state: $E_{\text{amorph}} = 0.9$ MPa and $\nu = 0.49993$. The needed mechanical properties of the crystalline phase are derived from *ab initio* calculations and from X-ray diffraction measurements. Their derivation is outlined in [158]. This phase presents a monoclinic behavior with a pronounced high stiffness in molecular chain direction.

6.2.4.1 Homogenization of the Lamellae at the Nanoscale

Based on the observations of Bedoui *et al.* [157], the geometry of the crux-like unit cell is specified for α-PP having a degree of crystallinity of 46% (see Figure 6.10).

The predicted effective Hooke matrix is nearly monoclinic in the local lamella axis system $(\vec{e}_1, \vec{e}_2, \vec{e}_3)$:

$$H^{\text{eff}}_{\text{bila}} = \begin{bmatrix} 1910 & 1780 & -0.007 & -185.7 & 0.004 \\ 2151 & 1923 & -0.006 & -76.18 & 0.008 \\ & 2469 & -0.008 & 342.1 & 0.009 \\ S & & 62.07 & 0.007 & -5.4 \\ & Y & & 795.5 & -0 \\ & & M & & 48.55 \end{bmatrix}$$

6.2.4.2 Effective Elastic Properties of an Injection-Molded 2 mm PP Plate

At first, the injection molding simulation is performed for hot process parameters ($T_{\text{wall}} = 80$°C and $T_{\text{melt}} = 280$°). Then, for five elements over the plate thickness, the spherulite microstructure evolution is predicted. Corresponding results can be found in [159]. The predicted effective Young modules in the structural axis over the plate thickness and experimental in-plane values are given in

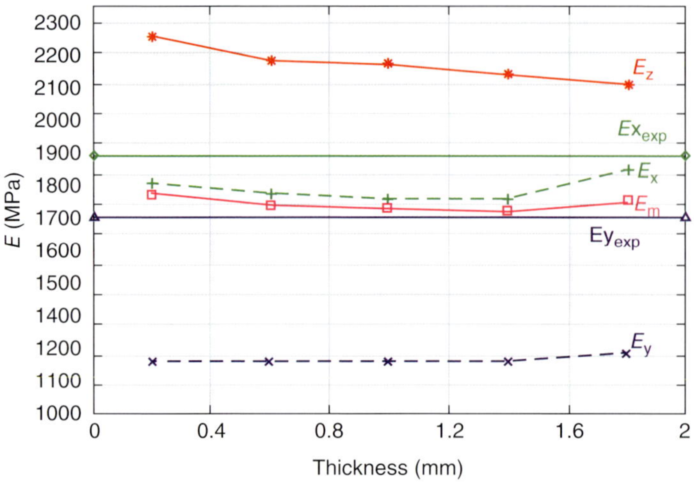

Figure 6.11 Variation of the predicted effective Young modulus over the plate thickness and comparison with experimental values.

Figure 6.11. Note that direction X is identical with the flow direction. The two-level homogenization scheme predicts a pronounced orthotropic effective behavior with: $E_Y < E_X < E_Z$. Predicted Young modulus E_X and the mean one are in a good agreement with the experimental value in flow direction. Only transverse modulus E_Y is significantly lower than the measured one, leading to a larger in-plane orthotropy. Young modulus E_Z decreases over the plate thickness, whereas E_X and E_Y are higher at the plate border, where the spherulites are smaller due to higher cooling rate, than in its center.

6.2.5
On the Effective Mechanical Properties of Discontinuous Fiber Composites (DFC): Application to a Ribbed Beam

Discontinuous fiber composites (DFCs) are heterogeneous systems made of strands chopped from a pristine unidirectional composite tape. The unidirectional composite tape being a combination of unidirectional long fibers (carbon or glass) embedded in thermoplastic (PEEK) or thermosetting (epoxy, phenolic, etc.) matrix.

This class of material continues to gain interest in aerospace applications as well as in other industrial sectors.

Such materials allow to achieve good stiffness performances [160, 161] while being suited for compression molding [162–165]. The latest being of utmost importance for the aerospace industry as it allows high production rates and the possibility of creating both thin and bulky/3D parts.

In this work we are interested in predicting the curing behavior of a ribbed beam (see Figure 6.12) of which interest is the prediction of the in-service mechanical

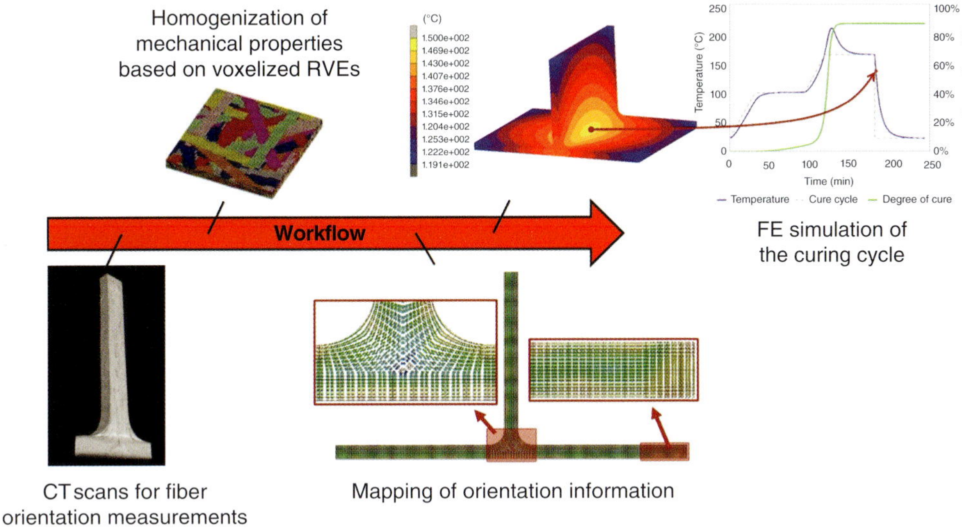

Figure 6.12 ICME workflow for DFC application. (Reproduced with the permission of e-Xstream Engineering.)

behavior. Thus, we are linking process-induced phenomena (e.g., microcracks) to the serviceability prediction of a given component.

6.2.5.1 Determination of the Distribution of Fibers Orientation

In order to predict the mechanical behavior of a component made of DFC material, such as the one presented in Figure 6.12, it is very important to be able to obtain information regarding the fibers, that is, strands, respective orientation.

For components obtained via injection molding of short-fiber-reinforced plastics (SFRP), numerical methods already exist for predicting the orientation of the short fibers. This methods are based on the Folgar–Tucker model (derived from the Jeffery equations) with various closure methods for obtaining the evolution of the orientation tensors. However, these methods are based on a list of assumptions which are too strong for application to the fiber orientation prediction of DFC.

We then relied on an experimental method for obtaining information regarding the strands orientation in the ribbed beam. Thanks to CT scans performed on the component and the postprocessing of the results, via the dedicated software VolumeGraphics®, we were able to retrieve information on the fibers orientation in the component at the end of the compression molding process.

The different orientations were then mapped to the structural mesh that we wanted to use for the curing cycle analysis.

6.2.5.2 Computation of the Homogenized Mechanical Properties

For the sake of reducing computation time, it is not possible to keep the detail of every strands in the FE model. We thus decided to rely on homogenization techniques for being able to work with bigger solid FEs which contain homogenized mechanical properties that are linked to the stress tensor associated to the element.

Many homogenization schemes can be found in the literature [44, 51, 78, 166] However, due to the high complexity of the mesostructure, we decided to use the full-field homogenization approach. This approach consists in analyzing RVEs as being low-scale FE problems, thus we ran thermomechanical FE simulations over a wide range of RVEs (one per orientation tensor) in order to obtain the homogenized properties, for example, Young's moduli, shear moduli, Poisson's ratios, and coefficients of thermal expansion (CTEs) of the material. Every homogenized mechanical property is dependent on both temperature and degree of cure.

The RVEs consists of a packing of voxelized strands of various lengths and widths which are oriented in such a way, inside the RVE, that we match the orientation tensor extracted from the CT scans. The variability, in terms of both length and width, comes from classical Gaussian distribution of which parameters come from experimental measurements.

6.2.5.3 Analysis of the Curing Cycle

The curing cycle is one of the most critical step in the compression molding of components made of DFC. Indeed, the various strands orientation in the component induces abrupt changes of mechanical properties over small distances. One of the most important property to take into account is the anisotropy of CTEs. These quick changes of CTEs will induce different expansion modes in the component, linked to the applied cure cycle, and thus produce residual stresses that can be correlated to the apparition of microcracks during the process.

In order to exhibit such behaviors and thus predict the apparition of defects in correspondence with the cure cycle, we ran a FE thermomechanical simulation on the whole structural mesh of which mechanical properties are set, FE per FE, as the one obtained from full-field homogenization over RVEs.

6.2.5.4 Analysis of the In-Service Mechanical Behavior

After performing the previously presented operations, it is then possible to run (thermal) mechanical structural analysis which aim at predicting the serviceability of the considered component under this service loading conditions.

For running these structural simulations, the mesh and orientations remain the same, the distribution of residual stresses is readily taken into account from the curing cycle analysis and the homogenized mechanical properties stay the same.

The "in-service" simulations give us the displacement, stress, and strain history that can further be used for failure prediction based on macroscopic failure criteria relying on the homogenized stresses and/or strains.

6.2.5.5 Outcome

A complete process-to-structure analysis of DFC-based ribbed beam was presented. The complete workflow allows accurate prediction in both terms of process influence on residual stresses appearance and defects development and serviceability of the component.

Full-field homogenization is a powerful tool for obtaining mechanical properties of highly heterogeneous materials. However, the high number of orientation tensors obtained from CT scans makes this method time costly for preprocessing of the structural model.

One of the main drawback of the presented workflow is the necessity to rely on CT scans for obtaining information regarding the orientation tensors. This method is costly in both time and material and thus the development of an accurate fiber orientation prediction model for compression molded DFC-based components will be a necessity in the near future.

6.3 Mean-Field Homogenization

6.3.1 Fiber-Reinforced Overmolded Composite Parts: An Industrial Application Example

One challenge of modern integrated material modeling methodologies is to understand the influences of the material structure, usually mostly induced by the process, over the heterogeneous material properties. In this subchapter we will present an application showing how it's possible to link short and continuous fiber-reinforced composites microstructures to material and structural behaviors. Being able to establish such a link allow to reach a better design for a given specific performance of a part under a constraint such as, for example, minimizing the overall weight. The following application was performed using Digimat [167].

6.3.1.1 Integrated Material Modeling of Polymer-Reinforced Composites

The targeted application is exposed in Figure 6.13 and represent an overmolded beam made of a short-fiber-reinforced polymer (SFRP) core within continuous fiber-reinforced polymer (CFRP) skins.

The SFRP and CFRP material microstructures are induced by the respective forming processes, for example, injection and draping. In the case of injection of SFRP, fiber orientation is diffused, varies spatially along the structures as well as strongly through the thickness. For CFRP, in this case a 2D woven material, the respective and relative orientation of the two yarns also varies from point to point. Such microstructural variations strongly influence material behaviors (e.g., elastic stiffness, nonlinear plastic behavior, failure, etc.). So the micromechanical material models are used to bridge the gap between the influence of processing and the performance of the final part in simulation via material microstructure and its influence over the material effective properties.

To identify the material parameters of the selected SFRP material model, one classical way to go is to inject a plaque from which one extracts three dumbbells (see Figure 6.14) in three different directions (usually 0°, 45°, and 90° with respect to the main injection flow direction). As the fibers are mainly oriented in flow direction at the skins of the plaque, and transversely at core, tensile tests over

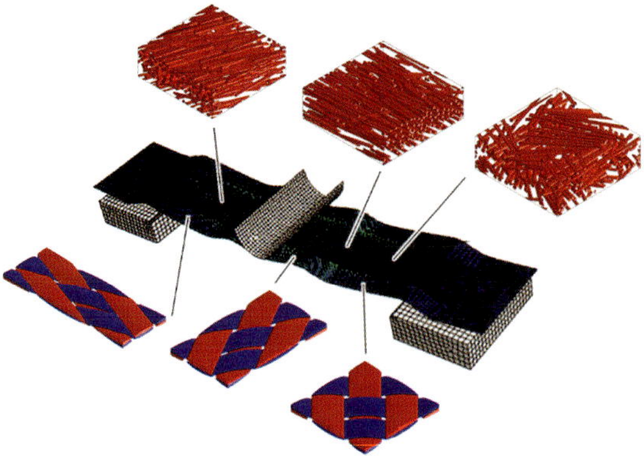

Figure 6.13 Representation of the composite beam under three-point bending. Illustration of the variability of the material (SFRP and CFRP) microstructures following the process (respectively, injection and draping).

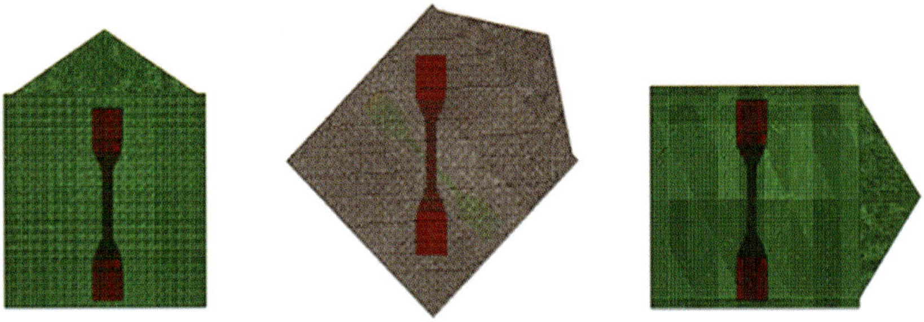

Figure 6.14 SFRP-injected plaque and extraction of dumbbells for material model parameter identification.

the three dumbbells will provide information about the anisotropy of the material behavior.

Concerning the CFRP the process is quite similar: A panel is manufactured with one or several fiber orientations, coupons are cut in different directions and submitted to tensile tests. In the current application the coupons contain a single family of yarns orientations, for example, 0°/90°.

The material models used in this application relies on mean-field homogenization which allows to model both the effects of material microstructure and constituents behaviors over the nonlinear composite response. Two different ways of using homogenization are here of application:

- *Sequential homogenization:* In such methods homogenization is performed prior to structural computations and used to predict all the material parameters required by an equivalent macrolevel material models, as well as their variations with local microstructures (or other varying fields).

- *Synergistic homogenization* (which also falls sometime in the category of concurrent homogenization techniques): Which resolves simultaneously the structural and homogenizations problems by using localization techniques (to compute micro fields starting from macrofields) and averaging techniques (to compute macrofields starting from microfields).

These kind of methodologies are implemented in Digimat [167] and are respectively called hybrid and micro solution procedures.

The material parameters of these material models are identified, using reverse engineering procedures, based on the mechanical responses of the dumbbells and coupons extracted from the plaques. Figure 6.15 shows both the experimental and numerical predictions for the SFRP and CFRP materials. Starting from these material parameters, the micromechanical model is then able to predict the material responses for any kind of microstructure when subjected to any mechanical loading conditions.

The composite beam was produced using both injection molding for the SFRP core (using Autodesk Moldflow [168]) and draping for the CFRP skins (using Dassault Systèmes Simulayt [169]). As schematically illustrated on Figure 6.13, this induces a varying microstructure which is handled by the model.

So the material models based on homogenization and the scale transition techniques plays a central role when coupling the processing steps to the computation of the structural performance. When such integrated methodology is applied one can compute the behavior of a structure following how it was processed. In this example the mechanical behavior of a pure SFRP-injected beam and the overmolded composite beam can be computed as illustrated on Figure 6.16.

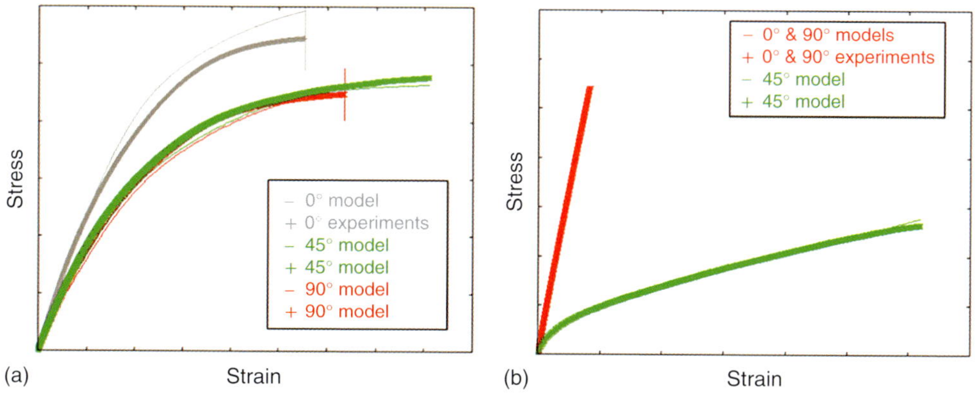

Figure 6.15 Experimental and numerical responses of the SFRP dumbbells (a) and CFRP coupons (b).

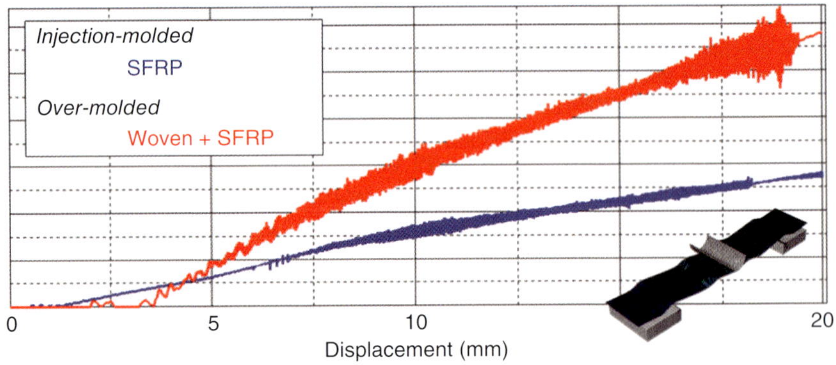

Figure 6.16 Mechanical responses of an SFRP-injected beam and an overmolded CFRP + SFRP beam subjected to three-point bending.

6.3.1.2 Conclusions

Integrated computational methodologies allowing to link process to structural behavior via the effective properties of materials open the doors for improved design. As the material, as well as the process, have become direct design variables in the computer analysis, a larger design space can be covered in a few variations. By changing the geometry, material and processing conditions of the part at the same time, an optimum design can be found more easily than by using classical computational methodologies.

6.4 Screening and Virtual Testing of Material Properties

6.4.1 Material Screening and Design Based on *n*th-Order Bounds

The effective potential related to linear physical properties of multiphase materials can be enclosed by bounds of different order which consider the single crystal behavior and n-point statistical information of the microstructure. The bounds enclose sets of properties of the material, which can be used to decide if a material is to be considered for an application with prescribed properties or to select materials from large data bases (material screening). In this section, we will present the concept for anisotropic linear elastic properties and give examples for polycrystals made of cubic materials.

6.4.1.1 *n*th-Order Bounds of Linear Anisotropic Elastic Properties

The effective linear elastic material behavior of multiphase materials is fully specified by the fourth-order effective stiffness tensor $\overline{\mathbb{C}}$, which is connected to the effective elastic energy density by $\overline{W} = \overline{\varepsilon}\cdot\overline{\mathbb{C}}[\overline{\varepsilon}]/2$ (quadratic form of $\overline{\mathbb{C}}$), where $\overline{\varepsilon}$ denotes the effective infinitesimal strain tensor (see [170]). The effective stiffness tensor can be bounded from above and below as follows (see [170]):

$$\mathbb{C}^{0-} \leq \mathbb{C}^{1-} \leq \mathbb{C}^{2-} \leq \cdots \leq \overline{\mathbb{C}} \leq \cdots \leq \mathbb{C}^{2+} \leq \mathbb{C}^{1+} \leq \mathbb{C}^{0+}$$

where all inequalities are to be interpreted as quadratic forms. The $\mathbb{C}^{n\pm}$ are tensorial nth-order upper and lower bounds, respectively. They take into consideration the single crystal material behavior and n-point statistical information of the microstructure. The important information is that, for example, if up to two-point statistical information of the microstructure is considered, then a set of properties is enclosed by $\mathbb{C}^{2\pm}$, which then present the properties hypothetically reachable by all variations of one-point and two-point statistics (volume fractions of constituents and morphology). This crucial information is used in [171] for an inverse design (through the usage of the "properties closure") and in [172, 173] for the inverse design of cubic materials incorporating a material selection based on single-phase material data bases. The aforementioned hierarchy of bounds exists also for other linear physical properties, for example, thermoelasticity and thermal conductivity.

6.4.1.2 Zeroth-, First- and Second-Order Bounds of Linear Elastic Properties of Cubic Materials

For a cubic material with single crystal stiffness $\mathbb{C}^c = \sum_{\alpha=1}^{3} \lambda_\alpha^c \mathbb{P}_\alpha^c$, see [173], the isotropic zeroth-order bounds discussed in [170] and proven in [174] are given by

$$\mathbb{C}^{0-} = \lambda_1^c \mathbb{P}_1 + \lambda_2^- \mathbb{P}_2\,, \quad \mathbb{C}^{0+} = \lambda_1^c \mathbb{P}_1 + \lambda_2^+ \mathbb{P}_2\,, \quad \lambda_2^- = \min\{\lambda_2^c, \lambda_3^c\},$$
$$\lambda_2^+ = \max\{\lambda_2^c, \lambda_3^c\},$$

with $\mathbb{P}_1$ and $\mathbb{P}_2$ being the identities on spherical and symmetric deviatoric second-order tensors. For the zeroth-order bounds of arbitrarily anisotropic tensor, see [174]. With the zeroth-order bounds one is able to give the range of effective properties hypothetically reachable independently of the microstructure statistics. That means that solely using the zeroth-order bounds we are able to go through material databases in a blink of an eye and select candidate materials for prescribed properties before even thinking about the microstructure. This is an efficient form of materials screening based on physical principles. Of course, the materials screening should consider further properties, see [175], but this is not the focus of the present example.

The first-order bounds take into consideration one-point statistics, that is, volume fractions of the microstructure constituents. For uniform polycrystalline materials the one-point statistics reflect the volume fractions of the material orientation. For large numbers this is represented by orientation distributions.

For cubic material the first-order bounds are given as follows (see [173]):

$$\mathbb{C}^{1-} = \left(\frac{1}{\lambda_1^c}\mathbb{P}_1 + \left(\frac{2}{5\lambda_2^c} + \frac{3}{5\lambda_3^c}\right)\mathbb{P}_2 + \sqrt{\frac{6}{5}}\left(\frac{1}{\lambda_2^c} - \frac{1}{\lambda_3^c}\right)\mathbb{V}'\right)^{-1},$$

$$\mathbb{C}^{1+} = \lambda_1^c\mathbb{P}_1 + \left(\frac{2}{5}\lambda_2^c + \frac{3}{5}\lambda_3^c\right)\mathbb{P}_2 + \sqrt{\frac{6}{5}}\left(\frac{1}{\lambda_2^c} - \frac{1}{\lambda_3^c}\right)\mathbb{V}',$$

where $\mathbb{V}'$ is the fourth-order texture coefficient representing the influence of the orientation distribution. If the two-point statistics are assumed to be isotropic, then the second-order Hashin–Shtrikman bounds are given by (see [173])

$$\mathbb{C}^{\text{HS}}(c_2) = (\lambda_1^c\mathbb{P}_1 + f_1(\lambda_i^c, c_2)\mathbb{P}_2 + f_2(\lambda_i^c, c_2)\mathbb{V}')(\mathbb{P}_1 + f_3(\lambda_i^c, c_2)\mathbb{P}_2 + f_4(\lambda_i^c, c_2)\mathbb{V}')^{-1},$$

$$\mathbb{C}^{2-} = \mathbb{C}^{\text{HS}}(c_2 = \lambda_2^-), \quad \mathbb{C}^{2+} = \mathbb{C}^{\text{HS}}(c_2 = \lambda_2^+),$$

where the scalars f_i are functions of the eigenvalues λ_i^c of the cubic materials and of the eigenvalue c_2 of an isotropic comparison material $\mathbb{C} = \lambda_1^c\mathbb{P}_1 + c_2\mathbb{P}_2$.

6.4.1.3 Application Example of Bounds for Material Screening and Design

In order to present a simple application example, we will take into consideration Young's modulus E in tensile direction $\boldsymbol{n}$ of a material with stiffness $\mathbb{C}$ defined as follows:

$$E(\mathbb{C}) = (\mathbb{C}^{-1} \cdot (\boldsymbol{n} \otimes \boldsymbol{n} \otimes \boldsymbol{n} \otimes \boldsymbol{n}))^{-1}.$$

Based on matrix algebra, it can be shown that this quantity obeys the bounds hierarchy for fixed $\boldsymbol{n}$, that is,

$$E^{0-} \leq E^{1-} \leq E^{2-} \leq \cdots \leq \overline{E} \leq \cdots \leq E^{2+} \leq E^{1+} \leq E^{0+},$$

where $\overline{E} = E(\overline{\mathbb{C}})$ and $E^{n\pm} = E(\mathbb{C}^{n\pm})$. It should be remarked that all bounds can also be used in order to enclose other quantities, for example, components of the effective stiffness.

In this example, we look for a material offering the following properties:

- $\overline{E} = 190$ GPa in $\boldsymbol{n} = \boldsymbol{e}_1$ direction with respect to sample basis directions $\{\boldsymbol{e}_i\}$
- $\overline{E} = 150$ GPa in $\boldsymbol{n} = (\boldsymbol{e}_1 + \boldsymbol{e}_2)/\sqrt{2}$ direction with respect to sample basis directions $\{\boldsymbol{e}_i\}$
- Orthotropic effective behavior in respect to sample basis directions $\{\boldsymbol{e}_i\}$

First, we consider the exemplary material data base presented in Table 6.1, in which the elastic constants of the single materials, the eigenvalues and Young's modulus of the corresponding zeroth-order bounds are given.

With the information given in Table 6.1 we are able to decide which of the materials is suitable for our future application. In this case aluminum is too weak, cubic silicon carbide is to stiff but copper suits our needs since as reflected by its zeroth-order bounds it offers us, in principle, ranges in which our prescribed properties lie. This crucial information embodied in the zeroth-order bounds can be embedded easily in any material data base allowing the designer to explore the properties offered by single materials and combination of materials. This allows an efficient, physically based material screening for systemic design methods using large material data bases. It should be remarked that this approach can also be used for composites, for example, the zeroth-order bounds of a weak aluminum matrix with stiff silicon carbide inclusions can be computed in order to see what possibilities are offered by this material combination.

The further design steps are to determine what kind of microstructure influence is needed in order to obtain properties as close as possible to the prescribed ones. The first-order bounds and second-order bounds can be taken into account in order to explore the sets of all possible properties offered by variation, for example, of all possible one-point statistics (orientation distribution of single material behavior), see, for example, [171 – 173]. The upper half of Young's modulus of the single crystal behavior of copper is plotted in gold as a polar plot in Figure 6.17(a), together with its zeroth-order bounds in orange. In Figure 6.17(b) and (c) analog plots for the first-order bounds and for the second-order bounds with two different texture coefficients showing orthotropic behavior are depicted. This reflects graphically the bounds hierarchy and the drastic properties changes and design variability for the effective material behavior through purposeful microstructure variation.

6.4.2
Comparison of *In Situ*/XCT Measurements with Virtual Testing of SFRP Materials

The realistic structural analysis of composite components (e.g., SFRP) has to include their complex and inhomogeneous local microstructural behavior. Typical approaches in the field of SFRP are using local output data from rheological simulations (i.e., fiber volume fraction and second-order orientation tensor) to develop appropriate material laws, either by analytical (e.g., MT) or numerical (FEM modeling of RVE) homogenization methods. In the latter case, the micromechanical FEM models are commonly based on generic realizations of RVEs, which show the according micromechanical statistical features, but do not resemble any real microgeometry. Available software tools (e.g., Digimat) allow for a semiautomatic application of these methods in the field of structural analysis of SFRP components in the elastic and (to some extent) inelastic regime. However, failure and damage mechanism are inherently local phenomena, taking place on the sub-RVE level (e.g., local fiber – matrix debonding, matrix cracking, etc.) and being strongly dependent on the local fiber arrangement. Hence, detailed information on damage- and failure-related material parameters (e.g., interface strength, etc.) may only be attained by studying the mechanical behavior of real SFRP samples on the sub-RVE level.

The proposed procedure aims at such a sub-RVE analysis by a direct comparison of experimental XCT-data of small-scale tensile SFRP specimens tested *in situ* inside a XCT and the corresponding "virtual testing" of its numerical FEM-counterpart, which includes all microstructural details in the area of

Table 6.1 Exemplary material database with linearly independent elastic constants, eigenvalues of zeroth-order bounds, and material ranges $[\mathbf{E}^{0-}, \mathbf{E}^{0+}]$ in GPa.

Material	Elastic constants $\{C_{1111}, C_{1122}, C_{2323}\}$	λ_1^-	λ_2^-	λ_1^+	λ_2^+	$[E^{0-}, E^{0+}]$
Aluminum	{107, 61, 28}	229	46	229	56	[63, 75]
Copper	{168.4, 121.4, 75.4}	411	47	411	151	[67, 191]
Silicon carbide	{228, 133.5, 110.9}	495	95	495	222	[272, 511]

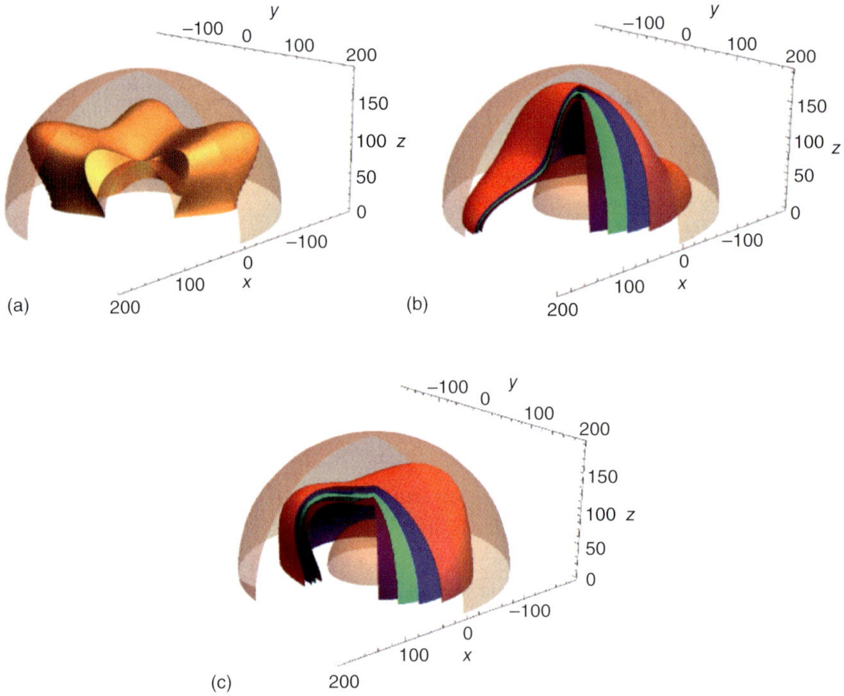

Figure 6.17 Polar plots of Young's modulus for copper: (a) single crystal behavior $\mathbb{C}^c$ in gold and $\mathbb{C}^{0\pm}$ in orange; (b) and (c) $\mathbb{C}^{1-}$ in purple, $\mathbb{C}^{2-}$ in green, $\mathbb{C}^{2+}$ in blue, and $\mathbb{C}^{1+}$ in red with two different orthotropic texture coefficients.

interest, which have been identified by a XCT fiber extraction analysis of the very SFRP sample under consideration.

6.4.2.1 Methods

Sample preparation: In the first step suitably sized small-scale flat specimens are cut out of a SFRP (PP-GF30) 2.5 mm plate (produced by injection molding) and are mounted in an *in situ* tensile testing device within a XCT (GE Phoenix/X-ray Nanotom 180NF).Due to the limited available tension force of the testing device (500 N) the reduced section of the specimen is limited to a width of 3 mm.

***In situ* tensile test**: The complete reduced section of the unloaded specimen is subject to high-resolution (3 μm voxel) XCT scans, which are taken in the unloaded state, as well as at specific increasing load levels up to the fracture of the specimen. During the operation of the CT scans the specimen elongation is kept constant.

Extraction of individual fibers from XCT scan: A quantitative individual fiber extraction pipeline, implemented in our CT evaluation software iAnalyse is used to identify each individual fiber in the XCT scan area [176]. The data extracted for each fiber are: starting and endpoint (plus all intermediate points for nonstraight fibers), fiber length, diameter, and orientation. In addition, iAnalyse provides statistical data for the fiber assembly (fiber volume fraction, orientation tensor) which may be directly used in an analytical homogenization procedure.

"Virtual Testing" FEM model from extraction data: The aforementioned fiber extraction data is automatically transferred into a specific 3D FEM model of the specimen under consideration. A 3D model of a RVE at a position of interest within the XCT scan area which holds all fiber arrangement details (set up by a CATIA script) is automatically embedded within a "global" model of the specimen. Outside the RVE a generalized MT approach is used to describe the material response (considering the local orientation distribution, aspect ratio, and volume fraction of the fibers). The FEM model is subject to axial loading conditions, which correspond to the actual *in situ* testing.

6.4.2.2 Results

The evaluation of the combined experimental/analytical setting is based on the comparison of the findings of deformation, local failure, and damage in the XCT scan data at specific load levels and the corresponding stress and strain fields within the detailed RVE-FEM model area. Specific postprocessing capabilities have been introduced to evaluate fiber- and matrix stress and strain results (e.g., normal and shear stresses at individual fiber–matrix interfaces or matrix stresses in the vicinity of individual fibers). Figure 6.18 gives an example of high tensional interface stresses at a specific fiber interface and the corresponding interface debonding at this very fiber found in the XCT scan data of the loaded condition.

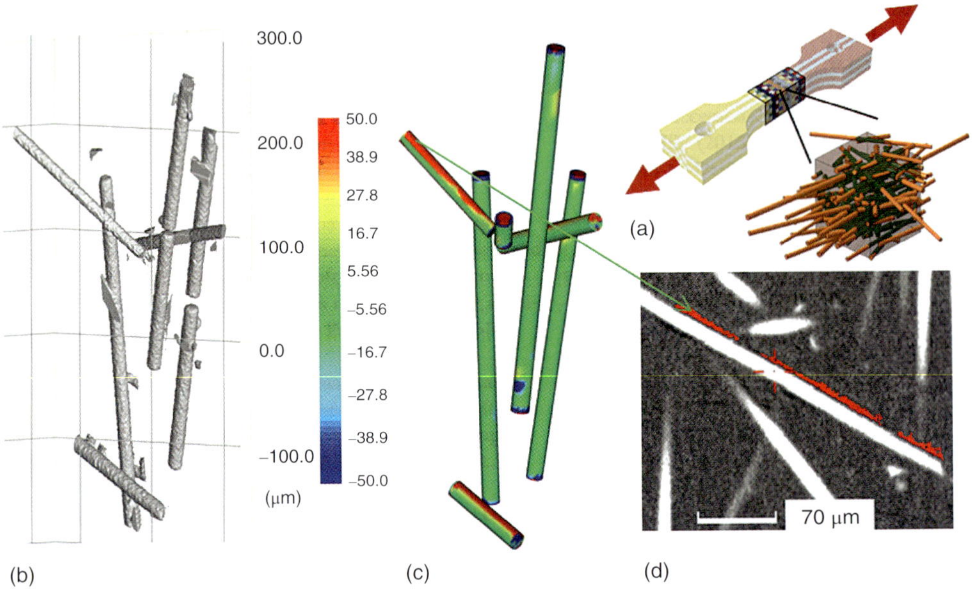

Figure 6.18 (a) FEM model of specimen with embedded RVE, (b) 3D visualization of XCT of critical area, (c) FEM results: normal stresses in interface (MPa), and (d) cutting plane in XCT scan showing interface debonding.

6.5
Software Tools for the Determination of Effective Properties

6.5.1
Software Categories

The software presented (nonexhaustive) here will be structured in the following categories:

6600 Composite modeling platform (including all following subcategories)

6610 Computation of effective properties based on different homogenization approaches (e.g., rule of mixture, mean field, etc.)

6620 Pre- (and post-) processing tool to support the computation of effective properties via FEA based on a realistic representation of the material microstructure in RVEs

6630 Mapping of effective properties based on results from manufacturing simulation

6640 Other

6.5.2
List of Software

Category	Name	Web	Type	Description
6600	Digimat	www.e-xstream.com	Commercial	Composite simulation platform that enables to predict material properties from microscale via FEA based and mean-field homogenization methods. Micro–macrocoupling is possible for a wide range of FE solvers. Mapping bridges the gap to manufacturing simulations and allows for integrative simulations. Full solutions for chosen tasks are available (e.g., injection-molded parts, screening of UD composite properties, design of honeycomb core plates)
6610	4aMicromec	http://www.4a-engineering.at/de/produkte/4amicromec.html	Commercial	Calculation of material properties for composite systems based on per-phase properties and microstructure information
6610	ESAComp	http://www.esacomp.com/overview/what-is-esacomp	Commercial	Micromechanical-based analysis of composites, from preliminary design of layered structures to the analyses of details
6610	MCQ http://analyswift.com/products/swiftcomp-vamuch-micromechanics-modeling-of-heterogeneous-materials/Composites	http://www.alphastarcorp.com/sections/Products/MCQ-Composites/index.jsp	Commercial	Characterization and qualification of composite material properties based on classical laminate theory (CLT) and micromechanics

Category	Name	Web	Type	Description
6610	Swift Comp	http://analyswift.com/products/swiftcomp-vamuch-micromechanics-modeling-of-heterogeneous-materials/	Commercial	General-purpose micromechanics analysis for heterogeneous materials for the prediction of effective multiphysical properties based on variational asymptotical method for unit cell homogenization (VAMUCH)
6610	HELIUS Composites	http://www.autodesk.in/products/helius-composite/overview	Commercial	Tool to understand the behavior of composite materials, laminates, and simple structures
6610	HELIUS MatSim	http://www.firehole.com/news/newsitem.aspx?id=111	Commercial	Web-based tool to simulate composite lamina properties based on constituent (fiber and matrix) level data and fiber volume
6610	MultiMech	http://multimechanics.com/	Commercial	Structural analysis of advanced composite parts and materials
6610	Multiscale.Sim	http://www.cybernet.co.jp/ansys/multiscale/multiscale_sim/	Commercial	Homogenization-method-based multiphysics analysis tool for the determination of equivalent material properties
6610	JMatPro	http://www.sentesoftware.co.uk/jmatpro.aspx	Commercial	Simulation software to calculate a wide range of material properties of (multicomponent) alloys
6610	VIPPER	http://www.materials.imdea.org/VIPPER	(Non)commercial	Simulation tool developed to predict ply properties of fiber-reinforced composite materials from the properties and spatial distribution of the different phases and interfaces in the composite

6610	Analyswift/ Swiftcomp	http://analyswift.com/2014/09/analyswifts-composite-simulation-tools-available-online-through-cdmhub-web-platform-at-purdue-university/	Commercial	General-purpose micromechanics analysis for heterogeneous materials with complex microstructure that predicts effective properties and recovers detailed multiphysical field distribution within the microstructure
6610	Genoa	http://www.ascgenoa.com/	Commercial	Engineering material characterization, durability, and damage tolerance and reliability software suite providing micro–macro progressive failure analysis (PFA) capability
6610	Cadec	http://en.cadec-online.com/	Noncommercial	Online application that performs composite materials analysis as described in the textbook *Introduction to Composite Materials Design*
6610	Zebulon/Z-set	http://www.zset-software.com/	Commercial	Simulation of mixed-mode delamination in composite laminate structures under complex combined tensile, compressive, and flexural loads
6610	4A Impetus Micromech	http://www.4a-engineering.at/de/produkte/4amicromec.html	Commercial	Computation of effective properties for fiber-reinforced composites based on per-phase material data and microstructure information

Category	Name	Web	Type	Description
6610	NASA-GMC	http://www.grc.nasa.gov/WWW/StructuresMaterials/MLP/software/mac-gmc/	Noncommercial	Micromechanics Analysis Code based on the Generalized Method of Cells for the prediction of elastic and inelastic thermomechanical and piezo/magnetoelectric responses of multiphased materials including an interface to Abaqus FEA software
6620	HOMAT		Noncommercial	Computation of effective properties of metals based on grain microstructure
6620	DAMASK	http://damask.mpie.de/	Noncommercial	Simulation of crystal plasticity within a finite-strain continuum mechanical framework
6620	MedTool	http://www.dr-pahr.at/software.php	Noncommercial	Fully automatic preparation and analysis of simulation models from 3D image data including material homogenization and interfaces to finite element solvers
6620	GeoDict	http://www.geodict.com/Solutions/aboutGD.php	Commercial	Complete software solution for multiscale 3D image processing, visualization, simulation-driven property characterization, material development, and process optimization

6620	Wisetex	https://www.mtm.kuleuven.be/Onderzoek/Composites/software	Commercial	Integrated tool for modeling textiles and textile composites with the purpose of predicting material properties
6620	Neper	http://neper.sourceforge.net/	Noncommercial	Software package for polycrystal generation and meshing
6620	DREAM.3D	http://dream3d.bluequartz.net/	Noncommercial	Generation of synthetic RVE
6620	SphäroSim	http://aixvipmap.iehk.rwth-aachen.de/index.php?id=2#	Noncommercial	Generation of mesoscale RVE based on spherolite growth in semicrystalline polymers
6620	OOF	http://www.nist.gov/mml/ctcms/oof/	Noncommercial	Calculation of macroscopic properties from images of real or simulated microstructures
6620	TexGen	http://texgen.sourceforge.net/index.php/Main_Page	Noncommercial	Geometry generation of textile structures used for the computation of properties (permeability and composite mechanical behavior)
6620	Palmyra	http://www.matsim.ch/products/palmyra.html	Commercial	Design of new composite materials by microstructure generation and calculation of physical properties
6630	Sundance	http://inthefold.autodesk.com/in_the_fold/2014/11/project-sundance-graduate-bolsters-autodesk-simulation-composite-analysis-offering.html	Commercial	Mapping of effective properties onto structural FEA models based on injection molding simulations

Category	Name	Web	Type	Description
6630	Converse	http://www.simcon-worldwide.com/wb/pages/en/products/converse.php?lang=EN	Commercial	Mapping of effective properties onto structural FEA models based on injection molding simulations
6630	Ultrasim	http://www.plasticsportal.net/wa/plasticsEU~pl_PL/portal/show/content/technical_resources/ultrasim/ultrasim	(Non)commercial	Mapping of effective properties onto structural FEA models based on processing simulations (mainly injection molding)
6630	SigmaLink	http://www.sigmasoft.de/sigmasoft-virtual-molding/products/add-ons/sigmalink/	Commercial	Mapping of effective properties onto structural FEA models based on injection molding simulations
6630	AnsysMoldSim	https://caeai.com/resources/getting-started-ansys-composite-preppost-acp	Commercial	Mapping of effective properties onto structural FEA models based on injection molding simulations
6630	ANSYS ACP	https://caeai.com/resources/getting-started-ansys-composite-preppost-acp	Commercial	Provides all the necessary functionality for the analysis of layered composite structures
6630	Simufact	http://www.simufact.de/en/index.html	Commercial	Simulation solution for metal forming, welding, and heat treatment with the ability to simulate the entire manufacturing process chain including the material behavior taken into account in structural analyses

6640	(Pre)VABS	http://analyswift.com/products/vabs-cross-sectional-analysis-tool-for-composite-beams/	Commercial	Decoupling 3D slender solids with complex microstructure into simple engineering beam models
6640	HyperSizer	http://hypersizer.com/products/pro.php	Commercial	Tool for composite optimization based on statistical methods
6640	Smart+	http://www.lem3.fr/chemisky/smartplus/	Noncommercial	C++ library with emphasis on speed and ease of use that facilitates the implementation of up-to-date constitutive model for materials in finite element analysis packages including tools to analyze the behavior of material considering load at the material point level
6640	GrantaDesign	http://www.grantadesign.com/products/mi/gateway.htm	Commercial	Material database providing reference data covering many categories of properties for metals, plastics, ceramics, composites, and natural materials

Note: Some of the given URLs might be out of date and thus non-functional.

References

1. Miehe, C. and Koch, A. (2002) Computational micro-to-macro transitions of discretized microstructures undergoing small strains. *Arch. Appl. Mech.*, **72** (4–5), 300–317. doi: 10.1007/s00419-002-0212-2.
2. Hill, R. (1963) Elastic properties of reinforced solids: some theoretical principles. *J. Mech. Phys. Solids*, **11** (5), 357–372. doi: 10.1016/0022-5096(63)90036-X.
3. Huet, C. (1990) Application of variational concepts to size effects in elastic heterogeneous bodies. *J. Mech. Phys. Solids*, **38** (6), 813–841. doi: 10.1016/0022-5096(90)90041-2.
4. Voigt, W. (1889) Ueber die beziehung zwischen den beiden elasticitätsconstanten isotroper körper. *Ann. Phys.*, **274** (12), 573–587. doi: 10.1002/andp.18892741206.
5. Reuss, A. (1929) Berechnung der fließgrenze von mischkristallen auf grund der plastizitätsbedingung für einkristalle. *ZAMM – J. Appl. Math. Mech./Z. Angew. Math. Mech.*, **9** (1), 49–58. doi: 10.1002/zamm.19290090104.
6. Hashin, Z. and Shtrikman, S. (1962) A variational approach to the theory of the elastic behaviour of polycrystals. *J. Mech. Phys. Solids*, **10** (4), 343–352. doi: 10.1016/0022-5096(62)90005-4.
7. Kanouté, P., Boso, D., Chaboche, J., and Schrefler, B. (2009) Multiscale methods for composites: a review. *Arch. Comput. Meth. Eng.*, **16** (1), 31–75. doi: 10.1007/s11831-008-9028-8.
8. Geers, M., Kouznetsova, V., and Brekelmans, W. (2010) Multi-scale computational homogenization: trends and challenges. *J. Comput. Appl. Math.*, **234** (7), 2175–2182. doi: 10.1016/j.cam.2009.08.077.
9. Llorca, J., González, C., Molina-Aldareguía, J.M., Segurado, J., Seltzer, R., Sket, F., Rodríguez, M., Sádaba, S., Muñoz, R., and Canal, L.P. (2011) Multiscale modeling of composite materials: a roadmap towards virtual testing. *Adv. Mater.*, **23** (44), 5130–5147. doi: 10.1002/adma.201101683.
10. Christman, T., Needleman, A., and Suresh, S. (1989) An experimental and numerical study of deformation in metal-ceramic composites. *Acta Metall.*, **37** (11), 3029–3050. doi: 10.1016/0001-6160(89)90339-8.
11. Tvergaard, V. (1990) Analysis of tensile properties for a whisker-reinforced metal-matrix composite. *Acta Metall. Mater.*, **38** (2), 185–194. doi: 10.1016/0956-7151(90)90048-L.
12. Brockenbrough, J., Suresh, S., and Wienecke, H. (1991) Deformation of metal-matrix composites with continuous fibers: geometrical effects of fiber distribution and shape. *Acta Metall. Mater.*, **39** (5), 735–752. doi: 10.1016/0956-7151(91)90274-5.
13. Wieckowski, Z. (2000) Dual finite element methods in homogenization for elastic–plastic fibrous composite material. *Int. J. Plast.*, **16** (2), 199–221. doi: 10.1016/S0749-6419(99)00070-4.
14. Segurado, J., Llorca, J., and González, C. (2002) On the accuracy of mean-field approaches to simulate the plastic deformation of composites. *Scr. Mater.*, **46** (7), 525–529. doi: 10.1016/S1359-6462(02)00027-1.
15. Ji, B. and Wang, T. (2003) Plastic constitutive behavior of short-fiber/particle reinforced composites. *Int. J. Plast.*, **19** (5), 565–581. doi: 10.1016/S0749-6419(01)00041-9.
16. Carrere, N., Valle, R., Bretheau, T., and Chaboche, J.L. (2004) Multiscale analysis of the transverse properties of Ti-based matrix composites reinforced by sic fibres: from the grain scale to the macroscopic scale. *Int. J. Plast.*, **20** (4–5), 783–810. doi: 10.1016/j.ijplas.2003.09.001.
17. Aboudi, J. (1996) Micromechanical analysis of composites by the method of cells – update. *Appl. Mech. Rev.*, **49** (10S), S83–S91. doi: 10.1115/1.3101981.
18. Dvorak, G., Wafa, A., and Bahei-El-Din, Y. (1994) Implementation of the transformation field analysis for

inelastic composite materials. *Comput. Mech.*, **14** (3), 201–228. doi: 10.1007/BF00370073.

19. Dvorak, G.J. and Benveniste, Y. (1992) On transformation strains and uniform fields in multiphase elastic media. *Proc. R. Soc. London, Ser. A*, **437** (1900), 291–310. doi: 10.1098/rspa.1992.0062.
20. Eshelby, J.D. (1957) The determination of the elastic field of an ellipsoidal inclusion, and related problems. *Proc. R. Soc. London, Ser. A*, **241** (1226), 376–396. doi: 10.1098/rspa.1957.0133.
21. Mori, T. and Tanaka, K. (1973) Average stress in matrix and average elastic energy of materials with misfitting inclusions. *Acta Metall.*, **21** (5), 571–574. Cited By (since 1996) 1814.
22. Benveniste, Y. (1987) A new approach to the application of Mori-Tanaka's theory in composite materials. *Mech. Mater.*, **6** (2), 147–157. doi: 10.1016/0167-6636(87)90005-6.
23. Kröner, E. (1958) Berechnung der elastischen konstanten des vielkristalls aus den konstanten des einkristalls. *Z. Phys. A: Hadrons Nucl.*, **151**, 504–518. doi: 10.1007/BF01337948.
24. Hill, R. (1965) A self-consistent mechanics of composite materials. *J. Mech. Phys. Solids*, **13** (4), 213–222. doi: 10.1016/0022-5096(65)90010-4.
25. Michel, J. and Suquet, P. (2003) Nonuniform transformation field analysis. *Int. J. Solids Struct.*, **40** (25), 6937–6955. doi: 10.1016/S0020-7683(03)00346-9. Special issue in Honor of George J. Dvorak.
26. Talbot, D.R.S. and Willis, J.R. (1985) Variational principles for inhomogeneous non-linear media. *IMA J. Appl. Math.*, **35** (1), 39–54. doi: 10.1093/imamat/35.1.39.
27. Talbot, D.R.S. and Willis, J.R. (1987) Bounds and self-consistent estimates for the overall properties of nonlinear composites. *IMA J. Appl. Math.*, **39** (3), 215–240. doi: 10.1093/imamat/39.3.215.
28. Ponte Castañeda, P. (1991) The effective mechanical properties of nonlinear isotropic composites. *J. Mech. Phys. Solids*, **39** (1), 45–71. doi: 10.1016/0022-5096(91)90030-R.
29. Ponte Castañeda, P. (1992) A new variational principle and its application to nonlinear heterogeneous systems. *SIAM J. Appl. Math.*, **52** (5), 1321–1341.
30. Talbot, D. and Willis, J. (1992) Some simple explicit bounds for the overall behaviour of nonlinear composites. *Int. J. Solids Struct.*, **29** (14–15), 1981–1987. doi: 10.1016/0020-7683(92)90188-Y.
31. Molinari, A., El Houdaigui, F., and Tóth, L. (2004) Validation of the tangent formulation for the solution of the non-linear Eshelby inclusion problem. *Int. J. Plast.*, **20** (2), 291–307. doi: 10.1016/S0749-6419(03)00038-X.
32. Hill, R. (1965) Continuum micromechanics of elastoplastic polycrystals. *J. Mech. Phys. Solids*, **13** (2), 89–101. doi: 10.1016/0022-5096(65)90023-2.
33. Pettermann, H.E., Plankensteiner, A.F., Böhm, H.J., and Rammerstorfer, F.G. (1999) A thermo-elasto-plastic constitutive law for inhomogeneous materials based on an incremental Mori-Tanaka approach. *Comput. Struct.*, **71** (2), 197–214. doi: 10.1016/S0045-7949(98)00208-9.
34. Doghri, I. and Ouaar, A. (2003) Homogenization of two-phase elasto-plastic composite materials and structures: study of tangent operators, cyclic plasticity and numerical algorithms. *Int. J. Solids Struct.*, **40** (7), 1681–1712. doi: 10.1016/S0020-7683(03)00013-1.
35. Doghri, I. and Tinel, L. (2005) Micromechanical modeling and computation of elasto-plastic materials reinforced with distributed-orientation fibers. *Int. J. Plast.*, **21** (10), 1919–1940. doi: 10.1016/j.ijplas.2004.09.003.
36. Molinari, A., Canova, G., and Ahzi, S. (1987) A self-consistent approach of the large deformation polycrystal viscoplasticity. *Acta Metall.*, **35** (12), 2983–2994.
37. Zaoui, A. and Masson, R. (2002) in *IUTAM Symposium on Transformation Problems in Composite and Active Materials*, Solid Mechanics and Its Applications, vol. **60** (eds Y.A. Bahei-El-Din, G.J. Dvorak, and G.M.L.

Gladwell), Springer, pp. 3–15, doi: 10.1007/0-306-46935-9_1.
38. Masson, R., Bornert, M., Suquet, P., and Zaoui, A. (2000) An affine formulation for the prediction of the effective properties of nonlinear composites and polycrystals. *J. Mech. Phys. Solids*, **48** (6–7), 1203–1227, doi: 10.1016/S0022-5096(99)00071-X.
39. Pierard, O. and Doghri, I. (2006) An enhanced affine formulation and the corresponding numerical algorithms for the mean-field homogenization of elasto-viscoplastic composites. *Int. J. Plast.*, **22** (1), 131–157. doi: 10.1016/j.ijplas.2005.04.001.
40. Mercier, S. and Molinari, A. (2009) Homogenization of elastic–viscoplastic heterogeneous materials: self-consistent and Mori-Tanaka schemes. *Int. J. Plast.*, **25** (6), 1024–1048. doi: 10.1016/j.ijplas.2008.08.006.
41. Doghri, I., Adam, L., and Bilger, N. (2010) Mean-field homogenization of elasto-viscoplastic composites based on a general incrementally affine linearization method. *Int. J. Plast.*, **26** (2), 219–238. doi: 10.1016/j.ijplas.2009.06.003.
42. Berveiller, M. and Zaoui, A. (1978) An extension of the self-consistent scheme to plastically-flowing polycrystals. *J. Mech. Phys. Solids*, **26** (5–6), 325–344. doi: 10.1016/0022-5096(78)90003-0.
43. Chaboche, J., Kanouté, P., and Roos, A. (2005) On the capabilities of mean-field approaches for the description of plasticity in metal matrix composites. *Int. J. Plast.*, **21** (7), 1409–1434. doi: 10.1016/j.ijplas.2004.07.001.
44. Pierard, O., Llorca, J., Segurado, J., and Doghri, I. (2007) Micromechanics of particle-reinforced elasto-viscoplastic composites: finite element simulations versus affine homogenization. *Int. J. Plast.*, **23** (6), 1041–1060. doi: 10.1016/j.ijplas.2006.09.003.
45. Wu, L., Noels, L., Adam, L., and Doghri, I. (2013) A combined incremental–secant mean–field homogenization scheme with per–phase residual strains for elasto–plastic composites. *Int. J. Plast.*, **51**, 80–102. doi: 10.1016/j.ijplas.2013.06.006.
46. Moulinec, H. and Suquet, P. (2003) Intraphase strain heterogeneity in nonlinear composites: a computational approach. *Eur. J. Mech. A. Solids*, **22** (5), 751–770. doi: 10.1016/S0997-7538(03)00079-2.
47. Ponte Castañeda, P. (1996) Exact second-order estimates for the effective mechanical properties of nonlinear composite materials. *J. Mech. Phys. Solids*, **44** (6), 827–862. doi: 10.1016/0022-5096(96)00015-4.
48. Suquet, P. (1995) Overall properties of nonlinear composites: a modified secant moduli theory and its link with Ponte Castañeda's nonlinear variational procedure. *C.R. Acad. Sci.*, **320**, 563–571.
49. Ponte Castañeda, P. (2002) Second-order homogenization estimates for nonlinear composites incorporating field fluctuations: I – Theory. *J. Mech. Phys. Solids*, **50** (4), 737–757. doi: 10.1016/S0022-5096(01)00099-0.
50. Ponte Castañeda, P. (2002) Second-order homogenization estimates for nonlinear composites incorporating field fluctuations: II – Applications. *J. Mech. Phys. Solids*, **50** (4), 759–782. doi: 10.1016/S0022-5096(01)00098-9.
51. Doghri, I., Brassart, L., Adam, L., and Gérard, J.S. (2011) A second-moment incremental formulation for the mean-field homogenization of elasto-plastic composites. *Int. J. Plast.*, **27** (3), 352–371. doi: 10.1016/j.ijplas.2010.06.004.
52. Lahellec, N. and Suquet, P. (2007) On the effective behavior of nonlinear inelastic composites: I. Incremental variational principles. *J. Mech. Phys. Solids*, **55** (9), 1932–1963. doi: 10.1016/j.jmps.2007.02.003.
53. Lahellec, N. and Suquet, P. (2007) On the effective behavior of nonlinear inelastic composites: II. A second-order procedure. *J. Mech. Phys. Solids*, **55** (9), 1964–1992. doi: 10.1016/j.jmps.2007.02.004.
54. Lahellec, N., Ponte Castañeda, P., and Suquet, P. (2011) Variational estimates for the effective response and

field statistics in thermoelastic composites with intra-phase property fluctuations. *Proc. R. Soc. London, Ser. A*, **467** (2132), 2224–2246. doi: 10.1098/rspa.2010.0609.

55. Brassart, L., Stainier, L., Doghri, I., and Delannay, L. (2011) A variational formulation for the incremental homogenization of elasto-plastic composites. *J. Mech. Phys. Solids*, **59** (12), 2455–2475. doi: 10.1016/j.jmps.2011.09.004.
56. Brassart, L., Stainier, L., Doghri, I., and Delannay, L. (2012) Homogenization of elasto-(visco) plastic composites based on an incremental variational principle. *Int. J. Plast.*, **36**, 86–112. doi: 10.1016/j.ijplas.2012.03.010.
57. Lahellec, N. and Suquet, P. (2013) Effective response and field statistics in elasto-plastic and elasto-viscoplastic composites under radial and non-radial loadings. *Int. J. Plast.*, **42**, 1–30. doi: 10.1016/j.ijplas.2012.09.005
58. Ponte Castañeda, P. (2012) Bounds for nonlinear composites via iterated homogenization. *J. Mech. Phys. Solids*, **60** (9), 1583–1604. doi: 10.1016/j.jmps.2012.05.004.
59. Agoras, M. and Ponte Castañeda, P. (2013) Iterated linear comparison bounds for viscoplastic porous materials with "ellipsoidal" microstructures. *J. Mech. Phys. Solids*, **61** (3), 701–725. doi: 10.1016/j.jmps.2012.11.003.
60. Ghosh, S., Lee, K., and Moorthy, S. (1995) Multiple scale analysis of heterogeneous elastic structures using homogenization theory and voronoi cell finite element method. *Int. J. Solids Struct.*, **32** (1), 27–62. doi: 10.1016/0020-7683(94)00097-G.
61. Ghosh, S. and Liu, Y. (1995) Voronoi cell finite element model based on micropolar theory of thermoelasticity for heterogeneous materials. *Int. J. Numer. Methods Eng.*, **38** (8), 1361–1398. doi: 10.1002/nme.1620380808.
62. Michel, J., Moulinec, H., and Suquet, P. (1999) Effective properties of composite materials with periodic microstructure: a computational approach. *Comput. Meth. Appl. Mech. Eng.*, **172** (1–4), 109–143. doi: 10.1016/S0045-7825(98)00227-8.
63. Feyel, F. (1999) Multiscale {FE2} elastoviscoplastic analysis of composite structures. *Comput. Mater. Sci.*, **16** (1–4), 344–354. doi: 10.1016/S0927-0256(99)00077-4.
64. Terada, K., Hori, M., Kyoya, T., and Kikuchi, N. (2000) Simulation of the multi-scale convergence in computational homogenization approaches. *Int. J. Solids Struct.*, **37** (16), 2285–2311. doi: 10.1016/S0020-7683(98)00341-2.
65. Miehe, C. (2002) Strain-driven homogenization of inelastic microstructures and composites based on an incremental variational formulation. *Int. J. Numer. Methods Eng.*, **55** (11), 1285–1322. doi: 10.1002/nme.515.
66. Kouznetsova, V., Brekelmans, W.A.M., and Baaijens, F.P.T. (2001) An approach to micro–macro modeling of heterogeneous materials. *Comput. Mech.*, **27** (1), 37–48. doi: 10.1007/s004660000212.
67. Feyel, F. and Chaboche, J.L. (2000) Fe2 multiscale approach for modelling the elastoviscoplastic behaviour of long fibre SiC/Ti composite materials. *Comput. Meth. Appl. Mech. Eng.*, **183** (3–4), 309–330. doi: 10.1016/S0045-7825(99)00224-8.
68. Kouznetsova, V., Geers, M.G.D., and Brekelmans, W.A.M. (2002) Multi-scale constitutive modelling of heterogeneous materials with a gradient-enhanced computational homogenization scheme. *Int. J. Numer. Methods Eng.*, **54** (8), 1235–1260.
69. Feyel, F. (2003) A multilevel finite element method (FE2) to describe the response of highly non-linear structures using generalized continua. *Comput. Meth. Appl. Mech. Eng.*, **192** (28–30), 3233–3244. doi: 10.1016/S0045-7825(03)00348-7.
70. Kouznetsova, V.G., Geers, M.G.D., and Brekelmans, W.A.M. (2004) Multi-scale second-order computational homogenization of multi-phase materials: a nested finite element solution strategy. *Comput. Meth. Appl. Mech. Eng.*, **193** (48–51), 5525–5550. doi: 10.1016/j.cma.2003.12.073.

71. Kaczmarczyk, L., Pearce, C.J., and Bicanic, N. (2008) Scale transition and enforcement of RVE boundary conditions in second-order computational homogenization. *Int. J. Numer. Methods Eng.*, **74** (3), 506–522. doi: 10.1002/nme.2188.
72. Moulinec, H. and Suquet, P. (1995) in *IUTAM Symposium on Microstructure-Property Interactions in Composite Materials*, Solid Mechanics and Its Applications, vol. **37** (ed. R. Pyrz), Springer, pp. 235–246, doi: 10.1007/978-94-011-0059-5_20.
73. Moulinec, H. and Suquet, P. (1998) A numerical method for computing the overall response of nonlinear composites with complex microstructure. *Comput. Meth. Appl. Mech. Eng.*, **157** (1–2), 69–94. doi: 10.1016/S0045-7825(97)00218-1.
74. Michel, J.C., Moulinec, H., and Suquet, P. (2001) A computational scheme for linear and non-linear composites with arbitrary phase contrast. *Int. J. Numer. Methods Eng.*, **52** (1–2), 139–160. doi: 10.1002/nme.275.
75. Kalamkarov, A.L., Andrianov, I.V., and Danishevskyy, V.V. (2009) Asymptotic homogenization of composite materials and structures. *Appl. Mech. Rev.*, **62** (3), 030802. doi: 10.1115/1.3090830.
76. Fish, J., Shek, K., Pandheeradi, M., and Shephard, M.S. (1997) Computational plasticity for composite structures based on mathematical homogenization: theory and practice. *Comput. Meth. Appl. Mech. Eng.*, **148** (1–2), 53–73. doi: 10.1016/S0045-7825(97)00030-3.
77. Aboudi, J., Pindera, M.J., and Arnold, S.M. (2003) Higher-order theory for periodic multiphase materials with inelastic phases. *Int. J. Plast.*, **19** (6), 805–847. doi: 10.1016/S0749-6419(02)00007-4.
78. Pierard, O., Friebel, C., and Doghri, I. (2004) Mean-field homogenization of multi-phase thermo-elastic composites: a general framework and its validation. *Compos. Sci. Technol.*, **64** (10–11), 1587–1603. doi: 10.1016/j.compscitech.2003.11.009.
79. Özdemir, I., Brekelmans, W.A.M., and Geers, M.G.D. (2008) Computational homogenization for heat conduction in heterogeneous solids. *Int. J. Numer. Methods Eng.*, **73** (2), 185–204. doi: 10.1002/nme.2068.
80. Coenen, E., Kouznetsova, V., Bosco, E., and Geers, M. (2012) A multi-scale approach to bridge microscale damage and macroscale failure: a nested computational homogenization-localization framework. *Int. J. Fract.*, **178** (1–2), 157–178. doi: 10.1007/s10704-012-9765-4.
81. Nguyen, V.P., Lloberas-Valls, O., Stroeven, M., and Sluys, L.J. (2012) Computational homogenization for multiscale crack modeling. Implementational and computational aspects. *Int. J. Numer. Methods Eng.*, **89** (2), 192–226. doi: 10.1002/nme.3237.
82. Mercatoris, B.C.N. and Massart, T.J. (2011) A coupled two-scale computational scheme for the failure of periodic quasi-brittle thin planar shells and its application to masonry. *Int. J. Numer. Methods Eng.*, **85** (9), 1177–1206. doi: 10.1002/nme.3018.
83. Coenen, E., Kouznetsova, V., and Geers, M. (2012) Novel boundary conditions for strain localization analyses in microstructural volume elements. *Int. J. Numer. Methods Eng.*, **90** (1), 1–21. doi: 10.1002/nme.3298.
84. Bosco, E., Kouznetsova, V.G., and Geers, M.G.D. (2015) Multi-scale computational homogenization–localization for propagating discontinuities using X-FEM. *Int. J. Numer. Methods Eng.*, **102** (3–4), 496–527. doi: 10.1002/nme.4838
85. Massart, T., Peerlings, R., and Geers, M. (2007) An enhanced multi-scale approach for masonry wall computations. *Int. J. Numer. Methods Eng.*, **69** (5), 1022–1059.
86. Belytschko, T., Loehnert, S., and Song, J.H. (2008) Multiscale aggregating discontinuities: a method for circumventing loss of material stability. *Int. J. Numer. Methods Eng.*, **73** (6), 869–894. doi: 10.1002/nme.2156.
87. Verhoosel, C.V., Remmers, J.J.C., Gutiérrez, M.A., and de Borst, R. (2010) Computational homogenization

for adhesive and cohesive failure in quasi-brittle solids. *Int. J. Numer. Methods Eng.*, **83** (8–9), 1155–1179. doi: 10.1002/nme.2854.

88. Wu, L., Tjahjanto, D., Becker, G., Makradi, A., Jérusalem, A., and Noels, L. (2013) A micro–meso-model of intra-laminar fracture in fiber-reinforced composites based on a discontinuous Galerkin/cohesive zone method. *Eng. Fract. Mech.*, **104**, 162–183. doi: 10.1016/j.engfracmech.2013.03.018.
89. Mulay, S., Becker, G., Vayrette, R., Raskin, J.P., Pardoen, T., Galceran, M., Godet, S., and Noels, L. (2015) Multiscale modelling framework for the fracture of thin brittle polycrystalline films: application to polysilicon. *Comput. Mech.*, **55**, 73–91. doi: 10.1007/s00466-014-1083-4
90. Voyiadjis, G.Z. and Deliktas, B. (1997) Damage in {MMCs} using the GMC: theoretical formulation. *Composites Part B*, **28** (5–6), 597–611. doi: 10.1016/S1359-8368(96)00074-1.
91. Oskay, C. and Fish, J. (2007) Eigendeformation-based reduced order homogenization for failure analysis of heterogeneous materials. *Comput. Meth. Appl. Mech. Eng.*, **196** (7), 1216–1243. doi: 10.1016/j.cma.2006.08.015.
92. Wu, L., Noels, L., Adam, L., and Doghri, I. (2012) A multiscale mean-field homogenization method for fiber-reinforced composites with gradient-enhanced damage models. *Comput. Meth. Appl. Mech. Eng.*, **233–236**, 164–179. doi: 10.1016/j.cma.2012.04.011.
93. Wu, L., Noels, L., Adam, L., and Doghri, I. (2013) An implicit-gradient-enhanced incremental-secant mean-field homogenization scheme for elasto-plastic composites with damage. *Int. J. Solids Struct.*, **50** (24), 3843–3860. doi: 10.1016/j.ijsolstr.2013.07.022.
94. Peerlings, R., de Borst, R., Brekelmans, W., and Ayyapureddi, S. (1996) Gradient-enhanced damage for quasi-brittle materials. *Int. J. Numer. Methods Eng.*, **39**, 3391–3403.
95. Geers, M. (1997) Experimental analysis and computational modelling of damage and fracture. PhD thesis. University of Technology, Eindhoven (Netherlands).
96. Peerlings, R., de Borst, R., Brekelmans, W., and Geers, M. (1998) Gradient-enhanced damage modelling of concrete fracture. *Mech. Cohes.-Frict. Mater.*, **3**, 323–342.
97. Peerlings, R., Geers, M., de Borst, R., and Brekelmans, W. (2001) A critical comparison of nonlocal and gradient-enhanced softening continua. *Int. J. Solids Struct.*, **38**, 7723–7746.
98. Wu, L., Sket, F., Molina-Aldareguia, J., Makradi, A., Adam, L., Doghri, I., and Noels, L. (2015) A study of composite laminates failure using an anisotropic gradient-enhanced damage mean-field homogenization model. *Compos. Struct.*, **126**, 246–264. doi: 10.1016/j.compstruct.2015.02.070.
99. Song, D., Agoras, M., and Ponte Castañeda, P. (2015) The evolution of pore shape and orientation in plastically deforming metals: implications for macroscopic response and shear localization. *Mech. Mater.*, **90**, 47–68. doi: 10.1016/j.mechmat.2015.01.011
100. Corcolle, R., Daniel, L., and Bouillault, F. (2008) Generic formalism for homogenization of coupled behavior: application to magnetoelectroelastic behavior. *Phys. Rev. B*, **78**, 214110. doi: 10.1103/PhysRevB.78.214110.
101. Miehe, C., Vallicotti, D., and Zäh, D. (2015) Computational structural and material stability analysis in finite electro-elasto-statics of electro-active materials. *Int. J. Numer. Methods Eng.*, **102** (10), 1605–1637. doi: 10.1002/nme.4855
102. Muliana, A. and Lin, C.H. (2011) A multi-scale formulation for predicting non-linear thermo-electro-mechanical response in heterogeneous bodies. *J. Intell. Mater. Syst. Struct.*, **22** (8), 723–738. doi: 10.1177/1045389X11406302.
103. Koutsawa, Y., Belouettar, S., Makradi, A., and Tiem, S. (2012) X-FEM implementation of VAMUCH: application

to active structural fiber multifunctional composite materials. *Compos. Struct.*, **94** (4), 1297–1304. doi: 10.1016/j.compstruct.2011.10.028.

104. Pham, K., Kouznetsova, V., and Geers, M. (2013) Transient computational homogenization for heterogeneous materials under dynamic excitation. *J. Mech. Phys. Solids*, **61** (11), 2125–2146. doi: 10.1016/j.jmps.2013.07.005.

105. Alali, B. and Lipton, R. (2012) Multiscale dynamics of heterogeneous media in the peridynamic formulation. *J. Elast.*, **106** (1), 71–103. doi: 10.1007/s10659-010-9291-4.

106. Hui, T. and Oskay, C. (2014) A high order homogenization model for transient dynamics of heterogeneous media including micro-inertia effects. *Comput. Meth. Appl. Mech. Eng.*, **273**, 181–203. doi: 10.1016/j.cma.2014.01.028.

107. de Souza Neto, E., Blanco, P., Sánchez, P., and Feijóo, R. (2015) An RVE-based multiscale theory of solids with microscale inertia and body force effects. *Mech. Mater.*, **80**, Part A, 136–144. doi: 10.1016/j.mechmat.2014.10.007.

108. Temizer, I. and Zohdi, T. (2007) A numerical method for homogenization in non-linear elasticity. *Comput. Mech.*, **40** (2), 281–298. doi: 10.1007/s00466-006-0097-y.

109. Temizer, I. and Wriggers, P. (2007) An adaptive method for homogenization in orthotropic nonlinear elasticity. *Comput. Meth. Appl. Mech. Eng.*, **196** (35–36), 3409–3423. doi: 10.1016/j.cma.2007.03.017.

110. Yvonnet, J., Monteiro, E., and He, Q.C. (2013) Computational homogenization method and reduced database model for hyperelastic heterogeneous structures. *Int. J. Multiscale Comput. Eng.*, **11** (3), 201–225.

111. Sparks, P. and Oskay, C. (2013) Identification of optimal reduced order homogenization models for failure of heterogeneous materials. *Int. J. Multiscale Comput. Eng.*, **11** (3), 185–200.

112. Legrain, G., Cartraud, P., Perreard, I., and Moës, N. (2011) An X-FEM and level set computational approach for image-based modelling: application to homogenization. *Int. J. Numer. Methods Eng.*, **86** (7), 915–934. doi: 10.1002/nme.3085.

113. Sonon, B., François, B., and Massart, T. (2012) A unified level set based methodology for fast generation of complex microstructural multiphase {RVEs}. *Comput. Meth. Appl. Mech. Eng.*, **223–224**, 103–122. doi: 10.1016/j.cma.2012.02.018.

114. Ostoja-Starzewski, M. and Wang, X. (1999) Stochastic finite elements as a bridge between random material microstructure and global response. *Comput. Meth. Appl. Mech. Eng.*, **168** (1–4), 35–49. doi: 10.1016/S0045-7825(98)00105-4.

115. Kanit, T., Forest, S., Galliet, I., Mounoury, V., and Jeulin, D. (2003) Determination of the size of the representative volume element for random composites: statistical and numerical approach. *Int. J. Solids Struct.*, **40** (13–14), 3647–3679. doi: 10.1016/S0020-7683(03)00143-4.

116. Gusella, V. and Cluni, F. (2006) Random field and homogenization for masonry with nonperiodic microstructure. *J. Mech. Mater. Struct.*, **1** (2), 357–386. doi: 10.2140/jomms.2006.1.357.

117. Trovalusci, P., Ostoja-Starzewski, M., De Bellis, M.L., and Murrali, A. (2015) Scale-dependent homogenization of random composites as micropolar continua. *Eur. J. Mech. A. Solids*, **49**, 396–407. doi: 10.1016/j.euromechsol.2014.08.010.

118. Yin, X., Chen, W., To, A., McVeigh, C., and Liu, W. (2008) Statistical volume element method for predicting microstructure–constitutive property relations. *Comput. Meth. Appl. Mech. Eng.*, **197** (43–44), 3516–3529. doi: 10.1016/j.cma.2008.01.008. Stochastic Modeling of Multiscale and Multiphysics Problems.

119. Yin, X., Lee, S., Chen, W., Liu, W.K., and Horstemeyer, M.F. (2009) Efficient random field uncertainty propagation in design using multiscale analysis. *J. Mech. Des.*, **131** (2), 021006. doi: 10.1115/1.3042159.

120. Clément, A., Soize, C., and Yvonnet, J. (2012) Computational nonlinear

stochastic homogenization using a nonconcurrent multiscale approach for hyperelastic heterogeneous microstructures analysis. *Int. J. Numer. Methods Eng.*, **91** (8), 799–824. doi: 10.1002/nme.4293.

121. Lucas, V., Golinval, J.C., Paquay, S., Nguyen, V.D., Noels, L., and Wu, L. (2015) A stochastic computational multiscale approach; application to MEMS resonators. *Comput. Meth. Appl. Mech. Eng.*, **294**, 141–167.

122. Baxter, S. and Graham, L. (2000) Characterization of random composites using moving-window technique. *J. Eng. Mech.*, **126** (4), 389–397. doi: 10.1061/(ASCE)0733-9399(2000)126:4(389).

123. Shinozuka, M. (1971) Simulation of multivariate and multidimensional random processes. *J. Acoust. Soc. Am.*, **49** (1B), 357–368. doi: 10.1121/1.1912338.

124. Shinozuka, M. and Jan, C.M. (1972) Digital simulation of random processes and its applications. *J. Sound Vib.*, **25** (1), 111–128. doi: 10.1016/0022-460X(72)90600-1.

125. Soize, C. (2001) Maximum entropy approach for modeling random uncertainties in transient elastodynamics. *J. Acoust. Soc. Am.*, **109** (5), 1979–1996. doi: 10.1121/1.1360716.

126. Soize, C. (2005) Random matrix theory for modeling uncertainties in computational mechanics. *Comput. Meth. Appl. Mech. Eng.*, **194** (12–16), 1333–1366. doi: 10.1016/j.cma.2004.06.038. Special Issue on Computational Methods in Stochastic Mechanics and Reliability Analysis.

127. Das, S. and Ghanem, R. (2009) A bounded random matrix approach for stochastic upscaling. *Multiscale Model. Simul.*, **8** (1), 296–325. doi: 10.1137/090747713.

128. Guilleminot, J., Noshadravan, A., Soize, C., and Ghanem, R. (2011) A probabilistic model for bounded elasticity tensor random fields with application to polycrystalline microstructures. *Comput. Meth. Appl. Mech. Eng.*, **200** (17–20), 1637–1648. doi: 10.1016/j.cma.2011.01.016.

129. Noshadravan, A., Ghanem, R., Guilleminot, J., Atodaria, I., and Peralta, P. (2013) Validation of a probabilistic model for mesoscale elasticity tensor of random polycrystals. *Int. J. Uncertainty Quantif.*, **3** (1), 73–100.

130. Cho, H., Venturi, D., and Karniadakis, G. (2013) Karhunen–Loève expansion for multi-correlated stochastic processes. *Probab. Eng. Mech.*, **34**, 157–167. doi: 10.1016/j.probengmech.2013.09.004.

131. Fish, J. and Wu, W. (2011) A nonintrusive stochastic multiscale solver. *Int. J. Numer. Methods Eng.*, **88** (9), 862–879. doi: 10.1002/nme.3201.

132. Bernhardt, R., Seyfarth, J., and Kilfoy, L. (2014) Benefits of standards for software providers. 1st International Workshop on Software Solutions for ICME, Aachen/Rolduc, June 24–27, 2014.

133. Seyfarth, J. and Assaker, R. (2014) Extracting effective properties from microstructures–current approaches. 1st International Workshop on Software Solutions for ICME, Aachen/Rolduc, June 24–27, 2014.

134. DIGIMAT (2015) DIGIMAT Users' Manual, Version 6.0.1, May 2015.

135. Taylor, G.I. (1938) Plastic strain in metals. *J. Inst. Metals*, **62**, 307–324.

136. Lebensohn, R. and Tomé, C. (1993) A self-consistent anisotropic approach for the simulation of plastic deformation and texture development of polycrystals: application to zirconium alloys. *Acta Metall. Mater.*, **41**, 2611–2624.

137. Miehe, C., Schroeder, J., and Schotte, J. (1999) Computational homogenization analysis in finite plasticity simulation of texture development in polycrystalline materials. *Comput. Meth. Appl. Mech. Eng.*, **171**, 387–418.

138. Lebensohn, R., Rollett, A., and Suquet, P. (2011) Fast Fourier transform-based modeling for the determination of micromechanical fields in polycrystals. *JOM*, **63**, 13–18.

139. Segurado, J. and Llorca, J. (2013) Simulation of the deformation of polycrystalline nanostructured Ti by computational homogenization. *Comput. Mater. Sci.*, **76**, 3–11.

140. Hill, R. (1966) Generalized constitutive relations for incremental deformation of metal crystals by multislip. *J. Mech. Phys. Solids*, **14**, 95–102.
141. Rice, J. (1971) Inelastic constitutive relations for solids: an internal-variable theory and its application to metal plasticity. *J. Mech. Phys. Solids*, **19**, 433–455.
142. Hill, R. and Rice, J. (1972) Constitutive analysis of elastic-plastic crystals at arbitrary strain. *J. Mech. Phys. Solids*, **20**, 401–413.
143. Roters, F., Eisenlohr, P., Hantcherli, L., Tjahjanto, D.D., Bieler, T.R., and Raabe, D. (2010) Overview of constitutive laws, kinematics, homogenization and multiscale methods in crystal plasticity finite-element modeling: theory, experiments, applications. *Acta Mater.*, **58**, 1152–1211.
144. Roters, F., Eisenlohr, P., Kords, C., Tjahjanto, D.D., Diehl, M., and Raabe, D. (2012) DAMASK: the Düsseldorf Advanced MAterial Simulation Kit for studying crystal plasticity using an FE based or a spectral numerical solver. *Procedia IUTAM*, **3**, 3–10.
145. Quey, R., Dawson, P.R., and Barbe, F. (2011) Large-scale 3D random polycrystals for the finite element method: generation, meshing and remeshing. *Comput. Meth. Appl. Mech. Eng.*, **200**, 1729–1745.
146. Dream.3d (2012) http://dream3d.bluequartz.net/.
147. Eiken, J., Böttger, B., and Steinbach, I. (2006) Multiphase-field approach for multicomponent alloys with extrapolation scheme for numerical application. *Phys. Rev. E*, **73**, 066122-1–066122-9.
148. Laschet, G. and Apel, M. (2010) Thermo-elastic homogenization of 3-D steel microstructures simulated by phase-field method. *Steel Res. Int.*, **81** (8), 637–643.
149. Read, W.T. and Shockley, W. (1950) Dislocation models of crystal brain boundaries. *Phys. Rev.*, **78** (1), 275–289.
150. Humphreys, F.J. (1997) A unified theory of recovery, recrystallization and grain growth, based on the stability and growth of cellular microstructures–I. The basic model. *Acta Mater.*, **45** (10), 4231–4240.
151. Laschet, G., Hul, P., and Apel, M. (2014) On the RVE size for quasi-isotropic macroscopic behaviour of a Fe-Mn-C steel microstructure simulated by the phase-field method. Proceedings of the WCCM-14, Barcelona, Spain, July 20–25, 2014.
152. Halton, J.H. (1960) On the efficiency of certain quasi-random sequences of points in evaluation multi-dimensional integrals. *Numer. Math.*, **2**, 84–90.
153. Chen, G., Ozden, U.A., Bezold, A., and Broeckmann, C. (2013) A statistics based numerical investigation on the prediction of elasto-plastic behavior of WC–Co hard metal. *Comput. Mater. Sci.*, **80**, 96–103.
154. Hachemi, A., Chen, M., Chen, G., and Weichert, D. (2014) Limit state of structures made of heterogeneous materials. *Int. J. Plast.*, **63**, 124–137.
155. Schmitz, G.J. and Prahl, U. (2012) *Integrative Computational Materials Engineering –Concepts and Applications of a Modular Simulation Platform*, Wiley-VCH Verlag GmbH.
156. Spina, P., Spekowius, M., and Hopmann, C. (2014) Multi-scale thermal simulation of polymer crystallization. *Int. J. Mater. Form.*, **8**, 497–504.
157. Bedoui, F., Diani, J., Regnier, G., and Seiler, W. (2006) Micromechanical modeling of isotropic behavior of semicrystalline polymers. *Acta Mater.*, **54**, 1513–1523.
158. Laschet, G., Kashko, T., Benke, S., Öte, M., and Bobzin, K. (2012) in *Integrative Computational Materials Engineering–Concepts and Applications of a Modular Simulation Platform*, Chapter 5 (eds G.J. Schmitz and U. Prahl), Wiley-VCH Verlag GmbH, pp. 81–115.
159. Michaeli, W., Hopmann, C., Bobzin, K., Arping, T., Baranowski, T., Heesel, B., Laschet, G., Schäfer, T., and Öte, M. (2012) Development of an integrative simulation method to predict microstructural influence on

the mechanical behaviour of semi-crystalline thermoplastic parts. *Int. J. Mater. Res.*, **103** (1), 120–130.

160. Feraboli, P., Peitso, E., and Cleveland, T. (2009) Characterization of prepreg-based discontinuous carbon fiber/epoxy systems. *J. Reinf. Plast. Compos.*, **28** (10), 1191–1214.

161. Feraboli, P., Cleveland, T., Ciccu, M., Stickler, P., and DeOto, L. (2010) Defect and damage analysis of advanced discontinuous carbon/epoxy composite materials. *Composites Part A*, **41** (7), 888–901.

162. Picher-Martel, G.-P., Levy, A., and Hubert, P. (2013) Squeeze flow of randomly oriented strands thermoplastic composites. The 19th International Conference on Composite Materials (ICCM-19), Montreal, QC, July 28 - August 2, 2013.

163. LeBlanc, D., Landry, B., Levy, A., Hubert, P., Roy, S., and Yousefpour, A. (2014) Compression moulding of complex parts using randomly-oriented strands thermoplastic composites. Proceeding of the International SAMPE Symposium and Exhibition, Seattle, WA, USA.

164. Harper, L.T. (2006) Discontinuous carbon fibre composites for automotive applications. PhD thesis. University of Nottingham.

165. Luchoo, R., Harper, L.T., Bond, M.D., Warrior, N.A., and Dodworth, A. (2010) Net shape spray deposition for compression moulding of discontinuous fibre composites for high performance applications. *Plast. Rubber Compos.*, **39** (3–5), 216–231.

166. Doghri, I. and Ouaar, A. (2003) Homogenization of two-phase elasto-plastic composite materials and structures. *Int. J. Solids Struct.*, **40** (7), 1681–1712.

167. Digimat (2014) The Multi-Scale Non-linear Material Modeling Platform, Product documentation, Version 5.1, www.e-Xstream.com (accessed 25 April 2016).

168. Autodesk Moldflow (2015) http://www.autodesk.com/products/simulation-moldflow/overview (accessed 25 April 2016).

169. Dassault Systèmes (2015) http://www.3ds.com/ (accessed 25 April 2016).

170. Kröner, E. (1977) Bounds for effective elastic moduli of disordered materials. *J. Mech. Phys. Solids*, **25** (3), 137–155.

171. Adams, B.L., Kalidindi, S.R., and Fullwood, D.T. (2013) *Microstructure Sensitive Design for Performance Optimization*, Butterworth-Heinemann, Waltham, MA.

172. Lobos, M. and Böhlke, T. (2014) Materials design for the anisotropic linear elastic properties of textured cubic crystal aggregates using zeroth-, first- and second-order bounds. *Int. J. Mech. Mater. Des.*, **10** (3), 1–20.

173. Böhlke, T. and Lobos, M. (2014) Representation of Hashin-Shtrikman bounds of cubic crystal aggregates in terms of texture coefficients with application in materials design. *Acta Mater.*, **67**, 324–334.

174. Nadeau, J.C. and Ferrari, M. (2001) On optimal zeroth-order bounds with application to Hashin-Shtrikman bounds and anisotropy parameters. *Int. J. Solids Struct.*, **38** (44), 7945–7965.

175. Ashby, M.F. (2010) *Materials Selection in Mechanical Design*, 4th edn, Butterworth-Heinemann.

176. Salaberger, D. *et al.* (2011) Evaluation of computed tomography data from fibre reinforced polymers to determine fibre length distribution. *Int. Polym. Proc.*, **26** (3), 283–291.

7
Numerical Methods

Carlos Agelet de Saracibar, Romain Boman, Philippe Bussetta, Juan Carlos Cajas, Miguel Cervera, Michele Chiumenti, Abel Coll, Pooyan Dadvand, Joaquin A. Hernández Ortega, Guillaume Houzeaux, Miguel Ángel Pasenau de Riera, and Jean-Philippe Ponthot

7.1
Overview

This chapter deals with numerical methods for the numerical simulation of a wide range of multiphysics and multiscale problems within an Integrated Computational Materials Engineering (ICME) framework. An outline of the chapter is as follows. Section 7.2 deals with preprocess and space discretization methods. Section 7.3 deals with an overview of different kinematic descriptions, solution methods for coupled problems, numerical methods for the solution of a wide variety of engineering problems governed by potential differential equation (PDE) – such as the finite difference method (FDM), finite element method (FEM), boundary element method (BEM), discrete element method (DEM), meshfree methods (MFMs), particle finite element method (PFEM), extended finite element method (XFEM), isogeometric analysis (IGA) method, and model order reduction (MOR) methods – and numerical methods for contact problems, such as Lagrange multipliers, penalty and augmented Lagrangian, direct elimination, mortar, and IGA methods. Section 7.4 deals with postprocess and visualization methods. Section 7.5 deals with an overview of numerical methods for mapping and data transfer in the FEM simulation of multiphysics/multiscale problems using either Lagrangian or arbitrary Lagrangian–Eulerian (ALE) formalisms. The need for efficient and accurate data transfer algorithms is a key issue in ICME and may play a central role in the numerical simulation of a manufacturing chain of processes, which may involve the use of different solution methods and software in the numerical simulation of each individual process of the process chain. Indeed, data transfer algorithms may be needed even in the numerical simulation of a single process, when Lagrangian or ALE formalisms are used and remeshing operations are needed. Section 7.6 deals with reduced-order models (ROMs), an efficient method to automatically reduce the inherent complexity of multiscale hierarchical models. Section 7.7 addresses different issues related to high-performance computing (HPC) and parallelization, and

Handbook of Software Solutions for ICME, First Edition. Edited by Georg J. Schmitz and Ulrich Prahl.

Section 7.8 includes a list of software for pre- and postprocessing. Finally, a list of references is provided.

7.2
Preprocess and Space Discretization Methods

7.2.1
Preprocess

Preprocessing is the preparation of a model for numerical simulation. Most of the numerical solutions are based on domain discretization called *mesh*. Preprocessing has the following main steps:

- Creating a geometry representing the real problem called *model*
- Applying the materials and boundary conditions
- Generating the mesh using space discretization methods

Creating the model is to convert the real geometry into a simplified one, which is suitable for the solver. An example would be a shell structure with a thickness that can be modeled with a surface and passing the thickness as a property to the solver. In practice, the models are usually imported from computer-aided design (CAD) systems. In many cases, these models are not suitable for simulations because they have some small defects such as overlapping surfaces or gaps between them, which should be cleaned up before the discretization step. For instance, in most computational fluid dynamics (CFD) simulations, a "watertight" model is required. For industrial simulations, this cleaning can be very time consuming and, therefore, state-of-the art meshing techniques are trying to relax this cleanness requirement.

Applying the material properties and boundary conditions is the next step of preprocessing. The use of graphical user interface (GUI) is highly recommended for this part. New trends of GUIs are trying to help as much as possible to define most of these parameters automatically for specific problems.

The discretization is another important part of preprocessing and consists of dividing the complex geometry of the model in simple geometric shapes, which can be used by the solver. The quality of the solution highly depends on the quality of this mesh. Generating a suitable mesh requires the knowledge about the solver and its acceptable discretization. These steps are shown in Figure 7.1. In this section, we focus on methods of discretization and their pros and cons.

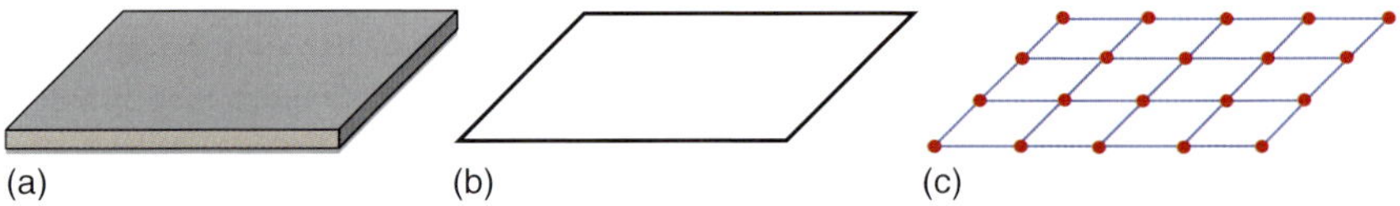

Figure 7.1 Preprocess example from a real geometry of a plate (a) to a surface representation of it without thickness (b) and the generated mesh (c).

7.2.2 Space Discretization Methods

As mentioned earlier, the discretization is one of the main steps in preprocessing. It is important to remark that the result of a numerical solution highly depends on the quality of the given mesh. Understanding the discretization methods and their correct use is important in order to generate a suitable mesh for the simulation to be run.

7.2.2.1 Structured and Semi-structured Meshing Algorithms

There are two main families of meshers depending on the kind of mesh they generate: structured and unstructured; see, for example, [1]. A structured mesh is defined as a mesh in which all inner nodes have the same degree (the degree of a node is the number of its incident elements), while the nodes of an unstructured mesh have different degrees. Informally, one can understand the structured mesh as a regular subdivision of the original shape. If a mesh is only structured in one direction, it is considered as a *semi-structured* mesh. Semi-structured meshes can only be applied to topologically prismatic geometries, and they basically repeat the structure of an unstructured mesh (in the base and top of the prismatic shape) in different layers following the structured direction. An example of these kinds of mesh is shown in Figure 7.2.

Structured and semi-structured meshers often receive the position of the nodes in the contours of the domain as input data and generate the positions of the inner nodes from a given interpolation; see, for example, [1–3]. These meshers are fast and reliable. Their main drawback is their sever restriction in the type of geometries that can be used. As an example, to generate a structured mesh of hexahedra, the input geometry must be previously divided into partitions of hexahedra. This makes them difficult to be used for arbitrary complex geometries.

7.2.2.2 Advancing Front

The advancing front method – see, for example, [4–7] – is a common technique for generating unstructured meshes. The algorithm starts creating elements on the

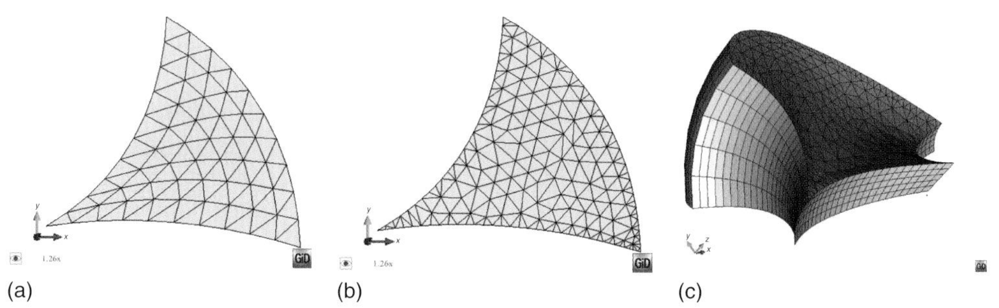

Figure 7.2 Examples of different types of mesh. (a) Structured triangle mesh, (b) unstructured triangle mesh, and (c) semi-structured prism mesh.

boundaries of the geometry and then advances to the interior of the volume until it is filled. The advancing front is mainly used for the generation of tetrahedral (or triangle in 2D) elements. Nevertheless some adaptations of the method have been done to generate other types of elements; see, for example, [8]. The high quality of the discretization and good control over the size transitions are the main characteristics of this technique. Generally speaking, the meshers based on advancing front technique can handle complex geometries but they should be clean geometries, that is, free of gaps, defects, and overlapping surfaces. This method is not particularly fast.

7.2.2.3 Delaunay

A *Delaunay* mesh is defined as a mesh in which all elements satisfy the Delaunay condition: the circumcircle (in 2D case) or the circumsphere (in 3D case) of any element has no node from the mesh inside; see, for example, [1]. Given a cloud of points, a Delaunay triangulation can always be created from their Dirichlet tessellation connecting these points with a set of triangles (in 2D) or tetrahedra (in 3D) without adding any extra node; see, for example, [1]. The Delaunay meshing methods – see, for example, [3, 9–11] – start from the contour of the domain and generate the Delaunay triangulation of its nodes. This mesh is the convex hull of the domain to be meshed. Then they improve the mesh by adding nodes while maintaining the Delaunay condition. The difficulty here is to respect the nonconvex boundaries of the model, which lead to the *constrained Delaunay* method. Unfortunately, in 3D, this method can lead to low-quality elements with almost zero volume, called *slivers*. For this reason, it is common to relax the Delaunay condition in some regions in order to avoid quality problems; see, for example, [12]. This method is faster than the advancing front technique for generating the mesh but generally with less quality. Many mesh generators are based on this technique but they need a cleaned geometry as input.

7.2.2.4 Space Decomposition Techniques

Space decomposition-based methods follow a different philosophy than the methods explained earlier. To generate the mesh, they basically subdivide the space into cells using a regular grid, like a *bin*, or using a hierarchical grid, like an *octree*, as can be seen in Figure 7.3. It is obvious that these grids do not represent accurately the model boundaries, so all methods in this category first find the part of the model they cover, and then try to fit their boundaries.

The bin structure is suitable for homogeneous discretization. It is common to use an octree structure instead of a regular grid (which gives the name of the family of methods), because it is more flexible for mesh generation purposes, as domains to be meshed and required elements' sizes are usually nonhomogeneous. Octree-based meshers were pioneered by Yerry and Shephard [13] and, since then, several approaches have been proposed; see, for example, [14–20]. These mesh generators are robust and also very fast and parallelizable, which make them a good choice for generating extremely large meshes. The drawback of these methods is the poor mesh quality in the boundaries and problems in preserving the

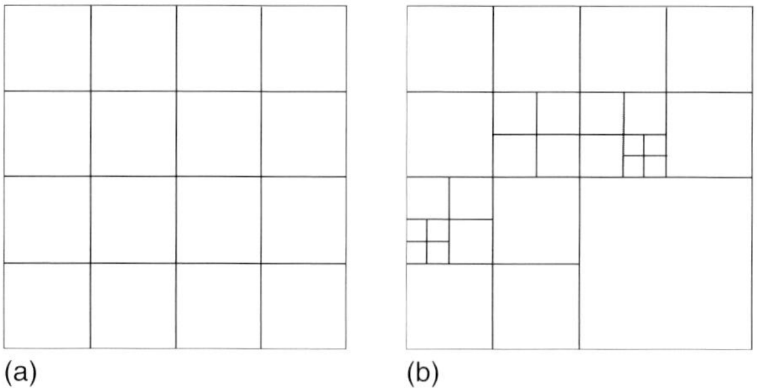

Figure 7.3 2D examples of typical structures used in space decomposition-based meshing algorithms. (a) Bin; (b) quadtree.

geometrical features (ridges or corners) and the topology of the domain. Nevertheless, recent works have been done in order to preserve perfectly the geometry and even to handle geometries that are not cleaned; see, for example, [15]. The last one is very important in order to work with very large and complex CAD inputs with many defects and in order to reduce the time significantly from CAD modeling to simulation.

7.3 Numerical Methods for Engineering Problems

7.3.1 Kinematic Frameworks

A fundamentally important consideration when developing a computer code is the choice of an appropriate kinematic description of the continuum. The choice of the kinematic framework determines the relationship between the deforming continuum and the underlying grid or mesh, conditions the ability of the numerical method to deal with large distortions, and provides an accurate resolution of material interfaces and mobile boundaries. The solution of a continuum mechanics problem involves a spatial discretization of the continuous fields in order to transform an infinite-dimensional problem into a finite-dimensional one. This discretization can also be regarded as a discretization of a frame of reference, in the sense that, once discretized, the physical properties of the continuum are only known at a finite number of mesh points.

The formulations of continuum mechanics usually make use of two classical descriptions of motion: the *Lagrangian* and the *Eulerian* description.

In a material or *Lagrangian* description, the frame of reference, known as material frame of reference, is attached to the material particles of the continuum, and any physical property is described in terms of the material particles. Particles are identified by their position at the reference configuration. Lagrangian descriptions, in which each individual node of the computational mesh follows the associated material particle during motion, are mainly used in solid mechanics. The Lagrangian description allows an easy tracking of free surfaces and interfaces between different materials, and avoids convective terms. It also facilitates the treatment of history-dependent material models. Its weakness is its inability to follow large distortions of the computational domain, as is the case for fluids or solids with large deformations, without recourse to frequent remeshing operations.

In a spatial or *Eulerian* description, the frame of reference, known as spatial frame of reference, is fixed in the space. Here, the computational mesh is fixed, mesh points are fixed to spatial points, and the continuum moves with respect to the grid. Physical properties are only known at those fixed spatial points. In the Eulerian description, remeshing procedures are avoided making this formulation especially suitable for fluid or large deformation solid problems. Nevertheless, two main drawbacks arise. First, convective terms appear due to the relative motion between the material (particles) and the computational (mesh points) frames of reference. These convective terms may introduce numerical difficulties and suitable numerical stabilization techniques have to be introduced. The second drawback is related to the description of moving boundaries and free surfaces.

In order to avoid the shortcomings of Lagrangian and Eulerian descriptions, an ALE description was developed in an attempt to combine the advantages of both of them, while minimizing their respective drawbacks, as far as possible; see, for example, [21]. In the ALE description, a new frame of reference, known as the computational frame of reference, is introduced. The computational frame of reference, which is neither attached to the particles nor fixed in the space, moves with an arbitrary velocity with respect to the fixed space. Both Lagrangian and Eulerian descriptions can be seen as particular cases of the ALE formulation. Taking as computational frame of reference the material one, the mesh is attached to the particles and one recovers the Lagrangian formulation. Taking as computational frame of reference the spatial one, the mesh is fixed in the space and one recovers the Eulerian formulation. Making an appropriate choice for the velocity of the computational frame of reference, the ALE allows the use of a pure Lagrangian formulation at the boundary (avoiding the problems associated to the Eulerian formulation) and, simultaneously, maintains a regular shape in the rest of the domain (avoiding severe distortion problems associated with the Lagrangian formulation). Then, the ALE description incorporates advantages from both Lagrangian (accuracy in the description of moving boundaries) and Eulerian (avoiding extreme distortions) formulations. Convective terms, however, are still present in the ALE formulation.

7.3.2 Computational Strategies for Coupled Problems

The numerical solution of the coupled problems involves the transformation of an infinite-dimensional dynamical system, governed by a system of quasilinear partial differential equations (PDEs), into a sequence of discrete nonlinear algebraic problems by means of the following two steps:

Step 1. The infinite-dimensional space is approximated by a finite-dimensional one via a Galerkin finite element (FE) projection. The spatial projection of the dynamic weak form of the governing equations leads to a nonlinear coupled system of ordinary differential equations (ODEs), which describe the time evolution of nodal degrees of freedom in the time interval of interest.

Step 2. The coupled system of nonlinear ODEs describes the time evolution in the time interval of interest, of the nodal degrees of freedom, and the internal variables associated with the FE Galerkin projection. A time discretization of this problem involves a partition of the time interval of interest. Within a typical time subinterval $[t_n, t_{n+1}]$, a time marching scheme for the advancement of the primary variables is introduced.

With regard to the time-stepping scheme, different strategies are possible, but they can be collected in two categories: *simultaneous* (monolithic) and *staggered* (*block-iterative* or *fractional-step*) time-stepping algorithms.

Simultaneous or *monolithic* time-stepping algorithms solve all the governing equations at once, thus advancing all the primary nodal variables of the problem simultaneously. This invariably leads to large and, most likely, unsymmetric system of equations, usually prohibitively expensive to solve. Furthermore, the use of different standard time-stepping algorithms developed for the single uncoupled problems is not straightforward, and it is not possible to take advantage of the different timescales possibly involved in each one of the subproblems.

Block-iterative time-stepping algorithms solve each one of the subproblems separately iteratively. This leads to smaller system of equations to be solved. The study of the stability of the resulting algorithm may be complicated, as it depends on the tolerance used to assess convergence. The problem of stability in time is then linked to that of convergence within the time step. *Fractional-step* algorithms are based on the use of an operator split, applied to the coupled system of nonlinear ODEs, and a product formula algorithm, which leads to a scheme in which each one of the subproblems defined by the partition is solved sequentially. This leads to a partition of the original problem into smaller and typical symmetric (physical) subproblems. Furthermore, the use of different standard time-stepping algorithms for the uncoupled subproblems is now straightforward, and it is possible to take advantage of the different timescales involved.

7.3.3 Numerical Methods for PDE

The numerical solution of problems governed by PDE can be accomplished by using different numerical methods. Those numerical methods could be classified into the following categories:

- **FDM**. The FDM is one of the easiest technique to solve PDEs. Within the FDM, the solution of the PDEs is obtained using finite difference equations to approximate the derivatives. The FDM can be a good choice to solve PDEs on simple and regular domains, but it becomes quite involved in complex and irregular domains. On the other hand, it is not easy to carry out the mathematical analysis (solvability, stability, and convergence) for the FDM, especially for nonlinear PDEs.

- **FEM**. The FEM can be considered as the most popular method for solving various PDEs; see, for example, [22]. Compared with other methods, it has well-established mathematical theory for various PDEs and is a good choice for solving PDEs over complex domains. Within the FEM, the PDEs to be solved are conveniently rewritten in weak form using variational methods. The continuum model domain is discretized using an FE mesh. Using a linear basis function over each element, an approximate solution is looked for at the mesh nodes.
- **BEM**. The BEM is a numerical method that is used to solve PDEs, which can be formulated as integral equations; see, for example, [23]. This induces a severe restriction on the applicability of the BEM, as nonlinear problems cannot be written as integral equations. The BEM can be thought of as the application of a FEM on the boundary of the spatial domain, rather than in the whole spatial domain. From a computational point of view, the BEM may be more efficient than other numerical methods for problems, which have a small surface-to-volume ratio. However, while the FEM typically yields sparse and banded matrices, resulting in storage requirements and computational time that grow linearly with the problem size, the BEM usually yields fully populated matrices, resulting in storage requirements and computational time which that grow quadratically with the problem size.
- **DEM**. The DEM assumes that materials can be represented by an assembly of rigid particles interacting among themselves. The overall behavior of the system is determined by the cohesive/frictional contact laws. The contact law can be seen as the formulation of the material model on the microscopic level. Cohesive bonds can be broken, which allows us to simulate the fracture of materials and its propagation. The motion of each individual particle in the DEM is computed by integrating in time the dynamic equilibrium equations of the particle under the interacting forces from adjacent particles. The DEM is widely recognized as a suitable tool to model geomaterials and advantages of this method can be clearly seen in the numerical modeling of geotechnical problems. A basic formulation of the DEM, using spherical or cylindrical particles, was first proposed by Cundall and Strack [24]. Further applications of the DEM in the numerical simulation of granular materials can be found in [25, 26]. A DEM formulation with enhanced features has been developed in CIMNE during the last years. The DEM formulation has been implemented in the explicit dynamic code DEMPack, which has several original features, such as a coupled DEM/FEM formulation, modeling of tool wear in rock cutting, thermomechanical coupling, and other capabilities, which are not available in commercial DEM codes. This unified DEM/FEM approach, developed by Oñate and Rojek [27], has been used for the numerical simulation of geotechnical problems, underground excavations,

rock cutting problems, and sand mold manufacture for lost foam casting processes; see, for example, [28–30].

- **MFMs**. MFM arose as a suitable solution to overcome some of the drawbacks shown by mesh-based methods, such as the FEM. The underlying structure of the classical mesh-based methods is not well suited to deal with discontinuities that do not coincide with the original mesh edges. With a mesh-based method, the most viable strategy to deal with moving discontinuities is to remesh whenever it is necessary. The remeshing process is expensive and nontrivial in 3D problems, and projection of quantities of interest between successive meshes usually leads to degradation of accuracy and often an excessive computational cost. The objective of MFM is to eliminate at least part of this mesh dependency by constructing the approximation entirely in terms of nodes (usually called particles in the context of MFMs). Moving discontinuities or interfaces can usually be treated without remeshing, with minor cost and accuracy degradation. The range of problems that can be addressed by MFMs is much wider than mesh-based methods. The following numerical methods are generally considered to fall within the general class of MFM: smoothed particle hydrodynamics (SPH) (1977), diffuse element method (DEM) (1992), dissipative particle dynamics (DPD) (1992), element-free Galerkin method (EFG/EFGM) (1994), reproducing kernel particle method (RKPM) (1995), finite pointset method (FPM) (1998), natural element method (NEM), material point method (MPM), meshless local Petrov Galerkin (MLPG), moving particle semi-implicit (MPS), generalized finite difference method (GFDM), particle in cell (PIC), moving particle finite element method (MPFEM), finite cloud method (FCM), boundary node method (BNM), meshfree moving Kriging interpolation method (MK), boundary cloud method (BCM), method of fundamental solution (MFS), method of particular solution (MPS), method of finite spheres (MFS), discrete vortex method (DVM), finite mass method (FMM) (2000), smoothed point interpolation method (S-PIM) (2005), meshfree local radial point interpolation method (RPIM), local radial basis function collocation method (LRBFCM), viscous vortex domains (VVD) method, discrete least-squares meshless method (DLSM) (2006), repeated replacement method (RRM) (2012), and radial basis integral equation method.
- **PFEM**. The PFEM, a new class of particle method developed by Oñate *et al.* [31], is an innovative Lagrangian method, which combines a particle-based method and the FEM. The PFEM treats the mesh nodes in the equivalent domains as Lagrangian particles that can freely move. An FE mesh connects the nodes defining the discretized domain where the continuum mechanics equations are solved via the standard FEM. The PFEM is a suitable approach for the numerical simulation of a wide range of problems, including solid mechanics, fluid dynamics, free-surface particle flows, fluid–solid interactions (FSIs), and so on; see, for example, [31–34]. The PFEM has been recently integrated with the DEM, leading naturally to a new particle-discrete finite element method (PDFEM), which seems ideal for the study of multiscale heterogeneous free-surface flows; see, for example, [28].

Mainly within the framework of the FEM, the following numerical methods can be considered:

- **XFEM**. The XFEM is a numerical technique that extends the classical FEM approach by enriching the solution space for solutions to differential equations with discontinuous functions. The XFEM was developed in 1999 by Belytschko and coworkers to help alleviate the shortcomings of the FEM to solve problems with localized features that are not efficiently resolved by mesh refinement. The idea behind is to retain most of the advantages of MFM, while alleviating their drawbacks. One of the initial applications was the modeling of fracture in a material. In this original implementation, discontinuous basis functions are added to standard polynomial basis functions for nodes belonging to elements that are intersected by a crack to provide a basis that includes crack opening displacements. A key advantage of XFEM is that in such problems the FE mesh does not need to be updated to track the crack path. XFEM has been also used for problems involving singularities, material interfaces, regular meshing of microstructural features, such as voids, and other problems where a localized feature can be described by an appropriate set of basis functions. It was shown that for some problems convergence rates and accuracy can be significantly improved. Moreover, treating problems with discontinuities with XFEM suppresses the need to mesh and remesh the discontinuity surfaces, reducing the computational costs and projection errors associated with conventional FEM, at the cost of restricting the discontinuities to mesh edges. There exist several research codes implementing this technique to various degrees, such as GetFEM++, xfem++, and openxfem++. XFEM has also been implemented in codes such as Altair Radioss, ASTER, Morfeo, and Abaqus. It is increasingly being adopted by other commercial FE software, with a few plug-ins and actual core implementations available (ANSYS, SAMCEF, OOFELIE, etc.).
- **IGA Method**. IGA is a recently developed computational approach that offers the possibility of integrating finite element analysis (FEA) into conventional NURBS-based CAD tools; see, for example, [35]. IGA employs complex NURBS geometry (the basis of most CAD packages) in the FEA application directly. This allows models to be designed, tested, and adjusted in one go, using a common data set. A reference free software implementation of some IGA methods is GeoPDEs, a suite of software tools written in Octave and fully compatible with MATLAB. The suite consists of a set of interrelated packages. The main package, geopdes_base, defines the basic data structures and methods and should also serve as an entry point for understanding the implementation of an IGA code. Other packages deal with applications in linear elasticity, fluid mechanics, and electromagnetism. A package specifically meant to allow handling multipatch NURBS geometries is also available. Likewise, other implementations can be found online. For instance, PetIGA is an open framework for high-performance IGA heavily based on PETSc. In addition, IGAFEM is another IGA code, which is implemented in MATLAB and supports Partition of Unity Method (PUM) enrichment IGA for 2D and 3D fractures.

- **MOR Method**. Developed from well-established mathematical theories and robust numerical algorithms, MOR has been recognized as very efficient for reducing the simulation time of large-scale systems; see, for example, [36] for an up-to-date complete review. Through MOR a small system with reduced number of equations (reduced model) is derived. The reduced model is simulated instead and the solution of the original PDEs or ODEs can then be recovered from the solution of the reduced model. As a result, the simulation time of the original large-scale system can be shortened by several orders of magnitude. The reduced model as a whole can also replace the original system and be reused repeatedly during the design process, which can further save much time. Parametric model order reduction (PMOR) methods are designed for MOR of parameterized systems. The parameters could be the variables describing geometrical measurements, material properties, the damping of the system, or the component flow rate. The reduced models are constructed such that all the parameters can be preserved with acceptable accuracy.

7.3.4 Numerical Methods for Contact Problems

Frictional contact problems arise in almost any material forming process where there is an interaction between different bodies, typically workpiece and tools, such as stamping, bulk forming, casting, friction stir welding, extrusion, rolling, powder compaction, mold filling, and cutting. Mathematically, the numerical analysis of frictional contact problems amounts to finding the solution of an initial boundary value problem (IBVP) within a constrained solution space. The variational formulation of a frictional contact problem includes restrictions on the admissible variations in the tangent solution space induced by the contact constraints yielding to variational inequalities (VIs); see, for example, [37, 38].

- **Lagrange Multipliers Method**. Within the Lagrange multipliers method, the kinematic contact constraints are satisfied by introducing a new set of unknowns, denoted as Lagrange multipliers. The size of the system of equations to be solved will increase/decrease according to the number of active Lagrange multipliers introduced to satisfy the contact constraints. On the other hand, the implementation of the Lagrange multipliers method can be cumbersome. In order to avoid those drawbacks, regularization methods were introduced.
- **Penalty-based methods: penalty and augmented Lagrangian**. A regularization of the frictional contact constraints, using penalty-based methods, allows us to bypass the need to find a solution within a constrained solution space and provides a very convenient displacement-driven frictional contact formulation. Among the penalty-based methods, the penalty method can be considered as the standard and most popular regularization procedure for the computational modeling of frictional contact problems; see, for example, [39–47]. To avoid some well-known drawbacks exhibited by the penalty method, such as penalty sensitivity, unphysical penetrations, and ill-conditioning of the system

of equations, while retaining its advantages, the augmented Lagrangian method, using Uzawa's algorithm, has been introduced. Uzawa's algorithm allows the use of lower penalty parameters, introducing Lagrange multipliers that need to be updated within an additional iteration loop, and keeps the convenient displacement-driven feature of the penalty method; see, for example, [44–50]. A displacement-driven formulation of frictional contact problems allows us to widely exploit features of the well-established framework developed for computational plasticity. Typically, most of the discrete frictional contact problems are formulated using the underlying local parameterization induced by the FE triangulation of the contact surfaces and standard frictional time integration algorithms may turn out to be useless if large slips are involved. A new frictional time integration algorithm suitable for large slips, designed to avoid those drawbacks, was proposed by Agelet de Saracibar [39]. The idea behind was to introduce a new slip path parameterization using Hermite interpolation functions, which, remarkably, is independent of the local surface FE parameterization used in the spatial triangulation.

- **Direct elimination method**. A direct elimination method for quasistatic and dynamic full stick and frictionless contact problems has been recently proposed by Di Capua and Agelet de Saracibar [51]. Lagrange multipliers and regularization methods, such as penalty or augmented Lagrange multipliers, are avoided. Constraints arising in quasistatic and dynamic, full friction, and frictionless contact problems are imposed in a strong fashion by a direct elimination of the involved degrees of freedom of the resulting system of equations. Drawbacks inherent to penalty-based methods, such as the selection of a suitable penalty parameter, unphysical penetrations or ill-conditioning of the system of equations, or the Lagrange multipliers methods, such as the need to dynamically expand the system of equations for the Lagrange multipliers, are fully avoided. Remarkably, the contact constraints are easily imposed by a number of matrix transformations applied to the global tangent operator and residual force vector of the contactless case problem.
- **Mortar element method**. It is well known that node-to-segment (NTS) formulations do not pass the contact patch test. An alternative to NTS formulations is the segment-to-segment (STS) formulation, such as the mortar element. A mortar element formulation for large deformation solid mechanics was presented by Puso and Laursen [52, 53] and Yang [54]. Dual-active set strategies based on the mortar element method have been presented by Wohlmuth [55] and Hüeber and Wohlmuth [56].
- **IGA method**. Within the IGA framework, introduced by Hughes *et al.* [35], the same smooth and higher-order basis functions, for example, NURBS, are used for both the CAD geometry and the approximation of the FEM solution fields, leading to evident potential advantages in the description of interacting surfaces undergoing large displacements and large sliding. A detailed and up-to-date review of isogeometric contact formulations can be found in [57].

7.4 Postprocess and Visualization Methods

7.4.1 Postprocess

The results of a simulation are given in tables of data over nodes or elements of the mesh. Extracting the interesting information from these results or visualizing them is called *postprocess*. One important part of postprocessing is the *data mining*, which consists of extracting information from the raw results of the simulation. Getting the pressure evolution in a point of the model, the maximum deformation of the model, or the average of velocity in a region are examples of such a process. The visualization of the results helps the engineer and the scientist in analyzing and interpreting the results. So the visualization of the results is a key part of the postprocessing.

It is important to emphasize the different nature of the simulation results. Usually, there are two categories of the results:

- *Nodal* results: These results are considered to be continuous and in many cases are even smooth. The displacement field in solid mechanic application is an example of a nodal result.
- *Elemental* results: They are also referred as *gauss points*' results or *cell* results. These results are usually discontinuous across neighboring elements. The tension field in solid mechanics is an example of such discontinuous gauss points' result. An important postprocessing step of these results is to convert them to some smoothed nodal ones.

As mentioned before, in many cases, the results of the simulation have to be analyzed in order to extract engineering design information. Some of the more standard queries are as follows:

- Calculating a result in a given spatial point, which does not correspond to a node of the mesh. This process can emulate a virtual sensor.
- Statistical postprocess of results. Involves spatial or temporal analysis of the results. Finding the average, maximum, minimum, variance, and so on are the examples of these statistics.
- Calculating secondary results. In this type of processes, the original results of the simulation are used in order to calculate some secondary results for engineers. Examples are the calculation of security coefficient from the calculation tensions or the streamlines of a velocity field in a CFD simulation.

Note that the solver may incorporate tools to facilitate this information without external postprocessing. Other "data analytics" queries that are used may include the detection of features such as vortices, contacts, huge damages, deformations, or domain discontinuities, or even more sophisticated higher-order features.

Figure 7.4 Images corresponding to different stages of the simulation of the sea waves against a platform. The mesh of the sea surface is not present in the simulation output data but calculated from the result "distance field" of the tetrahedral mesh in the postprocessing visualization tool.

These queries can also be related to a precomputed result like the average values on different time steps.

7.4.2
Visualization Methods

The visualization is the most important part of the postprocessing and also the most complex one from the development point of view. The increase in size and complexity of the simulation results to be shown are the main difficulties. Some visualization modes include data mining on the simulation results, as can be seen in Figure 7.4.

The following list gives a brief description of the common visualization modes:

- **Contour lines**: A very common visualization mode to show colored isolines at automatic chosen values of the results.
- **Contour fill**: Where a color map is used to show the result gradients in color bands, as can be seen in Figure 7.5.
- **Vectors**: Displaying vectors with the direction of the results. The magnitude can be shown as color or as size of the vector. This visualization is very helpful for vectorial fields like velocities and especially in volumes. It is convenient to also depict the tensorial fields with three vectors oriented according to the principal stresses.
- **Isosurface**: Is the surface that connects points with the same result value. It is an important tool for visualizing volumetric data such as the failure surface of a mechanical part, the free surface of the water in a channel, or the advancing front of the material in casting simulations.
- **Deformation**: Applies the displacements of the nodes to move the mesh into deformed configuration applying some amplifying factor. This tool is of special interest for solid mechanic applications.
- **Streamlines**: Curves that are tangent to a vector field, typically velocities, at a certain time step, showing the direction of the vectors. This visualization is suitable to understand the flow of fluid or for tracing particles in stationary flows.

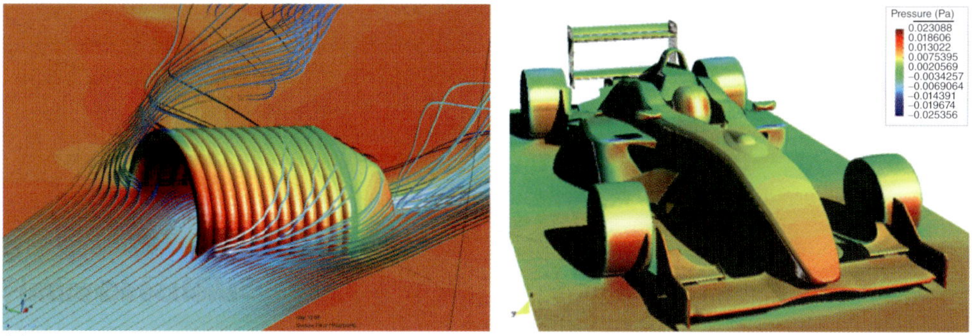

(a) (b)

Figure 7.5 (a) Streamline of the wind over structure. (b) Distribution of the nodal pressure in an air flow simulation of a racing car.

- **Pathlines or particle tracing:** Are curves representing the trajectories that individual particle follows in a vector field along the time steps of the simulation.

The recent increase of simulation model sizes results in large amount of data [58]. Postprocessing of such a huge data can be very time consuming or impossible in a small machine. So simplification of the geometry and results have become a part of the postprocessing [59, 60], as well as the use of advanced data analytics tools, like the ones based on Big Data technologies [61].

7.5 Mapping and Data Transfer Methods

Many numerical methods require more than one mesh to solve complex problems. For example, in the case of multiphysics models, it is sometimes useful to create one specific mesh dedicated to each physics model. As an illustration, in thermomechanics, a first mesh handles the mechanical part of the equations and a second one manages the thermal equilibrium since both problems have generally different length scales. In such staggered strategies, data transfers have to be frequently performed between the two meshes; see, for example, [62, 63]. Furthermore, in the numerical simulations of large deformations of solids with one single mesh, remeshing – and thus data transfer – is sometimes required during the solution procedure because of excessive mesh distortions. This new mesh generation can also be seen as an opportunity to optimize the mesh density with the aim of saving computational time (coarsening some part of the mesh) or with the aim of increasing the local accuracy of the solution (refinement of some part of the mesh). The new mesh density is then driven by some error estimation; see, for example, the review from [64]. During the remeshing operation, all the variables of the problem have to be transferred from the old mesh to the new one; see, for example, [65–67]. Two kinds of variables must be processed: the primary variables, which are stored at the nodes and allow an easy interpolation in order to approximate a continuous field, and the secondary variables, which are only defined at the discrete quadrature points of the FEs. In practice, data transfer methods are required in many computational techniques such as the ALE formalism using an operator-split procedure (see, for example, [21, 68–70] and many others), multigrid procedures (see, for example, [71, 72]), Arlequin method (see, for example, [73, 74]), global/local techniques (see, for example, [75, 76]), or the Chimera scheme (see, for example, [77, 78]). Data transfer algorithms are also an important step of the numerical simulation of a manufacturing chain of processes, which usually involves various solution methods and computer codes for the modeling of all the processes of the chain. The efficiency and the accuracy of the data transfer method play a central role in these complex simulations.

In the following sections, the methods are classified into four main types. The first type of methods calculates the values of the primary variables over the new mesh as an interpolation of nodal values of the containing element of the old mesh using shape functions. Secondary variables have to be extrapolated to nodes first in order to be processed. The second type calculates each new variable value from a smooth local reconstruction of a continuous field based on the neighbor values over the old mesh considered as a simple cloud of points. The two last presented sets of methods are projection operators: they are based on the fact that the integral of any quantity over the remeshed domain should remain unchanged by the transfer operation. They use either mortar elements or a discontinuous reconstruction of the transferred field to reach this goal.

7.5.1
Element Interpolation Methods

This family of method is the oldest, the simplest, and one of the most used data transfer methods. The new values of the variables are directly evaluated from a continuous field of this variable over the old mesh. Primary variables can be processed very easily by this method since they naturally define a continuous field over the mesh, thanks to the FE shape functions. The treatment of secondary variables is more complex because it requires an additional step for constructing a continuous field from the set of discrete values. This step usually involves the definition of auxiliary nodal values followed by the aforementioned interpolation. Several ways of defining these nodal values are proposed in the literature. Perić *et al.* [67] and Mediavilla *et al.* [66] use extrapolation using shape functions followed by the averaging of the values coming from adjacent elements. More elaborated techniques consist of using a moving least-squares approximation, for example, [79], the superconvergent patch recovery (SPR) technique, for example, [80, 81], or other numerical approximation methods, for example, [82, 83]. Once the continuous nodal field is constructed, the remaining problem is to find the FE of the old mesh, which contains each data point of the new mesh (see e.g., [65, 84, 85]). Then, the final value of the variable is obtained by using the shape functions of the corresponding element. This family of methods produces important diffusion errors, and they are only a good choice for problems where only a few remapping are necessary. In particular, using such methods in the frame of an ALE resolution procedure may lead to numerical instabilities.

7.5.2 Interpolation from Clouds of Points

These methods try to reconstruct a local smooth function from the pointwise values taken by the variable over the old mesh in the vicinity of the considered data point. Then, the value calculated at that particular data point is set equal to the value of the reconstructed function at the point location. Such remapping techniques completely ignore the connectivity of the FEs and only rely on a cloud of neighbor points that surround the point of interest. The aim of the reconstructed smooth function is to have a better approximation of the actual field than the one defined by the nodal values or the discrete Gauss point values. Many reconstruction techniques have been suggested in the literature. For example, [86] used an inverse distance algorithm. The diffuse approximation, which relies on moving least squares, can also be used to evaluate the transferred field at each data point; see, for example, [87, 88]. In addition, with the diffuse approximation, some admissible conditions can be imposed during the reconstruction process so that the transferred state of stress does respect equilibrium conditions as well as a yield criterion; see, for example, [88].

7.5.3 Projection using Mortar Elements

This kind of method is based on the conservation of the integral of the variable to be transferred over the spatial domain; see, for example, [62, 65, 89, 90]. This integral is computed over the new mesh using classical finite element shape functions. The coupling between values over the old mesh and values over the new mesh is evaluated, thanks to integrals over mortar elements; see [91]. Due to the structure of the resulting system of equations, the secondary variables can be easily computed by an element-by-element approach. As far as the primary variables are concerned, two ways can be employed to define the mortar elements. The first one consists of generating a "super mesh" to compute the coupling terms. Each element of this super mesh is included in one and only one element of the two other meshes (the old mesh and the new mesh). In other words, the geometry of one of the mortar elements that define the super mesh is given by the intersection between one element of the old mesh and one element of the new mesh. Especially in 3D, the calculation of this intersection is quite cumbersome. In the second way, the integrals over the mortar elements are computed by numerical integration. The advantage of this second technique is twofold. First, the implementation is much easier, especially in 3D, and second, its intrinsic simplicity saves a lot of computational time with respect to the previous method, which requires the generation and the management of the super mesh. Practically, the accuracy of the two methods is totally similar; see [65]. Unfortunately, the resulting transfer scheme is not monotonic (new maxima may appear during the procedure).

7.5.4
Projection using Discontinuous Reconstructions

This family of methods has been first developed for the transfer of secondary variables of elements using one single quadrature point. Similar to the mortar element method, the integral of the variable over the old mesh and the one over the new mesh are expected to be equal. Instead of relying on shape functions for the evaluation of this equality, it is assumed that the value of the variable to be transferred is constant over each element (and obviously equal to the value at the unique quadrature point); see, for example, [68]. This assumption leads to the well-known constant reconstruction technique used in the finite volume method. As a consequence, the reconstructed field is defined everywhere on the old mesh but it is discontinuous across element boundaries. Then, the previous integral equality can be rewritten as a simple explicit formula: the value of the variable in a particular element of the new mesh simply becomes a weighted sum of the values of all the intersecting elements of the old mesh. The weights are the ratios of the volumes of the intersections and the volume of the element. The resulting scheme is monotonic (no new extrema) and first-order accurate. Second-order accuracy can be achieved by using a linear reconstruction instead of a constant one: the variable is still discontinuous across element boundaries but it now varies linearly over each element. The interesting monotonicity property is lost but it can be recovered, thanks to appropriate gradient limiters. The method has been extended to elements with more than one quadrature point by Huerta and Casadei [92] and Rodriguez-Ferran *et al.* [93]. The underlying idea consists of splitting each element in a series of smaller elements surrounding each quadrature point for which the previous procedure can be applied. As an example, a quadrangle with four Gauss points is divided into four smaller elements using its median lines. Many variants dedicated to other types of elements may be found in the literature; see, for example, [94] for triangles. Another extension concerns the transfer of primary variables. In this case, the same procedure is applied to the dual mesh, the one with elements that surround each node of the initial mesh; see, for example, [65, 70, 95]. Moreover, the conservation of important physical quantities, such as mass and energy, or constraints on the internal variables of the material can be taken into account during the transfer; see, for example, [90].

7.5.5
Particular Case of ALE Remapping

The ALE formalism consists of decoupling the movement of the finite element mesh from the deformations of the underlying material; see [21] among many others. Compared with the classical Lagrangian formalism where the nodes follow the same material particles or to the Eulerian formalism where the nodes are fixed in space, the motion of the ALE computational grid can be arbitrarily chosen by the user of the FE code. ALE methods are very helpful in large deformation problems, when the quality of the FE mesh deteriorates rapidly during the simulation. For these particular models, the ALE mesh is constantly optimized and costly remeshing operations are delayed and sometimes completely avoided. The ALE equilibrium equations contain two main sets of unknowns defined at each node: the material velocity and the mesh velocity. Consequently, this system of equations must be completed with additional relationships describing the motion of the mesh. However, in the case of highly nonlinear problems including free boundaries and possibly contact between them, writing these second set of equations explicitly is very difficult. The common workaround is to solve the two sets of equations sequentially with an operator-split procedure. At each time increment, the nonlinear equilibrium equations are iteratively solved as in the Lagrangian case, that is, with a mesh that follows the material. Once the equilibrium is reached, the nodes are then repositioned using appropriate techniques; see [69]. The solution variables (e.g., the stress tensor and the internal state of the material) are eventually transferred from the old mesh configuration to the new one.

The transfer methods used in the ALE framework are very similar to the ones described in the general context of the previous sections. However, they are much faster because their implementation can take advantage of the fact that the topology of both meshes is exactly the same. In other words, each element keeps the same neighbors during the redefinition of the new mesh. Moreover, the distance between two corresponding elements is usually small because the transfer is executed very frequently. In practice, it means that the very CPU-expensive determination of the position of data points of the new mesh in the old mesh is completely avoided. For ALE models, this search can be restricted to the element itself and its direct neighbors. Some transfer methods can be even more accelerated. For example, projection methods using a discontinuous reconstruction can be formulated as a finite volume resolution of an advection problem, leading to a fast calculation of simple fluxes of the considered variable through the boundaries of the elements; see [93].

7.6 Reduced-Order Multiscale Models

7.6.1 Introduction

7.6.1.1 Simplifications in Hierarchical Multiscale Models

It is unarguable that *multiscale modeling* is a pivotal building block of the burgeoning ICME philosophy; see, for example, [96]. Indeed, one of the major goals of this top-down design philosophy is to try to meticulously modify and control features that span across a large spectrum of length scales so that the desired properties and performance characteristic at the "engineering" scale are achieved; therefore, it is crucial to have a computationally efficient way of modeling and linking the distinct phenomena that operate at each level in the material hierarchy.

Hierarchical multiscale approaches, the focus of this section, are based on the general principle of *divide and conquer*: the original domain is divided into smaller regions, with appropriate equations governing their interaction; then, if possible, these regions are further subdivided into smaller pieces and so on until arriving at the deepest level of the hierarchy, at which no more partitioning is possible, and recourse to experimentally derived relationships (or fundamental atomic principles) is to be made.

From the computational perspective, this divide-and-conquer strategy brings an obvious advantage: the problem becomes fully parallelizable (in space). However, this does not imply that the complexity of the problem is reduced – the total number of degrees of freedom remains the same. The actual feature that distinguishes a hierarchical multiscale model from a standard recursive domain decomposition approach is the introduction of *simplifying assumptions*, in either (or both) the equations that govern the *interaction* between contiguous domains at each scale or/and in the equations that govern the behavior of each domain under external actions. To illustrate this point, we succinctly describe in the following two well-known examples of hierarchical multiscale modeling, namely, *classical beam theory* and *homogenization of heterogeneous materials*.

Classical beam theory can be indeed legitimately regarded as a hierarchical two-scale model: to construct this theory, the beam is first ideally divided along its length into differential elements, and, then, the equilibrium of each of these elements is addressed by a continuum model (typically an elasticity model). The key feature that makes, in most cases, these elemental equilibrium problems amenable to analytical solution is that deformed cross sections are assumed to remain plane and normal to the deflected beam axis (Euler–Bernoulli hypothesis); in doing so, the number of degrees of freedom needed to describe the displacement of the interfaces is reduced from infinite to three (in-plane problems).

Another clear example of hierarchical multiscale modeling is the so-called first-order *homogenization* theory, employed for modeling materials that exhibit a clear heterogeneous composition at the micro- or mesoscale, but that can be regarded, for practical purposes, as homogeneous at the macroscale. In this case, the fundamental simplifying assumption is *scale separation*. This assumption presumes that the subdomains in which the body is divided – usually called representative volume elements (RVEs; see also Chapter 3) – are so small in comparison with the characteristic macroscopic length that they can be macroscopically viewed as points and, as a consequence, the only macroscopic information needed to characterize their state are the typical stress and strain measures employed in continuum mechanics. In fact, the only difference with a conventional continuum analysis is that the link between these macrostresses and macrostrains is established by solving a boundary value problem (BVP) at the RVE level (in this BVP, the macrostrain plays the role of "loading parameter" in the form of appropriate essential boundary condition). Of course, this RVE problem can be further streamlined by introducing simplifying assumptions regarding the topological arrangement and collective behavior of the heterogeneities in the RVE (this is what is done in, for instance, *mean-field* approaches; see, for example, [97]) or can be addressed directly, with no further simplifications, by standard numerical methods such as the FE method (this gives rise to computational homogenization approaches such as the FE^2 method; see, for example, [98]).

7.6.1.2 Physical Insight-Based Simplifications

Having argued that simplifications are the essential key to successful multiscale hierarchical modeling, we must come to grips with the question of how to arrive at such simplifications. How can one discern at each scale which are the *essential* features – that is, those features that significantly influence the response at the next, coarser scale in the material hierarchy – and which are not? In the "classical" physical modeling paradigm, simplifying assumptions are usually derived from empirical observations and then introduced in the formulation of the governing equations after arduous analytical manipulations. Needless to say, this task requires, aside from a solid theoretical background in the corresponding discipline, a good dose of inspiration and creativity. Much more often than not, the modeler becomes rapidly discouraged by the complexity of the task and, quite understandably, abandons any attempt of simplification. This inevitably translates into *unnecessarily* costly multiscale analyses. A clear example of this trend is the aforementioned multilevel FE method (FE^2): in this method, one has to solve at every increment and every Newton–Raphson iteration of the global, coarse-scale FE analysis –and for each Gauss point of the coarse-scale mesh – a local, nonlinear FE problem that, in turn, may involve several thousands of degrees of freedom. Such an enormous computational cost is the price paid for skipping the task of simplifying the RVE problem, and it explains why the routine use of this otherwise powerful and versatile method in the modeling of large structural systems featuring complex micro/mesostructures is still considered unpractical, especially when the system has to be analyzed for various configurations (as in design optimization).

7.6.1.3 Computer-Based Simplifications

Fortunately, introducing "physical insight-based" simplifications is not the only way of removing unnecessary complexity from a given computational model, multiscale or otherwise. In recent years, there has been a resurgence of interest in the so-called *MOR* techniques. These techniques allow us to significantly simplify, in a *systematic manner*, a given computational model using the information obtained from representative full-order simulations. The qualifier *systematic* here means that the task of discriminating what is essential and what is not is automatically accomplished by a *computer algorithm*, and, therefore, it is not contingent upon the depth of insight, knowledge, and "engineering judgment" of the modeler. But how can "engineering judgment" – the ability to discern the essential from the inessential, to detect *dominant* behavioral patterns – be replaced by a computer algorithm? Luckily, this challenging question has been already addressed in research areas such as data compression and data mining; see, for example, [99], and, as a result, many efficient *dimensionality reduction* (or data compression,

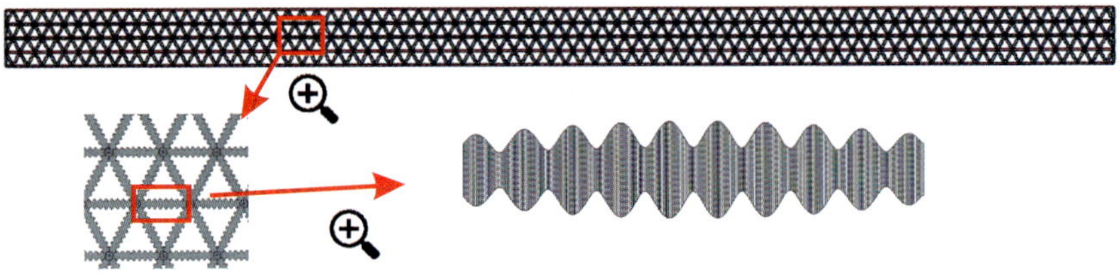

Figure 7.6 Geometry of both the truss and the bars ($n_{bar} = 1000$ and $n_{elem} = 250$).

in more common parlance) algorithms have been developed. The goal here is to elucidate how these techniques can be harnessed to systematically simplify a given multiscale hierarchical model. An elementary (yet insightful) structural problem – equilibrium of a planar truss formed by sinusoidally shaped bars – is used to convey the flavor of the proposed MOR strategy. Mathematical formalisms would require a lengthy exposition and hence are omitted. The reader interested in plunging into such formalisms is referred to a recent paper by Hernández *et al.* [100].

7.6.2 Problem Statement

Consider the planar (Warren) truss shown in Figure 7.6. The output of interest in the problem is the evolution of the bending moment at the left end, which is fixed, under arbitrary rotations of the right end. No external loads are applied. The truss is formed by $n_{bar} = 1000$ sinusoidally shaped bars; the geometry of a typical bar is displayed also in Figure 7.1. To properly capture the variation of cross-sectional area, each bar is, in turn, discretized into $n_{elem} = 250$ linear FEs. The stress–strain constitutive equation is an elastoplastic model endowed with linear hardening.

The equilibrium of this truss is naturally structured into a hierarchy of two nested equilibrium problems: on the one hand, the equilibrium of the assembly of $n_{bar} = 1000$ bar (the large scale) and, on the other hand, the equilibrium of the $n_{elem} = 250$ FEs of each bar (the small scale). The total number of integration points at which stresses, strains, and internal variables have to be computed (at each iteration and time step of the equilibrium at the largest scale) amounts, therefore, to $n_{bar} \times n_{elem} = 1000 \times 250 = 250\,000$. The aim of the MOR strategy described in the following is to drastically reduce this number of integration points. We shall proceed sequentially in the hierarchy, by first deriving a ROM for the small scale, and then, using this simplified model, constructing the pertinent ROM for the coarse scale.

7.6.3 Small-Scale ROM (Bar Equilibrium)

In the bar equilibrium problem, the displacements at the ends are the input parameters, while the outputs of interest are the corresponding reactive forces – these forces are eventually transferred to the larger scale to address the truss equilibrium problem.

The first step in the MOR process consists of performing FE analyses of the bar subjected to various, representative end displacement histories. The displacement solution fields resulting from these FE analyses are next subjected to dimensionality reduction in order to unveil and extract the dominant displacement modes. Here, we employ the arguably simplest and most popular of these dimensionality reduction algorithms: the singular value decomposition (SVD). In this case, an error truncation analysis reveals that the first six SVD dominant modes suffice to capture 95% of the information contained in the ensemble of displacement solutions.

In projecting the original FE equation onto the space spanned by these SVD modes, the number of degrees of freedom is automatically reduced from $n = 249$ to $n^* = 6$. However, the computation of the reduced internal forces still requires the evaluation of stresses, strains, and internal variables at the original nelem = 250 integration points; the complexity of the problem, thus, still depends on the resolution of the underlying FE mesh. Accordingly, the next step is to commensurately reduce the number of integration points. This reduction is carried out by approximating the stress field by a low-dimensional interpolant. The basis matrix for the interpolation should include both the dominant stress modes arising from an SVD of the corresponding FE stress solutions, as well as the matrix of the gradient of the SVD displacement modes (this last requirement stems from well-posedness considerations, as thoroughly explained in [100]). Concerning the reduced set of integration points, they must be chosen so as to minimize the interpolation error over the set of FE stress solutions. In this work, we have employed the point selection algorithm known as the empirical interpolation method (EIM); see, for example, [101, 102]. The $n^* = 8$ points chosen by this algorithm can be seen in Figure 7.7. Figure 7.7 shows also the plot of bending moment versus angular rotation for both the full-order multiscale analysis and the multiscale analysis using the ROM at the small scale. It can be readily appreciated that the impact, on the output of interest, of the introduced simplification is negligible, for both curves that are practically coincident, while the cost has been reduced by a factor of approximately 40.

7.6.4
Large-Scale ROM (Truss Equilibrium)

We now carry the simplifying endeavor a step further by deriving a ROM for the truss itself – one that involves the evaluation of axial forces at only a reduced number of bars, $n^*_{\text{bar}} \ll n_{\text{bar}}$. The procedure is exactly the same as the one outlined in the foregoing for the case of the bar problem: computation of SVD dominant displacement and force modes, projection onto the space spanned by the computed displacement modes, and replacement of the vector containing the end forces of all bars by a low-dimensional interpolant. We next examine the performance of this “doubly” reduced-order multiscale model in the case of

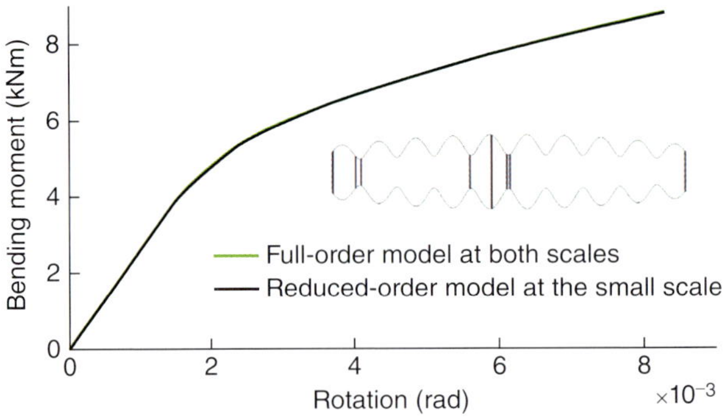

Figure 7.7 Bending moment versus angular rotation. Comparison between full-order model (250 integration points per bar) and ROM model (8 integration points per bar) at the small scale.

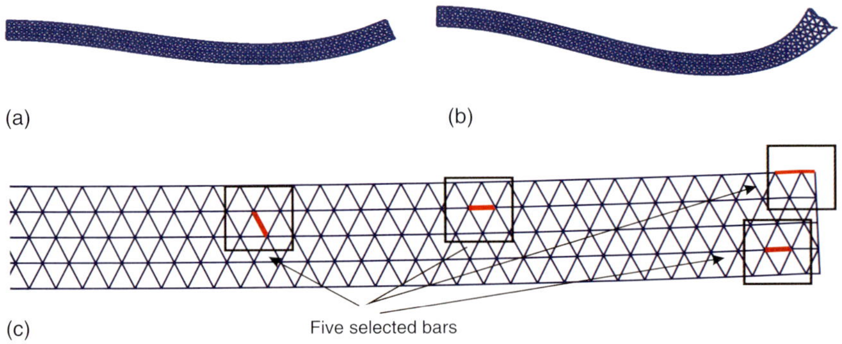

Figure 7.8 First two SVD dominant displacement modes of the truss (a) and bars selected for the reduced-order analysis at large scale (b). (c) Five selected bars.

using two SVD displacement modes and $n^*_{\text{bar}} = 5$ bar (see Figure 7.8) at the large scale and $n^*_{\text{elem}} = 8$ integration points at the small scale. This amounts to a total of $n^*_{\text{bar}} \times n^*_{\text{elem}} = 5 \times 8 = 40$ integration points and, consequently, implies a *reduction of the computational effort* of $n^*_{\text{bar}} \times n^*_{\text{elem}} / (n^*_{\text{bar}} \times n^*_{\text{elem}}) = 250\,000/40 = 6250$ (in terms of number of integration points). To assess the error, we conduct a simulation in which the right end of the truss is subjected to a load cycle: first, a clockwise rotation until $\vartheta = -8 \times 10^{-3}$ rad, then a counterclockwise rotation up to $\vartheta = 8 \times 10^{-3}$ rad, and finally, back to its initial, straight position.

In Figure 7.9, we plot the bending moment versus the angle of rotation computed with both the full-order and the doubly reduced-order multiscale models. Observe that the initial descending branches of both graphs are practically indistinguishable despite the marked nonlinearity of the process. This remarkable accuracy is due to the fact that the "calibration" of the large-scale ROM, that is, the determination of displacement and force dominant basis vectors, was carried out using solutions obtained in this range. Discrepancies with the high-fidelity

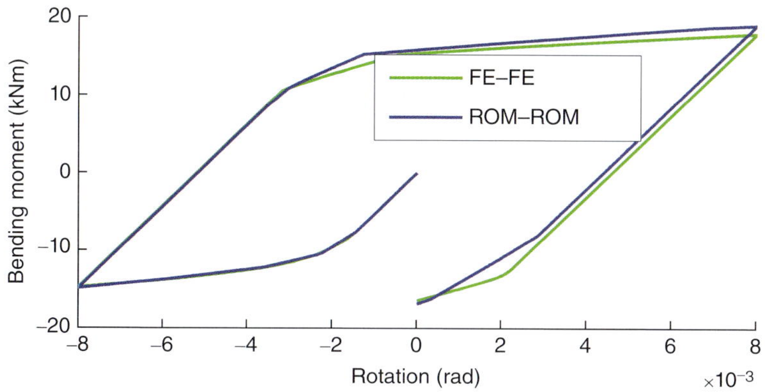

Figure 7.9 Bending moment versus rotation of the right end computed with both the full-order multiscale model (labeled FE–FE) and the doubly reduced-order model (labeled ROM–ROM). The FE–FE model uses $n_{bar} \times n_{elem} = 250\,000$ integration points, while the ROM–ROM model employs only $n^*_{bar} \times n^*_{elem} = 40$ integration points.

response are observed after plastic yielding in the reverse direction commences, yet they remain within reasonable bounds. Remarkably, the error in predicting the residual bending moment (at the end of the process) is less than 5%.

7.6.4.1 Acknowledgments

Research leading to these results has received funding from the European Research Council under the European Union's Seventh Framework Program (FP/2007-2013)/ERC Grant Agreement n. 320815, Advanced Grant Project COMP-DES-MAT, and the Spanish Ministry of Science and Innovation under grant BIA2011-24258.

7.7 HPC and Parallelization Methods

7.7.1 Introduction

"Theoretical and applied mechanics (TAM) is the branch of applied science concerned with the study of mechanical phenomena: the behavior of fluids, solids, and complex materials under the actions of forces […]" – Oden *et al.* [103]

Computational mechanics (CM) is that subdiscipline of TAM concerned with the use of computational methods and devices to study events governed by the principles of mechanics. High-performance computational mechanics is a CM subdiscipline. The use of HPC tools in CM gives a different perspective to mathematics, physics, and computer science involved and the engineering problems to solve. Nowadays, HPC is no longer an option when one pretends to solve engineering problems. It is therefore of primary importance to put HPC at the core of engineering, from education to software design.

This revolution is quite recent, although HPC has been present since more than six decades. It was triggered 10 years ago when out-of-the-shelves supercomputers were made available to the scientific community. This obliged the code developers to start working on an enabling phase to adapt computational mechanics codes to the parallel architectures. This phase was highly promoted in Europe by PRACE European project and PRACE RI as well.

At the beginning of this revolution, mainly two paradigms were at the center of the developments for both shared memory computers and distributed memory supercomputers. For the first one, OpenMP (www.openmp.org/wp) has been the standard, while for the last MPI (www.mpi-forum.org) has been the standard. However, the sudden appearances of new architectures like accelerators and new programming models (such as CUDA) have sewed confusion among code developers. Constant refactoring of codes is inviable and the lack of standard programming language was surely the reason why the success was not at the level of the possibilities provided by these architectures.

In this section, we focus particularly on message passing interface techniques for distributed memory supercomputers. The techniques for shared memory supercomputers are more standard, and those for accelerators are highly dependent on architecture and programming model. Their studies would require a single book to cover all the aspects. We decided to focus on MPI because, nowadays, this is the only option to communicate between the different nodes of a supercomputer. MPI is the primary requirement to run on supercomputers, and probably the most time-consuming technique for parallel code developers. We start by describing classical substructuring techniques, commonly used in codes involving iterative solvers. Then, we introduce different solvers and preconditioners to accelerate the convergence. The next section treats multiphysics and multidomain couplings in a parallel context.

7.7.2 Substructuring

Classical computational mechanics codes consist mainly of an assembly process, to compute the coefficients of a matrix A and a vector b, and a solution process to solve the system $Ax = b$. Although we treat mainly the FE method, we mention the finite volume (FV) and finite difference (FD) as well, as their parallelization is a bit different and for the sake of comparison. Let us consider the partition in two subdomains of a mesh shown in Figure 7.10. The unknowns in partitions 1 and 2 are referred to with subindices 1 and 2, respectively, and the common nodes on the interface with subindex 3.

Using this numbering, the matrix of global system A can be rewritten as

$$A = \begin{pmatrix} A_{11} & 0 & A_{13} \\ 0 & A_{22} & A_{23} \\ A_{31} & A_{32} & A_{33} \end{pmatrix}.$$

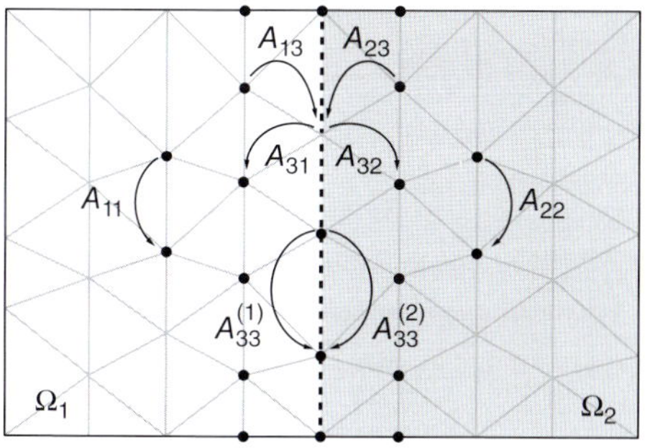

Figure 7.10 Decomposition into two subdomains and unknown renumbering.

We first observe that this matrix can be computed from the following two local matrices:

$$A_1 = \begin{pmatrix} A_{11} & A_{13} \\ A_{31} & A_{33}^{(1)} \end{pmatrix} \qquad A_2 = \begin{pmatrix} A_{22} & A_{23} \\ A_{32} & A_{33}^{(2)} \end{pmatrix},$$

where the blocks $A_{33}{}^{(1)}$ and $A_{33}{}^{(2)}$ represent the interactions between the interface degrees of freedom such that $A_{33} = A_{33}{}^{(1)} + A_{33}{}^{(2)}$. Now, let us consider the basic operation of an iterative solver that is the matrix–vector product $y = Ax$:

$$\begin{pmatrix} y_1 \\ y_2 \\ y_3 \end{pmatrix} = \begin{pmatrix} A_{11} & 0 & A_{13} \\ 0 & A_{22} & A_{23} \\ A_{31} & A_{32} & A_{33} \end{pmatrix} \begin{pmatrix} x_1 \\ x_2 \\ x_3 \end{pmatrix} = \begin{pmatrix} A_{11}x_1 + A_{13}x_3 \\ A_{22}x_2 + A_{23}x_3 \\ A_{31}x_1 + A_{32}x_2 + A_{33}x_3 \end{pmatrix}.$$

We note that this result can be obtained by performing two local matrix–vector products using the local matrices

$$\begin{pmatrix} y_1 \\ y_3^{(1)} \end{pmatrix} = \begin{pmatrix} A_{11}x_1 + A_{13}x_3 \\ A_{31}x_1 + A_{33}^{(1)}x_3 \end{pmatrix}, \qquad \begin{pmatrix} y_2 \\ y_3^{(2)} \end{pmatrix} = \begin{pmatrix} A_{22}x_2 + A_{23}x_3 \\ A_{32}x_2 + A_{33}^{(2)}x_3 \end{pmatrix},$$

and then, by performing the following assembly operation $y_3 = y_3{}^{(1)} + y_3{}^{(2)}$. This observation is the basis of substructuring methods. Matrix–vector products can be carried out in parallel on the local meshes of the partitions, and then an assembly step is required to obtain the results on these partitions.

The process is as follows:

- Assemble local matrices A_1 and A_2.
- Compute local matrix–vector products.
- Exchange through MPI the local results.

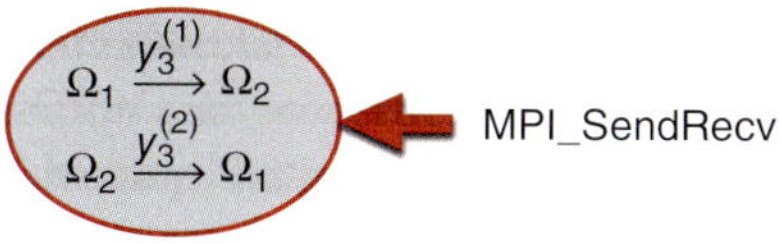

- Assemble the results:

$$\Omega_1 : y_3 = y_3^{(1)} + y_3^{(2)}, \qquad \Omega_2 : y_3 = y_3^{(2)} + y_3^{(1)}.$$

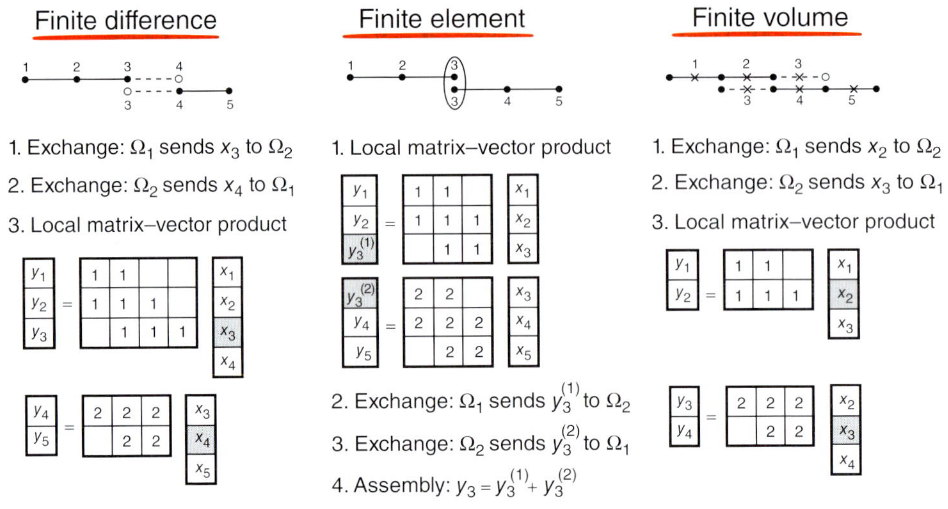

Figure 7.11 Parallel matrix–vector product for FD, FE, and FV.

The procedure is a bit different from other numerical methods, such as FV and FD, for which cells are duplicated. Figure 7.11 compares the usual implementations of parallel matrix–vector products for these three methods. Usually, in FD and FV, the matrices are not square due to the presence of halo nodes, and instead of exchanging the result of the product y, the halo values x are exchanged before the product.

7.7.3 Algebraic Solvers

On the one hand, the previous parallelization technique is almost sufficient when a simple iterative solver is robust enough to achieve convergence. Simple preconditioning techniques can be implemented as well, although most of them should be local to each partition to maintain the MPI communications only between neighbors. Among others, let us mention linelet; see, for example, [104] and Gauss–Seidel. On the other hand, very stiff problems require the use of direct solvers to ensure convergence. Direct solvers have been receiving more and more attention recently; let us mention, among others, the multifrontal (see, for example, [105]), and supernodal (see, for example, [106]) techniques. Multifrontal techniques are represented by the multifrontal massively parallel solver (MUMPS) (see, for example, [107]), and the Watson Sparse Matrix package (WSMP) (see, for example, [108]). SuperLU DIST (see, for example, [109]) is an MPI parallel version of SuperLU family of solvers for unsymmetric systems based on supernodal right LU factorization. A list of the direct sparse solvers is

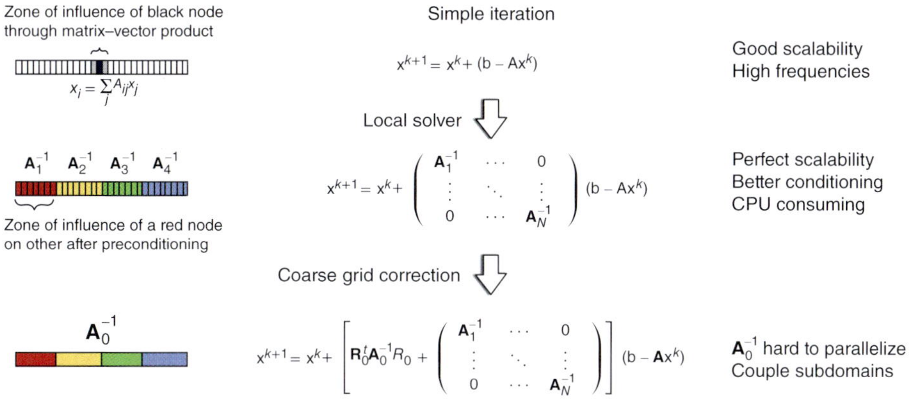

Figure 7.12 Solvers with domain decomposition preconditioners and coarse grid correction.

presented in [110]. However, these solvers usually lack scalability when using a very large number of CPUs and are still unaffordable for very large problems.

There exists a family of intermediate choices, usually referred to as domain decomposition methods in the literature. Figure 7.12 summarizes the effect of these domain decomposition preconditioners used to precondition classical Krylov solvers. On the left we depicted a one-dimensional mesh divided into four partitions.

On the top of the figure, we present the simplest iterative solver, called the simple iteration or Richardson method. We easily observe that from an iteration k to the next one, the zone of influence of a node consists of its neighboring nodes (left part of the figure). In fact, when performing the matrix–vector product, the update on a node will depend on the previous values of its neighboring nodes. This implies a local mechanism to propagate the error: if an error is introduced at one node, it will propagate from node to node until it disappears. This solver is more or less representative of common iterative solvers. It exhibits a good scalability, as only point-to-point communications are required after each matrix–vector product, and in the case of more complex solvers such as CG, BICGSTAB, and GMRES (see, for example, [111]), additional global communications are required to compute scalar products. Such solvers are good to damp high-frequency errors.

The idea to propose a more global mechanism for propagating the error is to consider a local direct solver to each partition. Then, the zone of influence of a partition will be its surrounding partitions. If the partitions are well balanced, the scalability is perfect and the algebraic system is better conditioned. As examples of such solvers, let us mention the additive Schwarz method, the block LU method, and the Schur complement method.

Having done this, we observe, however, that we are left with another grain of zone of influence. We lack a global mechanism to communicate between all the partitions at the same time. This is the role of the so-called coarse solvers. This problem appears when we face lots of partitions. Even if this is not yet very common, engineering codes are now prepared to run on thousands up to millions of partitions. Therefore, this global mechanism is getting more and more importance.

7.7.4 Efficiency

There are two important parameters to quantify the efficiency of a parallel code. They are the strong and weak scalabilities. Let us define dof as the number of degrees of freedom of a problem. Let $t_{np}(\text{dof})$ be the time required to solve this problem of size dof using np partitions. We define the strong and weak scalability efficiencies as

$$\text{Strong scalability} = \frac{t_1(\text{dof})}{(np \times t_{np}(\text{dof}))}, \quad \text{weak scalability} = \frac{t_1(\text{dof})}{t_{np}(np \times \text{dof})}.$$

On the one hand, strong scalability tells us how much faster we can go using more partitions. On the other hand, a unit weak scalability efficiency tells us that refining the mesh by a factor np and using np partitions the problem is solved with the same duration. Weak scalability is much harder to obtain and requires the use of complex domain decomposition solvers. First, the conditioning of a matrix degrades with increasing dof. Second, domain decomposition solvers require coarse grid corrections to mitigate the dependence on the number of partitions. These coarse solvers are usually a bottleneck in parallel, as they require a direct solver that hardly scale. Weak scalable solvers are, therefore, an active area of research and the main concern of the domain decomposition community (www.ddm.org).

7.7.5 The Challenges

In the previous sections, we presented the basic parallelization strategies for distributed memory supercomputers, which mainly concentrate on the algebraic solvers. However, engineering problems are more and more complex, and modern simulation codes must be adapted constantly to tackle new physical challenges, usually involving new numerical challenges. Some examples are multiphysics and multidomain problems (fluid–structure interactions (FSIs)), contact problems, fracture dynamics, multiscale coupling, and so on. All these problems require careful MPI implementation to maintain the performance of the original parallel code.

Let us take the example of FSI. Although monolithic approaches have been considered in the literature, they are not the rule. The coupling of fluid and structure codes has been envisaged recently in a multiple-code framework. This technique is very convenient as it enables the use of existing codes and is relatively simple to implement. Figure 7.13 illustrates the concept. On a

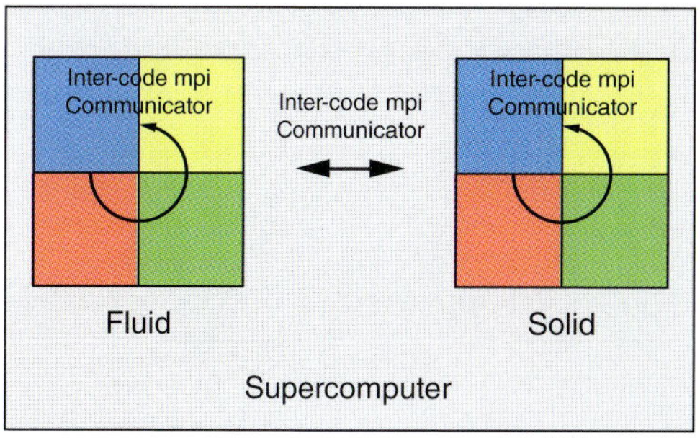

Figure 7.13 FSI interactions in a multiple-code framework using intra- and intercode communicators.

supercomputer, a parallel fluid solver and a parallel solid solver are running independently, each one within its MPI communicator, referred to as intracode communicator. Now, if these individual communicators were generated from the MPI_COMM_WORLD using the function MPI_COMM_SPLIT, then we have a mechanism to communicate forces and displacements between the two codes. This is the role of the intercode communicator. The main implementation difficulty involves the optimization of the preprocess, that is, setting up the interpolation and communications scheduling to be used each time the displacement is required by the fluid code or the force is required by the solid code.

7.8 Software Codes

Public domain, downloadable, and university codes (www.robertschneiders.de/meshgeneration/software.html):

- ADMesh (A.D. Martin): A program for processing triangulated solid meshes in STLformat.
- ANGENER (Vit Dolejsi): 2D triangulation using anisotropic mesh adaptation.
- AUTOMESH2D (Shandong University): A fully automatic adaptive quad mesh generator, especially suited for metal forming simulation.
- CAMINO (T. Chen): 3D/2D meshing program using a generalized octree/quadtree approach.
- Cart3D (M.J. Aftosmis): Preprocessing tools and mesh generator for Cartesian mesh generation.
- CGAL mesh generation (CGAL): Various packages for triangle and tetrahedral mesh generation.
- CGM (T. Tautges): A code library that provides geometry functionality used for mesh generation and other applications.

- Chimera Grid Tools (W.M. Chan): A software package containing a variety of tools for the Chimera overset grid approach for solving complex configuration problems.
- COG 2.0 (I. Schmelzer): A grid generation package for 2D/3D grid generation. Essential part is public domain. Its aim is to create Delaunay grids with few nodes for complex geometries.
- COMET (M. Cervera, C. Agelet de Saracibar, M. Chiumenti): A FEM software for the numerical simulation of coupled thermomechanical problems. Includes GiD as pre- and postprocessor.
- CQMesh (M. Siqueira): AC++ program for generating convex quadrilateral meshes of arbitrary polygonal domains.
- CSCMDO (B. Jones): A general-purpose multiblock 3D volume grid generator that is suitable for multidisciplinary design optimization.
- CUBIT (SANDIA, BYU): A 2D/3D finite element mesh generation tool that is being developed to pursue the goal of robust and unattended mesh generation (quads and hexa elements).
- Delaundo (J.-D. Müller): A 2D Delaunay mesh generator. Produces high-quality triangular grids.
- DelIso (T.K. Dey): Can mesh an isosurface from volume data with Delaunay triangles that have bounded aspect ratio.
- Delos (O. Stab): An automatic 2D mesh generator for "almost flat" surfaces.
- DelPSC (T.K. Dey): Produce quality Delaunay meshes (weighted) for a large class of 3D domains. The algorithm guarantees that almost all triangles and tetrahedra have bounded radius–edge ratio, except a few ones in the vicinity of small input angles or near the boundary.
- DIAMESH (A. Rassineux): 2D (triangular) and 3D (tetrahedral) meshing and surface remeshing.
- Discretizer (B. Bergqvist): An interactive block-structured mesh creation tool for CFD applications.
- DistMesh (P.-O.Perrson): MATLAB code for generation of unstructured triangular and tetrahedral meshes.
- enGrid (enGits): Tetrahedral grid generation with prismatic boundary layers.
- EZ4U (LACAN): Quadrilateral unstructured mesh generation, with GUI.
- femmesh (Medical Physics, UCL): A UNIX/OpenWindows program designed to interactively generate 2D FEM meshes composed of three-node triangular elements.
- FIST (M. Held): A robust polygon triangulation code. Can handle many kinds of degenerate data.
- GENIE++ (B. Soni): A collection of software packages that generates computational grids for internal–external flow configurations. GENIE++ generates 3D structured, multiblock grids.
- Geompack++ (B. Joe): A mathematical software package written in standard Fortran 77 for the generation of 2D and 3D triangular/tetrahedral finite element meshes using GEOMetric algorithms. GEOMPACK90, the substantially enhanced successor of GEOMPACK, is a comprehensive software package for

FE mesh generation (triangular, quadrilateral, surface, tetrahedral, and hexahedral dominant).

- GiD (Quantech): GiD is a universal, adaptive, and user-friendly pre- and post-processor for numerical simulations in science and engineering. It has been designed to cover all the common needs in the numerical simulations field from pre- to postprocessing: geometrical modeling, effective definition of analysis data, meshing, data transfer to analysis software, as well as the visualization of numerical results.
- Globegen (N. Ahmad): An unstructured prismatic grid generator for the entire globe.
- GMSH (J.-F. Remacle, C. Geuzaine): A Delaunay-based mesh generator that generates adapted meshes for lines, surfaces, and volumes.
- GNU Triangulated Surface Library (Source forge): Intended to provide a set of useful functions to deal with 3D surfaces meshed with interconnected triangles. Contains 2D Delaunay and constrained Delaunay triangulation, surface refinement and coarsening, and much more.
- GrAL (Guntram Berti): A generic library for grid data structures and operating algorithms.
- GridEx (GEOLAB): An interactive software system developed by GEOLAB of the NASA Langley Research Center for the generation of unstructured meshes. The software integrates native CAD geometry access, multiple unstructured meshing algorithms, and interactive 3D computer graphics through a GUI resulting in a package that is both powerful and easy to use.
- GRIDGEN (NASA): A software system for the generation of 3D, multiple blocks, structured grids. GRIDGEN is a visually oriented, graphics-based interactive code used to decompose a 3D domain into blocks, distribute grid points on curves, initialize and refine grid points on surfaces, and initialize volume grid points.
- Gridgen (P. Sakov): An orthogonal grid generator, based on the CRDT algorithm, that can easily handle elongated regions with a few hundred boundary points.
- Gridgen (USGS): A MATLAB-based tool to construct orthogonal curvilinear grids for NetCDF files for ECOM and SCRUM ocean circulation models.
- gridpak (from IMCS): 2D orthogonal grid generation for coastal engineering.
- GridTool (NASA Langley): Surface Modeling and Grid Generation Tool.
- GRUMMP (ANSLab): Quality generation and refinement of unstructured o meshes.
- GTS (SourceForge): GTS stands for the GNU Triangulated Surface Library. Open-source free software Library intended to provide a set of useful functions to deal with 3D surfaces meshed with interconnected triangles.
- G3D (IMI): A C, X/Motif, and OpenGl application, which generates 3D grids used in groundwater simulations.
- HypGrid (Riso Wind Energy Department): a grid generator based on hyperbolic equations for orthogonality and cell area for both two- and three-dimensional domains.

- IMTEK Mathematica Supplement (Freiburg Chair for Simulation): A Mathematica interface to various mesh generators.
- iso2mesh: A free MATLAB/octave-based mesh generation and processing toolbox. It can create 3D tetrahedral FE mesh from surfaces, 3D binary, and grayscale volumetric images such as segmented MRI/CT scans.
- JMesh (J.T. Hoffman): A tool for generating approximations of minimal surfaces given by the Weierstrass Formula.
- LaGriT (Los Alamos National Laboratory): An unstructured grid generation and optimization software package used for semiconductor device modeling, CFD, and porous flow modeling. This software is especially useful for 3D moving surface type applications.
- LBG (WIAS): A layer-based 3D structured (Cartesian) grid generator.
- LBIE-Mesher (Austin CCV): Level Set Boundary Interior and Exterior Mesher can extract adaptive and quality 2D (triangular or quadrilateral) meshes over isosurfaces and 3D (tetrahedral or hexahedral) meshes with isosurfaces as boundary surfaces directly from volumetric imaging data. LBIE-Mesher can generate 3D meshes for the volume interior to an isosurface, the volume exterior to an isosurface, or an interval volume between two isosurfaces.
- Gaëtan Compère (MAdLib): An open-source mesh adaptation library that performs global node repositioning and mesh adaptation by local mesh modifications on tetrahedral or triangular meshes.
- MAKROS-A (G. Boege): Quadrilateral surface meshing for AutoCAD data.
- Mefisto-maillages (A. Perronnet): Structured or nonstructured generation of 2D and 3D meshes, part of the Mefisto FE program.
- MegaCads (DLR Institute for Design Aerodynamics): Multiblock elliptic grid generation and CAD system.
- Meshing tools (P. Frey): Medit is an interactive 3D viewing program. It has been designed to allow easy and interactive manipulation of unstructured (2D, 3D, and surface) meshes.
- MeshLab: An open-source, portable, and extensible system for the processing and editing of unstructured 3D triangular meshes.
- Mesh Maker Pro (D. Keller): A tool to create 3D models for graphics programming.
- Mesh2D (F.X. Giraldo): An adaptive triangular mesh generator for unstructured CFD.
- MeshGenC++ (J. Rossmanith): A software package for generating unstructured meshes in 2D (triangles) and 3D (tetrahedrals).
- METAFOR (J.P. Ponthot and R. Boman): A C++ implicit/explicit finite element code for solids submitted to large deformations and frictional contact (http://metafor.ltas.ulg.ac.be/).
- MESQUITE (TSTT): A linkable software library that lets users improve the quality of their meshes. It uses advanced smoothing, optimization, and local swapping/splitting operations.
- MG (L.C.G. Coelho): A system for 3D FE mesh generation with interactive graphics capabilities.

- NETGEN (J. Schoberl): A 2D/3D mesh generator for CSG geometries (advancing front and Delaunay methods, hierarchical mesh refinement).
- MMESH3d (S. Marras): A semi-structured multiblock 2D/3D mesh generator for hexahedrons and prisms (wedges of triangular base) in 3D, and quads and tri in 2D, thought mostly for topography-based domains.
- NWGrid (PNL): Integrates automated grid generation, time-dependent adaptivity, applied mathematics, and numerical analysis for hybrid grids on distributed parallel computing systems.
- OpenMesh (Computer Graphics Group, RWTH Aachen): A generic and efficient data structure for representing and manipulating polygonal meshes.
- OpenVolumeMesh (Computer Graphics Group, RWTH Aachen): A generic and versatile index-based data structure for polytopal meshes.
- Overture (Center for Applied Scientific Computing): A library of C++ classes for the solution of PDE on complicated domains based on multiblock structured and overlapping grid technology. Also useful for grid generation.
- PAMGEN (Sandia): Hexahedral (3D) or quadrilateral (2D) FE meshes of simple shapes (cubes and cylinders) in parallel.
- PARMGRIDGEN (I. Moulitsas): Serial library written in ANSI C that implements (serial) algorithms for obtaining a sequence of successive coarse grids, well suited for geometric multigrid methods.
- QualMesh (T.K. Dey): Quality Delaunay volume mesh of a polyhedral domain.
- Qhull (B. Barber): A general dimension code for computing convex hulls, Delaunay triangulations, Voronoi vertices, and half-space intersections.
- QMG (S. Vavasis): FE mesh generation in 2D/3D (triangles/tetrahedra) integrated into MATLAB.
- QUIKGRID (J. Coultharq): A scattered data surface (grid) generator and viewer.
- SimLab (P. Chew): A tool for creating guaranteed-quality triangulations of planar areas.
- snappyHexMesh (OpenCFD): Generates 3D meshes containing hexahedra and split hexahedra automatically from triangulated surface geometries in stereolithography (STL) format.
- SolidMesh (MSU-ERC, MSU): Grid generation system to create both 2D and 3D unstructured grids. Surface grids can be created in parametric space on the NURBS or by using a 3D point insertion method iterating between parametric space and physical space.
- Stellar (B. Klingner): A tetrahedral mesh improvement program.
- SurfRemesh (T.K. Dey): Remesh, a polygonal surface with Delaunay triangles that have bounded aspect ratio. It maintains the topology and approximates the geometry of the original surface.
- SUS (ULPGC IUSIANI): Simultaneous untangling and smoothing of tetrahedral meshes.
- T3D (Daniel Rypl): A powerful mesh generator capable to discretize complex 3D domains into triangular and tetrahedral meshes of high quality.
- TCGRID (NASA Glenn Research Center): 3D grid generation code for turbomachinery blades.

- Tetgen (S. Hang): Generates exact Delaunay, constrained (conforming) Delaunay, and quality conforming Delaunay tetrahedralizations.
- TMG (M. Paolini): 2D automatic triangular mesh generator (advancing front).
- Triangle (J. Shewchuk): Generates exact, constrained, and quality conforming Delaunay triangulations (2D).
- UGRID (D. Hawken): Rapid generation of smooth unstructured, structured, or hybrid 2D grids about an airfoil or cylinder.
- Unamalla (P. Barrera-Sanchez): Easy-to-use grid generator designed to solve the grid generation problem over very irregular planar regions using rectangular structured meshes.
- Vcat2tets: Generates multimaterial tetrahedral meshes from segmented 3D images.
- VERDICT (Sandia National Laboratories): A library for computing the quality metrics of an FE mesh.
- VGRID (S. Pirzadeh): A robust, user-friendly computer program for the generation of 3D unstructured (triangular surface and tetrahedral volume) grids in geometrically complex domains.
- VGM (S.J. Alter): A powerful grid generation manipulation tool for 1D, 2D, or 3D structured grids.
- ViennaGrid (K. Rupp): A C++ library designed for the handling of structured and unstructured meshes in arbitrary spatial dimensions using different coordinate systems.
- Volume (NASA Geometry Lab): Interactive program written for SGI workstations to generate multiblock structured volume grids.
- Voro++ (C. Rycroft): Open-source software library for the computation of the Voronoi diagram.
- XGEN (P. Solin): A moving particle scheme for 2D unstructured triangle grid generation.
- xprob (UCD magnetics and machines group): Automatic triangular mesh generation for FE solution of static magnetic field problems.
- Yams (P. Frey): Given a surface triangulation, Yams allows the user to easily create a geometric mesh, a curvature-based mesh, or to adapt the mesh to a desired size map.
- 3DMAGGS (S.J. Alter): An elliptic volume grid generator (multiblock structured grids) used to generate computational domains for CFD analyses of aerodynamic vehicles.

Companies offering mesh generation software:

- ADINA: ADINA user interface is ADINA's mesh generator, capable of generating surface and volume meshes (tetrahedra) from CAD data.
- ANSYS: CFX-Mesh, highly automated surface and volume meshing. TurboGrid, a lightning fast, application-specific, interactive grid generation tool to create high-quality CFD meshes.
- Algor: Houdini is an unstructured hex meshing code for mechanical engineering and CFD. There are also modules for 2D, surface, and tet meshing.

- Altair: Hypermesh, a high-performance FE pre/postprocessing system for engineering simulation.
- AMPS Technologies: AMPSolid, a semiautomatic mesh generator. It is based on integrating the ACIS solid modeler and uses an overlay algorithm to generate mostly structured hexahedral meshes. The surface and tetrahedral elements are generated automatically.
- Analytical Methods Inc.: SURFGEN, PEP, surface, and volume gridding for CFD simulations.
- ANSYS Inc.: Pre- and postprocessing for the ANSYS product line.
- Argus Interware Inc.: MeshMaker, 2D triangular and quadrilateral FE mesh generation (structured and unstructured) for numerical simulations in computational geosciences. Geometry handling (generation, cleanup, watertight preparation, and defeaturing), surface and volume meshing, geometry and mesh morphing, and direct interfaces to several CFD and FEA solvers.
- Beaver Discovery, Inc.: GridCAD a powerful tool for 2D and 3D structured or unstructured grid generation, in-static computation of 3D grids, aeroelastic phenomena, and CAD modeling.
- BETA CAE Systems: ANSA, automatic surface mesh generation, and FEA preprocessing. Geometry handling (generation, cleanup, watertight preparation, and defeaturing), surface and volume meshing, geometry and mesh morphing, and direct interfaces to several CFD and FEA solvers.
- BUD: BudMesh2D, an all quadrilateral 2D mesh generator producing high mesh quality for rubber and tire manufacturers.
- CAE Software Solutions: SPIDER, an octree-based meshing method that produces boundary fitted, local refined, conformal meshes with one or more boundary layers.
- Calmar Research Corporation: AGPS, Aero-Grid Paneling System – a surface geometry system.
- Cambridge Flow Solutions: BOXERMesh, a mesh-generating system for CFD applications.
- Catalpa Research Inc.: TIGER, 3D structured grid generation for general turbomachinery configurations.
- Sharc: Harpoon, body-fitted Cartesian meshing (hex elements, hanging nodes).
- CentaurSoft: Centaur is a complete unstructured grid generation package, containing a CAD conversion engine, a surface, prismatic and tetrahedral grid generator, and a grid adapter.
- Compass: GiD, a universal user-friendly 2D/3D structured/unstructured mesh generation and visualization.
- Channel Consulting Ltd.: TriGrid is a modeling system for building triangular grids.
- Program Development Company: GridPro is an elliptic, two-dimensional, single block, structured grid generation, and display software package for IBM/PC and compatible computers.
- Computational Mechanics Australia Pty. Ltd.: QUAD-GEN, automatic generator of quadrilateral meshes in nonhomogeneous domains. ANISO-QUAD is

a quadrilateral mesher designed specifically for the generation of anisotropic quadrilateral meshes. Source code is provided.

- Computing Objects: CM2 MeshTools includes a triangle Delaunay mesher, a quadrangle Delaunay mesher, a tetrahedron Delaunay mesher, anisotropic meshers, and surface meshers.
- EDS: COSMOS/AccuStress, a meshing control tool, coupled to a stress calculation module.
- csimsoft: Trelis is csimsoft's high-end commercial preprocessor for FEA and CFD, based on the time-proven CUBIT software from Sandia National Laboratories.
- DHCAE Tools: CastNet, an automated hybrid mesher using hex, tets, prism, and pyramid elements with direct interfaces to CFD solvers, for example, OpenFOAM (complete model setup) and Fluent. CastNet reads CAD-kernel data as input, uses curvature-controlled element sizes, and has advanced boundary layer features.
- T-Systems Digital Engineering Solutions: MEDINA is a universal CAE pre- and postprocessor.
- Electricité de France: HOMARD, adaptation of 2D/3D meshes by refinement and unrefinement techniques.
- Edinburgh Petroleum Services: PanMesh, a mesh generator customized for modeling geometric structures.
- EKK: KENT, a series of mesh generation tools designed specifically for the casting industry.
- Elements Research: EleGrid generates structured/unstructured meshes on parametric surfaces.
- NISA: DISPLAY IV, automatic and mapped mesh generation for 2D and 3D geometries using shells and solids (structured and unstructured).
- EMRL: GMS is a comprehensive software package for computer simulations of groundwater problems. SMS is a surface water modeler. Both packages come with a meshing module.
- Siemens PLM: FEMAP, an FE preprocessor for 2D (triangles/quads) and 3D (tetrahedral).
- FEGS: CADfix, geometry repair, volume/surface meshing, based on the medial object technology.
- Femsys Ltd.: FEMGV, a general-purpose pre- and postprocessing program for FEA, CFD, and FD.
- Field Precision: MetaMesh, 3D structured conformal mesh, hexadron elements. Mesh 4.5, 2D structured conformal mesh, triangular elements.
- Fluent Inc.: GAMBIT has a single interface for geometry creation and meshing that brings together all of Fluent's preprocessing technologies in one environment. G/Turbo, unstructured meshing for turbomachinery. TGrid, advanced hybrid volume mesh generation using tetrahedra, pyramids, and prisms (wedges or hexahedra) and 2D meshing with triangles and quadrilaterals.
- IBM: CATIA Surface 2, CATIA's powerful surface mesher.

- ANSYS Inc.: ICEM CFD Hexa, a semiautomated hexahedral meshing module that provides rapid generation of multiblock structured or unstructured volume meshes; ICEM CFD Tetra is the object-based tetrahedral meshing module, while ICEM CFD Autohexa is the fully automatic object-oriented hexahedral mesh generator. There are also modules for unstructured triangular/quadrilateral surface meshing and hex meshing with semiautomated blocking.
- ISE: MESH ISE, a dimension-independent Delaunay grid generator for semiconductor devices.
- ITC: GEMS, geometric modeling and structured grid generation.
- MacNeal-Schwendler Corp.: MSC.AMS, tool set for manual meshing and geometry creation.
- Materialise: Mimics, medical image processing and mesh generation; 3-matic, forward engineering including mesh generation.
- MIDASoft: MIDAS/FX-Modeler, auto surface meshing for structural engineering.
- NUMECA: HexPress is NUMECA's new unstructured hexahedral mesh generator.
- PC-Progress: Meshgen 2D, a Windows application for FEM data preprocessing (triangles).
- Pointwise Inc.: Pointwise, surface and volume grid generation (3D, multiblock, structured grids).
- Program Development Company: Grid Pro, automatic surface and volume grid generator, multiblock structured approach.
- Quantech: GiD, universal pre- and postprocessor.
- SCOREC: AGMD, the Algorithm Oriented Mesh Database (AOMD), serves as a mesh management library (or database) that provides a variety of services for mesh users.
- Siemens PLM: NX, finite element preprocessing (surface and volume meshing).
- Simmetrix Inc.: MeshSim, fully automatic mesh generation direct from CAD models.
- Simpleware Ltd.: +FE, generation of FE models from scanned object data.
- Simulation Works. Inc.: Kubrix, all-hexahedral unstructured mesh generation based on a fuzzy logic approach. Also builds structured meshes.
- Distene: TetMesh-GHS3D, Yams, MeshAdapt, HEXOTIC. Most of the codes have been developed in the GAMMA project at INRIA.
- smile:)consult: Janet, an efficient software package to generate model meshes for different numerical methods.
- Symscape: SimLab Meshing Tools, integrated structured and unstructured volume meshing.
- Taitech: DRAGON Grid, hybrid grids for CFD simulation, unstructured at the boundary, structured in the domain.
- Transoft International: Fluidyn-CAD/GEN, a combined CAD package and mesh generator for CFD.
- XYZ Scientific Applications Inc.: Truegrid, multiblock structured grid generation.

References

1. George, P.L. (1991) *Automatic Mesh Generation: Application to Finite Element Methods*, John Wiley & Sons, Inc.
2. Farin, G. (1997) Curves and surfaces for computer-aided geometric design – a practical guide, in *Computer Science and Scientific Computing*, 4th edn, (ed. J.M Ortega) Academic Press, https://books.google.es/books?id=Y7rSBQAAQBAJ&pg=PR3&lpg=PR3&dq=computer+science+and+scientific+computing+academic+press&source=bl&ots=rRk1eoAZgs&sig=NoHQFuXhtaa5z9Vf14p3BCY57wM&hl=es&sa=X&ved=0ahUKEwiGmM-qgY_OAhWMbhQKHZfUCY8Q6AEIITAA#v=onepage&q=computer%20science%20and%20scientific%20computing%20academic%20press&f=false (accessed 27 July 2016).
3. Frey, P.J. and George, P.L. (2000) *Mesh Generation: Application to Finite Elements*, Hermès Science Publications.
4. Lo, S.H. (1985) A new mesh generation scheme for arbitrary planar domains. *Int. J. Numer. Methods Eng.*, **21** (8), 1403–1426.
5. Löhner, R. and Parikh, P. (1988) Generation of three-dimensional unstructured grids by the advancing front method. *Int. J. Numer. Methods Fluids*, **8**, 1135–1149.
6. Nguyen, V.-P. (1982) Automatic mesh generation with tetrahedron elements. *Int. J. Numer. Methods Eng.*, **18** (2), 273–289.
7. Peraire, J., Vahdati, M., Morgan, K., and Zienkiewicz, O.C. (1987) Adaptive remeshing for compressible flow computations. *J. Comput. Phys.*, **72** (2), 449–466.
8. Owen, S.J. and Saigal, S. (2000) H-morph: an indirect approach to advancing front hex meshing. *Int. J. Numer. Methods Eng.*, **49** (1–2), 289–312.
9. Cheng, S.-W., Dey, T.K., and Shewchuk, J. (2012) *Delaunay Mesh Generation*, 1st edn, Chapman & Hall, CRC Press.
10. Löhner, R. (2008) *Applied Computational Fluid Dynamic Techniques: An Introduction Based on Finite Element Methods*, John Wiley & Sons, Ltd.
11. Shewchuk, J.R. (2012) Unstructured mesh generation, in *Combinatorial Scientific Computing*, Chapter 10 (eds U. Naumann and O. Schek), Chapman & Hall, CRC Press, pp. 259–285, https://www.crcpress.com/Combinatorial-Scientific-Computing/Naumann-Schenk/p/book/9781439827352 (accessed 27 July 2016).
12. George, P.-L., Hecht, F., and Saltel, É. (1990) Fully automatic mesh generator for 3d domains of any shape. *IMPACT Comput. Sci. Eng.*, **2** (3), 187–218.
13. Yerry, M.A. and Shephard, M.S. (1984) Automatic three-dimensional mesh generation by the modified-octree technique. *Int. J. Numer. Methods Eng.*, **20** (11), 1965–1990.
14. Bern, M., Eppstein, D., and Gilbert, J. (1994) Provably good mesh generation. *J. Comput. Syst. Sci.*, **48** (3), 384–409.
15. Coll, A. (2014) Robust volume mesh generation for non-watertight geometries. PhD thesis. Universitat Politècnica de Catalunya (UPC) – BarcelonaTech, July 2014.
16. Labelle, F. and Shewchuk, J.R. (2007) Isosurface stuffing: fast tetrahedral meshes with good dihedral angles. *ACM Trans. Graphics*, **26** (3), 57.
17. Mitchell, S.A. and Vavasis, S.A. (1992) Quality mesh generation in three dimensions. Proceedings of the 8th Annual Symposium on Computational Geometry, ACM, pp. 212–221.
18. Qian, J. and Zhang, Y. (2010) Sharp feature preservation in octree-based hexahedral mesh generation for cad assembly models, in *Proceedings of the 19th International Meshing Roundtable* (ed. S. Shontz), Springer, pp. 243–262, http://www.springer.com/us/book/9783642154133 (accessed 27 July 2016).
19. Schneiders, R. (1996) A grid-based algorithm for the generation of hexahedral element meshes. *Eng. Comput.*, **12**, 168–177.
20. Shephard, M.S. and Georges, M.K. (1991) Automatic three-dimensional mesh generation by the finite octree

technique. *Int. J. Numer. Methods Eng.*, **32** (4), 709–749.

21. Donéa, J., Huerta, A., Ponthot, J.-P., and Rodríguez-Ferran, A. (2004) Arbitrary Lagrangian–Eulerian methods, in *Encyclopedia of Computational Mechanics* (eds E. Stein, R. De Borst, and T.J.R. Hughes), John Wiley & Sons, Ltd.
22. Zienkiewicz, O.C., Taylor, R.L., and Zhu, J.Z. (2013) *The Finite Element Method: It's Basis and Fundamentals*, 7th edn, Butterworth-Heinemann.
23. Brebbia, C.A. and Wrobel, L.C. (1996) *Boundary Element Techniques: Theory and Applications in Engineering*, 2nd edn, WIT Press, Southampton.
24. Cundall, P.A. and Strack, O.D.L. (1979) A discrete numerical method for granular assemblies. *Geotechnique*, **2**, 47–65.
25. Campbell, C.S. (1990) Rapid granular flows. *Annu. Rev. Fluid Mech.*, **2**, 57–92.
26. Williams, J. and O'Connor, R. (1999) Discrete element simulation and contact problem. *Arch. Comput. Methods Eng.*, **6** (4), 279–304.
27. Oñate, E. and Rojek, J. (2004) Combination of discrete element and finite element methods for dynamic analysis of geomechanics problems. *Comput. Methods Appl. Mech. Eng.*, **193**, 3087–3128.
28. Rojek, J. and Oñate, E. (2008) Coupled discrete/finite element modelling of geotechnical problems. 8th World Congress on Computational Mechanics (WCCM 2008), Venice, Italy, July 2008.
29. Rojek, J., Oñate, E., Labra, C., Kargl, H., and Akerman, J. (2009) Optimizing rock cutting through computer simulation, in *Technology Innovation in Underground Construction* (ed. G. Beer), CRC Press.
30. Rojek, J., Zarate, F., Agelet de Saracibar, C., Gilbourne, C., and Verdot, P. (2005) Discrete element modelling and simulation of sand mould manufacture for the lost foam process. *Int. J. Numer. Methods Eng.*, **62** (11), 1421–1441.
31. Oñate, E., Idelsohn, S., Del Pin, F., and Aubry, R. (2004) The particle finite element method. An overview. *Int. J. Comput. Methods*, **1** (2), 267–307.
32. Idelsohn, S., Oñate, E., and Del Pin, F. (2003) A Lagrangian meshless finite element method applied to fluid–structure interaction problems. *Comput. Struct.*, **81**, 655–671.
33. Idelsohn, S., Mier-Torrecilla, M., and Oñate, E. (2009) Multi-fluid flows with the particle finite element method. *Comput. Methods Appl. Mech. Eng.*, **198**, 2750–2767.
34. Oñate, E., Celigueta, M.A., and Idelsohn, S. (2006) Modeling bed erosion in free surface flows by the particle finite element method. *Acta Geotech.*, **1**, 237–252.
35. Hughes, T.J.R., Cottrell, J.A., and Bazilevs, Y. (2005) Isogeometric analysis: CAD, finite elements, NURBS, exact geometry and mesh refinement. *Comput. Methods Appl. Mech. Eng.*, **194**, 4135–4195.
36. Quarteroni, A. and Rozza, G. (eds) (2014) *Reduced Order Methods for Modeling and Computational Reduction*, Springer International Publishing.
37. Duvaut, G. and Lions, J.L. (1972) *Les Inequations en Mecanique et en Physique*, Dunod, Paris.
38. Kikuchi, N. and Oden, J.T. (1988) *Contact Problems in Elasticity: A Study of Variational Inequalities and Finite Element Methods*, SIAM, Philadelphia, PA.
39. Agelet de Saracibar, C. (1997) A new frictional time integration algorithm for large slip multi-body frictional contact problems. *Comput. Methods Appl. Mech. Eng.*, **142**, 303–334.
40. Agelet de Saracibar, C. (1998) Numerical analysis of coupled thermomechanical contact problems. Computational model and applications. *Arch. Comput. Methods Eng.*, **5**, 243–301.
41. Agelet de Saracibar, C., Cervera, M., and Chiumenti, M. (1999) On the formulation of coupled thermo-plastic problems with phase-change. *Int. J. Plast.*, **15**, 1–34.
42. Agelet de Saracibar, C., Cervera, M., and Chiumenti, M. (2001) On the constitutive modeling of coupled thermomechanical phase-change problems. *Int. J. Plast.*, **17**, 1565–1622.

43. Agelet de Saracibar, C. and Chiumenti, M. (1999) On the numerical modeling of frictional wear phenomena. *Comput. Methods Appl. Mech. Eng.*, **177**, 401–426.
44. Chiumenti, M., Agelet de Saracibar, C., and Cervera, M. (2008) On the numerical modeling of the thermomechanical contact for metal casting analysis. *J. Heat Transfer*, **130**, Art. No. 061301.1-10.
45. Laursen, T.A. (1992) Formulation and treatment of frictional contact problems using finite elements. PhD dissertation. Department of Mechanical Engineering, Stanford University.
46. Laursen, T.A. (2002) *Computational Contact and Impact Mechanics*, Springer.
47. Laursen, T.A. and Simo, J.C. (1993) A continuum-based finite element formulation for the implicit solution of multibody, large deformation frictional contact problems. *Int. J. Numer. Methods Eng.*, **36**, 3451–3485.
48. Simo, J.C. and Laursen, T.A. (1992) An augmented Lagrangian treatment of contact problems involving friction. *Comput. Struct.*, **42**, 97–116.
49. Wriggers, P. (1995) Finite element algorithms for contact analysis. *Arch. Comput. Methods Eng.*, **2** (4), 1–50.
50. Wriggers, P. and Zavarise, G. (1993) On the application of augmented Lagrangian techniques for non-linear constitutive laws in contact interfaces. *Commun. Numer. Methods Eng.*, **9**, 815–824.
51. Di Capua, D. and Agelet de Saracibar, C. (2015) A direct elimination algorithm for quasi-static and dynamic contact problems. *Finite Elem. Anal. Des.*, **93** (1), 107–125.
52. Puso, M.A. and Laursen, T.A. (2004) A mortar segment-to-segment contact method for large deformation solid mechanics. *Comput. Methods Appl. Mech. Eng.*, **193**, 601–629.
53. Puso, M.A. and Laursen, T.A. (2004) A mortar segment-to-segment frictional contact method for large deformations. *Comput. Methods Appl. Mech. Eng.*, **193**, 4891–4913.
54. Yang, B. (2006) Mortar finite element methods for large deformation contact mechanics. PhD dissertation. Duke University.
55. Wohlmuth, B. (2000) A mortar finite element method using dual spaces for the Lagrange multiplier. *SIAM J. Numer. Anal.*, **38**, 989–1012.
56. Hüeber, S. and Wohlmuth, B. (2005) A primal-dual active set strategy for non-linear multibody contact problems. *Comput. Methods Appl. Mech. Eng.*, **194**, 3147–3155.
57. de Lorenzis, L., Wriggers, P., and Hughes, T.J.R. (2014) Isogeometric Contact: A Review. ICES Report 14–09, The University of Texas at Austin.
58. Pasenau, M.A. *et al.* (2011) Scalable system for large unstructured mesh simulation. Conference Paper, May 2012, Parallel CFD 2011, Barcelona.
59. Pasenau, M.A. (2014) Detail-preserving mesh simplification. Master thesis. Universitat Politècnica de Catalunya (UPC) – BarcelonaTech.
60. Pasenau, M.A. and Andújar, C. (2014) Detail-preserving mesh simplification for scientific visualization. 11th World Congress on Computational Mechanics, Barcelona, July 2014.
61. Lange, B. and Nguyen, T. (2014) Big Data architecture for large-scale scientific computing. World Academy of Science – ABDA 2014, Las Vegas, NV, September 2014.
62. Dureisseix, D. and Bavestrello, H. (2006) Information transfer between incompatible finite element meshes: application to coupled thermoviscoelasticity. *Comput. Methods Appl. Mech. Eng.*, **195**, 6523–6541. doi: 10.1016/j.cma.2006.02.003
63. Ramadan, M., Fourment, L., and Digonnet, H. (2010) A parallel multimesh method for speeding-up multiphysics computations – application to the cogging process. IV European Conference on Computational Mechanics, Paris, France.
64. Grätsch, T. and Bathe, K.J. (2005) A posteriori error estimation techniques in practical finite element analysis.

Comput. Struct., **83** (4–5), 235–265. doi: 10.1016/j.compstruc.2004.08.011

65. Bussetta, P., Boman, R., and Ponthot, J.-P. (2014) Efficient 3D data transfer operators based on numerical integration. *Int. J. Numer. Methods Eng.*, **102** (3–4), 892–929. doi: 10.1002/nme.4821.
66. Mediavilla, J., Peerlings, R., and Geers, M. (2006) A robust and consistent remeshing-transfer operator for ductile fracture simulations. *Comput. Struct.*, **84**, 604–623. doi: 10.1016/j.compstruc.2005.10.007
67. Perić, D., Vaz, M. Jr.,, and Owen, D. (1999) On adaptive strategies for large deformations of elasto-plastic solids at finite strains: computational issues and industrial applications. *Comput. Methods Appl. Mech. Eng.*, **176**, 279–312. doi: 10.1016/S0045-7825(98)00342-9
68. Benson, D.J. (1989) An efficient, accurate, simple ALE method for nonlinear finite element programs. *Comput. Methods Appl. Mech. Eng.*, **72**, 305–350. doi: 10.1016/0045-7825(89)90003-0
69. Boman, R. and Ponthot, J.-P. (2012) Efficient ALE mesh management for 3D quasi-Eulerian problems. *Int. J. Numer. Methods Eng.*, **92**, 857–890. doi: 10.1002/nme.4361
70. Boman, R. and Ponthot, J.-P. (2013) Enhanced ALE data transfer strategy for explicit and implicit thermomechanical simulations of high-speed processes. *Int. J. Impact Eng.*, **53**, 62–73. doi: 10.1016/j.ijimpeng.2012.08.007
71. Biotteau, E., Gravouil, A., Lubrecht, A.A., and Combescure, A. (2010) Multigrid solver with automatic mesh refinement for transient elastoplastic dynamic problems. *Int. J. Numer. Methods Eng.*, **84** (8), 947–971. doi: 10.1002/nme.2927
72. Trottenberg, U., Oosterlee, C.W., and Schuller, A. (2001) *Multigrid*, Academic Press, London.
73. Ben Dhia, H. (1998) Multiscale mechanical problems: the Arlequin method. *C.R. Acad. Sci., Ser. IIb: Mec., Phys., Astron.*, **326** (12), 899–904. doi: 10.1016/S1251-8069(99)80046-5
74. Ben Dhia, H. and Rateau, G. (2005) The Arlequin method as a flexible engineering design tool. *Int. J. Numer. Methods Eng.*, **62** (11), 1442–1462. doi: 10.1002/nme.1229
75. Gould, P.L. and Hara, T. (2002) Recent advances in local–global FE analysis of shells of revolution. *Thin Walled Struct.*, **40** (7–8), 641–649. doi: 10.1016/S0263-8231(02)00017-4
76. Mote, C.D. (1971) Global–local finite element. *Int. J. Numer. Methods Eng.*, **3**, 565–574. doi: 10.1002/nme.1620030410
77. Renaud, T., Costes, M., and Péron, S. (2012) Computation of GOAHEAD configuration with chimera assembly. *Aerosp. Sci. Technol.*, **19** (1), 50–57. doi: 10.1016/j.ast.2011.07.001
78. Steger, J. and Benek, J. (1987) On the use of composite grid schemes in computational aerodynamics. *Comput. Methods Appl. Mech. Eng.*, **64** (1–3), 301–320. doi: 10.1016/0045-7825(87)90045-4
79. Alves, M., Fernandes, J., Rodrigues, J., and Martins, P. (2003) Finite element remeshing in metal forming using hexahedral elements. *J. Mater. Process. Technol.*, **141**, 395–403. doi: 10.1016/S0924-0136(03)00388-1
80. Zienkiewicz, O. and Zhu, J.Z. (1992) The superconvergent patch recovery (SPR) and adaptive finite element refinement. *Comput. Methods Appl. Mech. Eng.*, **101** (1–3), 207–224. doi: 10.1016/0045-7825(92)90023-D
81. Khoei, A.R. and Gharehbaghi, S. (2009) Three-dimensional data tranfert operators in large plasticity deformations using modified-SPR technique. *Appl. Math. Modell.*, **33**, 3269–3285. doi: 10.1016/j.apm.2008.10.033.
82. Bucher, A., Meyer, A., Görke, U.J., and Kreißig, R. (2007) A comparison of mapping algorithms for hierarchical adaptive FEM in finite elasto-plasticity. *Comput. Mech.*, **39**, 521–536. doi: 10.1007/s00466-006-0051-z
83. Larson, M.G. and Bengzon, F. (2008) Adaptive finite element approximation of multiphysics problems. *Commun. Numer. Methods Eng.*, **24**, 505–2500. doi: 10.1002/cnm.1087

84. Ortiz, M. and Quigley, J.J. (1991) Adaptive mesh refinement in strain localization problems. *Comput. Methods Appl. Mech. Eng.*, **90**, 781–804. doi: 10.1016/0045-7825(91)90184-8
85. Saksono, P. and Perić, D. (2006) On finite element modelling of surface tension. *Comput. Mech.*, **38**, 251–263. doi: 10.1007/s00466-005-0745-7
86. Yu, L., Liu, J., Kong, X., and Hu, Y. (2008) Three-dimensional RITSS large displacement finite element method for penetration of foundations into soil. *Comput. Geotech.*, **35**, 372–382. doi: 10.1016/j.compgeo.2007.08.007
87. Labergère, C., Rassineux, A., and Saanouni, K. (2011) 2D adaptive mesh methodology for the simulation of metal forming processes with damage. *Int. J. Mater. Form.*, **4**, 317–328. doi: 10.1007/s12289-010-1001-z
88. Villon, P., Borouchaki, H., and Saanouni, K. (2002) Transfer of admissible plastic fields (transfert de champs plastiquement admissibles). *C.R. Mec.*, **330**, 313–318. doi: 10.1016/S1631-0721(02)01457-2 (French, Abridged English version).
89. Farrell, P., Piggott, M., Pain, C., Gorman, G., and Wilson, C. (2009) Conservative interpolation between unstructured meshes via supermesh construction. *Comput. Methods Appl. Mech. Eng.*, **198**, 2632–2642. doi: 10.1016/j.cma.2009.03.004
90. Orlando, A. and Perić, D. (2004) Analysis of transfer procedures in elastoplasticity based on the error in the constitutive equations: theory and numerical illustration. *Int. J. Numer. Methods Eng.*, **60**, 1595–1631. doi: 10.1002/nme.1021
91. Bernardi, C., Maday, Y., and Patera, A. (1994) A new non conforming approach to domain decomposition: the mortar element method, in *Collège de France Seminar* (eds H. Brezis and J.L. Lions), Pitman.
92. Huerta, A. and Casadei, F. (1994) New ALE applications in non-linear fast-transient solid dynamics. *Eng. Comput.*, **11** (4), 317–345. doi: 10.1108/02644409410799317
93. Rodriguez-Ferran, A., Casadei, F., and Huerta, A. (1998) ALE stress update for transient and quasistatic processes. *Int. J. Numer. Methods Eng.*, **42**, 241–262.
94. Alauzet, F. and Mehrenberger, M. (2010) P1-conservative solution interpolation on unstructured triangular meshes. *Int. J. Numer. Methods Eng.*, **84** (13), 1552–1588. doi: 10.1002/nme.2951
95. Benson, D.J. (2008) Momentum advection on unstructured staggered quadrilateral meshes. *Int. J. Numer. Methods Eng.*, **75** (13), 1549–1580.
96. Panchal, J.H., Kalidindi, S.R., and McDowell, D.L. (2013) Key computational modeling issues in integrated computational materials engineering. *Comput. Aided Des.*, **45** (1), 4–25.
97. Bohm, H.J. (1998) A Short Introduction to Basic Aspects of Continuum Micromechanics. CDL-FMD Report, 3, https://www.ilsb.tuwien.ac.at/links/downloads/ilsbrep206.pdf (accessed 27 July 2016).
98. Feyel, F. and Chaboche, J.L. (2000) FE-2 multiscale approach for modelling the elastoviscoplastic behaviour of long fibre SiC/Ti composite materials. *Comput. Methods Appl. Mech. Eng.*, **183** (3), 309–330.
99. Salomon, D. (2004) *Data Compression: The Complete Reference*, Springer-Verlag, New York.
100. Hernández, J.A., Oliver, J., Huespe, A.E., Caicedo, M.A., and Cante, J.C. (2014) High-performance model reduction techniques in computational multiscale homogenization. *Comput. Methods Appl. Mech. Eng.*, **276**, 149–189.
101. Barrault, M., Maday, Y., Nguyen, N.C., and Patera, A.T. (2004) An empirical interpolation method: application to efficient reduced-basis discretization of partial differential equations. *C.R. Math.*, **339** (9), 667–672.
102. Chaturantabut, S. and Sorensen, D.C. (2010) Discrete empirical interpolation for nonlinear model reduction. Decision and Control 2009 Held Jointly with the 28th Chinese Control

Conference 2009, CDC/CCC 2009, Proceedings of the 48th IEEE Conference, pp. 4316–4321.

103. Oden, J.T., Belytschko, T., Babuska, I., and Hughes, T.J.R. (2003) Research directions in computational mechanics. *Comput. Methods Appl. Mech. Eng.*, **192**, 913–922.

104. Soto, O., Löhner, R., and Camelli, F. (2003) A linelet preconditioner for incompressible flow solvers. *Int. J. Numer. Methods Heat Fluid Flow*, **13** (1), 133–147.

105. Amestoy, P., Duff, I., and L'Excellent, J.Y. (2000) Multifrontal parallel distributed symmetric and unsymmetric solvers. *Comput. Methods Appl. Mech. Eng.*, **184** (2–4), 501–520.

106. George, A. and Liu, J.W. (1981) *Computer Solution of Large Sparse Positive Definite*, Prentice Hall, Englewood Cliffs, NJ.

107. Duff, I. and Reid, J. (1983) The multifrontal solution of indefinite sparse symmetric linear-equations. *ACM Trans. Math. Software*, **9** (3), 302–325.

108. Gupta, A., Koric, S., and George, T. (2009) Sparse matrix factorization on 1150 massively parallel computers. Proceedings of the Conference on High Performance Computing Networking, Storage and Analysis, SC' 09, ACM, New York, pp. 1:1–1:12.

109. Li, X. and Demmel, J.W. (2003) SuperLU DIST: a scalable distributed memory sparse direct solver for unsymmetric linear systems. *ACM Trans. Math. Software*, **29**, 110–140.

110. Davis, T.A. (2012) Summary of Available Software for Sparse Direct Methods.

111. Saad, Y. (2003) *Iterative Methods for Sparse Linear Systems*, SIAM.

8
Platforms for ICME

Adham Hashibon, Önder Babur, Mauricio Hanzich, Guillaume Houzeaux, and Bořek Patzák

8.1
Introduction

Numerical simulations are now routinely used in research and industry and are accepted as reliable analysis tools and, in many cases, are used as part of the design process. Existing knowledge in various disciplines has been integrated into advanced simulation tools, which have been developed over many years. In recent years, it has become increasingly clear that the focus will be on integrating existing knowledge from different disciplines to enable further progress in many fields, including structural and material engineering. Therefore, one of the key challenges in numerical modeling is to enable efficient use and interoperation of a wide variety of problem-specific computational tools and databases for simulating, analyzing, and understanding complex, multiphysics problems [1–3].

At present, the routine use of ICME tools is limited to small groups of researchers in large companies, universities, and software companies. To reach widespread acceptance and use of ICME tools by the research and engineering community, standardized, open integration tools have to be developed to integrate individual simulation tools along the process chain. The availability of such tools with commercial support is one of the challenges that have to be reached to extend ICME to all stakeholders in small and medium businesses, research institutes, and universities.

Software integration focuses on combining individually developed software components into an integrated product. Traditionally, researchers have tended to use single-purpose programs or shell scripts to integrate existing applications. This approach may initially look promising, but it later becomes clear that there are many restricting issues and limitations. This approach provides a very low level of interoperability – data is usually exchanged using input and output files, allowing interoperation at program level rather than at function level. While this allows different codes to be combined, substantial coding and deep understanding of the code structure is usually required. The level of code reuse is typically very low, and this is a barrier to extension and maintenance. More

Handbook of Software Solutions for ICME, First Edition. Edited by Georg J. Schmitz and Ulrich Prahl.

sophisticated solutions that allow connecting individual components to enable seamless communication and coordination are necessary.

Today, software integration use in industry is mainly made on the basis of continuum models describing large, that is, component-scale systems. Furthermore, numerical simulations in industrial applications are overly dominated by structural mechanics or computational fluid dynamics tools based on finite elements or finite volume numerical schemes as part of product life management or computer-aided engineering (PLM/CAE) processes. However, advances in nanotechnology demands that processes occurring on the smaller scales, that is, electronic, atomistic, and mesoscopic be also taken into account. Though in principle nanoscale systems can be modeled using single-scale models, for example, atomistic molecular dynamics, often the nanoscale system is only one part of a larger system or device exhibiting interconnected physical or chemical processes and components operating on microscopic, mesoscopic, and macroscopic scales. This demands extending the tools and approaches of ICME to include those on the smaller scales as well. In particular, integrated platforms are particularly attractive for nanoscale systems as the numerous advanced single-scale tools that are available readily to researchers and engineers can be effectively reused as building blocks for new multiscale and multiphysics simulations.

8.2 Integration Approaches

The reliable multiscale/multiphysics numerical modeling requires the inclusion of all relevant physical phenomena along the process chain and across multiple scales, requiring the combination of knowledge from multiple fields. The development of a new multiphysics tool for a particular problem would be extremely time and resource consuming. A more viable approach lies in combining existing, usually, single-physics tools to build a customized multiphysics simulation chain for a particular problem. The most important advantage of this modular approach stays in reusing of existing matured, validated, and verified tools. In order to achieve a full potential of such a modular approach, an integration platform is needed to provide the underlying infrastructure that enables to facilitate interoperability between different tools; see [4] for an excellent review.

An integration platform aims to provide an environment in which the individual components (applications or tools) can share and exchange information and are integrated together in workflows to solve the complex problems, providing interoperability between different hardware platforms, operating systems, and programming languages, and taking into account security and licensing issues.

The platform should be able to integrate *third-party* simulation tools provided with or without access to its source code. The general requirements include the ability to steer individual applications and provide a mechanism(s) for data exchange and persistent storage. Data import/export modules to well-established data formats are necessary for interoperability and to facilitate pre- and postprocessing.

The principal role of the framework is to steer individual components (applications) and to provide high-level data exchange services. Efficient steering and data exchange require some kind of control channel to be defined and implemented by individual simulation tools. This channel should allow the framework to call the individual codes at appropriate times, handle exceptional situations, and request/update application data. There are many common points within the field of computational steering [5, 6], which deals with interactive control over a computational process. Here, techniques for monitoring and run-time changes of simulation data and parameters have been developed, including data exchange in heterogeneous distributed environments.

Platforms should provide a convenient way to define simulation scenarios. This requirement can be achieved, for example, either by the user-defined script that interacts with the platform or is interpreted by the platform engine. Some platforms are developed using scripting languages, such as Python. Then the scripting language can be directly used to define simulation scenario, giving the user a full power of interpreted programming language. This allows to make complex simulations more interactive and flexible, approaching the idea of interactive visual computing [5]. Usually, on top of this rather low-level layer, a graphical workflow editor is developed, providing user-friendly environment. The workflow editor can perform type and compatibility checks even before the simulation scenario is executed. There are many existing tools for defining, managing, and executing workflows, such as FireWorks [7], Kepler [8], and OpenMole [9].

Regarding the development in hardware and software technology in the last decades, the integration platform should enable integration of distributed, parallel, and scattered applications over the network in an efficient and scalable way based on standardized communication protocols. Parallel and distributed applications typically come with a distributed parallel data structure and several application instances running on different processors. The platform should provide a communication layer allowing unified application data access and steering, allowing serial and parallel applications to be managed using the same interface, hiding the differences of communication with local and remote simulation tools. Standard methods, such as "naming services" or a "service discovery," that allow individual component lookup based on name and/or services should be provided, as well as services for job and resource allocation on remote computing resources.

The specific platform requirements are derived directly from the actual integration model used. For example, in case of object-oriented approaches, the interfaces to data access have to be implemented in the same way for all tools that are integrated. If the tools are already existing, then special software wrappers can be developed to create a unified interface for the incorporated or integrated tools. If a data-centric model is used for integration, then conversion tools or interfaces for each model have to be developed. However, regardless of the integration approach, there are some common high-level requirements that are expected for each platform.

In this section, an introduction to related software engineering issues and recommend best practices approaches to software integration is given. Later, an overview of algorithms for multiphysics coupling is presented, focusing on weak, intermediate, and strong solution strategies.

The integration involves, in general, two main aspects. One is obviously technical in nature, that is, how to achieve the interoperability in terms of the software tools and their respective data representation models. The other is of more fundamental scientific nature and pertains specifically to the coupling and linking of the underlying physical and chemical models. Achieving interoperability requires that methods and models are developed both to allow technical as well as scientific integration.

8.2.1
A Categorization of Software to be Integrated

Before discussing the integration approaches, it is essential to emphasize a determining factor for the whole process: the nature of the existing software to be integrated. For this, we use a similar but slightly changed terminology as used in software testing, and define three categories of software:

- Black-box software: Typically includes commercial software with no access to source code. This type of software offers limited flexibility, for example, by means of configuration options and file I/O, and has to be considered and integrated "as is" without any modification or knowledge on internals.
- Gray-box software: Includes software that offers some kind of a controlled interface, for example, a Python [10], MATLAB, COM [11], or DLL interface, but still hides its source code. The software itself is still not modifiable in principle but allows a more flexible communication with the outside.
- White-box software: Is the open source or in-house developed software, where full access to source code is available. This allows the most flexibility and modification, as the internals and interfaces can be freely adapted; even a piece of code can be (partially) translated into another programming language with the help of automatic wrappers such as SWIG [12], Boost [13], and Forthon [14]. Of course quality factors such as modularity, extendibility, and understandability play a big role for the ease of integration.

8.2.2
Object-Oriented Approaches

Modern software integration approaches are related to the concept of software component, a reusable program building block that can be combined with other components in the same or other computers in a distributed network to form a complex application. The object-oriented approach involves designing a system of interacting objects for the purpose of solving a software problem. Software objects (or components) consist of state, described by internal attributes and interface, representing the behavior of the object in terms of methods. Methods operate on an object's state and are the only mechanism for interacting with other objects.

Hiding internal state and requiring all interaction to be performed through an object's methods is known as *data encapsulation* – a fundamental principle of object-oriented programming.

Classes are a template definition of the methods and variables in a particular kind of objects. Thus, an object is a specific instance of a class. Classes can create hierarchies, and inheritance lets the structure and methods in one class pass down the hierarchy. Another key concept in object-oriented design is the concept of abstract classes (ACs) or as it is sometimes called abstract base classes (ABCs). Both ACs and ABCs are used to define an interface in terms of member functions and are intended to be inherited from by concrete classes. The ABCs normally have only declaration and absolutely no implementation, while ACs may have some basic methods implemented. It is a way of forcing a contract between the class designer and the users of that class. If one wishes to create a concrete class (a class that can be instantiated) from an AC, then one must declare and define a matching member function for each abstract member function of the base class. Such an interface concept allows using any derived class on a very abstract level, using a common interface for services, without being concerned with the implementation details of an individual software component.

Integration platform defines a set of ACs representing the entities in the model space. Typically, they define AC representing an application code, which defines an interface that needs to be implemented by application code to plug into the platform. This interface defines steering and data exchanges services. In this way, all application can be managed using the single, common interface. This concept can be extended to support creation of a hierarchy of cooperating applications, where a platform script representing a low-level problem can be transformed into a class implementing application interface, which permits its reuse in more complex simulation scenarios.

One of the key features is that ACs can be used to represent not only the applications but also the data in the model space, such as properties and fields. This approach focuses on services provided by high-level data and abstract from underlying data, their representation, and storage. In this way, such a platform does not depend on particular data format. Users are simply interested about how to get the data (how to evaluate the field at given position, for example) and not how the data are stored. As a consequence, this leads to natural independence on particular data storage format(s).

Object-oriented frameworks enjoy high popularity in multiscale/multiphysics communities. Prominent examples include Elmer [15], MOOSE [16], OpenFOAM [17], SU2 [18], Kratos [19], OOFEM [20], MuPIF [21, 22], CoolFluid3 [23], MBDyn [24], and SimPhoNy [25].

8.2.3 Component-Based Approaches

Component-based software engineering CBSE [26] extends the notion of component as defined in the earlier section, putting emphasis on several features. Firstly, components are considered as encapsulated semantically related executable pieces of software (in contrast to objects). Independent deployment is therefore an essential part of a component. Another feature is that components explicitly specify their interfaces (via some type of interface description language – IDL), including their context dependencies (what services it requires and offers). With the help of these features, components lead to a better "plug-and-play" integration, also at run time. The ultimate goal of CBSE is to build a catalog of available software components and transform the software development process into a commercial-off-the-shelf component evaluation, selection, and composition activity as much as possible, thus achieving maximum reuse. Note that component concept is very similar to the object concept, but it does not necessarily have to be implemented object-oriented. In OO implementations on the other hand, a component typically maps to a group of internal objects cooperating to provide/consume the specified services to/from the outside. CBSE has gained popularity in the early 1990s and has achieved a wide applicability with generic standards such as Microsoft's COM/.NET [11], OMG's CORBA [27], and Sun's JavaBeans [28] for enterprise interoperability.

Not surprisingly, it is considerably welcomed in the multiscale/multiphysics community. Common Component Architecture (CCA) [29], for instance, aims to provide a component model specific to high-performance computing (HPC) applications. For reference, several platforms adopting this approach are Salome [30], Uintah Computational Framework [31], Virtual Test Bed [32], Integrated Plasma Simulator [33], PUPIL [34], Cheops [35], APECS [36], Earth System Modeling Framework [37], and to some extent FLASH [38] and Roccom/ Rocstar [39].

8.2.4 Service-Oriented Approaches

Service-oriented architecture (SOA) [40] overlaps heavily with CBSE; it revolves around designing loosely coupled, interoperable pieces of software to be integrated and orchestrated. On the other hand, it differs by its emphasis on Internet technologies and service layer. Although a service is conceptually similar to a component, that is, a loosely coupled entity with clearly defined interface, it is better defined as the logical unit of functionality that encapsulates a repeatable activity. Typical examples are the banking operations, for example, withdraw money from an account, which involves the interaction of many components. So the service layer can be thought of being a layer above the components, the interplay of which realizes the services, likewise components are on top of objects as described earlier. Another important feature of SOA is that the services have to be remotely invokable, publishable, and discoverable. A very popular implementation of SOA is Web Services [41]. It contains an XML-based service description language WSDL [42], service registry and location facility (UDDI), and a messaging protocol SOAP [43] over HTTP. A lot of effort has been put on areas such as service orchestration and quality assurance in terms of service-level agreements;

SOA has been in great use for business-to-business (B2B) interoperability and cloud computing (software as a service (SAAS)). Being a relatively recent technology, however, SOA has not been used often in the ICME/multiphysics/multiscale community. RealityGrid Steering Toolkit [44], for instance, uses SOA for the grid computing infrastructure. There are also other efforts in the literature such as [45, 46] and more recently MSC SimEnterprise [47]. One can draw more examples for general simulation integration [48] from environmental/earth sciences [49, 50] and life/health sciences [51].

8.2.5 Data-Centric Approaches

While the approaches listed earlier mostly focus on the connection and interoperability of functionality, there are also data-centric approaches, which focus on common data representation and conversion. This is often done in an *ad hoc* way in various tools having hard-coded programs or scripts that convert the data formats into their own. This problem can be divided into two: format conversion and model conversion/integration. Firstly, the data has to be structurally converted into other formats. This is generally the trend in most applications, where they provide utilities for conversion to/from widely applicable common scientific data formats such as HDF [52], netCDF [53], and VTK [54] from/to their own formats. The more difficult problem is the conversion of data model rather than the format. This is usually done with the help of some metadata, whether it be models (or schemas, metamodels, etc.) or ontologies; both of which are summarized separately in the following sections. Note that data-centric approaches can be very beneficial and sometimes the most feasible approach when black-box integration is considered. DEEPEN [55], for instance, advocates this approach.

8.2.6 Model-Based Approaches

Model-based approaches, which are fundamental in model-driven software engineering [56] and model-based systems engineering [57], rely on a common data/component model and operations on the model level such as conversion/transformation/generation, verification, and optimization. A typical example for the model-based data integration is the integration of databases via schema mappings, which has a long history of application. Database integration is facilitated in the multiscale integration of data, for example, for climate, GIS, or bioinformatics databases. Other approaches involve a common data/component model, specified by a domain-specific language (DSL): a markup or modeling language for instance. Note that every API of a program/library already implicitly specifies an embedded DSL (EDSL); in object-oriented approaches, for example, this model can be partly represented by the class diagram of the program. There are also examples of explicit EDSLs such as in CoolFluid3 [23] and Elmer [15] for symbolic representation of equations. More interesting DSLs for integration and interoperability are the explicit external ones. In the multiscale/multiphysics domain, there have been a handful of examples of such approaches. Materials markup language (MatML) [58] is an XML-based markup language for describing and exchanging material properties. Multiscale Modeling Language (MML) [59] aims to specify the architecture of a multiscale simulation program based on the Complex Automata (CxA) formalism. MML supports graphical and XML-based specifications and is used in the MUSCLE 2 [60] framework for coupling different simulations. Model-based systems can very well support generative paradigms, for example, for generating code from models. An example is The Tensor Contraction Engine [61] for quantum chemistry, which can generate optimized Fortran code from a high-level description language. A better example is the VPH/Physiome project [62], which includes, among many other things, Systems Biology Markup Language, CellML, and FieldML with an extensive toolset for model conversion and code generation.

8.2.7 Ontology-Based Approaches

Ontology [63] is formalism from artificial intelligence and knowledge representation disciplines of computer science, gaining a high popularity in Web Engineering with the advent of Semantic Web. Ontologies model the concepts and relationships of a domain in a formal way. Thus, it gives a kind of "meaning" (semantics) to the concepts, as they are used in the domain. Usually ontologies have a layered structure: an upper ontology for describing the generic elements, a domain ontology derived from it for describing domain-specific elements, and possible extensions. With this hierarchical semantic data model at hand, individual model developers can achieve semantic interoperability by extensions and mappings. Note that this is basically another model-based approach with a unique line of development and standards such as Web Ontology Language OWL [64] and Resource Description Framework RDF [65]. The use of ontologies for expressing domain knowledge in materials science and engineering has been widely researched since the 1990s (e.g., [66, 67], and [68]). There are a few ontology-based multiphysics/multiscale integration frameworks as well. Simantics [69] and CAPE-OPEN [70] based on OntoCAPE [71] are such frameworks for chemical process engineering. A recent effort, MoDeNa [72], aims at developing a framework based on a multiscale mathematical ontology for the same domain, with a toolset of editors, code generators, and so on. Other examples are Gene Ontology [62, 73, 74].

8.2.8 Existing Standards for Integration

High-Level Architecture (HLA), defined under IEEE Standard 1516 [75], is a general-purpose architecture for distributed computer simulation systems. Originally, it has been developed by the US military and later has been adopted by the IEEE as standard. It was also used by the European Space Agency in the EODiSP

Open and Distributed Simulation Platform project [76]. HLA architecture enables several simulation systems to interact and work together regardless of the computing platforms. The HLA links individual tools (called federates) together into federation. The federates interact through the run-time infrastructure (RTI), which provides a set of software services that support federates in coordinating their operations and data exchange during a run-time execution. The federation object model (FOM) describes the data exchange in the federation, for example, the objects and interactions that will be exchanged. There are many commercial and open-source implementations of the HLA interface specification; see [77] for reference.

Functional Mock-up Interface (FMI) is a tool-independent standard to support both model exchange and cosimulation of dynamic models using a combination of XML files and compiled C code [78]. The FMI has been developed as a part of the ITEA2 MODELISAR project. It is based on separation of description of interface data (XML file) and functionality (C code or binary). FMI is supported by more than 35 tools and is used by automotive and nonautomotive organizations throughout Europe, Asia, and North America. The FMI specification is freely available [79].

The CCA Forum [29] is a group of researchers from national labs and academic institutions committed to defining standard component architecture for HPC. The objective of the CCA Forum is to define a minimal set of standard interfaces that a high-performance component framework has to provide to components and can expect from them, in order to allow disparate components to be composed together to build a running application. Such a standard will promote interoperability between components developed by different teams across different institutions.

CAPE-OPEN standards provide a set of interfaces that allows the integration of process simulation components from various sources into process simulation environments. The architecture is based on process modeling environment used to run the simulations and process modeling components, representing evaluation of physical properties, simulation of a particular unit operation, or numerical solver. The individual components are connected using concept of units representing a single-component calculation, streams that contain specification of information flow between units, ports that connect units to streams, and flow sheets that allow building a collection of units and streams. CAPE-OPEN is based on established software technologies, such as COM and CORBA. The standard is open, multiplatform, uniform, and available free of charge. At present, the standard is maintained by the CAPE-OPEN Laboratories Network (CO-LaN), the international user-driven organization [80].

8.2.9 Coupling and Linking Approaches

In general, there are different types of couplings between individual application codes. For example, a multiphysics problem may exhibit one-way coupling in which one physics affects another but not vice versa. However, more often the problems are mutually dependent, where the degree of physical coupling can vary and is often governed by one or more physical parameters.

Spatial coupling is concerned with information exchange in space. Data exchange between applications using different types of discretizations (FE, FV, FD) and/or nonmatching grids can be challenging, especially when adaptive, parallel, or distributed simulation scenarios are considered. Another interesting point is spatial coupling of the discrete and continuum descriptions, where information exchange can be done on the interface or over transitional zone, where weak compatibility conditions are enforced.

Temporal coupling refers to dependencies between components that involve time. A major difficulty is associated with different timescales required by individual physical processes considered, which may vary by orders of magnitude. Even if the timescales are approximately the same, the individual applications can have their own requirements related to time stepping, induced mainly by time discretization. The stability requirement typically leads to small time steps in explicit methods, while implicit methods may allow the use of much larger time steps. The particular choice depends on accuracy versus efficiency requirements.

In [81], an excellent overview of coupling algorithms ranging from weak coupling based on fixed-point iterations to strong coupling represented at the extreme by a pure Newton method is presented. The focus is on intermediate coupling strategies, which seek to provide the benefits of strong coupling but with little to no requirements imposed on applications codes beyond providing weak coupling data.

The *strong coupling* solution strategies are based on solution of the problem, described by a set of coupled differential equations, using Newton's method. This leads to an iterative solution scheme requiring evaluating Jacobian matrix, containing the sensitivities (derivatives) of each equation to each unknown. This is sometimes referred as monolithic solution scheme. Such task can be relatively simple when the complete system of equations is modeled in a single application but can be extremely complicated when coupling two or more independent applications. The need for implementing evaluation of missing sensitivities and for assembly and solution of the global system of linearized equations renders the whole coupling strategy very demanding. The strongly coupled solution scheme also poses additional requirements on the platform itself, as it should provide support to assemble and solve the resulting (sparse) linearized system, and this is often beyond the capabilities provided by existing platform solutions.

On the other side, the *weakly coupled* (sometimes referred also as partitioned) solution strategies are based on the idea of the fixed-point iteration (also known as the Picard or nonlinear Richardson iteration method). In this case, the individual subproblems are solved in a sequence while considering the external variables to be fixed. The updated values are transferred to the next subproblem, and this is repeated until all individual subproblems are updated and an overall convergence is reached. The key advantage is that individual applications update their own variables independently, so there is no need to form and solve global system of equations. Due to its simplicity not only in terms of implementation but also regarding the required functionality provided by individual simulation codes, the weak coupling is usually considered as the only option supported by the integration platform. However, weakly coupled physical models may experience convergence problems and often exhibit slow convergence.

The *intermediate coupling* algorithms try to preserve the simplicity of weak coupling with the performance of strong coupling. These approaches are based on Jacobian-free Newton–Krylov (JFNK) methods (see [82]) based on Newton method that employs a Krylov-based iterative linear solver. The advantage lies in the fact that the Krylov linear solver only requires evaluating the action of the Jacobian on a vector. Thus, the Jacobian matrix does not have to be formed, and the key consequence is that directional derivatives needed may be approximated. JFNK can be thought as an approximate, inexact Newton method with approximations due to evaluation of Jacobian action and potentially solution errors associated with linear iterative solver.

In general, in all approaches, the consistency of the internal values across applications is required at each global time step. A sequence of global time steps has to be established based, for example, on the global Courant–Friedrichs–Lewy (CFL) condition, a timescale of relevant physical phenomena, or accuracy constraints. Within the limits of the global time step, each application can choose its own local time stepping.

8.3 High-Performance and Distributed Computing

Complex simulations are extremely resource and time demanding. Distributed and parallel computing environments provide needed resources and computational power. Common features of these environments are a distributed data structure and concurrent processing on distributed processing nodes. This brings in an additional level of complexity that needs to be addressed in a platform design. The important role of any integration platform is providing a communication mechanism that will take care of the network communication between the objects.

Even sequential simulation scenarios can profit from distributed computing model. Individual simulation tools can be available as services remotely. This is called SAAS, a software licensing and delivery model in which software is licensed on a subscription basis and is centrally hosted. Integration of such applications into complex simulation scenarios requires at least remote communication mechanism to exist. Another example includes the multiscale simulations, where RVE simulations on a lower scale require only boundary and initial conditions from the upper scale and can be naturally executed concurrently.

However, increasing the number of existing applications can exploit parallel computing resources. The parallelization allows not only obtaining results in acceptable time by significantly speeding up the analysis but also performing large and complex analyses, which often do not fit into a single, even well-equipped, machine with one processor unit (regardless of achieved speedup). Sequential computer codes run on a computer with a single processing unit. The algorithm is translated into a sequence of instructions, which are executed by a processing unit one by one. On the other hand, parallel computing is based on premise that large problems can be divided into smaller tasks, which are solved concurrently by the simultaneous use of multiple computing resources. The design of parallel algorithms requires the partitioning of the problem into a set of tasks, the number of which is greater than or equal to the number of available processors. The partitioning of the problem can be fixed at run time (static load balancing) or can change during the solution (dynamic load balancing). The latter option is often necessary in order to achieve good load balancing of work among processors and thus optimal scalability.

Many existing platform solutions supporting distributed communication are based on service bus topology. Here, every application has a single connection to a service bus, usually called the RTI. The RTI provides information, synchronization, and data exchange services. At the same time, individual parallel simulation codes are regarded as a single entity, assuming that there is a centralized node (often called master node) that interfaces to the platform. This naturally hides the complexity of internal application parallelism and data decomposition. However, the existence of centralized component, either being represented by RTI on the platform side or master node in case of parallel application code, is a potential bottleneck for scalability. Such a central component is by design incompatible with massively parallel applications: in massively parallel applications, scalability is synonymous to data distribution and mutual point-to-point communication. The scalability of the centralized approach depends on the amount of interapplication data exchange. The scalable solution should make possible the fully distributed communication scheme, where all nodes can communicate concurrently. This is quite common pattern when considering communication within a single application, but in case of integration platform this is a challenging topic. To establish such a distributed data exchange can be a nontrivial task, as individual simulation tools may use different decompositions and discretizations that can even change over time (adaptivity, dynamic load balancing). There already exist successful approaches illustrating the importance of distributed communication topology; see, for example, [83].

8.3.1 HPC Hardware

In 1965, future Intel cofounder Gordon E. Moore published a paper entitled "Cramming more components onto integrated circuits" [84]. In that paper, he made a prediction about technology evolution that states "The number of transistors incorporated in a chip will approximately double every 24 months."

Today, there is an open discussion about the current validity of Moore's statement from 50 years ago. On the one hand, the frequency scaling of the technology being used for building computers is off [85], as can be observed in Figure 8.1, while packing an always growing amount of transistors in the

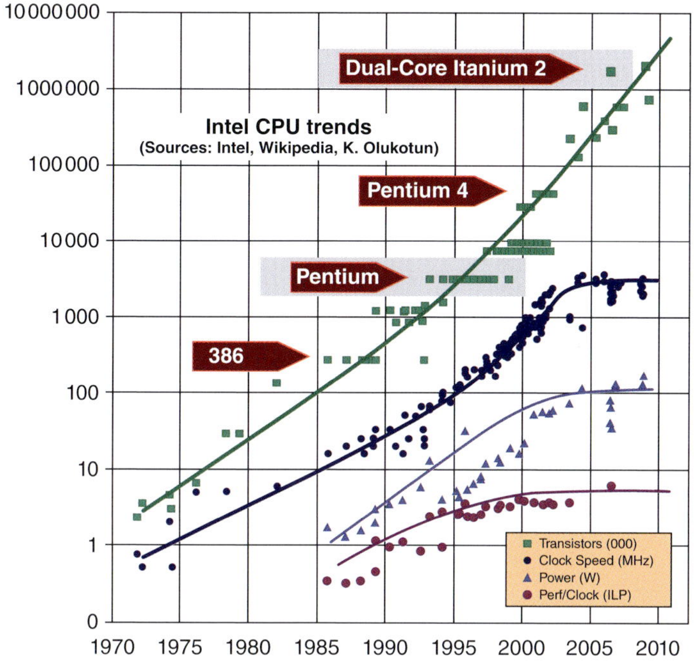

Figure 8.1 Transistors, clock speed, power consumption, and performance. (Reproduced with the permission of [85].)

same space is reaching physical limits that might not be overcome [86]. On the other hand, there are ways of improving computational power that extends the way CMOS technology is currently being used by combining it with other technologies [87] in the so-called "More-than-Moore" approach [88]. One way of doing that is packing more "processors" to do a single task, or build specialized devices for tackling specific problems.

This trend defined the modern processors' architectures sporting many cores in a single chip from vendors such as Intel's Xeon [89], AMD's Opteron [90], or IBM's Power8 [91], among others. Those processors, however, are still devoted to general-purpose computations and can be found in modern laptops, tablets, and even smartphones.

A newer trend to overcome Moore's Law is to build processing devices that are developed with a more specific goal in mind. Clear examples of such approaches are the (already extinct) Cell/B.E. from IBM [92] or the trending GPUs from NVIDIA [93] and AMD [94], which were originally developed for multimedia tasks and gaming and are now being used extensively for HPC. This fact can be observed in the TOP500 list ranking the biggest HPC system worldwide [95]. Intel has also bet on this field with its new Xeon Phi [96] brand of coprocessors.

During the history of HPC, and related to what Moore has said in the 1960s, there are around three orders of magnitude of improvement in the computational power of HPC machines every decade [95]. Currently, the most powerful HPC machines in the world rank in the order of tenths of PFlops1. The next challenge is the EFlop (1018 Flops), but the technology that should be used to build such machines is not clear enough. Mostly, this is due to the amount of computational resources needed (millions of cores) and the energy to be used for empowering the machine that could not be affordable [97].

For facing those challenges, many different alternatives are being explored by different research prototypes, such as Montblanc and Montblanc 2 [98] or DEEP and DEEP-ER [99].

8.3.2 HPC Programming Models

Most of the current HPC platforms are built using an interconnection of many computational resources, each of which can be called a computational node. Figure 8.2 shows such an architecture, where a computational node is composed of some sort of host that is a compute device that is able to run the operating system needed for the machine to work. In some cases, that host is also complemented with a device (or devices) that could be an accelerator (GPU) or coprocessor (Xeon Phi). In any case, inside a computational node, there is a common address space (CAS) memory that can be accessed both by the host or device. Also, for having a complete HPC environment, the computational nodes need to communicate in order to collaborate in achieving a given parallel task. To that end, a communication network (or many) interconnects the computational nodes for sharing information.

Considering the different hardware architectures, there is a clear need for tools that enable the development of applications that are able to profit from such specialized architecture. Hence, many of the so-called programming models [100–102] come to help the developer on this task. Such programming models consist of different kinds of tools, such as libraries invoked from traditional sequential languages, language extensions, or complete new execution models that can be used for developing parallel and distributed applications for HPC platforms.

Moreover, as the HPC hardware is complex enough, the level of parallelism that should be tackled can be found at many different levels: embarrassingly parallel applications, message-passing parallelism, shared memory systems, and/or vectorial code at core level.

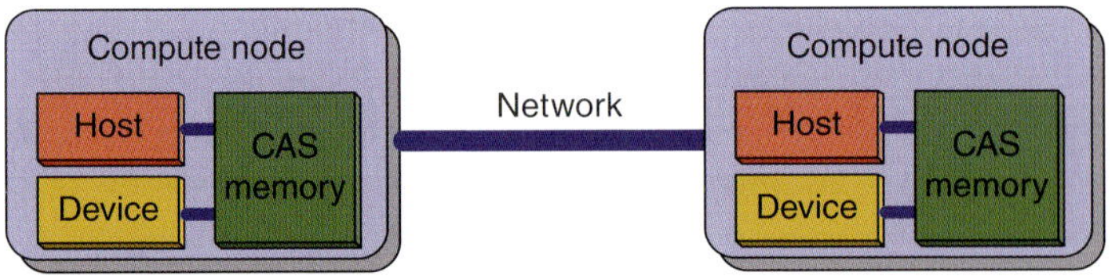

Figure 8.2 Typical HPC environment architecture.

The embarrassingly parallel level of parallelism could be found in those applications that have to provide their results as the condensation of the results of many tasks that are completely independent. Examples of this kind of applications are workflow oriented [103] or stochastic and Monte Carlo [104] frameworks, which need to explore a parametric space where each instance is independent of the rest. In such situations, the support provided by the programming model is related to the way the computational resources are being allocated, how the results from individual tasks are gathered and reduced to the final results, or even some sort of fault tolerance, among other concerns. Many workflow management systems such as Pegasus [105], Falkon [106], or Swift [107] are in use, among many others. The one to be used depends mostly on the problem at hand.

The next level of parallelism is typical in cases where the amount of computational resources for solving a single task is greater than can be obtained from a single computational node. In such cases, a middleware needs to provide the means for the different subtasks to cooperate (i.e., communicate) for reaching their goal. The current programming model standard for providing such communication mechanism is known as Message Passing Interface (MPI) [108, 109], whose first version was released in 1994. Currently, MPI standard is under version 3.0, but 3.1 is almost approved. It should be noticed that as MPI is a standard that has matured for more than 20 years, there are in the market many implementations that support the standard to a certain extent. Examples of MPI implementations are the open-source MPICH2 [110, 111] and OpenMPI [112]. Also, there are some from hardware vendors, such as Intel's IMPI [113] and IBM's POE [114].

Even if the resources inside a single computational node are enough for solving an HPC task, the developer needs some mechanism to profit from the many computational resources (i.e., cores, accelerators, coprocessors, etc.) that are available in modern HPC platforms. In order to do that, some programming models were developed that provide the means to use the memory in a computational node for sharing information, avoiding the necessity for an explicit exchange of information among the different subtasks of a parallel application. Such programming models ease the shared memory level of parallelism. The most extended programming model that provides a way for parallel subtasks to share data is OpenMP (OMP) [115]. Unlike MPI that defines a standard API and a subsequent library, OMP provides a series of pragmas2 that might be included in a sequential code that the compiler will understand, transforming such annotated regions in code that will run in parallel using the resources available in the computational node. Besides OMP, there are other alternatives, such as OMPSs [116] and OpenAcc [117], which are both based on pragmas like OMP. While the former is more general than OMP and deals with the code as if it were a workflow of tasks that can be mapped onto the computational resources, the latter try to profit from the capabilities of accelerators/coprocessors available into the computational node. Other alternatives, such as NVIDIA's CUDA [118] programming model, are oriented toward GPU development and are not based on pragmas but on some extensions to traditional C/C++ programming. Also, OpenCL [119] is a standard similar to

CUDA that extends the applicability not only to GPUs but also to any other kind of accelerator/coprocessor available in a computational node.

The last level of parallelism can be found in a single computational core or certain accelerators such as GPUs streaming multiprocessors (SM). In traditional processors, there are some instructions whose operands are not scalar values but vectors to which the same operation is applied at once. This is known as single instruction, multiple data (SIMD) instructions. For the SMs in a GPU, instead of a single instruction, there are many threads executing the same instruction at the same time over different data. This is known as single instruction, multiple thread (SIMT) scheme. Regarding the hardware architecture, different kinds of intrinsic instructions exist to provide with SIMD/SIMT capabilities. For example, Intel provides many different sets such as SSE [120], AVX, AVX-2, and AVX-512 [121].

8.3.3 On Major HPC/Distributed Computing Architectures

HPC/distributed computing has an orthogonal aspect in addition to the ones described earlier: system architecture and organization. Depending on factors such as underlying hardware/software, network connection, and organizational structure of ownership, there are three main technologies for distributed computing systems (DCS):

Cluster computing system (to be referred as cluster) is a collection of usually same type of computers that act together as single computing resource for parallel tasks.

Grid computing system (to be referred as grid) is a federation of heterogeneous computers that enable utilizing geographically distributed resources depending on their availability.

Cloud computing systems (to be referred as cloud) is a collection of virtualized computers that are dynamically provided as a single unified computing resource depending on user demand.

Although the distinction between these three technologies is not entirely black and white because of their coevolution in the recent years, there are significant differences. A cluster is composed of same type of commodity computers with high network interconnection (e.g., LAN). Typically, one organization owns and maintains the entire cluster, and resource and user management is operated centrally. Examples range from small clusters for in-house simulations to huge HPC workstation clusters and supercomputers. On the other hand, a grid consists of heterogeneous types of computers interconnected with varying speeds of network connection, possibly through the Internet. Multiple administrative organizations maintain their own subgrids and manage their users and resources. The scientific computing community has many success stories for grid computing such as European Grid Infrastructure (http://www.egi.eu/), BOINC (http://boinc.berkeley.edu/), and Open Science Grid (http://www.opensciencegrid.com/).

As many alternatives arise for supporting the grid computing concept, in 2006, the Open Grid Forum [122] association was created for providing standards for

Table 8.1 Differences between major distributed computing systems.

	Cluster	Grid	Cloud
Computers	Same type, commodity	Heterogeneous	Heterogeneous
Ownership	Single	Multiple	Single
Interconnection of computers	High speed (LAN)	Low speed (Internet)	High speed (LAN, fiber-optic, etc.)
User/resource management	Centralized	Distributed	Centralized with delegation possibility

those alternatives to cope with. Among the alternative platforms for providing grid computing resources are Globus [123], UNICORE [124], gLite [125], and ARC [126].

Finally, a cloud has a network of heterogeneous types of computers. There is usually a single cloud provider organization that maintains the resources and serves multiple customers, though it can delegate the management to third parties. The virtualization layer allows the easy dynamic allocation of resources to different customers according to their demand.

Among the current major cloud service providers are Amazon (http://aws.amazon.com/), Google (https://cloud.google.com/), Microsoft (http://azure.microsoft.com/en-us/), Apple (https://www.apple.com/icloud/), IBM (http://www-03.ibm.com/systems/x/solutions/cloud/), Verizon/Terremark (http://www.verizonenterprise.com/solutions/cloud/), and Cisco (http://www.cisco.com/web/solutions/trends/cloud/index.html). Table 8.1 briefly summarizes the points discussed.

The scientific community at the moment mostly uses clusters and grids; however, cloud has been an emerging trend since the 2000s and provides a scalable and cheap alternative for many other domains. The performance of cloud for the costly scientific tasks has been benchmarked and is reported to be lower than the previous alternatives [127, 128]. However, the cost-effectiveness, high utilization, and flexibility of cloud can be the driving factors of adopting it for ICME.

8.3.4 Fault Tolerance

As the Exascale era is only years apart, machines with millions of cores, many thousand nodes, complex networks, and huge I/O systems will become commodities. In such scenarios, distributed applications that run using a huge amount of computational resources will inexorably face some kind of failure that will prevent the normal execution of the workload [129]. This situation is even worse for grid or cloud environments that include even more failure points. Therefore, it is mandatory for any parallel applications to consider the fault tolerance as a design issue in order to be able to use the next generation of distributed environments in their full capacity.

Regarding the failures that may be produced while running a distributed application, they are due to three main reasons: data corruption, software errors, and hardware failures. No matter the reason, however, any fault tolerance system must be able to carry out two actions, detect the failure, and recover from it.

In order to detect a failure during the application execution, there are different strategies that should be used. For detecting a data corruption, some mechanism should be included in the application that checks the input data, before using it in the main processing. This strategy could be extended not only to the user data but to deeper levels of the application to check, for example, the information exchanged between computational domains or even between functions of a single thread. The level at which this detection is applied depends on how critical a certain process into the distributed application is. For detecting software errors, the best alternative is to have a proper unit testing policy during the development, jointly with good system and regression testing systems, which reduce the failure risk during execution [130]. That is, solving the problem before it arises in production. However, there will always be a situation not contemplated in the testing that will incur in a software error. For tackling those situations, a system integrated in the application that explicitly checks the information consistency through the application should be included [131]. Finally, the way for detecting hardware failures depends on the specific kind of failure being tackled; however, in general, it should be a supervised process from an external source. This is necessary, as a hardware failure may hinder the detection system if run using the faulty resource. Sometimes, there is the possibility to explicitly check the status of the hardware, but in most cases this can be solved by introducing timeouts for certain tasks, which will be assumed failed if the timeout expires without obtaining a result. In order to define the proper timeout for a certain task, a good knowledge of the application is mandatory [132].

Once the failure is detected, some measures should be taken for recovering a consistent state of the application and resuming the execution. If the failure is due to data corruption and depending on the critical path for such data, either it is skipped or the application is stopped and the user informed. If the failure is due to a software error found by analyzing the data throughout its flow into the application, the recovery action would be to relaunch the tasks for a limited number of times to avoid eventual errors [133] due to masked failures produced by undetected hardware failures or middleware malfunction. For those situations where the failure was produced by the underlying hardware, again to relaunch the affected task in a set of different resources is the most common strategy.

Finally, an important part of the recovery process is to be able to restore a consistent state; to that goal different techniques should be considered, such as checkpointing, message logging, rollback, message replaying, and message reordering [134].

8.4 Overview of Existing Platform Solutions

Many integration solutions are available nowadays. Many are provided by existing multiphysics simulation codes that integrate several simulation tools and may allow the integration of external, third-party modules. These capabilities can be found in commercial (ADINA [135], MSC Nastran [136], ANSYS Multiphysics [137], etc.), as well as open-source programs [15, 17]. A number of independent integration platforms have been developed in last decades, particularly for specific purposes to solve the problems in a specific domain. Despite this, many of these solutions have followed a generic approach permitting to use these tools in more general situations.

The major challenge for any platform remains the flexibility, allowing to easily integrate external simulation tools, supporting various data formats, and allowing to define complex simulation scenarios. The lack of standardization (in terms of data formats or communication interfaces) is probably the main limiting factor in the widespread usage of integrated solutions by commercial as well as academic users. Fortunately, there are ongoing standardization activities in ICME:

- The Minerals, Metals, and Materials society (TMS) within its Materials Processing & Manufacturing Division [138] is promoting the emerging area of integrated computational materials engineering.
- Integrated Computational Materials Engineering expert group (ICMEg) [139] is aiming at developing a global open standard for information exchange between the heterogeneous simulation tools in ICME.
- Consortium for Open Multiphysics [140] is forming a community of users dedicated to the design and implementation of a free and open interface and toolkit for application integration to aid in multiphysics development and simulation.
- European Materials Modelling Council (EMMC) [141] is a community-driven bottom-up action to connect all existing material modeling activities in Europe focusing on establishing the common language and standards. It has, for example, working groups on open simulation platform and on software interoperability.
- EU Multiscale Materials Modeling Cluster [142] brings together a number of EU-funded projects [139, 143–147], all developing open, integrated, and multipurpose numerical environments, for example, for nanodesign. The cluster is intended to enable knowledge exchange, foster adoption of novel approaches for multiscale modeling, and provide a platform for harmonizing standardization.

In the following platform directory, the important integration platforms that are collected from an ICMEg survey of simulation tools (http://www.icmeg.euproject.info/downloads.html), relevant reviews [148, 149], and Internet are listed.

Package name	Author/ company	www	Features	Short description
COMSOL Multiphysics	COMSOL, Inc.	http://www.comsol.com/comsol-multiphysics	LC, 3M	General-purpose software platform, for modeling and simulating physics-based problems. Add-on products can be used to add dedicated physics interfaces and tools for electrical, mechanical, fluid flow, and chemical applications
Simulia	Dassault Systemes	http://www.3ds.com/products-services/simulia	LC, 3M	Multiphysics Simulation from SIMULIA is based on the market-leading Abaqus technology and provides a complete suite of features for solving Multiphysics problems
modeFRONTIER	ESTECO	http://www.esteco.com/modefrontier	LC, 3M	modeFRONTIER® is an integration platform for multiobjective and multidisciplinary optimization. It provides a seamless coupling with third-party engineering tools, enables the automation of the design simulation process, and facilitates analytic decision making
Salome platform		http://www.salome-platform.org/	LO, PC, PP, 3M	SALOME is an open-source software that provides a generic platform for pre- and postprocessing for numerical simulation. It is based on an open and flexible architecture made of reusable components

Package name	Author/ company	www	Features	Short description
LIME	Consortium for Advanced Simulation of Light Water Reactors (CASL)	http://sourceforge.net/projects/lime1/	LO, PC, 3M	The Lightweight Integrating Multiphysics Environment (LIME) for coupling codes is a small software package for creating multiphysics simulation codes
OpenPALM	CERFACS	http://www.cerfacs.fr/globc/PALM_WEB/index.html	LO, 3M, DC	OpenPALM is a software allowing the concurrent execution and the intercommunication of programs based on in-house as well as commercial codes
PreCICE	TU München	http://www5.in.tum.de/wiki/index.php/PreCICE_Webpage	LO, 3M	Precise Code Interaction Coupling Environment (PreCICE) is a coupling tool for partitioned simulations of multiphysics scenarios
MCT	DOE	http://www.mcs.anl.gov/research/projects/mct/		MCT is a set of open-source software tools for creating coupled models. MCT is fully parallel and can be used to couple message-passing parallel models to create a parallel coupled model
DEEPEN	EU	http://www.nmp-deepen.eu/		The major objective of DEEPEN is to develop an integrated multiscale simulation capability for predictive design of novel materials and nanostructures, targeted at two of the most important present device challenges, namely future CMOS and future energy-efficient light sources

Package name	Author/ company	www	Features	Short description
SimPhoNy	EU	http://www.simphony-project.eu/		The SimPhoNy project aims to develop an easy-to-use integrated multiscale modeling environment for *in silico* discovery and design of nanoenabled systems and materials. The main concept of the SimPhoNy framework is to augment existing open-source and commercial simulation tools and supplement them with sophisticated interface software libraries
MMP/MuPIF	EU	http://mmp-project.eu	LO, PP, 3M, DC	MMP project will develop an integrated modeling platform, especially equipped to target multiscale and multiphysics engineering problems. The innovation of MMP lies in its generic and modular concept, supported by data standardization and proper definition of application interfaces
NanoSim	EU	http://www.sintef.no/Projectweb/NanoSim/		The objective of the NanoSim project is to create an efficient and cost-effective multiscale simulation platform based on free and open-source codes. This platform will connect models spanning a wide range of scales from the atomic scale through the particle and cluster scales, the industrial equipment scale, and the full system scale

Package name	Author/ company	www	Features	Short description
MoDeNa	EU	http://modena.units.it		MoDeNa aims at developing, demonstrating, and assessing an easy-to-use multiscale software modeling framework application under an open-source licensing scheme that delivers models with feasible computational loads for process and product design of complex materials
Cactus	NCSA,AEI, CCT	http://cactuscode.org/		Cactus provides computational scientists and engineers with a collaborative, modular, and portable programming environment for parallel high-performance computing
SimPARTIX	Fraunhofer IWM	http://simparix.com	LC, PC, 3M,DC	SimPARTIX® provides modeling tools for the mechanics of materials and fluids using mesh-free methods, therefore providing complexity in describing complex changing and dynamic geometries
Elmer	Helsinki CSC-IT	https://csc.fi/web/elmer	LO, PC, 3M, DC	Elmer is an object-oriented finite element multiphysics framework for mechanical engineering, fluid dynamics, electromagnetics, heat transfer, and acoustics

Package name	Author/ company	www	Features	Short description
OOFEM	Czech Technical University	http://www.oofem.org/	LO, PC, 3M, DC	OOFEM is a object-oriented finite element framework for mechanical engineering, transportation, and fluid mechanics
Moose	Idaho National Labs	http://mooseframework.org/	LO, PC, 3M, DC	Moose is a multiphysics object-oriented simulation environment for microstructure, chemistry, geomechanics, and superconductivity
OpenFOAM	OpenFOAM Foundation	http://www.openfoam.com/	LO, PC, 3M, DC	OpenFOAM is a multiphysics computational fluid dynamics framework
SU2	Stanford University	http://su2.stanford.edu/	LO, PC, 3M, DC	SU2 (Stanford University Unstructured) is a computational fluid dynamics package for aerospace
Kratos	CIMNE	http://www.cimne.com/kratos/	LO, PC, 3M, DC	Kratos is a framework for parallel multiphysics calculation for solid mechanics, fluid dynamics, and thermodynamics
CoolFluid3	Von Karman Institute	http://coolfluid.github.io/	LO, PC, 3M, DC	CoolFluid3 is a collaborative simulation environment for complex multiphysics for fluid dynamics, hydrodynamics, and aerodynamics
Trilinos	Sandia National Laboratory	http://trilinos.org/	LO, PC, 3M, DC	Trilinos is an open-source object-oriented framework for the solution of large-scale, complex multiphysics engineering and scientific problems

Package name	Author/ company	www	Features	Short description
CHIMPS	Stanford University	—	PP, PC, PO, DC	Coupler for High-Performance Integrated Multi-Physics Simulations library (CHIMPS) is a code coupling environment to facilitate the setup and execution of large-scale integrated simulations
SIERRA	Sandia National Laboratory	—	PO, DC	SIERRA is an object-oriented computational framework for developing complex multiphysics applications
MUSCLE 2	MAPPER project	http://www.qoscosgrid.org/trac/muscle	LO, PC, 3M, DC	Multiscale Coupling Library and Environment (MUSCLE 2) is a portable framework for multiscale modeling and simulation
Rocstar/ Roccom and IMPACT	University of Illinois, Illinois Rocstar LLC	http://www.csar.illinois.edu/rocstar/, http://www.illinoisrocstar.com/	LO/LC, PC, 3M, DC	Roccom/IMPACT is the software integration framework within the Illinois Rocstar suite for rocket simulation
MpCCI	Fraunhofer SCAI	http://www.mpcci.de/	LC, PC, 3M, DC	MpCCI (mesh-based parallel code coupling interface) is a commercial framework for the coupling of different simulation codes
OOF2	NIST	http://www.ctcms.nist.gov/oof/oof2/	LO, PC/PP, 3M, DC	OOF2, an object-oriented framework for finite element analysis of microstructures
PHYSICA	University of Greenwich	—	?, LF, 3M, DC	PHYSICA is a multiphysics computational framework for solidification/melting

Package name	Author/ company	www	Features	Short description
CMDF	California Institute of Technology	http://web.mit.edu/mbuehler/www/research/CMDF/CMDF.htm	PP	Computational Materials Design Facility (CMDF) is a framework for multiscale modeling and simulation of materials
AixVipMap	RWTH Aachen	http://aixvipmap.iehk.rwth-aachen.de/	LC, PO, DC	AixVipMap (the Aachen Virtual Platform for Materials Processing) is a platform for integrated computational materials engineering

Characteristics: License: *LO,LC* (Open Source, Commercial), major programming language (*PP* – Python, **PC** – C/C++, **PJ** – Java, **PO** – Other), **3M** – extensible to third-party modules, **DC** – support for distributed and/or parallel computing.

References

1. van der Velde, P. and Mallinson, G.D. (2007) The design of a component-oriented framework for numerical simulation software. *Adv. Eng. Software*, **38** (3), 182–192.
2. de Sturler, E., Hoeflinger, J., Kale, L., and Bhandarkar, M. (2001) A new approach to software integration frameworks for multi-physics simulation codes, in *The Architecture of Scientific Software* (eds R.F. Boisvert and P.T.P. Tang), Kluwer Academic Publishers, Boston, MA. ISBN: 0-7923-7339-1
3. Montarnal, P., Dimier, A., Deville, E., Adam, E., Gaombalet, J., Bengaouer, A., Loth, L., and Chavant, C. (2005) Coupling methodology within the software platform Alliances, in *International Conference on Computational Methods for Coupled Problems in Science and Engineering COUPLED PROBLEMS 2005* (eds E. Onate, M. Papadrakakis, and B. Schrefler), CIMNE, Barcelona.
4. Keyes, D.E. *et al.* (2012) Multiphysics Simulations: Challenges and Opportunities. Technical Report ANL/MCS-TM-321, Revision 1.1, Argonne National Laboratory.
5. Parker, S.G., Beazley, D.M., and Johnson, C.R. (1997) Computational steering software systems and strategies. *IEEE Comput. Sci. Eng.*, **4** (4), 50–59.
6. van Liere, R., Mulder, J.D., and van Wijk, J.J. (1997) Computational steering. *Future Gener. Comput. Syst.*, **12** (5), 441–450.
7. FireWorks (2015) Open-Source Code for Defining, Managing, and Executing Scientific Workflows, http://pythonhosted.org/FireWorks/ (accessed 26 April 2016).
8. Kepler (2015) Kepler Scientific Workflow System, https://kepler-project.org/ (accessed 26 April 2016).

9. OpenMOLE (2015) Open MOdeL Experiment (OpenMOLE), http://www.openmole.org (accessed 26 April 2016).
10. Rossum, G. and Drake, F.L. Jr. (eds) (2006) *An Introduction to Python*, Network Theory Ltd., 164 pp.
11. Williams, S. and Kindel, C. (1994) The Component Object Model: A Technical Overview. Microsoft Technical Report.
12. SWIG (Simplified Wrapper and Interface Generator), http://swig.org/ (accessed 26 April 2016).
13. Boost C++ Wrapper Generator, http://www.boost.org (accessed 26 April 2016).
14. Forthon Python Interface Generator for Fortran Based Codes, http://hifweb.lbl.gov/Forthon/ (accessed 26 April 2016).
15. Elmer (2011) http://www.csc.fi/english/pages/elmer (accessed 26 April 2016).
16. MOOSE Framework (2015) http://mooseframework.org/ (accessed 26 April 2016).
17. OpenFOAM (2011) http://www.openfoam.com, (accessed 26 April 2016).
18. SU2 (2015) http://su2.stanford.edu/ (accessed 26 April 2016).
19. KRATOS (2015) http://www.cimne.com/kratos/ (accessed 26 April 2016).
20. OOFEM.ORG (2015) http://www.oofem.org/ (accessed 26 April 2016).
21. OOFEM.ORG (2015) http://mech.fsv.cvut.cz/mupif (accessed 26 April 2016).
22. Patzák, B. *et al.* (2012) MuPIF – A distributed multi-physics integration tool. *Adv. Eng. Software*, **60–61**, 89–97, doi: 10.1016/j.advengsoft.2012.09.005
23. Coolfluid 3 (2015) http://coolfluid.github.io/ (accessed 26 April 2016).
24. MBDyn (2015) https://www.mbdyn.org/ (accessed 26 April 2016).
25. SimPhoNy http://www.simphony-project.eu/ (accessed 26 April 2016).
26. Heineman, G.T. and Councill, W.T. (2001) Component-based software engineering, in *Putting the Pieces Together* (eds I. Gorton, G.T. Heineman, I. Crnković, H.W. Schmidt, J.A. Stafford, C. Szyperski, and K. Wallnau), Addison-Westley.
27. Vinoski, S. (1997) CORBA: integrating diverse applications within distributed heterogeneous environments. *IEEE Commun. Mag.*, **35** (2), 46–55.
28. DeMichiel, L.G. (2002) Enterprise JavaBeans Specification, Version 2.1.
29. Common Component Architecture (2015) http://www.cca-forum.org/ (accessed 26 April 2016).
30. SALOME (2015) http://www.salome-platform.org/ (accessed 26 April 2016).
31. Uintah (2015) http://www.uintah.utah.edu/ (accessed 26 April 2016).
32. VTB (2015) http://vtb.engr.sc.edu/vtbwebsite/ (accessed 26 April 2016).
33. Cswim (2015) http://cswim.org/ips/ (accessed 26 April 2016).
34. PUPIL (2015) http://pupil.sourceforge.net/ (accessed 26 April 2016).
35. Schopfer, G. *et al.* (2004) Cheops: a tool-integration platform for chemical process modelling and simulation. *Int. J. Software Tools Technol. Trans.*, **6** (3), 186–202.
36. Zitney, S.E. (2010) Process/equipment co-simulation for design and analysis of advanced energy systems. *Comput. Chem. Eng.*, **34** (9), 1532–1542.
37. ESMF (2015) http://www.earthsystemmodeling.org/ (accessed 26 April 2016).
38. Flash Center (2015) http://flash.uchicago.edu/site/flashcode/ (accessed 26 April 2016).
39. Center for Simulation of Advanced Rockets (2015) http://www.csar.illinois.edu/rocstar/ (accessed 26 April 2016).
40. Erl, T. (2006) *Service-Oriented Architecture: Concepts, Technology, and Design*, Pearson Education India.
41. Alonso, G. *et al.* (2004) *Web Services*, Springer-Verlag, Berlin, Heidelberg.
42. Christensen, E. *et al.* (2001) Web Services Description Language (WSDL) 1.1.
43. Box, D. *et al.* (2000) Simple Object Access Protocol (SOAP) 1.1.
44. RealityGrid (2015) http://www.realitygrid.org (accessed 26 April 2016).
45. dos Santos, I.H.F., Raposo, A.B., and Gattass, M. (2006) A service-oriented architecture for a collaborative engineering environment in petroleum engineering. *Res. Interact. Des.*, **2**, 110.

46. Chen, X. *et al.* (2006) Soar-dsgrid: service-oriented architecture for distributed simulation on the grid. Proceedings of the 20th Workshop on Principles of Advanced and Distributed Simulation, IEEE Computer Society.
47. MSC.Software (2015) SimEnterprise, http://www.mscsoftware.com/assets/MSC_SimEnterprise_WP.pdf (accessed 26 April 2016).
48. Brutzman, D. et al. (2002) Extensible Modeling and Simulation Framework (XMSF) Challenges for Web-Based Modeling and Simulation.
49. Plale, B. *et al.* (2006) Casa and lead: adaptive cyberinfrastructure for real-time multiscale weather forecasting. *IEEE Comput.*, **39** (11), 56–64.
50. Smiatek, G. (2005) SOAP-based web services in GIS/RDBMS environment. *Environ. Modell. Software*, **20** (6), 775–782.
51. Arbona, A. *et al.* (2007) A service-oriented grid infrastructure for biomedical data and compute services. *IEEE Trans. Nanobiosci.*, **6** (2), 136–141.
52. The HDF Group (2015) http://www.hdfgroup.org/HDF5/ (accessed 26 April 2016).
53. UCAR (2015) http://www.unidata.ucar.edu/software/netcdf/ (accessed 26 April 2016).
54. VTK (2015) http://www.vtk.org/ (accessed 26 April 2016).
55. Deepen (2015) http://www.nmp-deepen.eu/ (accessed 26 April 2016).
56. Schmidt, D.C. (2006) Guest editor's introduction: model-driven engineering. *Computer*, **39** (2), 0025–0031.
57. Estefan, J.A. *et al.* (2007) Survey of model-based systems engineering (MBSE) methodologies. *Incose MBSE Focus Group*, **25** (8).
58. Kaufman, J.G. and Begley, E.F. (2003) MatML: a data interchange markup language. *Adv. Mater. Processes*, **161** (11), 35–39.
59. Falcone, J.-L., Chopard, B., and Hoekstra, A. (2010) MML: towards a multiscale modeling language. *Procedia Comput. Sci.*, **1** (1), 819–826.
60. Muscle2 (2015) http://www.qoscosgrid.org/trac/muscle (accessed 26 April 2016).
61. TCE (2015) http://www.csc.lsu.edu/~gb/TCE/ (accessed 26 April 2016).
62. Physiome Project (2015) http://physiomeproject.org (accessed 26 April 2016).
63. Staab, S. and Studer, R. (2010) *Handbook on Ontologies*, Springer Science & Business Media.
64. McGuinness, D.L. and Van Harmelen, F. (2004) OWL Web Ontology Language Overview. W3C Recommendation 10 October 2004.
65. Klyne, G. and Carroll, J.J. (2006) Resource Description Framework (RDF): Concepts and Abstract Syntax.
66. Sargent, P. Subrahmanian, E., Downs, M., Greene, R., and Rishel, D. (1992) Materials' information and conceptual data modelling, in *Computerization and Networking of Materials Databases: Third Volume*, ASTM STP 1140 (eds T.I. Barry and K.W. Reynard), American Society for Testing and Materials, Philadelphia, PA, p. 172.
67. Gruber, T.R. and Olsen, G.R. (1994) An ontology for engineering mathematics. 4th International Conference on Principles of Knowledge Representation and Reasoning.
68. Alberts, L.K. (1994) YMIR: a sharable ontology for the formal representation of engineering design knowledge, in *Formal Design Methods for Computer-Aided Design, IFIP Transactions* (eds J.S. Gero and E. Tyugu), North Holland, pp. 3–32.
69. Karhela, T., Villberg, A., and Niemistö, H. (2012) Open ontology-based integration platform for modeling and simulation in engineering. *Int. J. Model. Simul. Sci. Comput.*, **3** (02), 1250004.
70. Braunschweig, B.L. *et al.* (2000) Process modeling: the promise of open software architectures. *Chem. Eng. Prog.*, **96** (9), 65–76.
71. Morbach, J., Yang, A., and Marquardt, W. (2007) OntoCAPE—A large-scale ontology for chemical process engineering. *Eng. Appl. Artif. Intell.*, **20** (2), 147–161.

72. MoDeNa (2015) MOdelling of Morphology DEvelopment of Micro- and NAnostructures (MoDeNa), http://www.modenaproject.eu/ (accessed 26 April 2016).
73. Freitas, A., Asooja, K., Jares, J., Decker, S., and Sahay, R. (2014) A semantic web platform for automating the interpretation of finite element biosimulations. Proceedings of the 7th International Workshop on Semantic Web Applications and Tools for Life Sciences, Berlin, Germany, December 9–11, 2014.
74. Neal, M.L. et al. (2009) Advances in semantic representation for multiscale biosimulation: a case study in merging models. Pacific Symposium on Biocomputing, NIH Public Access.
75. IEEE Standard Association 1516–2010. (2015) *IEEE Standard for Modeling and Simulation (M&S) High Level Architecture (HLA)– Framework and Rules*, http://standards.ieee.org/findstds/standard/1516-2010.html (accessed 26 April 2016).
76. EODiSP (2015) EODiSP Open and Distributed Simulation Platform, http://www.pnp-software.com/eodisp/home.html (accessed 26 April 2016).
77. Wikipedia Contributors Run-Time Infrastructure (simulation), Wikipedia, The Free Encyclopedia, http://en.wikipedia.org/w/index.php?title=Run-time_infrastructure_(simulation)&oldid=641294065 (accessed 21 March 2015).
78. FMI (2015) Functional Mock-up Interface, https://www.fmi-standard.org/start (accessed 26 April 2016).
79. FMI-STANDARD (2015) Functional Mockup Interface Standard, Version 2.0, https://www.fmi-standard.org/downloads (accessed 26 April 2016).
80. COLaN (2015) CAPE-OPEN Laboratories Network, http://www.colan.org/ (accessed 26 April 2016).
81. Hooper, R., Hopkins, M., Pawlowski, R., Carnes, B., and Moffat, H.K. (2007) Final Report on LDRD Project: Coupling Strategies for Multi-Physics Applications. SANDIA REPORT SAND2007-7146, Sandia National Laboratories.
82. Knoll, D.A. and Keyes, D.E. (2004) Jacobian-free Newton-Krylov methods: a survey of approaches and applications. *J. Comput. Phys.*, **132** (2), 357–397.
83. Jaure, S., Duchaine, F., and Gicquel, L.Y.M. (2011) Comparisons of coupling strategies for massively parallel conjugate heat transfer with large eddy simulations, in *IV International Conference on Computational Methods for Coupled Problems in Science and Engineering, COUPLED PROBLEMS* (eds M. Papadrakakis, E. Onate, and B. Schrefler), International Centre for Numerical Methods in Engineering (CIMNE).
84. Moore, G.E. (1965) Cramming more components onto integrated circuits. *Electronics*, **38** (8), 114–117.
85. Sutter, H. (2005) The Free Lunch Is Over. HP, 2009. Application Lifecycle Management (ALM) | HP® Official Site, http://www8.hp.com/us/en/software-solutions/application-lifecycle-management.html (accessed 13 February 2015).
86. Feldman, M. (2015) Life Beyond Moore's Law.
87. Cavin, R.K., Lugli, P., and Zhirnov, V.V. (2012) Science and engineering beyond Moore's law. *Proc. IEEE*, **100**, 1720–1749. doi: 10.1109/JPROC.2012.2190155
88. Arden, W., Brillouët, M., Cogez, P., Graef, M., Huizing, B., Mahnkopf, R., Pelka, J., Pfeiffer, J.-U., Rouzaud, A., Tartagni, M., Van Hoof, C., and Wagner, J. (2011) *Towards a "More-than-Moore" Roadmap*, CATRENE.
89. Intel Corporation (2013) Intel® Xeon® Processor E7 Family: Supporting next generation RAS Servers, The Intel ® Xeon® Processor E7 Family.
90. Shainer, G., Lui, P., Hilgeman, M., Layton, J., Stevens, C., Stemple, W., Schultz, S., Ludden, G., Mora, J., and Kresse, G. (2013) Maximizing application performance in a multi-core, NUMA-aware compute cluster by multi-level tuning, in *Supercomputing* (eds J.M. Kunkel, T. Ludwig, and H.W. Meuer), Springer-Verlag, pp. 226–238.

91. Sinharoy, B., Van Norstrand, J.A., Eickemeyer, R.J., Le, H.Q., Leenstra, J., Nguyen, D.Q., Konigsburg, B., Ward, K., Brown, M.D., Moreira, J.E., Levitan, D., Tung, S., Hrusecky, D., Bishop, J.W., Gschwind, M., Boersma, M., Kroener, M., Kaltenbach, M., Karkhanis, T., and Fernsler, K.M. (2015) IBM POWER8 processor core microarchitecture. *IBM J. Res. Dev.*, **59**, 2:1–2:21. doi: 10.1147/JRD.2014.2376112
92. Chen, T., Raghavan, R., Dale, J.N., and Iwata, E. (2007) Cell broadband engine architecture and its first implementation—a performance view. *IBM J. Res. Dev.*, **51**, 559–572.
93. NVIDIA Corporation (2014) NVIDIA G eForce GTX 980.
94. Rupley, J., King, J., Quinnell, E., Galloway, F., Patton, K., Seidel, P., Dinh, J., Bui, H., and Bhowmik, A. (2013) The floating-point unit of the Jaguar x86 Core. IEEE 21st Symposium on Computer Arithmetic, pp. 7–16, doi: 10.1109/ARITH.2013.24.
95. Dongarra, J., Strohmaier, E., and Meuer, H.W. (1993) TOP500, http://www.top500.org (accessed 26 April 2016).
96. Jeffers, J. and Reinders, J. (2013) *Intel Xeon Phi Coprocessor High-Performance Programming*, Newnes.
97. Bergman, K., Borkar, S., Campbell, D., Carlson, W., Dally, W., Denneau, M., Franzon, P., Harrod, W., Hill, K., Hiller, J. *et al.* (2008) Exascale Computing Study: Technology Challenges in Achieving Exascale Systems.
98. Ramirez, A. (2010) Montblanc, http://www.montblanc-project.eu/ (accessed 12 February 2015).
99. DEEP Project – Homepage (2011) http://www.deep-project.eu/deep-project/EN/Home/home_node.html (accessed 12 February 2015).
100. Darlington, J., Ghanem, M., and To, H.W. (1993) Structured parallel programming. Programming Models for Massively Parallel Computers, 1993. Proceedings, IEEE, pp. 160–169).
101. Shan, H. and Singh, J.P. (1999) A comparison of MPI, SHMEM and cache-coherent shared address space programming models on the SGI Origin 2000. Proceedings of the 13th International Conference on Supercomputing, ACM, pp. 329–338.
102. Chen, L.T. and Bairagi, D. (2010) Developing Parallel Programs–A Discussion of Popular Models. Technical Report, Oracle Corporation.
103. Hanzich, M., Rodriguez, J.E., Gutierrez, N., and Cela, J.M. (2014) BSIT: a flexible platform for development of efficient simulation software. Presented at the IACM – ECCOMAS, Barcelona.
104. Angelikopoulos, P., Papadimitriou, C., and Koumoutsakos, P. (2012) Bayesian uncertainty quantification and propagation in molecular dynamics simulations: a high performance computing framework. *J. Chem. Phys.*, **137**, 144103. doi: 10.1063/1.4757266
105. Deelman, E., Blythe, J., Gil, Y., Kesselman, C., Mehta, G., Patil, S., Su, M.-H., Vahi, K., and Livny, M. (2004) Pegasus: mapping scientific workflows onto the grid, in *Grid Computing* (ed. M.D. Dikaiakos), Springer-Verlag, pp. 11–20.
106. Raicu, I., Zhao, Y., Dumitrescu, C., Foster, I., and Wilde, M. (2007) Falkon: a fast and light-weight tasK executiON framework, in *Proceedings of the 2007 ACM/IEEE Conference on Supercomputing, SC '07* (ed. B. Verastegui), ACM, New York, pp. 43:1–43:12, doi: 10.1145/1362622.1362680.
107. Wilde, M., Hategan, M., Wozniak, J.M., Clifford, B., Katz, D.S., and Foster, I. (2011) Swift: a language for distributed parallel scripting. *Parallel Comput.*, **37**, 633–652.
108. Dongarra, J., Hempel, R., Hey, T., and Walker, D. (1991) Message Passing Interface (MPI) Forum Home Page, http://www.mpi-forum.org/ (accessed 13 February 2015).
109. Snir, M., Otto, S., Huss-Lederman, S., Walker, D., and Dongarra, J. (1998) *MPI-The Complete Reference: The MPI Core*, vol. **1**, 2nd (Revised) edn, MIT Press, Cambridge, MA.
110. Gropp, W., Lusk, E., and Thakur, R. (1999) *Using MPI-2: Advanced Features of the Message-Passing Interface*, MIT Press, Cambridge, MA.

111. Gropp, W. (2006) MPICH | High-Performance Portable MPI, http://www.mpich.org/ (accessed 13 February 2015).
112. Gabriel, E., Fagg, G.E., Bosilca, G., Angskun, T., Dongarra, J.J., Squyres, J.M., Sahay, V., Kambadur, P., Barrett, B., Lumsdaine, A., Castain, R.H., Daniel, D.J., Graham, R.L., and Woodall, T.S. (2004) Open MPI: goals, concept, and design of a next generation MPI implementation. Proceedings, 11th European PVM/MPI Users' Group Meeting. Budapest, Hungary, pp. 97–104.
113. Gergana, S. (2012) Intel® MPI Library, https://software.intel.com/en-us/intel-mpi-library (accessed 13 February 2015).
114. IBM Corporation (2014) Platform MPI (No.SC27-5319-00), IBM.
115. Dagum, L. and Menon, R. (1998) OpenMP: an industry standard API for shared-memory programming. *IEEE Comput. Sci. Eng.*, **5**, 46–55.
116. Duran, A., Ayguadé, E., Badia, R.M., Labarta, J., Martinell, L., Martorell, X., and Planas, J. (2011) OmpSs: a proposal for programming heterogeneous multi-core architectures. *Parallel Process. Lett.*, **21**, 173–193, doi: 10.1142/S0129626411000151, Extrae user guide, 2014. Barcelona Supercomputing Center.
117. OpenACC Consortium, 2013 (2015) The OpenACCTM Application Programming Interface, http://www.openacc-standard.org/ (accessed 26 April 2016).
118. Kirk, D.B. and Hwu, W.W. (2010) *Programming Massively Parallel Processors: A Hands-on Approach*, 1st edn, Morgan Kaufmann Publishers Inc., San Francisco, CA.
119. Munshi, A., Gaster, B., Mattson, T.G., Fung, J., and Ginsburg, D. (2011) *OpenCL Programming Guide*, 1st edn, Addison-Wesley Professional.
120. Intel Corporation (2007) Intel® SSE4 Programming Reference (No. D91561-003).
121. Intel Corporation (2011) Intel® Advanced Vector Extensions Programming Reference (No. 319433–011).
122. Open Grid Forum (2015) www.ogf.org (accessed 26 April 2016).
123. Foster, I. and Kesselman, C. (1997) Globus: a metacomputing infrastructure toolkit. *Int. J. High Perform. Comput. Appl.*, **11**, 115–128.
124. Streit, A., Erwin, D., Lippert, T., Mallmann, D., Menday, R., Rambadt, M., Riedel, M., Romberg, M., Schuller, B., and Wieder, P. (2005) UNICORE—from project results to production grids. *Adv. Parallel Comput.*, **14**, 357–376.
125. Laure, E., Edlund, A., Pacini, F., Buncic, P., Barroso, M., Di Meglio, A., Prelz, F., Frohner, A., Mulmo, O., Krenek, A. *et al.* (2006) Programming the grid with gLite. *Comput. Methods Sci. Technol.*, **12** (1), 33–45.
126. Eerola, P., Ekelöf, T., Ellert, M., Grønager, M., Hansen, J.R., Haug, S., Kleist, J., Konstantinov, A., Kónya, B., Ould-Saada, F. *et al.* (2007) Roadmap for the ARC grid middleware, in *Applied Parallel Computing. State of the Art in Scientific Computing* (eds B. Kågström, E. Elmroth, J. Dongarra, and J. Waśniewski), Springer-Verlag, pp. 471–479.
127. Kondo, D., Javadi, B., Malecot, P., Cappello, F., and Anderson, D.P. (2009) Cost-benefit analysis of cloud computing versus desktop grids. IEEE International Symposium on Parallel & Distributed Processing, 2009. IPDPS 2009, IEEE, pp. 1–12.
128. Iosup, A. *et al* (2011) Performance analysis of cloud computing services for many-tasks scientific computing. *IEEE Trans. Parallel Distrib. Syst.*, **22** (6), 931–945.
129. Cappello, F. (2009) Fault tolerance in petascale/exascale systems: current knowledge, challenges and research opportunities. *Int. J. High Perform. Comput. Appl.*, **23**, 212–226. doi: 10.1177/1094342009106189
130. Myers, G.J., Sandler, C., and Badgett, T. (2011) *The Art of Software Testing*, John Wiley & Sons, Inc.
131. Qin, F., Lu, S., and Zhou, Y. (2005) SafeMem: exploiting ECC-memory for detecting memory leaks and memory corruption during production runs.

11th International Symposium on High-Performance Computer Architecture, 2005. HPCA-11, IEEE, pp. 291–302.
132. Goel, A.L. and Okumoto, K. (1979) Time-dependent error-detection rate model for software reliability and other performance measures. *IEEE Trans. Reliab.*, **28**, 206–211.
133. Fuchs, W.K., Huang, Y., and Wang, Y.-M. (1996) Progressive retry method and apparatus for software failure recovery in multi-process message-passing applications. Google Patents EP 0691610 A1.
134. Wang, Y.-M., Huang, Y., and Kintala, C. (1997) Progressive retry for software failure recovery in message-passing applications. *IEEE Trans. Comput.*, **46**, 1137–1141.
135. ADINA (2011) http://www.adina.com (accessed 26 April 2016).
136. MSC Nastran (2011) http://www.mscsoftware.com (accessed 26 April 2016).
137. ANSYS (2011) http://www.ansys.com (accessed 26 April 2016).
138. The Minerals, Metals, and Materials Society (TMS) (2015) Materials Processing & Manufacturing Division, http://www.tms.org (accessed 26 April 2016).
139. Integrated Computational Materials Engineering Expert group (ICMEg) (2015) http://www.icmeg.euproject.info/imceg-project.html (accessed 26 April 2016).
140. Illinois Rocstar (2015) Consortium for Open Multiphysics, http://www.illinoisrocstar.com/engineering-services/hpc/openmultiphysics (accessed 26 April 2016).
141. EMMC (2015) European Materials Modelling Council (EMMC), http://emmc.info/ (accessed 26 April 2016).
142. Goldbeck Consulting Limited (2015) EU Multiscale Materials Modelling Cluster, http://www.goldbeck-consulting.com/home.html/27120/25020.html (accessed 26 April 2016).
143. DEEPEN (2015) DEEPEN: From Atom-to-Device Explicit simulation Environment for Photonics and Electronics Nanostructures, www.nmp-deepen.eu (accessed 26 April 2016).
144. MMP (2015) Multiscale Modelling Platform: Smart Design of Nano-Enabled Products in Green Technologies, www.mmp-project.eu (accessed 26 April 2016).
145. MoDeNa (2015) MOdelling of Morphology DEvelopment of Micro- and NAnostructures, www.modenaproject.eu (accessed 26 April 2016).
146. NanoSim (2015) A Multi-Scale Simulation-Based Design Platform for Cost-Effective CO2 Capture Processes Using Nano-Structured Materials, http://www.sintef.no/Projectweb/NanoSim/ (accessed 26 April 2016).
147. SimPhoNy (2015) Simulation Framework for Multi-Scale Phenomena in Micro- and Nanosystems, http://www.simphony-project.eu/ (accessed 26 April 2016).
148. Babur, Ö., Smilauer, V., Verhoeff, T., and van den Brand, M. (2015) A survey of open source multiphysics frameworks in engineering. *Procedia Comput. Sci.*, **51**, 1088–1097, doi: 10.1016/j.procs.2015.05.273.
149. Babur, Ö., Verhoeff, T., and van den Brand, M. (2015) Multiphysics and Multiscale Software Frameworks: An Annotated Bibliography. Computer Science Reports 15-01, Technische Universiteit Eindhoven, 37 pp., https://pure.tue.nl/ws/files/14461807/CSR_15_01.pdf (accessed 11 July 2016).

9
Future Directions

Ulrich Prahl and Georg J. Schmitz

9.1
Lessons Learned

When compiling the *Handbook of Software Solutions for ICME*, a number of topics requiring needs for further development and actions showed up. Future workflows simulating materials and components will increasingly aim at describing and predicting the components performance, that is, their functionality varying over time for a given load history (Figure 9.1).

Desired component functionalities in general will require a suitable combination of a number of properties of the material. These effective properties are determined by the microstructure and by the properties of the different phases within the microstructure and can be calculated by homogenization tools (h-tools). While the properties of the individual phases for a given status in time may be derived from electronic/atomistic and mesoscopic (e/a/m) models, their arrangement within the microstructure evolves in time and is described by continuum (c) models at the mesoscale. This evolution is influenced by the boundary conditions imposed by the process conditions on the process scale. This large number of correlations can be described by coupling and linking a variety of different models describing phenomena across length scales and timescales.

Having a look at the mere statistics of the number of software tools in the different fields, it is obvious that the number of codes per field is rather different. First of all, the continuum mechanical FE-based codes on macroscale are the most established group of tools with many different variants and details, while the thermodynamic CALPHAD-based tools are smaller in number. This reflects the fact that the FE approach has been invented and developed over many decades since the 1960s, while the CALPHAD approach started about 20 years later in the 1980s. In addition, the engineering community being responsible for the development and being interested in the applications of FE codes aims at merging and combining codes. A consolidation of the market toward a few big players can currently be observed.

Handbook of Software Solutions for ICME, First Edition. Edited by Georg J. Schmitz and Ulrich Prahl.

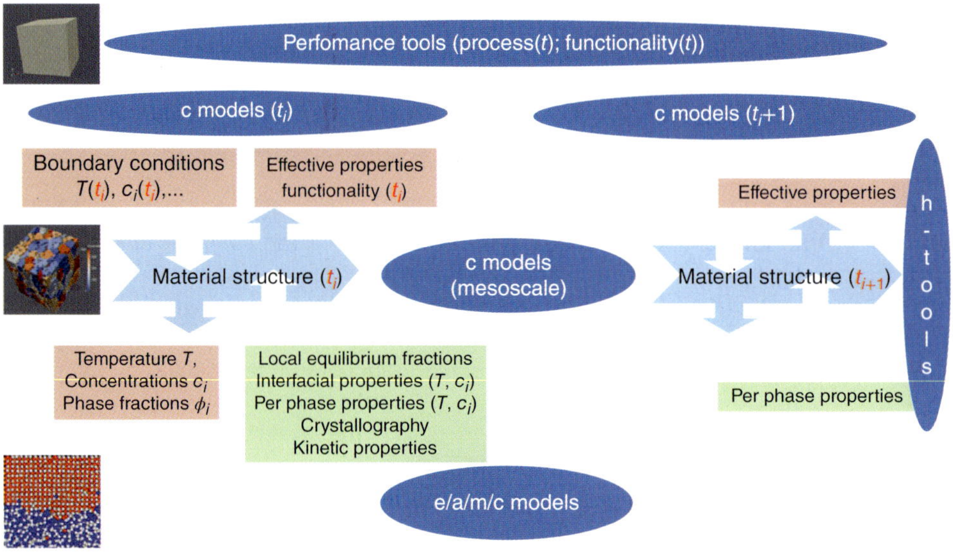

Figure 9.1 Future simulation workflow for materials aims at predicting component performance based on material scientific knowledge on the correlation between "properties, microstructure, and processing" (see text).

In contrast, the number for electronic, atomistic, and/or mesoscopic codes – at least on an academic software level – is very high, indicating that the development seems still in an exploration stage. Scientists from physics and chemistry developing this kind of codes do not yet show a tendency to consolidate into a small number of "surviving codes."

On the macroscopic scale of the processes (Chapter 2), a large number of codes could be identified for the area of primary shaping. One reason for this large number may be that different material classes were considered here (single crystals, polymers, ceramics, metals, and alloys), many of which already represent final products and do not need any further processing. Another reason may be that primary shaping simulations in general can start from well-defined initial conditions, whereas subsequent simulations downstream the processing chain in general rely on assumptions for the initial conditions.

Forming simulations represent a second important class of simulation tools. The number of tools listed for – mechanical – joining, machining, coating, fatigue, and corrosion is quite small and is frequently based on solidification and/or forming concepts using general-purpose tools. Besides the complex physics and chemistry describing these processes, a more critical argument might be the relative market share of the industrial communities in these areas, at least in comparison to primary shaping and forming markets.

A large number of codes and approaches operating on the scale of the process and a huge number of codes operating in the e/a/m area are contrasted by a limited number of codes describing microstructures and thermodynamics. Furthermore, tools dedicated to bridging the scales are not very frequent, reflecting gaps and missing bridges between the different model worlds.

Looking at the variety of data formats, a huge number of different formats were identified for both input and output. Some, but by far not all, are given in the following listing: ASCII, AVS UCD, BYU, CML Molecule, COMSOL Multiphysics, COSMOS, CSV, DEM, EnSight, Enzo, Exodus II Facet, Flash, Fluent Case, FEM Neutral Format, Femap, Gaussian Cube, GiD, Gmsh, HDF, JPEG, LS-DYNA, MFiX, Metafile Series, NetCDF, Nrrd, OBJ, OpenFOAM, OpenDX, PDB, PLOT3D, proSTAR (STAR-CD), PLY Reader, PNG, PVD, RTXMLPolyData, Sim Exodus, Sim Spy Plot, SESAME, SLAC, STL, Tecplot, TIFF, UNV Universal File Format, VTK, VPIC, VRML, WindBlade, XDMF, XML, XYZ, and others.

The list of these input/output (I/O) formats – let their number be N – is most probably still far from being complete. However it becomes obvious that coupling and linking of different simulation tools in an ICME approach will strongly be hindered by the necessity of creating and maintaining $N \times (N+1)$ conversion tools if each tool is to communicate with each other. Besides the different file formats, even for the same format the different nomenclatures for the entities being described are compatible or even identical only in very few cases. This clearly indicates the need for a single common metadata description offering to limit the number of necessary conversion tools to N again. Each format allows for the conversion of I/O into the single and unique metadata framework.

Thus, the main learning of this book might be summarized as follows:

> There is a need for establishing and fostering a metadata structure promoting a plug-and-play-type interoperability between codes based on widely agreed communication standards and on workflows on common ICME platforms.

Some considerations about this metadata description and the implications are discussed in the following sections.

9.2 Interoperability and Communication Standards

Further efficient exploitation of the benefits of ICME requires communication standards to be implemented enabling a plug-and-play-type operation when combining ICME simulation tools within a simulation workflow (Figure 9.2).

General considerations then relate to what shall be considered as a standard. In modern technological developments, de facto standards arising from best practices usually evolve much faster as compared to legal standards defined by any authority. It is of utmost importance to identify already existing data standards and best practices on which the further development can be build. For example, within the continuum mechanical FE-based approaches, a rather limited number of multipurpose programs such as Abaqus, ANSYS, and Marc reveal very similar data formats and data structures which may be used or even already be considered as

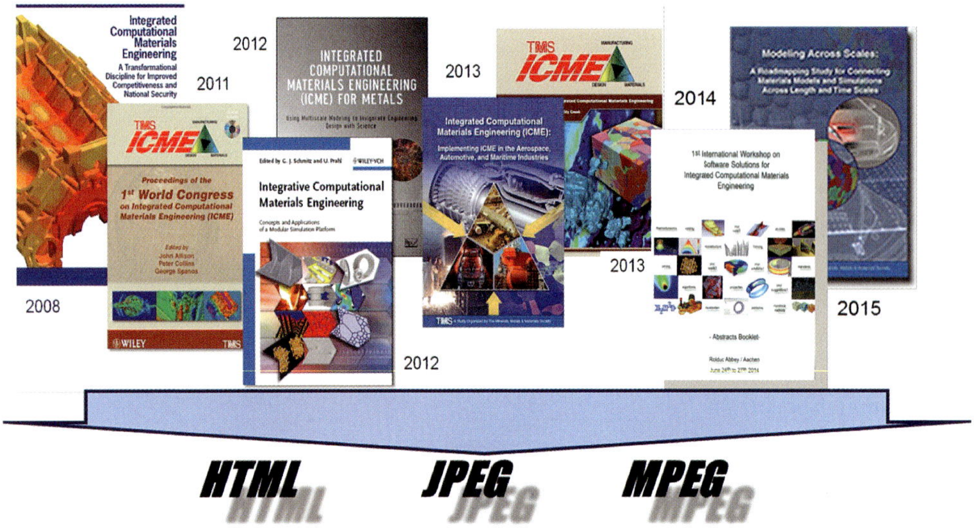

Figure 9.2 The further development of ICME needs a data standard allowing the exchange of simulation information including all types of data and metadata in a simple way similar to exchanging pictures in the jpeg/mpeg or HTML formats.

a basis for a de facto standard for metadata on properties and continuum mechanical results. Within the crystallographers' community, the CIF [1] has evolved into a de facto standard describing crystallographic and molecular structures in many applications.

For data in the area of thermodynamics, a fundamental challenge is to avoid problems related to the use of different definitions of thermodynamic quantities. For this purpose, use of a common definition is encouraged, and the definitions provided by IUPAC in *Quantities, Units and Symbols in Physical Chemistry* (Green Book) should be considered for this purpose. A well-known issue is that different phase names are used by different communities. There is a need for a unique identification of "phases" that can be used by different software tools internally for translation into the common names. In addition to a common terminology, there is a need for a protocol facilitating the exchange of thermodynamic information between different software codes.

In a next step the target of standardization has to be specified. The description of engineering-type initial state data, boundary conditions, properties, and other calculation results has to be surely in the focus of such standardization activities. A further enrichment of these data by the origin of the data, their history, and their trustworthiness is often necessary and beneficial providing respective metainformation. Questions arise such as follows:

- What are relevant metadata, and how can they be structured or classified?
- Which metadata are relevant with respect to interoperability?
- What is a material, and which metadata are needed for the generic description of a material state?

- What is a property, and which metadata are needed for the generic description of a property?

The third general consideration is about the sustainability of the further development of standards and metadata schemes once they are developed. Comparable to the NAFEMS association [2] being responsible for orchestrating the development of information exchange between FE tools, a body is needed to orchestrate the further development of ICME-type communication protocols. The evolution of NAFEMS over the decades provides a template for a meaningful sequence to proceed with the standardization process also in ICME. The ICMEg project [3] has initiated the discussion on best practices for information exchange between the heterogeneous variety of commercial and academic simulation tools.

The only common denominator of all tools is that they all act or draw on a material. Any standard proposed will thus only be sustainable if it follows the generic nature of materials and processes. The standardized description of a material covering data at multiple scales and providing data for – and integrating data of – different models and tools on a physically sound basis thus is the most important step toward interoperability. Interoperability will be achieved between models, across scales, and along the evolution of a material and its properties during a number of processing steps once a metadata keyword list for the description of a material is available.

To find a widespread acceptance, any analysis of ICME results and their visualization must be possible on the basis of open and freely available postprocessing tools. Here, the graphical visualization community provides a high level of knowledge and best practices on how to operate large datasets in short time and how to identify and to get those data that are relevant for a specific exploitation task.

9.3 Hierarchical Description of a Material

A huge variety of models and tools has been collected in the present book. All these tools operate at different scales, with different objectives, aiming at different properties, different materials, and so on. Their essential common denominator is the material. To further discuss the generic structure of a best practice material description, it is essential to formulate a common answer to the question: What is a material?

9.3.1 What Is a Material?

The most general answer probably is shown in Figure 9.3.

Knowing the positions respectively the arrangement of all atoms of all the elements accordingly fully describes the material for a fixed moment in time. This may practically be possible for a small number of atoms, for example, in molecular

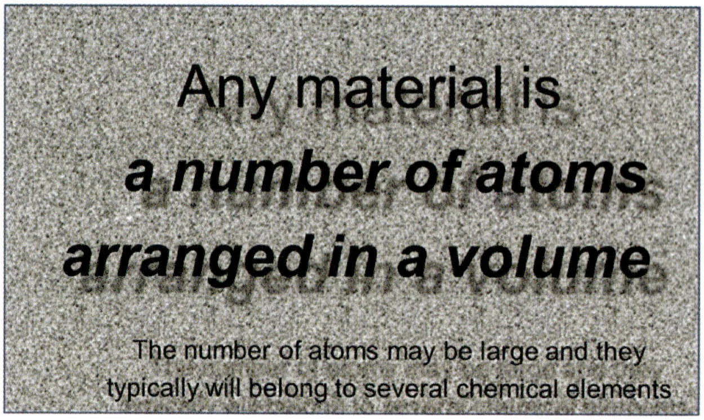

Figure 9.3 Attempting a generic definition of a material.

Figure 9.4 Hierarchical description of a material starting from a macroscopic RVE being the smallest discretization element in a process-scale simulation (a) via ensembles/phases (b), grains/features (c), and continuous fields around or within the grains (d) down to atomistic-scale arrangements at interfaces (e) (see text).

dynamics simulations. For most technical materials with some 10^{23} atoms, however, such a description is obsolete. Nature fortunately helps by "self-assembly" of atoms, for example, into crystals, grains, phases, and other objects or features (including defects). The material can thus be described with a reasonable effort on the basis of the positions of such features.

Materials thus reveal a hierarchical structure comprising atoms, grains, phases, and eventually the multicomponent, multiphase, and polycrystalline microstructure itself (Figure 9.4). Any approach to define a technical material thus should be based on a hierarchical description. Depending on the scale being described, the number of atoms can be

- Small (individual atoms or molecules)
- Medium (particles, clusters of atoms, precipitates)
- Large (grains, phases, multiphase, polycrystalline materials)

Materials then are ensembles of objects or features in a volume where the objects or features consist of atoms (electrons, protons, and neutrons). In general, atoms may be atoms of different chemical elements, and objects thus may be atoms, molecules, particles, precipitates, grains, grains of different phases, and so on. Objects/features (Figure 9.4c) each have an identity and different properties or attributes, for example, position, orientation, volume, shape, composition, and defect structure. Within this generic interpretation, interfaces are treated as objects and can also be assigned with attributes for interface properties.

Ensembles (Figure 9.4b) each have different properties and attributes that are typical for all objects in the ensemble, for example, phase type, lattice parameters, crystal symmetry, and chemical composition of the phase. The description of the crystal structure of such an ensemble might follow the CIF standards [1].

The next higher level of description is the volume containing a defined amount of material. If the volume is large enough to host a relevant number of atoms and to be representative for the material, it might be called a representative volume element (RVE) (Figure 9.4a). The RVE is assumed to be a homogeneous piece of material having effective, that is, homogenized or averaged properties. Different RVEs at different positions in a component reveal macroscopic local properties that by now typically are taken from tables (e.g., for similar materials). As the microstructure determines the properties, statistical information about microstructure features within the RVE thus is highly desirable. Examples for such statistical information are average grain sizes or grain size distributions to describe the material strength following the Hall–Petch relationship. Further examples are texture to explain anisotropy, secondary arm spacing to give permeability or strength, phase fractions to quantify precipitation hardening, phase composition to control solution hardening, or dislocation densities to yield work hardening.

Objects within a material in general interact via fields that are described by scalar, vector, or tensor fields in a continuum description (Figure 9.4d). Such fields accordingly are continuously defined in real space, each field thus being a function of (x,y,z) and a function of time. For atoms and molecules, these fields might be magnetic fields, dipole fields, and so on, while for grains and phases, concentration fields, stress fields, and temperature fields would be typical examples. The distribution of a large number of discrete objects in a volume can also be described by a field such as the concentration field of atoms of a specific element. Continuous fields in general are stored in a discretized numerical representation called a grid. The simplest grid is a regularly structured grid of simple cubes called voxel (volume element cell).

In general, data have to be defined within the context and the scale they are used. An instructive example is the keyword "concentration" of alloy elements. For an RVE of a material, this concentration corresponds to the nominal or averaged composition of the overall material. In general the alloy elements however are not evenly distributed in the RVE but form phases and grains/features. Multiple grains or features of the same phase form an ensemble. The concentration of the ensemble thus corresponds to the nominal average composition of the phase. The concentration of a grain corresponds to the average composition within the grain. The concentration within or between the grains may further continuously vary in space being discretized, for example, into the cells. The concentration of a cell thus depicts the highest level of detail. Such a cell itself can, however, be again considered as an entire RVE at a smaller scale, where, for example, discrete atoms of different chemical elements and their motion in space are described (Figure 9.4e).

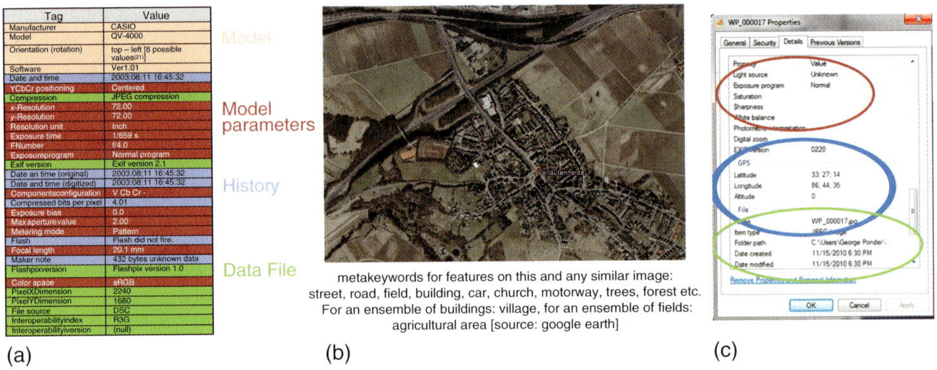

Figure 9.5 Metadata for a .jpeg image. In addition to the metadata describing its content (a small village close to Aachen, b) a number of metadata ("data about this picture") are nowadays typically stored. They may comprise information about the model being used to capture the data ("the camera model"), the model parameters (exposure time, resolution; a), data about the history (where and when the picture has been recorded, and c) eventually information about its archiving location.

Such a hierarchical structure allows for both homogenization and averaging and the seamless integration of more detailed information from smaller scales.

9.4 Metadata

Material modeling must be capable of describing properties of complex materials and components as a function of time starting from their processing and ending with recycling at the end of their application life cycle. To justify any effort being spent in ICME, the user needs information about costs, capacities, and efforts necessary in relation to the quality of simulation results, the possible benefit, and many more.

Furthermore, the user must be capable of archiving, documenting, and retrieving the results. This also includes information about the history of all the steps taken and models being used to obtain the results. Additional information in the form of metadata will thus be needed to support the design of future simulation tasks similar to the metainformation available for jpeg files (Figure 9.5).

Thus, in any future ICME context, metadata (i.e., "data about data") will play a crucial role. Examples of metadata might be (list not complete):

- Metadata for data history/data origin
- Metadata for models
- Metadata for numerical representations
- Metadata for solvers
- Metadata for workflows
- Metadata for processes
- Metadata for materials

- Metadata for properties
- Metadata for costs

While all of such metadata seem relevant, important, and beneficial, setting priorities for their definition seems necessary and will be decisive to set up the ICME approach effectively and efficiently. Although recognizing the importance of metadata, for example, on data origin and on documentation, priority has to be given to metadescriptions of materials, their properties, and models for their evolution in view of interoperability between the different simulation tools. A further challenge is the need for thermodynamic data and, in particular, phase-based property data. Often the relevant data needed for practical applications of an ICME approach is simply incomplete.

The approaches to establish a structure to the keyword list of metadata describing a material are:

(i) To start from descriptions of simple materials (e.g., some atoms in a volume) and increase complexity up to the description of polycrystalline multiphase materials comprising defects.

(ii) To have an intrinsic hierarchy within the metadata keyword list. The hierarchy in this context is essentially given by using the same keyword at different levels of detail (RVE, Ensemble, Object/Feature, and Cell).

9.5 Metadata Schemata

Following the hierarchical nature of a material, a hierarchical description scheme is needed for a thorough description of material. The HDF5-type hierarchical data format approach [4], which has recently been customized to the description of several aspects of material microstructures in the DREAM.3D software [5], seems most promising in this context. This scheme and also other schemes such as MongoDB [6] seem worthy to be further developed on the basis of a defined keyword list to categorize and structure the collected keywords including all their attributes.

Each of the metadata keywords specified for the description of a material will require and profit from a number of further attributes. Examples of beneficial attributes are on data type (integer, real, character), dimensionality (scalar, vector, tensor), model domain by which it can be tackled (e.g., electronic, atomistic, mesoscopic, mesoscale continuum, continuum), software tools allowing its determination, and application area (structural materials, functional materials, high-temperature materials, etc.). Helpful are further attributes for the minimum and maximum of data validity range (e.g., the interval between 0 and 1 for "fractions") and error estimates. Especially error/uncertainty estimates on all types of data are a major issue in view of uncertainty propagation along the entire ICME workflow.

The next basic steps to be taken toward the specification of a communication standard for ICME are the compilation of structured metadata keyword lists for:

(i) The description of materials with respect to their microstructure (geometry/topology)
(ii) The description of the energetic state of a material (thermodynamics, potentials)
(iii) The description of the kinetics of material transformations
(iv) The description of properties of materials
(v) The numerical representation of continuum fields

A compilation of a metadata keyword list for the description of polycrystalline, multiphase, and multicomponent materials with respect to their microstructure has been published recently [7]. This metadata keyword list has to be condensed, consolidated, and structured for the description of a material by proposing names and unique identifiers and by adding short but sound descriptions. Especially the specification of unique identifiers and the sound description of their meaning and their format will provide the basis for the specification of an ontology of the field. This ontology will allow for the creation of lookup tables for respective translations at a later stage. In particular, the metadata keywords for the description of a microstructure will require proposals for a naming scheme and the definition of respective wordings, for example, with respect to the definition and understanding of the term microstructure.

Beyond conventional static descriptions of a material, detailed descriptions of three-dimensional (3D) microstructures and their evolution will become possible. As the microstructure is the carrier of all material properties, such a sound description of a microstructure (in the sense of describing a state variable of the system) will allow for future predictions of material properties and their evolution or variation, for example, with composition and temperature. This will tremendously advance the field of materials science and engineering for technical materials.

Further development depends on unique and useful metadata keyword specification also for other domains. This standardized nomenclature has to be specified by domain experts. In several areas best practices are already established, for example, crystallographers have developed the information file CIF. Keywords on mechanical properties as used in FE simulations are converging to a few best practice solutions already revealing a very similar nomenclature such as Abaqus [8],Marc [9]/ANSYS [10], and Z-set [11]. Future format standards for analysis, visualization, and workflows such as HDF5 [4], XDMF [12], JSON [13], and VTK [14] are discussed and developed by computer scientists, and all can be considered as best practices in their fields.

From a practical viewpoint, the resulting structured metadata keyword lists have to be challenged and revised on the basis of scenarios with increasing complexity. What are the suitable and viable scenarios for data exchange? Scenarios may range from simple scenarios for internal validation of the basic keyword concept up to scenarios for complex multiscale process chains in industrial settings. However – by intention – it should not be foreseen to actually perform time-consuming simulation runs as test-case scenarios.

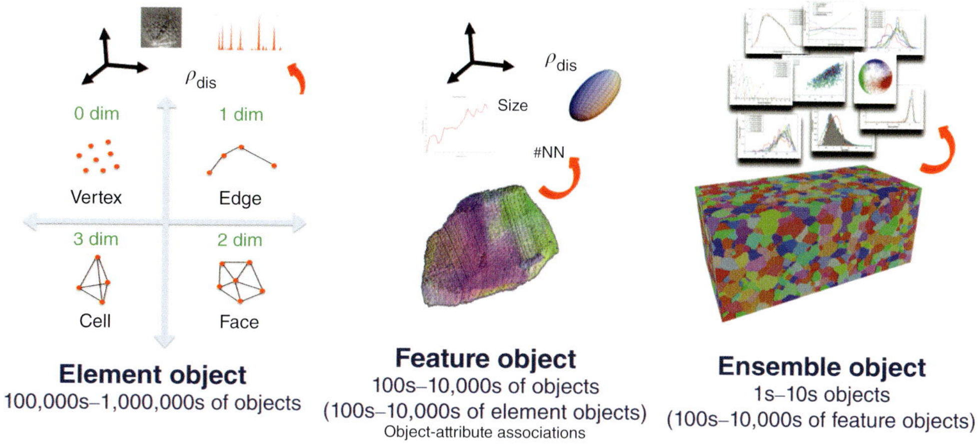

Figure 9.6 The HDF5 data structure as customized for materials engineers in the .dream3D format. This format offers a hierarchical description for the storage and easy access to data and metadata on each scale ranging from spatially fully resolved data (elements) information about discrete features (feature) up to statistical information (ensembles). (Reproduced with the permission from BlueQuartz.)

Sandbox scenarios only verifying the integrity of the information flow should be used instead. These sandbox scenarios have to especially account for requirements of industrial users. Additionally, within sandbox scenarios, issues such as economic or ecological impact can easily be discussed as well.

A hierarchical approach will also be helpful to design the visualization approach (Figure 9.6). It turns out that 3D spatially resolved data could be described in vtk only. Description of per-phase data, RVE statistics, and RVE effective properties needs an additional structure as, for example, provided by HDF5 with vtk being something such as a subset. The description of workflows and the temporal evolution of materials will need combinations of HDF5 (and other data and file types) and probably is best achieved on the basis of xdmf with HDF5 being a subset then.

9.6 Platforms: Orchestration of Simulation Tools

Interoperability between the heterogeneous variety of currently available and future simulation tools will be achieved on the basis of aforementioned metadata concepts. To further exploit interoperability, there will be a need to steer and to orchestrate the workflow and the information flow. This holds not only for information flow between the different models and simulation tools in the simulation chain but also for data exchange with databases.

Some conceptual ICME approaches are under development utilizing platform concepts. Examples of ongoing activities are nanoHUB [15], the AMP2 package [16], or AIIDA [17]. nanoHUBis an example for a platform providing access to a huge number of software tools which can be run remotely. It provides also access functionalities from the Web, for example, via user login and user administration. A further example is VFab, a customized graphical user interface to control and steer an ICME platform for electronics [18].

When operating a platform, different types of information exchange can be distinguished: linking and coupling. In a linking type of information exchange, results from tool 1 are conveyed to tool 2 at the end of processing by tool 1, for example, based on a file-type information exchange. In a coupling mode of operation, information exchange between tools 1 and 2 proceeds at each time step. Such coupling is extremely important in the case of nonlinear systems of equations. Orchestration of the different tools in this case becomes even more important, for example, with respect to timing the execution of the codes for each time step. Respective platforms allowing to orchestrate different tools on a per-time-step basis are currently under development in Europe in a number of ongoing projects [3, 19–23]. These activities are based on concepts such as API wrappers and metadata schemata.

Based on a technical interoperability and an orchestration of different tools by platform concepts, the commercial exploitation of such platform concepts has to be envisaged as well. Questions arise such as the following: How can commercial packages be included? How can accounting schemes be generated and implemented? How can confidentiality be assured for commercial use?

A portability of platform concepts, that is, the option to install such platforms anywhere, will be a crucial point to be met in this context. Such portability will allow industrial users to run the platform at their own premises and including their own proprietary tools while at the same time avoiding unsecured communication of confidential data via the Internet.

9.7 Databases: Storage and Retrieval of Information

Materials science and engineering increasingly draws on simulation. New material developments and scientific findings are created by intensively applying simulations on the basis of the ICME approach. The reliability of results from data-driven materials science and thus the publication of simulation data along with the scientific papers are most important to maintain transparency and trustworthiness in this concept. Recent schemata such as open data repositories offer options for further use of published numerical data by external scientists for comparison with own data or for further analysis. Thus, numerical datasets themselves have both a scientific and a commercial value. Their sustainable, readable, and retrievable storage and handling become an important issue.

The materials simulation community is learning and increasingly committing that publications without data are fraud. The related necessity of handling huge data amounts – which will be necessary to follow the accepted rules of good scientific practice – still requires some work to be done. Ensuring data standardization, the availability of metadata, and interoperability are only some but important and crucial tasks to be tackled.

The Big Data Initiative represents a first step into this direction. Data available from public sources are collected, stored, and analyzed. The effective search for desired and reliable data and information turns out to be a bottleneck as well as the quality of data. Based on a metadata keyword structure, the storage and the retrieval of information on material properties from databases will become largely facilitated and much more effective. A metadata curation scheme may help to overcome obstacles of data inconsistencies. Discussions on how to handle, include, and also exploit commercial datasets are ongoing.

The question will be how to search and how to make sure to find the required and meaningful data in a reasonable time. What are actually the data of interest? How reliable and how complete is the available information?

When searching for a material, already the specification of the composition bears imponderability and uncertainties. Even tiny impurities of some elements – frequently even not being specified in details of experimental data – may play a significant role. In electronic materials, the effect of tiny amounts of dopants making a qualitative difference between a semiconductor and a conductor is well known. The exploitation and the assessment of effects of such tiny amounts of impurities on the formation sequence, the morphology of phases, and accordingly the properties of structural materials, however, are still in its infancy.

Big data-type storage comprising all types of data and literature available about a material and exploiting it on the basis of statistical evaluation thus can only be one of the approaches. Complementary storage of 3D digital microstructures and of per-phase properties seems to be a most interesting approach. Simulating a microstructure or compiling a respective dataset from experimental information is quite an effort. Available microstructure datasets can however be used to derive the properties of the materials, and especially they can also serve as initial conditions for the description of the further evolution of the material. The per-phase properties – that is, the properties of single crystals without defects in contrast to the properties of polycrystalline, multicomponent, and multiphase materials – are physical constants depending on the composition and temperature only and thus do not vary over time. Once calculated, such data can be reused again and again if they are properly stored in databases and easily retrievable. One of the major future tasks therefore is the definition and compilation of suitable data structures, allowing simple and fast search and retrieval operations.

Currently available commercial materials databases comprise information about material composition and properties and in some cases also information about the processing history. Entire processing sequences have to be fully documented and to be made retrievable to recover the history of the material. In contrast, the microstructure integrates all processing history and also provides the basis for the determination of the material properties. Material properties/data of materials can be calculated on the basis of the microstructure and the known per-phase properties, for example, by mathematical homogenization, by virtual testing, or even by simple averaging.

It thus seems meaningful to establish and maintain microstructure databases. Microstructure characteristics can be used as search parameters for data retrieval. Such characteristics could be the number of phases present in the microstructure, number and names of chemical elements and their relative amounts in microstructure, statistical data such as average grain sizes, and many more based on a future metadata keyword list.

9.8 Sustainability

The materials simulation community is rapidly evolving and generates a large number of ideas, models, codes, and tools each year. Quite a number of new members join this community annually. Sustainability of the newly developed structures for standards and metadata has to be adequately secured. For this purpose, the ICMEg e.V. was founded in 2014 in Aachen, Germany. ICMEg aims at the creation and especially the further maintenance of an open, global standard for information exchange between the heterogeneous variety of commercial and academic simulation tools.

Based on a first structured metadata keyword list for the description of a material including short explanations, procedures to challenge the standard formulations have to be iteratively developed. The challenges and their outcomes have to be documented in an evolving "White Book" along with a documentation of the challenging procedures. This White Book will be available for public discussions on the websites of the European Materials Modelling Council (EMMC) [24] and ICMEg [3].

The White Book "Communication standards for successful ICME" will document all rationale and arguments for the specification of communication standards for ICME as well as the standard formulations themselves. The first version of this book will be compiled by the ICMEg consortium. It then will be further developed interactively and iteratively in discussions with all interested stakeholders. The interactive compilation of the White Book and its open availability to a general public, however, shall ensure further harvesting and collection of ICME-related modeling software by establishing a registry of codes. The collection within this book provides a valuable basis for such a registry. It will offer incentives for software providers to register their codes and assign a unique code ID and to allow for easy retrieval of the codes and their inclusion into metadata schemata.

Printed versions of any book are rapidly outdated. Accordingly, there is an urgent need for an electronic version that will be made available at a global one-stop shop maintained, for example, by EMMC marketplace and by mirror sites. ICMEg then might be the nucleus for a future noncommercial organization devoted to maintain the standards and their further development within an open but orchestrated public discussion within the ICME community.

9.9 Outlook

Future materials and component engineering will require an orchestrated interaction between the heterogeneous varieties of software tools being depicted in this book. The properties of a material and thus eventually the performance of a component are determined by the material's microstructure, by the properties of the individual phases and interfaces in the microstructure, and eventually by the evolution of the microstructure along the entire processing and load history. The microstructure – in a very general formulation – is an arrangement of a large number of atoms of various chemical elements in space. Controlling and describing the properties accordingly requires the knowledge of the spatial distribution of the chemical elements. Process simulation software tools in future have to account for this need and should provide information on the distribution of the chemical elements beyond the statistical information currently being provided on, for example, grain size distributions. Such information – once available on the component scale and along the entire process chain – will allow tackling phenomena downstream the production cycle such as welding or especially corrosion processes, both being strongly dependent on local chemical composition.

Modern materials engineering is closely related to the visual inspection and analysis of microstructures being observed by different types of microscopes. The documentation of respective observations has always been based on photographs which – only few decades ago – were classical black-and-white hard copies being processed from negative films in a complex chemical procedure. In the mid-1980s, first devices became available to digitize such analog pictures and allowed for digital image analysis and digital evaluation of microstructure features. First digital cameras entered the mass market in the 1990s and allowed for a much easier recording of image data related to microstructures. Only few data formats for such image information have widely spread and became a de facto standard. Microstructure characterization and documentation accordingly have made a significant similar progress. Nowadays the entire stacks of images resulting from, for example, serial sectioning in optical microscopy or from EBSD measurements are compiled into 3D datasets providing full 3D information about the microstructure. Information originating from different characterization technologies is increasingly integrated into comprehensive file structures. Thus, the description methodology for microstructures becomes not only colorful but also hierarchical, enriched, and multidimensional.

At the same pace, computers and modeling tools in the area of material modeling have evolved. Many of them are detailed in this book, and many have already found applications in industrial settings. A numerical description of 3D microstructures has been shown to be possible even for complex alloy systems in some cases, and first steps have been taken to integrate respective information along different process steps and to link different phenomena affecting microstructure evolution.

Hierarchical data structures such as the HDF5 scheme provide a structured framework for an integration of both experimental and simulation-based microstructure and property information originating from various sources. The customization of HDF5 for purposes of the materials engineering community has been kick-started by the open-source software tool DREAM.3D [5].

Missing information may be generated by electronic, atomistic, or mesoscopic models and may enter into the same data structure. Examples could be the composition- and temperature-dependent anisotropic properties of individual phases and the properties of interfaces, triple, and quadruple junctions. The term "property" in this context covers a broad range of properties such as thermomechanic, thermodynamic, electromagnetic, or optical properties, to name a few. While these properties essentially refer to the application of the final product, kinetic properties such as diffusion data or electrical or thermal conductivities are the most interesting properties during the production of materials and components.

The future of the black-and-white 2D micrograph in early metallography will be a colorful 3D (or even 4D) digital dataset. Colorful in this context means that its data are recorded not only in visible light (LOM) but also in the light of electrons (EDX/EBSD), X-rays (tomography), neutrons (neutron diffraction), and other types of characterization. Furthermore, all types of data generated by any of the software tools listed in this book or their combinations may be added to such datasets.

It is like adding the ultrasound information of a bat to the normal light information in a classical jpeg picture. Such a dataset may comprise all information that is necessary and valuable to describe a material and to derive its properties. Eventually, the file ending ".dream3d" may turn from dream to reality and be renamed to ".DM3D – Digital Microstructures in 3D." Such files may then serve as a hub to exchange information between all types of models, across all scales, and along all processes.

The editors do hope that this book provides a helpful basis to approach this vision of complex 3D multidata simulation results being exchanged as simply as jpeg pictures.

References

1. Hall, S.R., Allen, F.H., and Brown, I.D. (1991) The crystallographic information file (CIF): a new standard archive file for crystallography. *Acta Crystallogr., Sect. A: Found.*, **A47**, 655–685, http://www.iucr.org/__data/iucr/cif/standard/cifstd1.html (accessed 10 May 2016).
2. NAFEMS http://www.nafems.org/ (accessed 25 April 2016).
3. Schmitz, G.J. and Prahl, U. (2014) ICMEg the Integrated Computational materials engineering expert group - a new European coordination action. *Integr. Mater. Manuf. Innovation*, **3**, 2, and www.icmeg.eu (accessed August 2015).
4. The HDF Group www.hdfgroup.org/HDF5/ (accessed August 2015).
5. Groeber, M.A. and Jackson, M.A. (2014) DREAM.3D: a digital representation environment for the analysis of microstructure in 3D. *Integr. Mater. Manuf. Innovation*, **3**,

5, http://www.immijournal.com/content/3/1/5 and dream3d.bluequartz.net/ (accessed August 2015).

6. Membrey, P., Plugge, E., and Hawkins, D. (2010) *The Definitive Guide to MongoDB: The noSQL Database for Cloud and Desktop Computing*, Apress, and https://www.mongodb.org/ (accessed October 2015).
7. Schmitz, G.J., Böttger, B., Apel, M., Eiken, J., Laschet, G., Altenfeld, R., Berger, R., Boussinot, G., and Viardin, A. (2016) Towards a metadata scheme for the description of materials – the description of microstructures, *Sci. Technol. Adv. Mater.*, (2016) DOI: 10.1080/14686996.2016.1194166.
8. Hibbitt, Karlsson, and Sorensen (2001) *ABAQUS/Standard User's Manual*, vol. **1**, Hibbitt, Karlsson & Sorensen, and www.3ds.com/products-services/simulia/products/abaqus/ (accessed August 2015).
9. MSC Marc (2010) Volume A: Theory and User Information. MSC Software Corporation, and http://www.mscsoftware.com/ (accessed 25 April 2016).
10. Fluent, Ansys (2009) 12.0 User's Guide. Ansys Inc., and http://www.ansys.com/ (accessed 25 April 2016).
11. Z-set http://www.zset-software.com/ (accessed 25 April 2016).
12. Clarke, J.A. and Mark, E.R. (2007) Enhancements to the eXtensible Data Model and format (XDMF). DoD High Performance Computing Modernization Program Users Group Conference, IEEE, 2007, and http://xdmf.org/ (accessed October 2015).
13. JSON http://json.org/ (accessed October 2015).
14. Schroeder, W.J., Lorensen, B., and Martin, K. (2004) The Visualization Toolkit. Kitware, VTK-The Visualisation Toolkit, www.vtk.org (accessed August 2015).
15. Klimeck, G. *et al.* (2008) nanohub. org: advancing education and research in nanotechnology. *Comput. Sci. Eng.*, **10** (5), 17–23, and https://nanohub.org/ (accessed 10 May 2016).
16. MTCG http://mtcg.snu.ac.kr (accessed 25 April 2016).
17. Pizzi, G., Cepellotti, A., Sabatini, R., Marzari, N., and Kozinsky, B. (2016) AiiDA: automated interactive infrastructure and database for computational science. *Comput. Mater. Sci.*, **111**, 218–230, http://www.aiida.net/ (accessed 10 May 2016).
18. KIST http://nano.vfab.org/ (accessed 25 April 2016).
19. EU Project MoDeNa http://modena.units.it/ (accessed August 2015).
20. DEEPEN http://www.nmp-deepen.eu/ (accessed August 2015).
21. Simphony http://www.simphony-project.eu (accessed August 2015).
22. MMP http://www.mmp-project.eu/ (accessed August 2015).
23. NanoSim http://www.sintef.no/Projectweb/NanoSim (accessed August 2015).
24. The European Materials Modelling Council emmc.info (accessed 10 May 2016).

Index

Handbook of Software Solutions for ICME, First Edition. Edited by Georg J. Schmitz and Ulrich Prahl.

n

o

p